Makhana (Euryale ferox) as a Functional Food

Makhana (Euryale ferox) as a Functional Food

Cultivation, Chemistry, Processing, and Commercialization

Edited by

Tanmay Sarkar
Department of Food Processing Technology, Malda Polytechnic,
West Bengal State Council of Technical Education,
Government of West Bengal, Malda, West Bengal, India

Manoj Kumar
ICAR–National Research Centre for Makhana, Darbhanga, Bihar, India

ELSEVIER

ACADEMIC PRESS
An imprint of Elsevier

Academic Press is an imprint of Elsevier
125 London Wall, London EC2Y 5AS, United Kingdom
50 Hampshire Street, 5th Floor, Cambridge, MA 02139, United States

Notices

ISBN: 978-0-443-44421-0

For information on all Academic Press publications visit our website at https://www.elsevier.com/books-and-journals

Publisher: Mara Conner
Acquisitions Editor: Palak Gupta
Editorial Project Manager: Sudip Saha
Production Project Manager: Aishwarya R.
Cover Designer: Mark Rogers

Typeset by TNQ Tech

Dedication

To my beloved mother, Arati Sarkar; and father, Bholanath Sarkar; and my sister, Tanusree Sarkar, whose unwavering love, support, and encouragement have shaped my journey.
And to God, for His strength and guidance throughout every step of the way.

Dr. Tanmay Sarkar

To the farmers and laborers of Bihar and across India engaged in Makhana cultivation and processing, whose resilience and indigenous knowledge sustain a living tradition and a vital agro-ecosystem.
Your skill and dedication have transformed Mithila's Makhana into a globally recognized superfood.

Dr. Manoj Kumar

Contents

2. Taxonomy, morphology, and growth ecology of *Euryale ferox*

Zahid Ahmad Thoker, Lovleen Marwaha, Anand Mohan and Ajit Prakash

11. Makhana plant protection measures and nutrient requirements for successful crop production

Vinod Kumar Padala, Akhileshwar Vishwakarma, N. Ramya and Manoj Kumar

12. Makhana processing waste valorization

Dharmender, Rahul Kumar Rout, Anupam Sahoo, Akhileshwar Vishwakarma, Shivani Jha and Manoj Kumar

14. Value added product development using makhana (snacks, flour, bars, etc.)

Ritu Kukde, Vaishali Thakre, Pakkiranna Sivamma, Thongam Sunita, Priyabrat Kapri, Pratiksha Puranik, Vikram Kad and Ganesh Shelke

15. Packaging technologies and smart packaging for makhana-based products

Aastha Bhardwaj, Bisma Jan, Shazan Khan, Nitya Sharma, Praveen Kumar Dubey and Tanweer Alam

Contributors

Abdul Momin Rizwan Ahmad, Department of Health Sciences, University of York, York, United Kingdom

Tanweer Alam, Indian Institute of Packaging (IIP), New Delhi, India

Aastha Bhardwaj, Indian Institute of Packaging (IIP), New Delhi, India

Anindita Ray Chakravarti, Department of Food and Nutrition, Maharani Kasiswari College, University of Calcutta, Kolkata, West Bengal, India

Mahesh Chandak, Wockhardt Research Center, Wockhardt Limited, Chhatrapati Sambhajinagar, Maharashtra, India

Simran Choudhury, Department of Food Science and Nutrition Management, J.D. Birla Institute, Kolkata, West Bengal, India

Ana Flávia Coelho Pacheco, Cândido Tostes Dairy Institute (EPAMIG), Lieutenant Luis de Freitas, Juiz de Fora, Brazil

Paulo Henrique Costa Paiva, Cândido Tostes Dairy Institute (EPAMIG), Lieutenant Luis de Freitas, Juiz de Fora, Brazil

Chinmayee Dash, Pharmacology, Department of Pharmaceutical Sciences and Technology, Birla Institute of Technology, Ranchi, Jharkhand, India

Sidharth Dash Sharma, Jyoti and Bhupat Mehta School of Health Sciences and Technology, Indian Institute of Technology Guwahati, Guwahati, Assam, India

Ishita Debnath, Pharmacology, Department of Pharmaceutical Sciences and Technology, Birla Institute of Technology, Ranchi, Jharkhand, India

Dharmender, ICAR-National Research Centre for Makhana, Darbhanga, Bihar, India

Praveen Kumar Dubey, Indian Institute of Packaging (IIP), Lucknow, Uttar Pradesh, India

Priti Duhan, K.L. Mehta Dayanand College for Women, Faridabad, Haryana, India

Mounika Edde, College of Agricultural Engineering, PJTAU, Sangareddy, Telangana, India; Sri Krishnadevaraya College of Agricultural Sciences, Anantapur, Andhra Pradesh, India

Walid Elfalleh, Department of Biology, College of Science, Imam Mohammad Ibn Saud Islamic University (IMSIU), Riyadh, Saudi Arabia

Beenish Israr, Institute of Home Sciences, Faculty of Food Nutrition and Home Sciences, University of Agriculture, Faisalabad, Pakistan

Nizwa Itrat, Institute of Home Sciences, Faculty of Food Nutrition and Home Sciences, University of Agriculture, Faisalabad, Pakistan

Bisma Jan, Department of Biotechnology, IILM University, Greater Noida, Uttar Pradesh, India

Vidyanath Jha, Department of Botany, L N Mithila University, Darbhanga, Bihar, India

Shivani Jha, ICAR-National Research Centre for Makhana, Darbhanga, Bihar, India

Vikas Joon, ICAR-Central Plantation Crops Research Institute Kasaragod, Kasaragod, Kerala, India

Vikram Kad, Dr. A.S.C.A.E & T, MPKV, Rahuri, Maharashtra, India

Priyabrat Kapri, ICAR-Central Institute of Post Harvest Engineering and Technology, Ludhiana, Punjab, India

Ayda Kargar-Ghavidel, Department of Food Science and Technology, Faculty of Agricultural Sciences, University of Guilan, Rasht, Iran

Reza Karimi, Department of Food Science and Technology, Faculty of Agricultural Sciences, University of Guilan, Rasht, Iran

Shazan Khan, Indian Institute of Packaging (IIP), New Delhi, India

Ritu Kukde, ICAR-Central Institute of Post Harvest Engineering and Technology, Ludhiana, Punjab, India

Anand Kumar, College of Food Science and Technology, Guangdong Ocean University, Zhanjiang, China

Ankit Kumar, ICAR-National Research Centre on Litchi, Rohua, Bihar, India

Chaman Kumar, ICAR-National Research Centre on Litchi, Rohua, Bihar, India

Lokendra Kumar, ICAR-Indian Institute of Wheat and Barley Research, Karnal, Haryana, India

Manoj Kumar, ICAR-National Research Centre for Makhana, Darbhanga, Bihar, India

Naresh Kumar, Eternal University, Baru Sahib, Himachal Pradesh, India

Sriranjani Kumar, Institute of Plant and Microbial Biology, Academia Sinica, Taipei City, Taiwan; Department of Life Sciences, National Central University, Taoyuan City, Taiwan

Kalyani Kumari, ICAR-National Institute of Seed Science & Technology, Mau, Uttar Pradesh, India

Chandra Bhushan Kumar Sinha, Department of Biotechnology, L N Mithila University, Darbhanga, Bihar, India

Selva Kumar T., Vel Tech Rangarajan Dr. Sagunthala R&D Institute of Science and Technology, Chennai, Tamil Nadu, India

H.L. Kushwaha, ICAR-Central Arid Zone Research Institute, Jodhpur, Rajasthan, India

Charu Lata, ICAR-Indian Institute of Wheat and Barley Research, Karnal, Haryana, India

Naveen Kumar Mahanti, Post-harvest Technology Research Station, Dr. YSR Horticultural University, Venkataramannagudem, Andhra Pradesh, India

Manukonda Maheswari, Dr. NTR College of Food Science & Technology, Acharya N.G. Ranga Agricultural University, Bapatla, Andra Pradesh, India

Muhammad Rizwan Majeed, Department of Nutrition, University of Tennessee, Knoxville, TN, United States

Lovleen Marwaha, Department of Zoology, School of Bioengineering and Biosciences, Lovely Professional University, Phagwara, Punjab, India

Luiza Oliveira Mello, Cândido Tostes Dairy Institute (EPAMIG), Lieutenant Luis de Freitas, Juiz de Fora, Brazil

Anand Mohan, Department of Botany, School of Bioengineering and Biosciences, Lovely Professional University, Phagwara, Punjab, India

Rezvane Nategh-Khoshnoud-Roohi, Department of Food Science and Technology, Faculty of Agricultural Sciences, University of Guilan, Rasht, Iran

Aliakbar Nategh-Khoshnoud-Roohi, Department of Cellular and Molecular Biology, Faculty of Basic Sciences, Rasht Branch, Islamic Azad University, Guilan, Iran

Wilson de Almeida Orlando Júnior, Cândido Tostes Dairy Institute (EPAMIG), Lieutenant Luis de Freitas, Juiz de Fora, Brazil

Flaviana Coelho Pacheco, Cândido Tostes Dairy Institute (EPAMIG), Lieutenant Luis de Freitas, Juiz de Fora, Brazil; Federal University of Viçosa, Viçosa, Brazil

Vinod Kumar Padala, ICAR-National Research Centre for Makhana, Darbhanga, Bihar, India

Palak, Pharmacology, Department of Pharmaceutical Sciences and Technology, Birla Institute of Technology, Ranchi, Jharkhand, India

Ajit Prakash, Department of Biochemistry and Biophysics, University of North Carolina, Chapel Hill, NC, United States

Om Prakash, ICAR-Central Arid Zone Research Institute, Jodhpur, Rajasthan, India

Subrata Pramanik, Department of Biological Sciences, Indian Institute of Science Education and Research Kolkata, Mohanpur, West Bengal, India

Gopal Pramanik, Pharmacology, Department of Pharmaceutical Sciences and Technology, Birla Institute of Technology, Ranchi, Jharkhand, India

Bogala Pravallika, Department of Processing and Food Engineering, Dr. NTR College of Agricultural Engineering, Bapatla, Andhra Pradesh, India

Pratiksha Puranik, Dr. A.S.C.A.E & T, MPKV, Rahuri, Maharashtra, India

N. Ramya, ICAR-Indian Institute of Agricultural Biotechnology, Ranchi, Jharkhand, India

Shailendra M. Raut, ICAR-National Research Centre for Makhana, Darbhanga, Bihar, India

Isabella de Andrade Rezende, Cândido Tostes Dairy Institute (EPAMIG), Lieutenant Luis de Freitas, Juiz de Fora, Brazil

Rahul Kumar Rout, ICAR-National Research Centre for Makhana, Darbhanga, Bihar, India

Parnasree Sadhukhan, Department of Food Science and Nutrition Management, J.D. Birla Institute, Kolkata, West Bengal, India

Vijeta Sagwal, ICAR-Indian Institute of Wheat and Barley Research, Karnal, Haryana, India

Anupam Sahoo, ICAR-National Research Centre for Makhana, Darbhanga, Bihar, India

Mathangi Raja Sekhar, ICAR-Central Arid Zone Research Institute, Jodhpur, Rajasthan, India

Sreemoyee Sensharma, Jyoti and Bhupat Mehta School of Health Sciences and Technology, Indian Institute of Technology Guwahati, Guwahati, Assam, India

Javaria Shabbir, Department of Nutrition and Dietetics, The University of Faisalabad, Faisalabad, Pakistan

Nitya Sharma, Food, Land and Water Program, WRI India, New Delhi, India

Arpita Sharma, ICAR-Indian Institute of Wheat and Barley Research, Karnal, Haryana, India

Ganesh Shelke, Dr. A.S.C.A.E & T, MPKV, Rahuri, Maharashtra, India

Basant Kumar Singh, Botanical Survey of India, Kolkata, West Bengal, India

Tushadri Singh, ICAR-Indian Institute of Wheat and Barley Research, Karnal, Haryana, India

Pakkiranna Sivamma, ICAR-National Institute of Seed Science & Technology, Mau, Uttar Pradesh, India

Saranya Devi Subburaj, Computational Biology and Therapeutics Lab, Saveetha College of Physiotherapy, Saveetha Institute of Medical and Technical Sciences (SIMATS), Saveetha University, Chennai, Tamil Nadu, India

Thongam Sunita, ICAR-Central Institute of Post Harvest Engineering and Technology, Ludhiana, Punjab, India

Vaishali Thakre, ICAR-DoCA Pulse Storage Project, Dr. PDKV, Akola, Maharashtra, India

Soumit Thakur, Pharmacology, Department of Pharmaceutical Sciences and Technology, Birla Institute of Technology, Ranchi, Jharkhand, India

Zahid Ahmad Thoker, Department of Zoology, School of Bioengineering and Biosciences, Lovely Professional University, Phagwara, Punjab, India

Ganjahalli Vani, ICAR-Central Institute of Agricultural Engineering, Bhopal, Madhya Pradesh, India

Akhileshwar Vishwakarma, ICAR-National Research Centre for Makhana, Darbhanga, Bihar, India

About the editors

Dr. Tanmay Sarkar is currently working as a Lecturer in the Government of West Bengal and Visiting Professor, Center for Research and Development, Sri Eshwar College of Engineering, Coimbatore, Tamil Nadu, India. He has received his Bachelor of Technology, Master of Technology, and PhD in Engineering from Jadavpur University, Kolkata, India. With 9 years of experience in both teaching and research, his areas of interest include bioactive components from natural sources, ethnic foods, mathematical modeling, and phytochemicals. He is an editor for several SCI, SCIE, and Scopus-indexed journals and has edited more than 30 books with Elsevier, Springer, CRC Press, and Wiley. Dr. Sarkar has been classified among the top 2% of global highly cited researchers published by Stanford University, USA, in the years 2024 and 2025.

Dr. Manoj Kumar is a Senior Scientist at ICAR–National Research Centre for Makhana, Darbhanga, Bihar, India. He earned his BSc (Agriculture) (Hons) from Uttar Banga Krishi Viswavidyalaya, West Bengal, and completed his MSc and PhD in Soil Science & Agricultural Chemistry at the Indian Agricultural Research Institute, New Delhi, receiving a gold medal. He secured All India Rank 1 in both the Agricultural Research Service (ARS) and ICAR-JRF examinations in his discipline. With over 16 years of research experience in soil science and plant nutrition, Dr. Kumar has contributed to micronutrient management for soil–crop–human health, climate change impacts on nutrient dynamics in soil–plant systems, makhana nutrition and quality improvement, and sustainable natural resource management. He has authored over 125 scientific publications and has guided/coguided 15 postgraduate and doctoral students. He is a recipient of several prestigious awards, including the Jawaharlal Nehru Award (ICAR), Dr. S.P. Raychaudhuri Gold Medal, Dhiru Morarji Memorial Award, and Young Scientist Award from reputed professional societies.

Preface

Makhana (*Euryale ferox*) has been pivotal in the amalgamation of tradition and nutrition for ages. It is widely revered in many parts of Asia for its cultural and religious significance, while it is gradually capturing the limelight at the global level as a functional food. The medicinal values of this aquatic crop have positioned it among plant-based nutraceuticals, reflecting its significance in therapeutic roles as well. These multidisciplinary perspectives and the emerging role of makhana as a functional food are integrated in this book to provide a comprehensive understanding of this underutilized yet promising crop.

The chapters of this volume extensively carve out the details of makhana from the basic to the advanced level. It starts with the historical background, taxonomy, and ecological characteristics and seamlessly progresses to cultivation practices, genetic improvement, and crop protection strategies that are essential for sustainable production. Postharvest management, processing technologies, and physicochemical attributes have been equally emphasized along with the fundamentals of nutritional composition, starch and protein functionality, and lipid stability. This helps to provide an academic holism about makhana, while catering to its practical relevance.

The terminal section explores areas like the development of value-added products, waste valorization, packaging innovation, safety regulations, and market dynamics. These domains address the potential of makhana as a crop of growing industrial importance that can support agronomic development and food security. It also provides insight into the unexplored opportunities and the related contemporary challenges within the sector.

The book mainly caters to researchers, food technologists, nutritionists, agricultural scientists, policymakers, and industry professionals with a holistic overview and an exhaustive compilation of information regarding this unexplored crop. This integrated compendium on makhana meticulously bridges traditional knowledge with modern scientific mechanisms and contributes to the advancement of functional foods and sustainable agronomic development.

Chapter 1

Introduction to Makhana (*Euryale ferox* Salisb.): Botanical, historical and cultural significance

Vidyanath Jha[1], Chandra Bhushan Kumar Sinha[2], Basant Kumar Singh[3] and Manoj Kumar[4]

[1]*Department of Botany, L N Mithila University, Darbhanga, Bihar, India;* [2]*Department of Biotechnology, L N Mithila University, Darbhanga, Bihar, India;* [3]*Botanical Survey of India, Kolkata, West Bengal, India;* [4]*ICAR-National Research Centre for Makhana, Darbhanga, Bihar, India*

1.1 Introduction

Makhana is cultivated as a crop in stagnant water bodies of Mithila area in northern Bihar, India. It is fast emerging as a superfood. Till about 5 decades back, it was called an orphan crop grown in about a dozen districts on the Indo-Nepal border. But the pace of liberalization and globalization unveiled the nutraceutical potentials of this under explored crop that has now attracted a wider scientific attention. All parts of the plant have been investigated for their nutritional and medicinal potentials. Even the seed coat of *E. ferox,* which is simply burnt as fuel during the popping process, has been found to be a repository of useful phytochemicals. There is an urgent need to remove the ambiguity regarding its pops being a product of lotus (*Nelumbo nucifera*). The indigenous fishing community in Mithila thrives upon its cultivation and processing for which its members are hired in non-traditional areas.The growers are getting all possible government incentives for its cultivation. A number of entrepreneurs have joined the business of production of processed items and their global trade. The crop is experiencing mechanization at all possible stages. It is now also possible to store the first roasted seeds for a few months before the second roasting is taken up. Recently the Govt. of India has set up a Makhana Board, and the crop has been granted a GI tagging as well. ICAR National Research Center for Makhana was established in 2002 at

Makhana (Euryale ferox) as a Functional Food. https://doi.org/10.1016/B978-0-443-44421-0.00020-2

Darbhanga. Rising pace of global warming has now led to its cultivation being taken up in knee-dip water bodies. It is time now to raise both production and productivity of Makhana crop. Farmers are now making all efforts to raise the crop productivity through the use of organic manures and synthetic fertilizers and pesticides. They are also using drones for this purpose. More and more of land area is being brought under Makhana cultivation. Arable plots are being transformed into shallow water holding plots by raising their ridges. Farmers are also selling the topsoil to convert the arable plots into Makhana plots having only 1.5–2 feet water. In a way this is a step that helps retention of the surface water and facilitates Makhana cultivation. Rural women who were earlier confined to their role in popping and retail marketing business have now stakes in the processes of weeding and transplantation as well. Water bodies in non-calcareous zones (generally between the Kosi that earlier flowed on Bengal border) were earlier considered to be more suitable for the crop. Huge number of water bodies in calcareous zones of even the state of Bihar were considered unsuitable for this purpose. But the growing mandate in favor of raising the overall crop production, districts such as Khagaria, Sitamarhi, Katihar, Purnea, Madhubani, Kisanganj, Supaul, Araria, Madhepura, Saharasa, Samastipur, Bhagalpur, Gopalganj, Muzaffarpur, Vaishali, Saran, Siwan, Patna, East and West Champaran have also been included in the list of providing incentives to the Makhana growers. Investors are now purchasing raw seeds from the state of Bihar and making pops through the popping machines installed outside the state.

The crop does have a sort of innate association with a submerged weed called *Ceratophyllum demersum* (Sinha and Jha, 2025a; Sodi and Kumar, 2020). *E. ferox* belonging to the order Nymphaeales and *C. demersum* belonging to the order Ceratophyllales have a significant role in the evolutionary history of angiosperms. High quality genes of the two aquaphytes greatly help to clarify phylogenetic relationships of early diverging angiosperms. These genomic resources are essential for future comparative investigations of genic evolution that underpin the morphological, physiological and ecological diversification of early angiosperms (Yang et al. 2020). The recently set up National Makhana Board has made an allocation in the annual action plan for the year 2026–27 under central sector scheme for development of Makhana. States such as Bihar, Uttar Pradesh, Chhattisgarh and Madhya Pradesh have got major allocations. Smaller allocations have been made for states such as Assam, Manipur, Odisha, Uttarakhand, Jammu & Kashmir and Andhra Pradesh. Thus altogether 10 states have come under this purview. The aim of this study is to provide a comprehensive overview of *E. ferox* (Makhana) by documenting its morphology and anatomy, evolutionary history, phytochemical and "omic" (genomic, transcriptomic, metabolomic, proteomic) characteristics, cultivation and post harvest practices, pests and breeding challenges and cultural significance -with the goal of highlighting its nutritional, medicinal and agronomic potential and identifying research and

TABLE 1.1 Names of Makhana in different Indian and foreign languages.

S.No.	Languages	Names
1.	Hindi	Makhana/Phool Makhana.
2.	Sanskrit	Makhanna Padmbijabh, Ankalodya.
3.	Maithili	Makhaan, Bherund, Phoka Makhaan.
4.	Odia	Kantapadma.
5.	Bengali	Makhan
6.	Manipuri	Thangjing.
7.	Assamese	Nikhori.
8.	Telugu	Melluni padmanu.
9.	Punjabi	Juwar/Jaivar
10.	Kashmiri	Pambuchh/Pumbach,/Jewar'go'tch.
11.	English	Gorgon nut, white ball, black nut, prickly water lily.
12.	Chinese	Qian Shi
14.	Japanese	Oni basu/Hasu ho mi.
15.	Korean	Ga si yeon
16.	Spanish	Semilla de loto.
17.	Russian	Lisiyorekh/gorgonovyy orekh.

development priorities for improved cultivation, value addition and conservation. This chapter contains 1 table and 5 plates incorporating altogether 34 photographs. Table 1.1 represents names of Makhana in different Indian and foreign languages.

1.2 Botanical studies

1.2.1 Morphological and anatomical details

The primary distribution of *E. ferox* in the Russian far East, particularly in Primorsky Krai considered a rare relict plant in these areas of and found in the Amur river basin lakes of Manchuria (bordering region) and specific localities such as Lachwin remains identified in Russia, Ukrain and Belarus. They are known for having low genetic diversity suggesting they are relicts of warming climatic periods. *E. ferox* is native to a range stretching from northern India through China, Korea and Japan to the Russian Far East. Koren et al. (2012) have reported low level of allozyme polymorphism in relict aquatic plants of the Far East *Nelumbo komarovii* Grossh and *E. ferox*. Although sometimes described as a perennial aquatic herb (Haines 1926; Biswas and Calder 1936; Subramaniam 1962). *E. ferox*, as observed in cultivation in north Bihar, behaves as a long seasonal annual crop. The plant is a large, prickly acaulescent, aquatic herb with floating leaves which are rugulose especially while young, prominently nerved and net-veined beneath. The plant grows in standing shallow water (0.3–1.5 m depth) and has a rhizomatous stem that is deeply rooted in sediment by means of thick, fleshy roots in clusters. The huge spiny, elliptic or orbicular mature leaves are peltate (0.3–1.2 m diameter). The leaves are green above and red or purple on the ventral side.

The rhizome of *E. ferox* is a short, thick, and erect, serving as a compact anchor in muddy substrates. Unlike long, creeping types, this structure supports a dense rosette of leaves and fibrous roots, often cultivated annually despite its basic perennial nature. The fruits mature and burst between June and August and the arillated globular seeds drop to the pond bed. To facilitate the collection of seeds from the water, most plants are removed during September–October. Normally, with the onset of winter, the remaining plants undergo natural death and decay. From December onwards, the seeds left over on the pond bed, start their hypogeal germination. The thick fibrous roots normally comprise three to four clusters, each having about 17 rootlets. The cluster emerges one after another at intervals of about 1 week. The foliage also follows a definite sequence in appearances. Early leaves are quite different from older ones. The first set of leaves are membranous, hastate to deeply sinuate with a dimension of 9×4 cm, and deep pinkish in color on both the dorsal and ventral sides. The stalks are correspondingly tender whilst the laminae have few or no prickles. The second set appears after about 15 days. The sagittate curled leaf buds later develop into a more orbicular shape (1.5–2.0 cm in length), purple below and green above. They are 6 cm long and when fully opened have dimensions of 15×14.5 cm. Their veins gradually become more prominent and develop spines. The petioles also become stiffer and more spiny than those of the first set. Mature leaves, which are rugose, orbicular and corrugate with a slight notch, start appearing a fortnight later. These persistent leaves vary in length and breadth from 30 to 120 cm and 25–110 cm, respectively. The veins are highly swollen on the ventral side and have sharp spines which are also profuse on the leaf stalk. Banerji (1972) considered the foliaceous chronology in *E. ferox* to be an ecological adaptation to its survival as the early clusters of roots can lose anchorage even after moderate surface disturbance. The rhizome of *E. ferox* is a short, thick, and erect, serving as a compact anchor in muddy substrates. This structure supports a dense rosette of leaves and fibrous roots, often cultivated annually despite its perennial nature. The absence of gluten in Makhana indeed has significant taxonomic value, as it serves as a biochemical indicator that distinguishes it from gluten-containing monocots such as wheat, barley, and rye.

E. ferox is famous for having gigantic leaves, with a diameter of up to 2.42 m. These rival even those of *Victoria amazonica* in the dimension of leaf surface

(Okada, 1928, 1935). The giant leaves of *E. ferox* along with that of *Victoria* sp. have been held as a botanical enigma. These plants may have evolved as a natural engineering solution to selection pressures that favored both a low leaf biomass while maintaining a large surface area available for photosynthesis (Box et al. 2022). Flowering in *E. ferox* starts in the second week of April when average air-temperatures are 30°C and the photoperiod is 12.60 h. Most flowers appear in the first week of May when air temperatures are 33°C and the photoperiod is 13.15 h, with available solar energy 243 cal/cm^2/day (Laal and Datta Munshi, 1985). The flowers are solitary, submerged and epigynous with four persistent, spiny sepals inserted on the torus above the level of the ovary, together with many seriate petals. Most flowers are cleistogamous, but chasmogamous flowers may also be produced (Okada, 1938). The latter open on the water surface for 2–3 days, then sink under water until the fruit is about to ripen and then reappear on the water surface. The cleistogamous flowers behave similarly, except that they do not open their sepals and petals. The fertilization of the cleistogamous flowers seems to take place before the flower buds appear on the water surface. The development of the embryo is far more advanced in the cleistogamous flowers than in the chasmogamous ones (Okada and Otaya, 1930). However, seed setting in both types is equally complete. The inferior, multicarpellary ovary develops into a spongy berry-like fruit which is densely prickly and contains 30–40 pea-sized seeds with hard black seed coats and a mucilaginous aril. The pulpy aril keeps the seeds floating for a few days after they dehisce, before they finally settle down to the bottom of the water. Dehiscence takes place either loculicidally or septicidally. Fruiting usually starts by the middle of May and continues until the ponds are cleared of plants for harvest during August–September. Each plant produces 15–20 fruits, each with an average of 30–40 seeds. The growth stages of *E. ferox* can be identified as seeding, seedling, grand growth and harvest phases occurring respectively from October–December, January–March, April–June and July–September with some overlapping depending upon variations in climatic conditions.

E. ferox exhibits characteristics typical of primitive angiosperms. The pollen grains are relatively large and generally monosulcate with a single elongated aperture. This feature is considered ancestral among flowering plants. The exine (outer wall) of the pollen is often reticulate or granulate, providing structural strength and protection. Unlike other water lilies, the pollen grains of *Euryale* have three nuclei (Gavhane et al., 2023). *E. ferox* has mostly the cleistogamous flowers which are suitable for commercial cultivation. The plant shows an alternate or spiral arrangement of leaves, where each leaf arises singly from nodes on a short, thick, submerged rhizome. The leaves are large, circular, and peltate, with long, often prickly petioles attached centrally to the lamina, an adaptation that facilitates efficient floating and light capture in aquatic environments. Since the rhizome remains embedded in mud, the phyllotactic pattern is best understood by observing the sequential emergence of petioles from the rhizome.

Blade of leaves floating on the water surface in *E. ferox* is of the epistomatic type. Various outgrowths of the epidermal origin include hydropots, coverts, and capitate trichomes occur in *E. ferox*. It has numerous long-pointed emergences (spines) of the epidermal and subepidermal origin. The stomata are large, the guard cells are reniform, and the stomatal slits are wide open. The petiole of *E. ferox* is characterized by the primary microstructure of the axial organ with typical histological and topographic complexes. The petiole epidermis of the species is single-layered and performs a secreting function. The zone of the primary cortex is differentiated into the external and internal areas. The outer cortex in the species remains equally occupied by five to seven layers of angular collenchymas. There is presence of paired concentric bundles with a common air cavity. The leaves exhibit a spectrum of micromorphological adaptations characteristic of the hydromorphic evolution of the plant. General aerenchymatization, reduction of conductive elements, weak tissue sclerization are adaptive features of the species. Assimilating, normally developed leaves are characterized by typical traits of organs found in their habitats. A common feature in their structure is the dorso-ventrality of the mesophyll and the difference between the upper and lower sides of the leaf blade. Secretory formations of the epidermal origin provide waterproof protection to plant organs in the aquatic environment (Tsyrenova, 2021).

The abaxial surface of *E. ferox* leaf is dark purple that is attributed to being rich in anthocyanins including delhinidin, pelargonidin, cyanidin, malvidin and peonidin and petunidin, all with strong antioxidant potential have been identified in *E. ferox* leaves (Jing et al., 2021).

The pedicel of *E. ferox* possesses closed, scattered vascular bundles and contains no cambium. Four main air canals are well developed. Mesophylls of sepal is differentiated into palisade and spongy parenchyma. Petal is simpler in structure than that of sepal with no differentiation of palisade tissue. On the basis of anatomical and developmental studies of the floral organs, Chen et al. (1991) suggested that *E. ferox* exhibits a number of the most primitive features, such as petaloid stamens, carpel with ovarian canal, elongated receptacle, prominent residual floral apex and laminar placentation. The development of floral parts and characteristics of ovary indicate that genus *Euryale* is much more similar to *Victoria*, *Nymphaea* and *Nuphur* than to *Nelumbo* and *Brasenia*. *E.ferox* is understood to have undergone an adaptive evolution. Its leaves float on the water surface and thus have limited living space. It develops large, rapidly expanding leaves to compete for space on the water surface. Wu et al. (2022a,b) reported that the cell proliferation of leaves is almost completed under water while the cell expansion occurs rapidly after they grow above water. They analyzed the genome and transcriptome dynamics of the plant through reconstruction of the three subgenomes generated from the paleo-hexaploidization event in *E. ferox* and revealed that one subgenome was phylogenetically close to *Victoria cruziana* that led to the development of gigantic floating leaves. Their finding provide novel insights into the genome evolution through polyploidization as well as the adaptive evolution regarding the leaf development accomplished through biased gene retention and expression sub-functionalization of multicopy genes in *E. ferox.*

1.2.2 The fibonacci angle in *E. ferox*

In *E. ferox*, this angle is most visible in the spiral arrangement of its numerous petals and stamens. The 137.5 degrees angle ensures that each new petal or seed is placed without overlapping the one directly below it, maximizing space efficiency and sunlight exposure. While many modern botanical and mathematical texts cite *E. ferox* (prickly water lily) as a classic example of Fibonacci phyllotaxis with a divergence angle of approximately 137.5 degrees, the original scientific reference for this specific observation is generally attributed to the pioneering work of Alexander Braun in the 1830s.

PLATE 1.1 Morphological features. 1. All three/four types of leaves, 2. Trimerous and Tetramerous flowers, 3. Ventral view of *E. ferox* leaf, 4. Harvest of Thangjing fruits in Manipur lakes by angling method, 5. A watercock *Gallicrex cinerea* associated with Makhana plants, 6. Visit of Prof. Naama Goren Inbar to Darbhanga in the year 2011.

1.2.2.1 Perisperm: The edible component

A key anatomical feature of *E. ferox* seeds is the presence of a well-developed perisperm, a diploid nutritive tissue derived from the nucellus that persists after fertilization and surrounds the embryo. Unlike endosperm, the perisperm serves as the principal storage tissue and constitutes the main edible portion of Makhana. It is rich in carbohydrates (approximately 75%–80%), along with moderate protein, minerals, and bioactive compounds, making it nutritionally significant. The perisperm is structurally stable even after drying, contributing to its commercial value. From a functional perspective, Makhana is recognized as a low glycemic index (GI) food and has been traditionally used in managing blood glucose-related disorders (Singh et al. 2025).

PLATE 1.2 Different parts of the plant. 7. Plantlets in young stage showing the fibrous roots, 8. Fully mature leaves of *E. ferox*, 9. A chasmogamous flower in full bloom, 10. A trimerous flower of *E. ferox*, 11. A tetramerous flower of *E. ferox*, 12. (a) Longitudinal section of a flower in young stage, 12. (b) Sepals and petals in a flower of *E. ferox*.

Perisperm is a layer of nutritive tissue that surrounds the embryo of a seed. An ovule consists of three parts: the integument that forms the outer layer, the nucellus and the female gametophyte in its center. The nucellus is the inner structure of the ovule that develops into the perisperm that feeds the embryo. Perisperm is the remains of the nucellus in seeds and is a diploid tissue. The nucellus develops into the perisperm after the process of fertilization. It provides nutrition to the developing embryo and is formed at the micropylar end of the ovule. Perisperm appears to be reddish and has a paper-like composition. This perisperm remains persistent after the egg fertilization. It surrounds the endosperm and is prominent in monocot seeds. *E. ferox* (Makhana) seeds are characterized by the presence of a prominent **perisperm**, which constitutes the main edible part of the seed. The perisperm is a diploid, maternal, starch-rich nutritive tissue that surrounds the embryo and is derived from the nucellus.

It is the white, starchy, and nearly fatless part of the seed that is consumed. It is rich in carbohydrates (roughly 75%–80%) and provides significant amounts of protein, minerals, and bioactive compounds. It is characterized as a "well-differentiated" layer that remains stable even after the seed is dried, enabling it to be a key component in the nutritional profile of the seed. It serves as a primary, persistent storage tissue for the developing embryo (Singh et al., 2025).

Makhana has been widely employed in traditional foods to cure a number of diseases associated to blood glucose. The retrogradation of starch is not a straightforward process, it manifests itself in completely different processes such as inflated firmness and recrystallization of gelatinized starch gels with time. The elements of starch, amylopectin and amylose, have completely different roles in retrogradation (Gupta et al. 2021).

1.2.3 Embryological studies

Embryological studies in *E. ferox* by Khanna (1964) have revealed that microsporogenesis results in the formation of tetrads which are tetrahedral and isobilateral. Pollen grains are two celled when shed. The ovules are crassinucellate, anatropous, and bitegmic. Double fertilization is simultaneous. The massive nucellus stores the food reserves. Embryo sac is of polygonum type. Development of endosperm is of cellular type which is distinct from other members of the family Nymphaeaceae. Suspensor develops along with the mature embryo that is globular. Two saucer shaped cotyledonary lobes arise from the ridge. Embryo development conforms to the asterad type with Penea variation. In some cases, polyembryony is also seen. The specific presence of three nuclei in the pollen grains of *E. ferox* is a distinguishing feature that sets it apart from other members of the family Nymphaeaceae. This characteristic is historically significant as it led to

Euryale sometimes being placed in a distinct family *Euryalaceae*, separate from Nymphaeaceae. The presence of three nuclei indicates a specific, more advanced developmental stage of the male gametophyte compared to 2-nucleate pollen found in many others. Some studies indicate that *E. ferox* can exhibit a nuclear type of endosperm development.

PLATE 1.3 Stages of Makhana cultivation. 13. A Makhana water body showing stunted growth of the crop due to weeds, 14. Weeds obstructing the crop growth, 15. Makhana competing with the growth of *Trapa* sp. (locally called Singhara), 16. Weeding performed by women farmers in a Makhana water body, 17. A traditional pond now being used for Makhana cultivation, 18. Makhana grown in a wetland created by brick kiln.

1.2.4 Phytochemical constituents

The identified cerebrosides, specifically ferocerebrosides A and B, along with complex tocopherol trimers (e.g., ferotocotrimers C and D, ferotocodimer A), are often found in basal dicot species within the order Nymphaeales. These compounds represent specialized, complex secondary metabolites (Kumar et al., 2025; Sinha and Jha, 2025b). *E. ferox* produces biofuels primarily through starch biosynthesis, with its 80% starch content serving as a feedstock for bioethanol, alongside minor lipid production for biodiesel. While not specializing in a unique pathway, the plant's metabolic pathways for starch and phenylpropanoid biosynthesis hold potential for biofuel development. *E. ferox* exhibits significant potential as a source of natural antimicrobial compounds. Kumar et al. (2024) shows strong antioxidant (Lee et al. 2002) hypoglycemic (Zhang et al. 2019), and anti-cancer (Nam et al. 2019) activities. Seeds may serve as a promising natural therapeutic for anxiety and depression as found in a pre-clinical study on mice by Yameen et al. (2026). It is also used in the preparations of Ayurvedic medicines for treating many diseases such as kidney disorders, persistent diarrhea, rheumatic disorders, excessive leucorrhea, bile disorders, hepatic dysfunctioning, etc. (Nehal et al. 2015). Current Chinese Pharmacopeia described it as "sweet, astringent, and flat, attributed to the spleen and kidney meridians, benefiting the kidney and consolidating sperm, tonifying the spleen and inhibiting diarrhea, eliminating dampness and arresting leucorrhea (Jiang et al. 2023)."

The compounds isolated from *E. ferox* seeds are a combination of glycosphingolipids (cerebrosides) and polyphenol-like tocopherol polymers. Studies further indicate that its glycemic response can be modified through processing techniques and incorporation of protein (gluten) and hydrocolloids. Additionally, the starch composition of Makhana, primarily amylose and amylopectin, plays a crucial role in retrogradation, a process involving recrystallization of gelatinized starch that affects texture and digestibility. Retrogradation leads to increased firmness over time and influences the functional and nutritional properties of the food (Gupta et al. 2021). Thus, the integration of botanical structure and biochemical properties highlights the importance of *E. ferox* as both a nutritionally valuable and functionally significant aquatic crop. Jiang et al. (2023) found the shell of *E. ferox* to be rich in polyphenols. They evaluated their in vitro antitumor activity. Ellagic acid, corilagin and geraniin were identified as the responsible compounds. Corilagin showed the strongest inhibitory effect on He La cell proliferation. Geraniin was found to inhibit He La cell invasion. Their study determined the chemical structures of polyphenolic compounds from *E. ferox* shell and their impact on tumor cell proliferation, migration and invasion, offering evidence for developing antitumor drugs and gaining insights for modernization of traditional Chinese medicinal resources (Jian et al., 2026).

E. ferox leaves accumulate high levels of diverse flavonoid compounds including the Flavonoid C-Glycosides (FCG) such as vitexin, isovitexin, and orientin that function as structural stabilizers. Its leaves undergo specialized development due to epidermal cells differentiating in a way to create protective stable layer that allows its large, thin leaves to survive in aquatic environment. The leaves expand rapidly even up to 2 m diameter within a span of 20 days. This is due to vigorous cell division while under submergence and accelerated cell expansion on reaching the surface. High concentration of flavonoids acts as a stress resistance mechanism. This facilitates the delicate giant leaves withstand environmental stimuli (Liu et al., 2025). As per the findings of Liu et al. (2025) *E. ferox* leaves accumulate high levels of diverse flavonoid compounds including the flavonoids C-glycosides such as vitexin, isovitexin, and orientin that function as structural stabilizers. *E. ferox* could serve as a promising natural therapeutic for anxiety and depression as well. A recent study by Yameen et al. (2026) reports the presence of polyphenolic flavonoids to have antioxidant properties that facilitate these potentials in the plant.

Devi et al. (2022) In this age of nanotechnology *E. ferox* leaves have been found to have the potential to facilitate green synthesis (bio fabrication) of silver nano particles named as ELAgNPs (*E. ferox* leaf silver nanoparticles), that could be used for biomedical applications. Biosynthesized ELAgNPs have been used as antibacterial agents against pathogenic bacteria. The greater free redical scavenging properties of ELAgNPs could be due to phytomoieties such as anthocyanins glycosides and phenolic compounds present on their surface during bio fabrication and stabilization. On account of its pronounced cytotoxic properties ELAgNPs hold the potential to be effective against cancer.

E. ferox as an aquatic crop suffers the attack of a number of fungal and insect pests of major and minor nature. Verma and Jha (1999) reported the development of fungal gall over the leaf of *E. ferox* from June onwards. The mode of transmission of the gall organism (*Doassansiopsis euryalis* Verma and Jha) needs to be explored in detail. The crop needs intervention to tackle these adversities through interventions at genetic level. The crop suffers losses due to the association of algal blooms as well. Losses inflicted by these blooms and association of mollusks, aphids nematodes etc. are yet to be assessed. A major challenge lies with development of its spineless forms on the pattern of *Trapa*, another aquatic crop being cultivated in the region. Spines on the plant body have been an age old adaptation against herbivory (Dwivedi et al., 1995, Mahto et al., 1993, Haider et al., 1995, Haider and Nath 1987; Jha et al., 1989, Banerji 1972, Mishra et al., 1989, 1990; Saraswati et al., 1990 etc.). TIFY genes are a family of plant specific transcription factors that play a role in leaf growth and development. TIFY (T stands for Threonine, I stands for Isolucine, F stands for Phenylalanine and Y stands for Tyrosine). Xu et al. (2023) identified 23 TIFY genes clustered into three groups. Molecular breeding in *E. ferox* is an emerging specialized approach designed to overcome the limitations of conventional breeding in this monotypic aquatic crop that is primarily self-pollinated and has limited natural genetic diversity. Pollination in this plant occurs underwater in hermatically sealed flowers. Traditional crossing is difficult. Molecular tools are therefore focused on accelerating the dert of spine free, high yielding and dense varieties. TIFY family has recently been analyzed in several sequenced plant genomes. Functional information regarding various TIFY proteins has also been obtained demonstrating their regulatory role. TIFY proteins regulate plant architecture including petiole elongation, hypocotyl growth, leaf development and spikelet development. It has putative transcription factors that is increasingly believed to play vital roles in control of various biological pathways including plant growth and development. They recently took up characterization and co-expression analysis of the TIFY family genes in *E. ferox* with reference to leaf development.

1.2.5 Resistant starch: A specific feature of *E. ferox*

Presence of resistant starch is generally more characteristic of dicot roots and tubers or specific high amylose variants. Another uniqueness of Makhana lies in the presence of resistant starch in its perisperm. It is a type of carbohydrate that escapes digestion in the small intestine. Its function is similar to soluble fiber through fermentation in the colon to feed beneficial gut bacteria. Its specialty lies in improving metabolic health. It boosts safety for weight management. It is a probiotic that enhances digestive health by feeding good intestinal bacteria and creating short chain fatty acids that nourish the colon. This is due to its property of resisting breakdown into glucose. It does not spike blood sugar and improves insulin sensitivity and thus aids in diabetes management. Makhana has a high compact starch structure with high amylose content. It is an excellent raw material for producing type-3 resistant starch (RS3) by adopting enzymolysis and retrogradation. It has been possible to raise the RS content in Makhana to approximately 15%–36% through enzymatic treatment with the use of amylopullulanase enzyme. This facilitates the reduction of readily digestible starch. Makhana helps manage obesity related to diabetes, oxidative stress and hypothyroidism.

1.2.6 "Omic" studies in *E. ferox*

E. ferox is fast emerging as a potential plant for "omic" studies including proteomics, transcriptomics, metabolomics etc. Wu et al. (2022a,b) studied the proteome of *E. ferox* seeds at different developmental stages. They screened and identified the key enzymes for biosynthesis of starch and flavonoids in its seeds. Abscisic acid (ABA) signal transduction pathway was also analyzed by them. Their study is likely to provide insights into the growth and development of *E. ferox* seed and lay a theoretical formulation for the subsequent study of the prevalent molecular mechanism. Metabolomics today has become an updated research tool for identification and quantification of secondary metabolites. Liu et al. (2023) undertook a comparative metabolomic profiling of secondary metabolites in different tissues of *E. ferox*. They obtained 589 secondary active substances from different tissues of *E. ferox* of which polyphenols occupy 87%. They made a functional characterization of phenylalamine ammonia lyase. They found leaves accumulating significant amounts of flavonoids with widespread glycosylation. Seed shell was found to be abundant in polyphenols while the fruit peel was rich in phenolic acids. Seed shell had good amount of antioxidant materials. Wu et al. (2021) Metabolomics and transcriptome has revealed analysis of biosynthesis mechanism of flavonoids in the seeds of *E. ferox* at different developmental stages.

1.2.7 Genomic studies

NBS-LRR (nucleotide-binding site-leucine-rich repeat) genes are the largest gene family in plants conferring resistance to pathogens. At present, studies on the evolution of NBS-LRR genes in angiosperms mainly focused on monocots and eudicots, while studies on NBS-LRR genes in the basal angiosperms are limited. *E. ferox* represents an early-diverging angiosperm order, Nymphaeales and confronts various pathogens during its lifetime, which can cause serious economic losses in terms of yield and quality. Qian et al. (2022) performed a genome-wide identification and analysis of NBS-LRR genes in *E. ferox* and identified 131 NBS-LRR genes that could be divided into three subclasses according to different domain combinations. These included 18 RNLs, 40 CNLs, and 73 TNLs. The *E. ferox* NBS-LRR genes are unevenly distributed on 29 chromosomes; 87 genes are clustered at 18 multigene loci, and 44 genes are singletons. Gene duplication analysis revealed that segmental duplications acted as a major mechanism for NBS-LRR gene expansions but not for RNL genes, because 18 RNL genes were scattered over 11 chromosomes without synteny loci, indicating that the expansion of RNL genes could have been caused by ectopic duplications. Ancestral gene reconciliation based on phylogenetic analysis revealed that there were at least 122 ancestral NBS-LRR lineages in the common ancestor of the three *Nymphaeaceae* species suggesting that NBS-LRR genes expanded slightly during speciation in *E. ferox*. Transcriptome analysis showed that the majority of NBS-LRR genes were at allow level of expression without pathogen stimulation. Their study characterized the profile of NBS-LRR genes in *E. ferox* and that could serve as a valuable resource for disease resistance breeding in *E. ferox*. Identification of NBS- LRR genes, analysis of features and revelation of chromosomal distribution pollen are likely to provide important references for molecular breeding in this plant (Qian et al., 2022). The genus *Euryale* is understood to have undergone two polyploidization events that occurred approximately 94–106 million years ago. *Euryale* passed through an ancient whole genome duplication (WGD) followed by one whole genome triplication (Yang et al. 2020).

1.2.8 Study of chloroplast genome

Guo and Minn (2020) reported complete chloroplast genome of *E. ferox* to be of 159,930 bp in length and have four sub-regions. The complete chloroplast genome sequence of *E. ferox* contains 89,678 bp of a large single copy region (LSC), 22,202 bp of small single-copy region (SSC) and 25,025 bp of two inverted repeat regions (IRs). The complete chloroplast genome of *E. ferox* is also consistent with other chloroplast genomes found in most plant species. The overall nucleotide composition of chloroplast genome sequence has: A (30.1%), T (30.8%), C (19.9%), G (19.2%) and the total GC content of 39.1%. Its sequence contains 127 genes, including 84 encoding genes, 35 transfer RNA genes and eight ribosomal RNA genes. As per the analysis result, the maximum likelihood (ML) tree presents that *E. ferox* clustered with *Nuphar advena* belonging to the family Nymphaeaceae in genetic evolution relationship by this method. Genetic diversity study of *E. ferox* through molecular markers depends on both quality and quantity of DNA. Kumar et al. (2017) standardized the process of isolation of DNA from its leaf tissue having high amount of polysaccharides, polyphenols and proteins. They optimized the CTAB (acetyl trimethyl ammonium bromide) protocol that yielded up to 4081 ng per gram good quality DNA from its leaf tissues. The method ensured the avoidance of DNA degradation contamination and low yield due to irreversible binding of phenolic compounds and coprecipitation with polysaccharides. The isolated DNA has excellent spectral qualities and is efficiently digested by restriction endonucleases and is suitable for development of molecular markers such as RAPD and ISSR.

1.2.9 Transcriptomic studies

Liu et al. (2018) took up transcriptome sequencing analysis during seed growth and development of *E. ferox*. They are of the opinion that increasing expression of PAL (Phenylalanineammonia lyase) and P450 encoded genes in phenylpropanoid biosynthesis (PPB). PPB could promote the maturation of *E. ferox* seed size, color, hardness and accumulation of medicinal components. PPB was found to be associated with the synthesis of secondary metabolites during seed growth and development. Phenylalanineammonia lyase (PAL) was the first key enzyme that catalyzed the conversion of phenylalanine to trans-cinnamic acid leading to the synthesis of flavonoids, lignins, and alkaloid. Its leaves undergo specialized development due to epidermal cells getting differentiated in a way to create protective stable layer that allows its large, thin leaves to survive in aquatic environment. The leaves expand rapidly even up to 2 m diameter within a span of 20 days. This is due to vigorous cell division while under submergence and accelerated cell expansion on reaching the surface. High concentration of flavonoids acts as a stress resistance mechanism. This facilitates the delicate giant leaves withstand environmental stimuli.

Using single cell technology Liu et al. (2025) constructed transcriptional maps to decipher the environmental adaptation strategies of *E. ferox* leaves that exhibit significant differences transitioning from submergence to

PLATE 1.4 Plant parts of *E. ferox* and certain stages of cultivation. 19. Thinning out of Makhana plants for transplantation elsewhere, 20. Makhana leaves charred due to non-availability of water during summer, 21. A mature fruit of *E. ferox*, 22. Arillated seeds floating on water surface after natural dehiscence of the fruit, 23. Inner placental mass after manual removal of the fruit sheath, 24. Arillated seeds collected after manual removal of the fruit sheath, 25. Raw seeds ready for popping, 26. Broken kernel after removal of seed coat, 27. Broken seed coat after removal of the kernel, 28. Fungal galls over the thick leaves of *E. ferox*.

emergence from water. Epidermis cells preferentially differentiate during the submerged stage involving the expression of numerous ribosomal proteins and plant immunity genes. They found that a significant number of mesophyll cells and hydathode cells, during the floating stage, undergo differentiation with energy metabolism genes exhibiting high activity. *E. ferox* leaves exhibit another adaptation in the form of specific accumulation of flavonoid C-glycosides in epidermal cells. The genes involved in flavonoid C-glycosides biosynthesis demonstrated specific expression and a crucial C-glycosyltransferase gene was characterized. This led to the construction of a transcriptional regulatory network that elucidates the molecular mechanisms by which *E. ferox* adapts to aquatic ecology at a single cell resolution.

1.3 Historical studies

Historical studies on *E. ferox* refer to its taxonomic and fossil histories.

1.3.1 Etymology & nomenclature history

Genus: *Euryale* Salisb. [Ann. Bot. (König and Sims) 2: 73 (1805)] named after one of the three Gorgons, the monstrous daughter of the sea God, who had venomous viper for hairs (Nayar, 1985); this is an allusion to the prickly character of the monotypic genus Euryale [*E. ferox* Salisb. Ann. Bot. (König and Sims) 2: 74 (1805)]. However, some unreferenced source mentions its origin from the Greek words *eurys* ("wide") and *ale* ("wandering" or "roaming"), translating to "far-roaming" or "wide-wandering", referring to its widespread leaves. The specific epithet '*ferox*' meaning 'fierce' likely refers to the plant's sharp prickles. Salisbury, the British Botanist first provided the description of the plant in modern botany, the type of which has been conserved in Banks Herbarium in British Museum (BM). The plant has been very nicely depicted in Curtis' Botanical Magazine 35: t, 1447, 1812 and Roxburgh's plants of the Coast of Coromandal 3:t. 244. 1815. There is also a good place of the plant under the replaced name, *Anneslia spinosa* Roxb. ex-Andrews in Botanist's Repository 10:t. 618. 1811.

This curious plant is a native of the East-Indies, and has been cultivated in China, according to the pretentions of the Chinese writers, more than a 1000 years before the commencement of the Christian era (Curtis's Botanical Magazine, 1811–12).

1.3.2 Enumeration

Euryale ferox Salisb. in Konig & Sims, Ann. Bot. 2: 74. 1805; Hook. F. & Thomson in Fl. Brit. India 1: 115. 1872; Mitra, R.L. in Sharma et al., (eds.) Fl. India 1: 427. 1993; Mitra in A.A. Mao & S.S. Dash, Fl. Pl. India Annot. Checkl. 1: 66. 2020.

E. indica Planch., Ann. Sci. Nat. Bot. 3, 19: 29. 1853. *Anneslia spinosa* Roxb. ex-Andrews, Bot. Rep. 10: t. 618. 1811.

Asm., Beng. & Hindi: *Makhana, Makhna*; Eng.: Fox Nut, Prickly water lily, Gorgon plant; Kh.: *Shriew-u-puban*; Or.: *Kunta-padam.*

Aquatic, annual, rhizomatous herb with fibrous rootstock. Leaves sagittate or elliptic-orbicular, 20–100 cm across, pale-pinkish beneath, green above, cordate or sinuate at base, leathery; floating leaves sub-leathery, glabrous above, veins abaxially strongly ribbed, prickly; submerged leaves not prickly; petioles long, prickly. Flowers cleistogamous, 2–5 cm in diam., solitary; torus prickly; peduncle stout, densely prickly. Sepals 4, connate, 2.2–3.5 × 1–2 cm, erect, triangular-ovate, green, prickly outside. Petals 20, in multiple whorls, oblong-lanceolate, 1–2 × 0.2–0.7 cm; outer purple-violet; inner whitish. Stamens many, 0.8–1 cm long; anthers oval, yellow. Ovary inferior, 8-carpelled, 8-loculed; style absent; stigma discoid, depressed, concave with as many rays as carpels. Berry dark purple, globose, spongy, densely prickly, crowned with the persistent calyx. Seeds eight to many, globose, black.

Fl. & *Fr.*: April–Dec.

Its Chromosome number is $2n = 58$ (Gupta and Roy 1973).

1.3.3 Fossil records of *Euryale* in Asia and Europe

All over the world, *Euryale* exists as a single species. In past geological periods, it grew in Europe, but it is completely extinct there at present (Okada, 1935). Miki (1960b) gave an account of Nymphaeaceae remains in Japan. The *Euryale* species obtained were classified as *E. ferox* Salisb, *E. nodulosa* Reid, *E. lissa* Reid and *E. europaea* Weber. Seeds of *E. ferox* from the Pleistocene stratum were obtained from 12 localities of different prefectures. *E. europaea* remains in Japan correspond to the species obtained from interglacial beds in Europe while those of *E. lissa* are similar to the species described from the Pliocene of Europe. The only report from India is by Dr. H. Falconer who discovered fossil seeds of *E. ferox* belonging to the Tertiary from Calcutta (north-east India) (cf. Hooker and Thompson, 1855). Closely related fossil species discovered in several localities of Europe include: (1) *Euryale limburgensis* Reid from the Pliocene strata of Tegelen clay in Holland, as well as from the Billstedt locality of Germany (Gripp and Beyle, 1937); (2) *Euryale lissa* Reid from the Pleistocene of Puentes de Garcia Rodriguez (Menendez Amor, 1975); (3) *Euryale europaea* Weber from the Pleistocene interglacial deposit at Likhvin in Kaluga (Weber, 1907) and from the Miocene of Konin (Ranicka-Bobrowska, 1959); (4) *Euryale carpatica* Szafer from the Pliocene sediments of Kroscienko (Szafer, 1947, 1954) and from the younger Tertiary of Poland. Fossil remains of *E. ferox* were obtained from the interglacial Likhvin sediments of European Pleistocene by Sukatscheff (1906), the Stanowice Pleis-tocene of Poland by Sobolewska (1970) and the Tonisberg

Holsteinian inter-glacial by Kempf (1966). Another account of its fossil record from the Oligocene of Scotland was given by Simpson (1936). Dorofeev (1963) suggested an evolutionary lineage of *E. ferox* with the Oligocene *Nikitinella* and *Tomskiella*, Miocene *Eoeuryale* and *Irtyshemia*, Pliocene *Pseudoeuryale* and Pleistocene *Palaeoeuryale*, all separate department from a succession leading to extant *Euryale*. Godwin (1975) studied the fossil and evolutionary history of *E. ferox* in temperate Europe and described it as a "discontinuous" species. Further accounts of the distribution in the past and evolutionary history of *E. ferox* have been given by Reid and Reid (1915), Miki (1933, 1937, 1960a), Nikitin (1957), Rochow (1957), Ushko (1959), Tralau (1963), Dorofeev (1972), Gall et al. (1974), Collinson (1980) and Jiffney and Collinson (1985).

Ancient exploitation of *E. ferox* at the Acheulian has been reported by Goren-Inbar et al. (2014). Israel dating back to approx. 790,000 years ago, their study documents the intentional harvesting and processing of *E. ferox* nuts, indicating a complex multistage and organized aquatic diet of early humans. They identified the plant remains via specific seed characteristics such as raphe, operculum and hilum. Their study connects the nuts to specific tools including stones and anvils, used for nut cracking. Prof. Inbar had an occasion to visit Darbhanga (Bihar, India) in the year 2011 to have a first-hand idea of the current use of the *E. ferox* nuts.

1.4 Key historical references

Early investigations into plant morphology and development laid the foundation for understanding phyllotaxis and the classification of aquatic plants such as *Euryale ferox*. Alexander Braun (1831/1835) made pioneering contributions to the study of phyllotaxis, the spatial arrangement of leaves and floral organs. Through detailed observations, he recognized recurring spiral patterns in several plant groups, including members of the Nymphaeaceae. His work highlighted the use of mathematical relationships, particularly fractional patterns, which approach the **golden angle** (**~137.5** degrees), which is now understood as a fundamental principle in plant organ arrangement. Wilhelm Hofmeister (1868) extended these ideas by proposing a developmental mechanism underlying phyllotaxis. He suggested that new organs (primordia) arise at positions of minimal spatial competition on the growing apex. This concept, often summarized as formation in the "least crowded space," naturally results in spiral arrangements consistent with the Fibonacci sequence and the golden angle. Richard Anthony Salisbury (1805) played a crucial role in the taxonomic history of *E. ferox*. He formally described the genus *Euryale*, distinguishing it within the family Nymphaeaceae. His classification work provided the systematic framework upon which later botanical and developmental studies have been built.

1.4.1 Systematic position

Kingdom: Plantae.
Phylum: Streptophyta
Class: Equisetopsida
Subclass: Magnoliidae
Order: Nymphaeales.
Family: Nymphaeaceae.

APG IV system of classification places Nymphaeaceae under the order Nymphaeales (Chase et al., 2016).

1.5 Cultural studies

Makhana is deeply entrenched with Mithila culture. It is sold as "prasad" all over India at religious shrines. Its white color makes it ideal for use during rituals related with last rites. There is a centuries old custom of sending its pops from the bride's side to the groom's place in the first year of their marriage. The festival of **Kojagara** is solemnized on the occasion of Ashwin Poornima (full moon day of the Indian calendar month of Ashwin, Oct–Nov.). In Mithila region it is essentially a marital ritual. The Ayurvedic contention of Makhana as aphrodisiac and spermatogenic properties have been experimentally substantiated through animal feeding trials at the Department of Zoology of the Banaras Hindu University, Varanasi, India (Shaw et al. 2025). Makhana pops forming part of the Ghunesh paag worn by the groom during marriage of Shrotriya Brahmins. Makhana garlands are offered to deities in temples and to distinguished guests.

Makhana (*E. ferox*) holds deep cultural, religious, and ethnic significance in the Mithila region as well as in other parts of India. Its uses are closely tied to rituals, traditions, and daily religious practices. In Mithila, during the sacred festival of Kojagari Purnima (observed on the full moon night of Ashwin in October), Makhana plays a special role in marital customs. The bride's family presents Makhana pops to the groom's family, a tradition unique to this region. Additionally, Makhana is offered as *prasad* to Goddess Lakshmi during this auspicious occasion.

On Somvati Amavasya, devotees use Makhana pops to count their circumambulations (*parikrama*) around temples or sacred trees, reflecting its role in ritual observance. During *Shraaddha Karma* (last rites rituals), Makhana is included among the offerings made to ancestors, symbolizing purity and faith. Makhana is also used in the preparation of non-flower garlands (*mala*), which are presented to deities and honored guests during religious ceremonies and public gatherings. Throughout the year, it is widely distributed as *prasad* at temples and religious shrines, reinforcing its sacred status. In the sacred thread ceremony (*Yajnopaveet*), starch derived from soaked Makhana pops is applied to strengthen the thread, highlighting its practical as well as ritualistic value.

During Navratri, sweet preparations such as *Makhaan Paagal* and *Makhaan Tasmai* are prepared and offered as part of the traditional *56 Bhog* to Goddess Durga. Kamla mai (the river Goddess) in formally worshipped before the Makhana harvest by the traditional harvesters, belonging to the fishing community.

Makhana also forms an essential component of *hawan samagri*, used in fire rituals along with sesame, barley, ghee, coconut, and honey. Furthermore, it is included in *Panchamewa*, a mixture of five dry fruits offered as *prasad* during various religious ceremonies. Thus, Makhana is not merely a food item but a culturally and spiritually significant element deeply embedded in the traditions and rituals of Mithila and Indian society. Makhana is commonly consumed during special occasions such as Diwali and Ashwin Purnima, where it is often prepared in the form of sweet dishes and desserts. An interesting physical property of Makhana is observed when its popped seeds are crushed into powder. This powder tends to float on the surface of water and does not mix easily unless it is stirred continuously. This floating behavior has attracted scientific attention due to its potential application in pharmaceutical science. **Makhana Dal Makhni (MDM)** is a creamy, protein rich twist on the classic Punjabi slow cooked black lentil dish incorporating roasted Makhana seeds for extra texture. Makhana roasted in ghee until crispy then added to dal during the final simmer. The unique property of *E. ferox* in the development of drug delivery systems. Notably, a study by Negi et al. (2011) utilized this characteristic to design a non-effervescent floating matrix tablet. Such tablets are based on the principle of floatability, allowing the drug to remain buoyant in the stomach for a longer duration, thereby improving drug release and absorption.

The National Research Center for Makhana, Darbhanga hosted a national seminar on this day on the October 17, 2024. A proposal has been mooted to declare this instant as National Makhana Day as per the Indian calendar. It is on this single day that there is maximum sale, purchase and distribution of Makhana.

1.5.1 A vehicle for culinary diplomacy

Makhana is fast emerging as a vehicle of culinary diplomacy in India. India recently showcased its rich agricultural heritage by serving premium Makhana dish during the 28th CSPOC (Conference of Speakers and Presiding Officers of the Commonwealth, Jan 2023) and the 18th G-20 Leaders' Summit, Sept. 2023) both held in the national capital, New Delhi. The initiative was supported by the Manigachhimidas Farmer Producer Company Ltd., Darbhanga. Out of the 38 key meetings, Makhana dish was served in 13 major events. The endeavor has been instrumental in opening global visibility for farmers and FPOs working in the field of Makhana. This is likely to open new avenues for export of value added agricultural products made of Makhana. Govt. of Bihar has organized "Makhana Mahotsav" (Festival of Makhana) at Patna and Bengaluru to popularize the crop and its diverse products.

PLATE 1.5 Cultural aspects of Makhana. 29. *Kojagara* festival - distribution of makhana amongst villagers, 30. *Ghunesh paag* strewn with makhana worn by a groom, 31. Makhana garland being offered to guests at RPCAU, Pusa on the occasion of inauguration of the Center for Excellence on makhana (26th February 2026), 32. Makhana *Kheer* (traditional/the one served in hotels with visible makhana pops), 33. Makhana *Panjiri* offered to Lord Krishna in Mathura, UP, India, 34. Makhana garland offered to Lord Shiva at Mahakaleshwar temple of Ujjain, M.P., India. Makhana garland offered to Goddess Kali in the Mithila region.

1.6 Conclusion

Makhana (*E. ferox* Salisb.) emerges as a plant of exceptional botanical, historical, and cultural importance. Botanically, it represents an evolutionarily significant basal angiosperm within the Nymphaeaceae, exhibiting a unique blend of dicot and monocot characteristics along with remarkable adaptations to aquatic environments. Its specialized features—such as floating leaves, cleistogamous reproduction, and a nutritionally rich perisperm—highlight both ecological resilience and economic value. Historically, fossil records across Europe and Asia indicate a much wider past distribution, emphasizing its status as a relict and monotypic genus today. Advances in genomic and phytochemical research further reveal its adaptive evolution, disease resistance potential, and bioactive compounds of medicinal importance. Culturally, Makhana holds deep-rooted significance, especially in the Mithila region of Bihar, where it is closely associated with rituals, traditions, and livelihoods. Its role in religious practices, traditional medicine, and local economies underscores its socio-cultural relevance.

In the modern context, *E. ferox* has gained recognition as a nutraceutical and "superfood," supported by its low glycemic index, antioxidant properties, and health benefits. Thus, integrating traditional knowledge with contemporary scientific research, Makhana stands as a promising crop for sustainable agriculture, nutritional security, and future biomedical applications.

References

Banerji, S.R., 1972. Infestation of *Euryale ferox* Salisb. by larvae of *Nymphula crisonalis* Walker and trials on its control. J. Bombay Nat. Hist. Soc. 69, 79–90.

Biswas, K., Calder, C.C., 1936. Handbook of Common Water and Marsh Plants of India and Burma. Govt. of India, New Delhi.

Box, F., Alexander, E., Jian, G., Chris, T., 2022. Gigantic floating leaves occupy a large surface area at an economical material cost. Sci. Adv. 8, eabg3790. https://doi.org/10.1126/sciadv.abg3790.

Braun, A. (1835). *On the Arrangement of Leaves in Plants (Phyllotaxis).* Leipzig, Germany: Leopold Voss.

Chase, M.W., Christenhusz, M.J.M., Fay, M.F., Byng, J.W., Judd, W.S., Soltis, D.E., Mabberley, D. J., Sennikov, A.N., Soltis, P.S., Stevens, P.F., 2016. An update of the angiosperm phylogeny group classification for the orders and families of flowering plants: APG IV. Botanical J. Linnean Soc. 181 (1), 1–20. https://doi.org/10.1111/boj.12385.

Chen, W.P., Zhang, S., Sheng, X.X., Weiming, Z., Guo, S., 1991. The anatomical structures and floral development of *Euryale ferox* Salisb. Acta Bot. Sin. 33 (2), 104–109.

Collinson, M.E., 1980. Recent and tertiary seed of the Nymphaeaceae *sensu lato* with a revision of *Brasenia ovula* (Brong.) Reid and Chandler. Ann. Bot. 46, 603–632.

Curtis's Botanical Magazine 1811-12. Euryale Ferox. Prickly Euryale, London, Academic Press, vols. 35–36:1447.

Devi, N., Rani, K., Kharb, P., Kaushik, P., 2022. Bio-fabrication of *Euryale ferox* (Makhana) leaf silver nanoparticles and their antibacterial, antioxidant and cytotoxic potential. Plants 11 (20), 2766. https://doi.org/10.3390/plants11202766.

Dorofeev, P.I., 1963. The Tertiary Flora of West Siberia. Moscow and Leningrad, p. 345 (in Russian).

Dorofeev, P.I., 1972. The taxonomy of Euryale ancestral forms. Bot. Zh. (Leningrad) 57, 1047–1054 (in Russian).

Dwivedi, A.K., Shekhar, R., Sharma, S.C., 1995. Ultrastructural studies of *Euryale ferox* leaf infected by *Alternaria alternata*. Indian Phytopathol. 48 (1), 61–65.

Gall, H., Jung, W., Dehm, R., 1974. Vorbericbt fiber die Tier und Pflanzenreste aus den obermiozaenen Riessee-Ablagerungen in der Forschungsbohrung Nordlingen. Geol. Bavarica 72, 53–58.

Gavhane, N.P., et al., 2023. Review on: prickly water lily. Int. J. Creat. Res. Thoughts 11 (1), 555–565.

Godwin, H., 1975. The History of the British Flora, second ed. Cambridge Univ. Press, Cambridge, p. 514.

Goren-Inbar, N., Melamed, Y., Zohar, I., Akhilesh, K., Pappu, S., 2014. Beneath still waters - multistage aquatic exploitation of Euryale ferox (Salisb.) during the Acheulian. 'human exploitation of aquatic landscapes' special issue. In: Ricardo (Ed.), Fernandes and John Meadows, Internet Archaeology. https://doi.org/10.11141/ia.37.1.

Gripp, K., Beyle, M., 1937. Das lnterglazial von Billstedt (Ojendorf). Mitt. Geol. Staat- sinst. Hamburg 16, 19–36.

Guo, Z., Min, J., 2020. Characterization the complete chloroplast genome of *Euryale ferox* (Nymphaeaceae), an medicinal plant species in China. Mitochondrial DNA B Resour. 5 (3), 2389–2390. https://doi.org/10.1080/23802359.2020.1775516.

Gupta, P.P., Roy, S.K., 1973. Karyotype of *Euryale ferox* Salisb. Curr. Sci. 42 (3), 106.

Gupta, U., Sonawan, S., Bhushette, P., Dabade, A., 2021. Low glycemic index simulated flour formulation from retrograded fox nut (*Euryale ferox*) starch. Ann. Food Sci. Technol. 22 (1), 76–84.43.

Haider, M.G., Nath, R.P., 1987. Chemical control of *Alternaria* leaf blight of makhana , *Euryale ferox*. Natl. Acad. Sci. Lett. 10 (9), 301–302.

Haider, M.G., Mahto, A., Mishra, R.K., 1995. Association of *Hirschmanniella oryzae* with *Euryale ferox* and its seasonal fluctuation. Indian J. Nematol. 25 (1), 115–116.

Haines, H.H., 1926. The Botany of Bihar and Orissa, Part II. Botanical Survey of India, Calcutta.

Hofmeister, W. 1868. *Allgemeine Morphologie der Gewächse (General morphology of plants)*. Leipzig, Germany: Wilhelm Engelmann.

Hooker, J.D., Thompson, T., 1855. Flora Indica. Royal Botanic Gardens, Kew, London.

Jha, V., Mishra, R.K., Jha, U.N., 1989. Storage pests of makhana (*Euryale ferox* Salisb. Geobios New Reports 8, 184–185.

Jian, X., Shan, T., Yang, Y., Lu, X., Ge, H., Li, L., Wang, H., Wang, L., 2026. Isolation, identification, and in-vitro antitumor activity of polyphenols from Euryale ferox Salisb. shell. Chem. Biodivers. 1, e03188. https://doi.org/10.1002/cbdv.202503188.

Jiang, J., et al., 2023. The ethnopharmacological, phytochemical, and pharmacological review of Euryale ferox Salisb.: A Chinese medicine food homology. Molecule 28 (11), 4399.

Jiffney, B.H., Collinson, M.E., 1985. Nymphaeaceae from the Oligocene. Biol. Abstr, 107847.

Jing, Z.H., Yin, M.J., Wang, Q., Bao, K., Zhou, P., Liu, C., Wu, Q., 2021. Expression profiling and functional verification of flavonoid 3′-hydroxylase gene from the leaves of *Euryale ferox*. China J. Chin. Times 46 (18), 4712–4720.

Kempf, E.K., 1966. Das Holstein Interglazial von Tonisberg in Rahmen des niederrheinischen Pleistoz~ins. Eiszeitalt. Ggw. 17, 5–60.

Khanna, P., 1964. Morphological and embryological studies in Nymphaeaceae *Euryale ferox* Salisb. Proc. Ind. Acad. Sci. 59 (4), 237–243.

Koren, O.G., Yatsunskaya, M.S., Nakonechnaya, O.V., 2012. Low level of allozyme polymorphism in relict aquatic plants of the Far East *Nelumbo komarovii* Grossh. and *Euryale ferox* Salisb. Russ. J. Genet. 48 (9), 912–919.

Kumar, N., Kumar, H., Kumari, S., 2017. Extraction of high quality genomic DNA for genetic diversity and molecular marker studies from leaves of makhana (*Euryale ferox*). J. Biodivers. 8 (2), 87–92.

Kumar, R., Sanduja, M., Painuly, N., 2024. A research on phytochemical and antimicrobial investigation of *Euryale ferox* seeds. Educ. Administr. Theory Pract. 30. https://doi.org/10.53555/kuey.v30i6.5884.

Kumar, R., Singh, I.S., Jha, V., 2025. Bioactive compounds in *Euryale ferox* Salisb. and their nutritional and therapeutic actions-a review. Vegetos 39. https://doi.org/10.1007/s42535-025-01209-x.

Laal, A.K., Datta Munshi, J.S., 1985. Phenology of *Euryale ferox* Salisb. Biol. Bull. India 7 (1), 13–20. 312.

Lee, S.E., Ju, M.E., Kim, J.H., 2002. Antioxidant activity of extracts from *Euryale ferox* seed. Exp. Mol. Med. 34 (2), 100–106.

Liu, A., Qu, C., Ling, W., Wang, T., Zhang, Y., Fang, Y., Guo, Y., He, M., Zou, J., Zhao, S., Feng, K., Wu, P., Li, L., 2025. A model for the adaptation of *Euryale ferox* leaves to aquatic environments through EfCGT1-controlled flavonoid C-glycoside-specific accumulation in epidermis cells. Plant Biotechnol. J. 23 (8), 3333–3348. https://doi.org/10.1111/pbi.70155.

Liu, A.L., Wang, Y.H., Wang, T.Y., Zhu, Y., Wu, P., Li, L.J., 2023. Comparative metabolomic profiling of secondary metabolites in different tissues of *Euryale ferox* and functional characterization of phenylalanine ammonia-lyase. Ind. Crop. Prod. 195, 116450. https://doi.org/10.1016/j.indcrop.2023.116450.

Mahto, A., Haider, M.G., Mishra, R.K., Jha, V., 1993. Nematode associations in pond growing *Euryale ferox* Salisbury (makhana) in Darbhanga (North Bihar) , India. J. Freshw. Biol. 5 (4), 299–304.

Menendez Amor, J., 1975. Paleobotanical analysis of some lignite samples from puentes do garcia Rodriguez (ka Coruna). Bol. R. Soc. Esp. Hist. Nat. Secc. Geol. 73 (1–4), 121–124.

Mishra, R.K., Saraswati, K.C., Jha, V., Kumar, R., 1989. Behavioral observations of Rhopalosiphum nymphaeae Linn. on the leaves of *E. ferox* Salisb. in Darbhanga, North Bihar. Biojournal 1 (2), 123–128.

Miki, S. 1933. On the Pleistocene flora in Prov. Yamashire etc. Bot. Mag. (Tokyo), 47, 619–631.

Miki, S., 1937. Plant fossils from the Stegodon bed and Elephas beds near Akashi. Jpn. J. Bot. 8, 303–341.

Miki, S., 1960. Nymphaeaceae remains in Japan with few fossil genus *Eoeuryale*. J. Inst. Polytech. Osaka City Univ. D11:, 63–78.

Miki, S., 1960. Seed remains of *Euryale* in Japan. J. Geobot. 8, 53–57.

Mishra, R.K., Jha, V., Kumar, R., Jha, B.P., 1990. The pests of *Euryale ferox* Salisb. in North Bihar. Environ. Ecol. 8, 133–136.

Nam, G.H., et al., 2019. *In vitro* and *in vivo* induction of p53-dependent apoptosis by extract of *Euryale ferox* Salisb in A549 human caucasian lung carcinoma cancer cells is mediated through akt signaling pathway. Front. Oncol. 9, 13406.

Nayar, M.P., 1985. Meaning of Indian Flowering Plant Names. Bishen Singh Mahendra Pal Singh, Dehradun.

Negi, J.S., Jugran, V., Kasliwal, N., 2011. Development of non-effervescent floating matrix tablets based on *Euryale ferox* seeds. Asian J. Pharm. 5 (2), 93–100.

Nehal, N., Mann, S., Gupta, R.K., 2015. Two promising under-utilized grains: a review. Indian J. Tradit. Knowl. 14 (3), 416–422.

Nikitin, P.A., 1957. Pliocenovye I Cet Vertienye Flory Voronezskoj Oblasti. Bot. Inst. Ira. V.L. Komarova A K Nauk SSSR, Moskva-Leningrad.

Okada, Y., 1928. Study of *Euryale ferox* Salisb. I. On the size of leaves, fruits etc. with some remarks on the mode of expansion of the leaf blade. II. On the variation in the shape of the seed. Sci. Rep. Tohoku Imp. Univ. Ser. IV (Biol.) 3, 271–279.

Okada, Y., 1935. Study of *Euryale ferox* Salisb. VIII. Miscellany. Sci. Rep. Tohoku Imp. Univ. Ser. IV (2), 455–459.

Okada, Y., 1938. On chasmogamous flowers of *Euryale ferox* Salisb. Ecol. Rev. 4, 159–163.

Okada, Y., Otaya, T., 1930. Study of *Euryale ferox* Salisb. VI. Cleistogamous versus chasmogamous flowers. Bot. Mag. (Tokyo) 44, 369–373.

Qian, L.H., Wu, J.Y., Wang, Y., Zou, X., Zhou, G.C., Sun, X.Q., 2022. Genome- wide analysis of NBS-LRR genes from an early-diverging angiosperm *Euryale ferox*. Front. Genet. 13, 880071. https://doi.org/10.3389/fgene.2022.880071.

Raniecka-Bobrowska, J., 1959. Trzeciorzedowa flora nasienna z Konina. Inst. Geol. Biul. 130 Zbadan trzeciorzedu w Polsce 2, 159–252.

Reid, C., Reid, E.M., 1915. The Pliocene Floras of the Dutch-Prussian Border. Gravenhage.

Rochow, M.V., 1957. Stacheln yon *Euryale* in norddeutschen Pleistoz~in, ein neues Interglazial Fossil. Geobot. Inst. Rtibel, Ziirich 34, 136–142.

Salisbury, R. A. (1805). Paradisus Londinensis. London, England: W. Hooker.

Saraswati, K.C., Mishra, R.K., Kumar, R., Jha, V., 1990. *Rhopalosiphum nymphaeae* Linn. infestation on the leaves of *Euryale ferox* in Darbhanga. J. Aphidol. 4 (1 & 2), 89–92.

Shaw, R., Deep, S., Jha, V., Chaube, R., 2025. *Euryale ferox* Salisb. ameliorates cadmium-induced testicular and hepatic impairments in male mouse model. Int. J. Biol. Innov. 7 (1), 54–65. https://doi.org/10.46505/IJBI.2025.7107.

Simpson, J.B., 1936. .Fossil pollen in Scottish tertiary coals. Proc. R. Soc. Edin. 56, 90–108.

Singh, B., Kumar, S.R., Harun, M., et al., 2025. Elucidating the seed storage behaviour of *Euryale ferox*: implications for seed genebanking. Genet. Resour. Crop. Evol. 72 (Suppl 1), 457–466. https://doi.org/10.1007/s10722-025-02604-8.

Sinha, C.B.K., Jha, V., 2025a. Impacts of climate change on cultivation of makhana (*Euryale ferox* Salisb.) and bio control of weeds- a case study of Darbhanga district in Bihar, India. Int. J. Plant Environ. 11 (2), 388–394.

Sinha, C.B.K. and Jha, V. 2025b. Identification of nutraceutically important metabolites in raw and popped makhana (*Euryale ferox* Salisb.) seeds using GC-MS technique. Int. J. Plant Environ. 11 (3), 514–524. https://doi.org/10.18811/ijpen.v11i03.09.

Sobolewska, M., 1970. *Euryale ferox* Salisb. in the Pleistocene of Poland. Acta Palaeobot. 11, 13–20.

Sodi, M., Kumar, S., 2020. Association of makhana (*Euryale ferox* Salisb.) with macrophyte weeds in Katihar district of Bihar, India. Sch. J Agri. Vet. Sci. 7, 231–239.

Subramaniam, K., 1962. Aquatic Angiosperms. C.S.I.R, New Delhi.

Sukatscheff, W., 1906. Ueber des Vorkommen der Samen von *Euryale ferox* Salisb. in einer inlerglazialen Ablagerung in Russland. Ber. Dtsch. Bot. Ges. B26, 132–137.

Szafer, W., 1947. Flora Pliocenska z Kroscienka n Dunajcem. Rozpr. Wydz. 3 Nauk Mat. Przyr. PAU. 72, 1–123.

Szafer, W., 1954. Pliocenska flora okolic Czorsztyokadna. Inst. Geol. Prace 1 (l).

Liu, X., He, Z., Yin, Y., Xu, X., Wu, W., Li, L., 2018. Transcriptome sequencing and analysis during seed growth and development in *Euryale ferox* Salisb. BMC Genom. 19 (1), 343. https://doi.org/10.1186/s12864-018-4707-9.

Tralau, H., 1963. Asiatic dicotyledonous affinities in Cainozoic Flora of Europe. K. Sven. Vetensk. Handl. 9, 7–87.

Tsyrenova, D.J., 2021. *Euryale ferox* Salisb. and *Nuphar japonica* DC. (Nymphaeaceae) micromorphology study from the Amur region. IOP Conf. Ser. Earth Environ. Sci. 895.

Ushko, K.A., 1959. Lichvinkskij (Czekalinskij) Razrez Mezlednikovych Ozernych Otlozenij. Kednikovyj Period, Moskwa.

Verma, R.A.B., Jha, V., 1999. New *Doassansiopsis* sp. associated with *Euryale ferox* Salisb. (makhana) in north Bihar (India). J. Freshw. Biol. 11 (1–2), 7–10.

Weber, C.A., 1907. Euryale europaea. Nov. Sp. Foss. Ber. Dtsch. Bot. Ges. 25, 150–157.

Wu, P., Wang, L., Liu, S., Liao, B., Xu, Y., Ma, D., Yang, M., 2021. Metabolomics and transcriptome analysis of the biosynthesis mechanism of flavonoids in the seeds of *Euryale ferox* Salisb. at different developmental stages. Mol. Genet. Genom. 296 (4), 953–970.

Wu, P., Zhang, L., Zhang, K., Yin, Y., Liu, A., Zhu, Y., Fu, Y., Sun, F., Zhao, S., Feng, K., Xu, X., Chen, X., Cheng, F., Li, L., 2022a. The adaptive evolution of *Euryale ferox* to the aquatic environment through paleo-hexaploidization. Plant J. 110, 627–645. https://doi.org/10.1111/tpj.15717.

Wu, P., et al., 2022b. E_fAB14 Transcription Factor Is Involved in the Regulation of Starch Biosynthesis in *Euryale ferox* Salisb (seeds).

Xu, L., Liu, A., Wang, T., Wang, Y., Li, L., Wu, P., 2023. Characterization and coexpression analysis of the TIFY family genes in Euryale ferox related to leaf development. Plants (Basel) 12 (12), 2323. https://doi.org/10.3390/plants12122323.

Yameen, M., et al., 2026. Anxiolytic and antidepressant potential of *Euryale ferox* Seeds: a preclinical study. Ann. ASH & KMDC 31 (1), 21–31.

Yang, Y., Sun, P., Lv, L., et al., 2020. Prickly waterlily and rigid hornwort genomes shed light on early angiosperm evolution. Nat. Plants 6, 215–222. https://doi.org/10.1038/s41477-020-0594-6.

Zhang, W.N., et al., 2019. Structural characterization and in vitro hypoglycemic activity of a glucan from *Euryale ferox* Salisb. seeds. Carbohydr. Polym. 209, 363–371.

Further reading

Boyd, C.E., 1968. Freshwater plants: a potential source of protein. Econ. Bot. 23, 123–127.

Guo, L., You, C., Zhang, H., Wang, Y., Zhang, R., 2022. Genome-wide analysis of NBS-LRR genes in Rosaceae species reveals distinct evolutionary patterns. Front. Genet. 13, 1052191 (doi.org).

Huang, Z., et al., 2020. Small auxin up RNAs in future distribution of indole 3 acetic acid play a potential role in increasing seed size in *Euryale ferox* Salisb. BMC Plant Biol. 2020.

Jha, V., Sinha, C.B.K., 2024. Economic and medicinal significance of makhana (*Euryale ferox* Salisb.). CGES Newsletter 9 (1), 5–11.

Jha, V., Sinha, C.B.K., 2025. Enhancing biomass production and utilization for sustainable energy in India. In: Yadav, A.K. (Ed.), Powering future: clean energy solutions for a sustainable planet. Clean energy production technologies. Springer, Singapore. https://doi.org/10.1007/978-981-96-5055-2_5.

Jha, V., Sinha, C.B.K., Das, D., 2026. Role of Makhana (*Euryale ferox* Salisb.) in shaping the economy and culture of Mithila region in Bihar, India. J. Exp. Environ. Res. 1 (1), 50–53.

Jha, V., Verma, A.M., Jha, A.K., 2003. Indigenous contrivances utilized in Makhana cultivation in North and North-Eastern India. In: Mishra, R.K., Jha, V., Dehadrai, P.V., MAKHANA, D.I.P. A. (Eds.), Indian Council of Agricultural Research, New Delhi, pp. 241–248.

Jha, V., Shalini, R., Kumari, A., Jha, P., Sah, N.K., 2018. Aquacultural, nutritional and therapeutic biology of delicious seeds of *Euryale ferox* Salisb.: a mini review. CurrPharm. Biotechnol. 19 (7), 545–555. https://doi.org/10.2174/138920101966618080816005844.

Kadono, Y., Schneider, E.L., 1987. The life history of *Euryale ferox* Salisb. in Southwestern Japan with special reference to reproductive ecology. Plant Species Biol. 2, 109–115.

Laal, A.K., 1981. Studies on the Ecology and Productivity of Swamps in North Bihar in Relation to Production of Fishes and Other Commodities. Ph.D. thesis. Bhagalpur Univ., Bhagalpur, India.

Sinha, C.B.K., 2024. Economic and medicinal significance of makhana (*Euryale ferox* Salisb.). CGES Newsletter 9 (1), 5–11.

Chapter 2

Taxonomy, morphology, and growth ecology of *Euryale ferox*

Zahid Ahmad Thoker[1], Lovleen Marwaha[1], Anand Mohan[2] and Ajit Prakash[3]

[1]*Department of Zoology, School of Bioengineering and Biosciences, Lovely Professional University, Phagwara, Punjab, India;* [2]*Department of Botany, School of Bioengineering and Biosciences, Lovely Professional University, Phagwara, Punjab, India;* [3]*Department of Biochemistry and Biophysics, University of North Carolina, Chapel Hill, NC, United States*

2.1 Introduction

Freshwater ecosystems are one of the most ecologically diverse and most biologically productive environments on earth that accommodate various communities of plants, microorganisms, and animals (Dudgeon et al., 2006). Aquatic macrophytes are among the leading producers in these ecosystems that are important in maintaining the ecological stability, nutrient cycle, and biodiversity (Lesiv et al., 2020). These plants consist of a very diverse array of submerged, floating, and emergent species adapted to live in wet water and are spread throughout lakes, ponds, wetlands, and slow-moving rivers. Macrophytes play an important role in regulating freshwater ecosystems because they create not only a food supply but also shelter and breeding sites for many aquatic life (Zedler and Kercher, 2005; Dudgeon et al., 2006; Thomaz and Cunha, 2010; Bornette and Puijalon, 2011; Alahuhta et al., 2021; Thiemer et al., 2021). Besides their ecological roles, aquatic plants also play a role in water purification, carbon sequestration, and sediment control, thus improving the well-being and productivity of the aquatic environments in general (Chaurasia, 2022; Rather and Sharma, 2023).

Euryale ferox Salisb (also fox nut or makhana) is a distinctive species of the *Nymphaea*ceae family of aquatic Macrophytes. The plant is especially remarkable because of its very specific morphological characteristics, ecological adaptation, and economic significance. *Euryale ferox* is a floating-leaf water-loving plant that is distributed extensively in Asia (in India, China,

Makhana (Euryale ferox) as a Functional Food. https://doi.org/10.1016/B978-0-443-44421-0.00010-X

Bangladesh, Japan, and Korea). In South and East Asia, the seeds of this plant have a long history of usage as a source of food and as a traditional medicine (Kapoor et al., 2022; Liaquat et al., 2022; Kumar et al., 2025). They are rich in nutrients and bioactive compounds, including polysaccharides, polyphenols, and minerals; these nutrients and compounds make the seeds antioxidants and therapeutic (Kapoor et al., 2022). The species has received growing interest over the past decades due to its nutritional and medicinal value, drawing attention from researchers in diverse disciplines, such as plant biology, aquatic ecology, agriculture, and food science. Knowledge of taxonomy, morphology, and growth ecology of *Euryale ferox* is thus necessary in addition to botanical and ecological research, as well as sustainable management and cultivation of this economically significant aquatic plant.

2.1.1 Importance of aquatic macrophytes in freshwater ecosystems

Aquatic macrophytes are a vital part of freshwater ecosystems, and they play various ecological roles that aid the stability and productivity of the ecosystem. These are primary producers, which change solar energy into biomass by a process called photosynthesis, and they are at the foundation of many aquatic food webs. They affect nutrient cycles, sedimentation, and habitat organization in the water body (Bornette and Puijalon, 2011; Lesiv et al., 2020; Alahuhta et al., 2021). Macrophytes also significantly act as a source of ensuring water quality by absorbing excessive nutrients, such as nitrogen and phosphorus in the surrounding water, thus setting in check the eutrophication and algal blooms (Chaurasia, 2022; Rather and Sharma, 2023).

Besides the control of nutrients, aquatic macrophytes serve as an important habitat and refuge for a diverse range of aquatic organisms, such as fish, invertebrates, amphibians, and microorganisms. Thick macrophyte beds may decrease turbulence of water, stabilize sediments, and increase oxygen supply in water bodies. They therefore facilitate intricate ecological interactions and add to the total biodiversity of freshwater ecosystems (Thomaz and Cunha, 2010).

Macrophytes also serve as valuable bioindicators of the environmental conditions. Macrophyte diversity, abundance, and community structure changes have a strong tendency to indicate changes in water quality, nutrients, or hydrological regimes. This is why aquatic macrophytes are often adopted in ecological control programs and freshwater ecosystem evaluation (Heegaard et al., 2001). In addition to ecological functions, a good number of aquatic plants can be of economic importance as food, fodder, medicinal sources, and raw materials to the different industries. Consequently, the biology and ecology of aquatic macrophytes are important to the conservation of the environment and long-term management of resources.

2.1.2 Economic and ecological significance of *Euryale ferox*

Euryale ferox is a particular plant among the various aquatic macrophytes found in freshwater wetlands because of the nutritional, medicinal, and economic value it has. The seeds of the plant are large and starchy in nature, which are refined to make edible puffed kernels commonly referred to as makhana or fox nuts. These seeds contain high carbohydrates, proteins, dietary fiber, and other bioactive compounds, which is why they make good functional food with possible health benefits (Kapoor et al., 2022; Kumar et al., 2025). It has been stated that *Euryale ferox* seeds are rich in polyphenols, flavonoids, and essential minerals, which can explain their antioxidant, antidiabetic, and cardioprotective effects (Liaquat et al., 2022; Kumar et al., 2025).

Makhana is a significant agro-aquatic farming system in India, and mostly in the state of Bihar, which supports the livelihoods and economies of the rural areas. Ponds, wetlands, and low water bodies favor the growth of the crop because it can grow well in shallow freshwater (Jha and Kargupta, 1991). The economic value of the *Euryale ferox* has grown significantly over the past few years because of the rise in consumer demand for nutritionally rich and plant-based food items.

Euryale ferox plays a role in the ecology of wetlands in a number of ways. Its extensive floating foliage provides cover and protection to aquatic organisms, its root system is involved in interacting with sediment and microbial communities, and it affects nutrient cycling and sediment stabilization (Wetzel, 2001; Thomaz and Cunha, 2010). The plant is also intertwined with other complex wetland plant communities since it interacts with other macrophytes and aquatic organisms. Consequently, *Euryale ferox* is significant in the ecological stability of freshwater wetlands in which it is naturally found.

2.1.3 Historical background and domestication

The cultivation of the *Euryale ferox* can be traced back several centuries in certain areas in Asia, where the plant has been traditionally cultivated to provide its edible seeds. It is recorded that the species has been a source of both food and medicine in the ancient Chinese and Indian traditional medical systems. According to the Chinese medical literature on classical Chinese medicine, the seeds of *Euryale ferox* were reported to be a functional food, which is therapeutic, especially in strengthening the digestive system and enhancing kidney functions (Kumar et al., 2025).

Makhana farming emerged as an ancient wetland crop in India, especially in the eastern Indo Gangetic plains. With time, local farmers came up with specially designed methods of planting the plant in ponds and swamps that were shallow in nature. These conventional methods of cultivation include seed broadcasting, seed harvesting by hand, and seed processing into puffed kernels (Jha and Kargupta, 1991). Despite the long-term farming of the crop, scientific studies of the biology, ecology, and agronomy of the crop have only begun to grow in the last few decades.

Recent research has started to investigate the genetic variability, metabolic makeup, and the ecological change of *Euryale ferox*. New understanding of the evolutionary adaptation and biochemical features of this species has been attained through advances in plant physiology, molecular biology, and metabolomics (Wu et al., 2022). This kind of research is necessary to enhance cultivation, save on genetic materials, and improve the productivity of the makhana production system.

2.1.4 Scope and objectives of the chapter

Still, in spite of the ecological and economic significance, no extensive scientific data about the taxonomy, morphology, and ecological traits of *Euryale ferox* have been accumulated, even though the taxon belongs to the fields of botany, ecology, agriculture, and food science. The species requires a unified knowledge base so that additional studies and sustainable use can be done. This chapter mainly aims to give an overall view of the taxonomy, morphology, and growth ecology of *Euryale ferox*. The chapter opens by discussing in some detail the taxonomic position and classification of the species in its family, Nymphaeaceae. This is then accompanied by the analysis of its vegetative and reproductive morphology, as well as the structural characteristics that allow it to survive in the water. The ecological reasons that affect the growth, distribution, and productivity of the species in both natural and cultivated habitats are also discussed in the chapter. This chapter should add to the existing knowledge in the fields to gain a clearer picture of the biological and ecological properties of *Euryale ferox*. This kind of knowledge is necessary to enhance cultivation practices, to preserve the biodiversity of wetlands, and to ensure sustainable utilization of aquatic plant resources.

2.2 Taxonomic classification and systematics

The systematic system of plant species classification gives a fundamental paradigm of the evolutionary association, morphological variety, and ecological adjustments. *Euryale ferox* Salisb. is an aquatic angiosperm plant, which belongs to the family *Nymphaea*ceae, a common group of freshwater plants commonly referred to as water lilies. The species is very interesting since it is the sole species of the genus *Euryale* that is still extant and which has several distinct morphological features that make it distinct among the other species of the family (Schneider and Williamson, 1993).

The *Nymphaea*les order, to which *Euryale ferox* belongs, is one of the earliest lineage splits of flowering plants. The representatives of this order are mostly water-based herbs with floating leaves, rhizomatous forms of growth, and with specific reproductive organs that are adapted to live in water (Borsch et al., 2008; Endress and Doyle, 2009; Soltis et al., 2018). Due to their low position in the evolutionary history of angiosperms, species in the *Nymphaea*les have been the focus of many studies on the origin and diversification of flowering plants (Borsch et al., 2008; Zhang et al., 2020).

Traditionally, morphological characteristics were used as the main criteria to classify *Euryale ferox*, including floral arrangement, leaf structure, and fruit morphology. Nevertheless, recent discoveries in molecular phylogenetics and genomic studies have presented new information on the evolutionary relationships among genera in the *Nymphaea*ceae (Löhne et al., 2007; Zhang et al., 2020; Wu et al., 2022). These studies have served to explain the formal position of *Euryale* and its connection with other water genera, such as *Nymphaea* and *Victoria*.

2.2.1 Taxonomic hierarchy of *Euryale ferox*

The taxonomic grouping of *Euryale ferox* indicates its accommodations in the wider grouping of flowering plants. Being part of angiosperms, the species is a big family of plants with closed seeds and special reproductive organs. *Euryale ferox* belongs to the order Nymphaeales, which has a number of aquatic plant families that are early-diverging angiosperm lineages (Soltis et al., 2018; Zhang et al., 2020).

The family *Nymphaea*ceae includes perennial water herbs, with floating leaves and conspicuous, usually insect-pollinated flowers. *Euryale ferox* is considered the only species of the genus *Euryale* that exists in this family (Schneider and Williamson, 1993; Borsch et al., 2008). Table 2.1 summarizes the hierarchical classification of the species by giving the systematic position of the species at the level of the kingdom to the species.

2.2.2 Family *Nymphaea*ceae: Diagnostic characteristics

*Nymphaea*ceae is a family of aquatic perennial herbs that is heavily dispersed in tropical and temperate freshwater habitats. The members of this family are usually attached to the lower sediments with the leaves positioned on the water surface. The leaves are usually wide and circular and are held by long petioles, which are born out of rhizomes that are embedded in the substrate (Sculthorpe, 1967; Schneider and Williamson, 1993). The large and visually prominent flowers that typically possess a high number of petals, stamens, and carpels are another hallmark of the *Nymphaea*ceae, with a spiral arrangement. These flower characteristics are deemed as being relatively primitive in angiosperms and serve as valuable indications of the evolutionary early history of flowering plants (Borsch et al., 2008; Soltis et al., 2018). The reproductive structures of the family are usually created above the water or close to the water, and this enables insects, such as beetles and flies to pollinate them. The fruits that grow after fertilization tend to be full of seeds that are suited to be dispersed in

TABLE 2.1 Taxonomic classification of *Euryale ferox*.

Taxonomic rank	Classification
Kingdom	Plantae
Clade	Angiosperms
Clade	Basal angiosperms
Order	*Nymphaeales*
Family	*Nymphaeaceae*
Genus	*Euryale* Salisb.
Species	*Euryale ferox* Salisb.

water. In this family, a number of genera are identified, such as *Nymphaea*, *Victoria*, *Barclaya*, and *Euryale*, whose morphological features are unique (Borsch et al., 2008; Zhang et al., 2020).

2.2.3 Genus Euryale: Evolutionary position

Euryale is a distinct genus in the family *Nymphaea*ceae in terms of evolution. *Euryale* is also a monotypic genus, unlike other genera in the family, which are represented by several species, and as a result, has only one species, *Euryale ferox* (Schneider and Williamson, 1993). Morphological studies have indicated that *Euryale* has a few structural similarities with the genus *Victoria* that includes giant Amazon water lilies. Both genera have enormous floating leaves and strong reproductive organs that have aquatic adaptations. Nonetheless, the abundance of defensive spines on the leaves, petioles, and fruits of *Euryale ferox* is a distinctive feature that separates it from the species of *Victoria* (Schneider and Williamson, 1993). The evolutionary meaning of the genus is that it represents a mixture of both primitive and specialized traits of the genus. Although the floral structure indicates characteristics that are usually found in early angiosperms, the large evolution of spines indicates evolutionary adaptations to environmental factors, such as herbivory and environmental stress (Sculthorpe, 1967).

2.2.4 Botanical nomenclature and synonyms

The English botanist Richard Anthony Salisbury, whose name is given in the abbreviation used in botanical authorities Salisb, first formally described the species *Euryale ferox* after the name of the species. The genus *Euryale* is a Greek mythological name of one of the Gorgons and probably has reference to the spiny look of the plant. Another name that displays the aggressive nature of the plant is the species name ferox, which translates to fierce or formidable; this is concerning the multiple sharp spines that cover the vegetative and reproductive parts of the plant. Several changes in taxonomic interpretation have been used to refer to the species in the course of the history of botanical classification. However, as per the International Code of Nomenclature of algae (as well as fungi and plants), *Euryale ferox* Salisb. is now the official scientific name (Turland et al., 2018). It is also commonly referred to by other vernacular names in different parts of the world, in addition to its scientific nomenclature. In India, it is also known as makhana or fox nut, but in China, it is widely known as gorgon fruit.

2.2.5 Phylogenetic relationships with related genera (*Nymphaea* and *Victoria*)

Euryale ferox shows strong phylogenetic ties with other aquatic genera within the *Nymphaea*ceae, for example, *Nymphaea* and *Victoria*. These genera have a number of morphological characteristics, such as floating leaves, rhizomatous growth forms, and large structures of reproductive organs, which are designed to support aquatic pollination systems (Borsch et al., 2008; Soltis et al., 2018). In spite of such similarities, the genera differ significantly in important morphological features. The species of *Nymphaea* usually have smooth leaves without protective spines and usually have typical radial notches in the leaf blade. However, species of Victoria have very massive floating leaves with raised edges that can hold a lot of weight on the water surface (Schneider and Williamson, 1993). *Euryale ferox*, though, has large floating leaves heavily densely covered with spines on both vegetative and reproductive structures. These characteristics give a solid morphological foundation on which the species is differentiated from other members of the family.

2.2.6 Molecular phylogeny and genetic studies

The development of molecular phylogeny has greatly helped to understand the evolutionary relationships of *Nymphaea*ceae. Contemporary research makes use of chloroplast, mitochondrial, and nuclear genomes' DNA sequences to evolve phylogenetic connections among plant taxa. Genomic studies have indicated that *Euryale ferox* has many genetic adaptations relating to aquatic living, which are genes that deal with leaf buoyancy, aquatic photosynthesis, and environmental resistance (Wu et al., 2022). Comparative genomic analysis has also given information on the evolutionary split of *Euryale* with other genera of the family. Massive phylogenomic studies of basal angiosperms have supported the claim that the representatives of the *Nymphaea*les constitute one of the first to evolve in flowering plants (Soltis et al., 2018; Zhang et al., 2020). These results demonstrate the value of *Euryale ferox* as a model species to study the behavior of aquatic vegetation and the early evolutionary development of angiosperms.

2.3 Botanical description and diagnostic features

Euryale ferox Salisb. is a floating-leaf water macrophyte, a species of the family *Nymphaea*ceae, and is very widely spread in the freshwater wetlands in South and East Asia. The plant also has a number of morphological adaptations that help survive in a shallow aquatic environment, such as floating leaves, long petioles, and a rhizomatous anchoring mechanism internalized in the bottom sediments (Kadono and Schneider, 1987; Conran, 2007). Such structural characteristics are common to most of the representatives of the order Nymphaeales, which is one of the first lines of angiosperms to diverge (Borsch et al., 2008; Soltis et al., 2018).

Euryale ferox, in comparison to other genera of the *Nymphaea*ceae, has some characteristic morphological features, the most prominent being the sharp spines in the leaves, petioles, and fruits. These spines are believed to be defensive mechanisms that might be used to guard the plant against the herbivore activities of aquatic animals. The enormous floating leaves also grow in the plant and spread over the water surface to enable a good amount of light to be absorbed and gases to be exchanged in the water bodies (Sculthorpe, 1967; Kadono and Schneider, 1987). The botanical description and diagnostic characters of *Euryale ferox* are necessary to identify the species and be able to differentiate it from other aquatic species that belong to the family *Nymphaea*ceae (Borsch et al., 2008).

2.3.1 General habit and growth form

Euryale ferox is traditionally defined as an aquatic herb growing in the annual cycle in shallow freshwater ecosystems, such as wetlands, ponds, and lakes. Seed germination normally takes place in the bottom sediments where the seeds are put into a dormant state until the environment creates favorable conditions that arouse growth (Kadono and Schneider, 1987). After germination, the plant grows underwater juvenile leaves, and then it grows to have large floating leaves, which extend above the water surface. The plant is rooted in the sediment with a short rhizome and fibrous root system, which takes in nutrients from the substrate. Long petioles protrude through the column of water and hold the floating leaves on the surface of the water. In good situations, the leaves may grow to very large sizes that extend up to about one - meter in diameter (Schneider and Williamson, 1993). This structure of leaves in a rosette form optimizes the sunlight exposure and at the same time shades the surrounding water surface, hence affecting the microhabitat conditions within the aquatic ecosystem (Wetzel, 2001; Thomaz and Cunha, 2010).

2.3.2 Distinctive morphological traits

Several characteristic features that obviously distinguish it among other aquatic plants can describe the morphology of *Euryale ferox*. The most

FIGURE 2.1 Diagnostic morphological features of *Euryale ferox*.

obvious feature is the fact that there are many stiff spines that cover the leaves, petioles, and reproductive organs. It is believed that these defensive structures help the plant against herbivorous animals and are thought to help in the support structure in water bodies. The leaves are large, circular, and floating in nature, with the upper surface being smooth and the lower surface being spiny with distinct venation on the lower side. The petioles are long, flexible, and cylindrical in nature so that the leaves can also adjust to water changes in level. The flowers are solitary, generally purple or violet in color, and contain many petals around the carpels or stamens in the center. The flowers grow into spiny fruits that have huge starchy seeds that are harvested as makhana after pollination (Wu et al., 2022; Kumar et al., 2025). The main morphological characteristics of species identification are represented in Fig. 2.1 that summarizes the vegetative and reproductive organs of *Euryale ferox*.

2.3.3 Comparison with closely related aquatic species

Euryale ferox, belonging to the family *Nymphaea*ceae, has a few morphological similarities to genera like *Nymphaea* and *Victoria*. Such genera have floating leaves, rhizomatous growth forms, and enormous reproductive structures that have adapted to aquatic systems of pollination (Borsch et al., 2008; Soltis et al., 2018). There are, however, critical morphological disparities that exist between these taxa. *Nymphaea* species are usually soft-leaved and show a notch in the leaf blade that is usually radial in shape. However, species of *Victoria* have very large floating leaves whose margins are raised and able to bear significant weight. Though the leaves of the *Victoria* species might have spines under the leaf surface, they have a very different leaf architecture, unlike *Euryale ferox* (Schneider and Williamson, 1993). The heavy spines on both the vegetative and reproductive organs, its globose fruits, and the peculiar form of the leaves render *Euryale ferox* easy to recognize among other aquatic representatives of the *Nymphaea*ceae.

2.4 Morphology of vegetative structures

The vegetative morphology of *Euryale ferox* Salisb. is an indication of a series of structural adaptations that help the plant to survive and reproduce effectively in freshwater aquatic habitats. Similar to other family structures in the *Nymphaea*ceae, the species have specialized vegetative components, such as a sediment-held root system, rhizomatous stem structure, long petioles, and very big floating leaves. They are used to absorb nutrients, ensure floatation, photosynthesize, and support their structure in the water (Sculthorpe, 1967; Colmer, 2003; Evans, 2004; Soltis et al., 2018). Aquatic macrophytes have vegetative organs that tend to have morphological and anatomical adaptations that are largely different when compared with the vegetative organs of terrestrial plants. Such adaptations involve the presence of reduced mechanical tissues, massive aerenchyma development, which is an internal gas transport and specialized epidermal features that control water exchange and guard plant tissues (Sculthorpe, 1967; Thomaz and Cunha, 2010; Bornette and Puijalon, 2011). These adaptations are especially intense in *Euryale ferox* because the plant has large floating leaves and is ecologically related to the shallow wetlands and ponds. Morphology of plants, hence, is important in ensuring that they stay afloat, capture as much light as possible, and that they can perform physiological roles when under water (Wetzel, 2001).

2.4.1 Root system and rhizosphere interactions

Root system *Euryale ferox* is mostly linked with the bottom sediment of the freshwater bodies where it has been planted. The roots are a result of a short rhizome that is set in the substrate and stretches in the surrounding sediment to absorb water and other vital nutrients. These roots tend to be fibrous and rather fragile relative to the roots of terrestrial plants, resulting from adaptation to the fragile and mostly anaerobic environment of wetland sediments (Sculthorpe, 1967; Wetzel, 2001). Rhizosphere, an area of soil that is affected by root activity, is significant in nutrient cycling and the interaction of microbes in aquatic macrophytes. *Euryale ferox* plants have their roots, which emit oxygen and organic matter into the subsurface sediments, which may alter microbial niches and increase nutrient availability (Wetzel, 2001; Thomaz and Cunha, 2010). The interactions such as these will help fill the overall productivity of wetland ecosystems and will help the plant to obtain necessary nutrients, such as nitrogen and phosphorus (Chaurasia, 2022; Rather and Sharma, 2023). The rhizome is in itself an anchoring structure that keeps the plant stable in the sediment and acts as the starting point of the leaves and reproductive structures. Although the rhizome is not very long, it plays a very important role in ensuring that water currents and sediment movement do not displace the plant structure, which is common in the aquatic environment.

2.4.2 Stem structure and adaptations

The stem is mainly a rhizomatous short structure that is found in the sediment of *Euryale ferox*. The rhizome is mostly submerged and serves as a central organ in which roots, petioles, and reproductive structures grow, unlike the elongated aerial stems of most terrestrial plants. The rhizome has a number of adaptations related to aquatic vegetation, such as decreased mechanical tissues and aerenchyma—air-filled spaces in plant tissues involved in the internal movement of gases. These airways enable oxygen generated in the leaves to be carried to the submerged tissues, thus facilitating respiration in conditions where the oxygen supply is low (Sculthorpe, 1967; Wetzel, 2001). The rhizome gives rise to long petioles, which stretch vertically through the column of water to help the floating leaves be held up by water. These petioles are highly flexible and cylindrical so that they can adapt to changes in the water depth and movement. Their elasticity ensures that the leaves are not damaged by the water currents, still keeping the leaves in an optimum position to facilitate photosynthesis (Thomaz and Cunha, 2010).

2.4.3 Leaf morphology and anatomical adaptations

One of the most characteristic vegetative characteristics of the plant is its leaves of *Euryale ferox*. They are massive, circular, and floating in nature, creating a massive canopy over the surface of the water. With good circumstances, it is possible that individual leaves can be up to 1 meter in diameter, giving them a large area to perform photosynthesis and gaseous exchange (Kadono and Schneider, 1987; Schneider and Williamson, 1993). The leaf surface is usually smooth and green, and its waxy cuticle is such that it prevents water repellence and provides buoyancy. On the other hand, the lower surface is usually reddish or purplish in hue and has a large number of veins running radially out of the center of the leaf blade. This type of venation gives the leaf structural support and helps in the transportation of water, nutrients, and photosynthetic material (Sculthorpe, 1967). Floating aquatic leaves also have special anatomies that are used to promote effective gas exchange. The stomata are usually found on the upper part of the leaf as opposed to the lower part of the leaf, which helps the leaf to be in direct contact with the atmosphere as the leaf floats on the water surface (Sculthorpe, 1967). In addition, the leaf has large aerenchyma tissue that enables buoyancy and facilitates internal

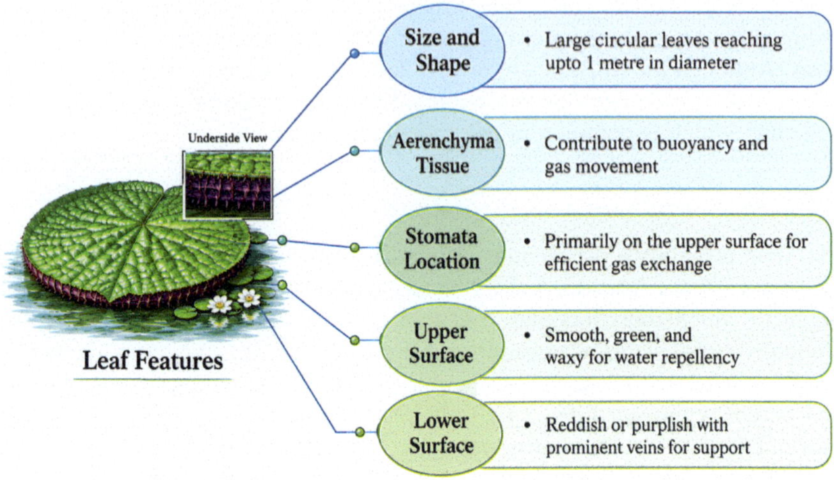

FIGURE 2.2 Morphological and anatomical features of the floating leaf of *Euryale ferox*.

movement of gases between submerged and aerial parts of plants (Wetzel, 2001). Fig. 2.2 summarizes the major morphological and anatomical characteristics of the floating leaves of *Euryale ferox*, such as the size of the leaf, stomata distribution, aerenchyma development, and surface characteristics.

2.4.4 Surface spines and defensive adaptations

One of the most striking vegetative characteristics of *Euryale ferox* is the presence of numerous sharp spines covering the leaves, petioles, and other plant parts. These spines are particularly dense on the lower surface of the leaves and along the petioles, forming an effective protective barrier. The development of these spines is believed to represent an evolutionary defense mechanism against herbivorous animals, such as fish, birds, and aquatic mammals, that might otherwise consume the plant's tissues (Schneider and Williamson, 1993). By deterring herbivory, the spines increase the likelihood that the plant can complete its life cycle and successfully produce seeds. In addition to their defensive role, the spines may also contribute to structural stability by reinforcing plant tissues exposed to mechanical stress caused by water movement.

2.4.5 Leaf anatomy and physiological significance

The next feature of *Euryale ferox* that is most remarkable is the sharp spines that can be found covering the leaves, petioles, and other parts of the plant. The spines are especially thick upon the lower surface of the leaves and along the petioles, as to which they furnish an effective protective covering. It is thought that the formation of these spines is a form of evolution in defense against herbivorous creatures, such as fish, birds, and water mammals, which would otherwise eat the tissues of the plant (Schneider and Williamson, 1993). The spines reduce the chances of the plant dying before it can reach its life cycle by preventing herbivory, which means that the plant is more likely to survive and be able to bear fruit. Besides the fact that the spines are used as a defense mechanism, they can also play a structural stability consideration in supporting the plant tissues undergoing mechanical stress due to the movement of water.

2.4.6 Leaf anatomy and physiological significance

The adaptation to the aquatic environment is evident in the anatomical design of *Euryale ferox* leaves, which have evolved features that facilitate physiological processes. The mesophyll of the leaf has extensive aerenchyma tissue, consisting of interconnected air spaces that aid the movement of internal gases between aerial and submerged plant organs. This is an internal system of aeration that enables oxygen produced in the process of photosynthesis to reach the submerged tissues, which is used in supporting cellular respiration (Sculthorpe, 1967; Wetzel, 2001). Large air spaces also play a role in the buoyancy of the leaves, so that the leaf blade is maintained at the water surface, where it is most likely to find light. In addition, vascular tissues in the leaf are positioned to facilitate the effective movement of water and nutrients through the rhizome and the dissemination of photosynthetic products in the plant. All these anatomical adaptations improve the physiological performance of *Euryale ferox* in the aquatic environments, enabling the plant to sustain high levels of photosynthesis and growth against these environmental constraints, such as limited oxygen availability and changes in water levels (Thomaz and Cunha, 2010). Table 2.2 summarizes the main vegetative organs of *Euryale ferox* and their functions.

2.5 Reproductive morphology

Reproductive structures have a central role in the life cycle and reproduction of aquatic vegetation. The reproductive morphology in *Euryale ferox* Salisb. adapts to a freshwater environment by allowing successful pollination, fertilization, and seed production. The plant is a family member of *Nymphaea*ceae, and its flowers are rather large and conspicuous, forming the water surface and enabling pollination through insects (Schneider and Williamson, 1993; Kadono and Schneider, 1987; Borsch et al., 2008). The *Euryale ferox* reproductive system comprises solitary flowers, spiny fruits, and a large-sized seeds, which are the main propagule of the species. These extremely specialized structures have a number of morphological features that differentiate the species among other aquatic plants of the order *Nymphaea*ceae (Schneider and

TABLE 2.2 Morphological characteristics of vegetative organs of *Euryale ferox*.

Vegetative structure	Key morphological characteristics	Ecological function
Roots	Fibrous roots embedded in sediment	Nutrient absorption and anchorage
Rhizome	Short rhizomatous stem	Structural support and vegetative growth
Petiole	Long flexible petiole connecting the leaf to the rhizome	Supports floating leaves
Leaves	Large circular floating leaves with spines	Photosynthesis and surface coverage
Spines	Sharp defensive structures on leaves and petioles	Protection from herbivory

Williamson, 1993). These reproductive characteristics are vital in research concerning plant taxonomy, plant ecology, and in the planting of makhana crops in the wetland environment (Jha and Kargupta, 1991).

2.5.1 Flower morphology and floral architecture

Euryale ferox has solitary flowers, which grow on long peduncles that run through the rhizome to the surface through the water column. These flora normally grow above the water surface and thus can easily be used by the insects in the process of pollination (Kadono and Schneider, 1987; Schneider and Williamson, 1993). The flora is divided into various parts with a number of distinct parts, such as the sepals, petals, stamens, and carpels of concentric whorl arrangement. The sepals tend to be protective and mostly green in color, covering the floral bud in the initial stages of development. The sepals consist of plenty of petals with the color of purple or violet, which makes the flowers look aesthetically pleasing to the pollinators (Borsch et al., 2008). The androecium of the flower has many stamens that surround the center gynoecium. These stamens give rise to the grains of pollen, which are carried to the stigma during the process of pollination. The gynoecium has more than one carpel, which later turns into the fruit after successful fertilization (Schneider and Williamson, 1993). Fig. 2.3 depicts the key flora of *Euryale ferox*.

2.5.2 Pollination mechanisms

Euryale ferox is mainly pollinated through insects attracted to the color, structure, and production of nectar in its flowers. Beetles and other insects are

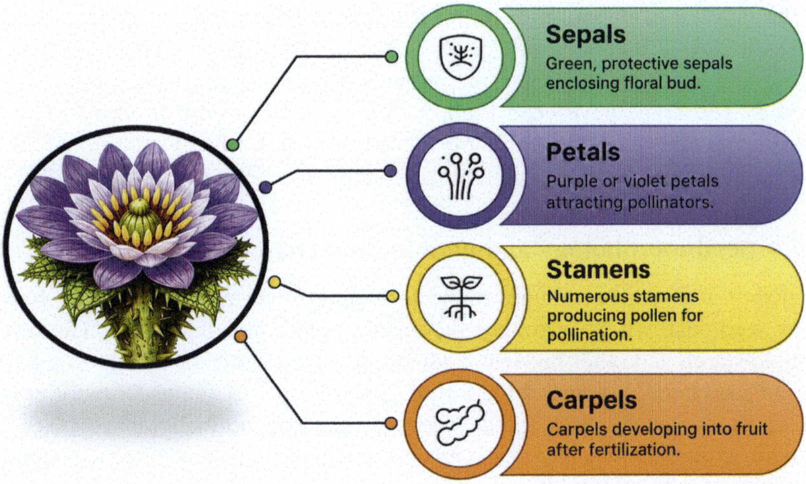

FIGURE 2.3 Floral structure of *Euryale ferox* showing major reproductive organs, including sepals, petals, stamens, and carpels.

also significant in the process of pollinating the stigma in most aquatic plants that are members of the *Nymphaea*ceae (Schneider and Williamson, 1993; Borsch et al., 2008). The formation of the flower on top of the water is one of the crucial adaptations that help the flower to interact with the terrestrial pollinators. When the flower opens, the grains of pollen are shed from the stamens and can be carried away by visiting insects or through direct contact with floral structures (Kadono and Schneider, 1987). Along with insect-mediated pollination, some amount of self-pollination is possible in *Euryale ferox*, especially in cases where pollinator abundance is low. This is a reproductive plasticity which increases the plant's capacity to generate viable seeds under fluctuating environmental conditions (Kadono and Schneider, 1987; Schneider and Williamson, 1993). The efficient pollination finally results in the fertilization of the ovules in the carpels, starting the growth of the fruits and seeds.

2.5.3 Fruit development and seed formation

Upon fertilization, *Euryale ferox* forms a spiny fruit structure inside the carpels and has many seeds. The fruit is normally globose-shaped and is enveloped with short spines similar to those occurring in the vegetative regions of the plant. These spines might be used as a defense mechanism against herbivorous or mechanical damage. As the fruit develops, the ovules in the carpels develop into seeds. The fruit will take some time to be detached by the plant, after which the mature seeds will be released to the surrounding water or sediment. In the natural ecology, seed spreading can be performed through the water flow or through the degradation of the fruit tissues, releasing the seeds to the rock (Wetzel, 2001). In agricultural systems, as in makhana production systems, the full-grown seeds are collected in wetlands and ponds and milled into edible kernels. The seeds have been found to be especially useful in terms of their nutritional value and are commonly used as a part of any traditional food system in South and East Asia (Jha and Kargupta, 1991; Kapoor et al., 2022).

2.5.4 Seed morphology and germination characteristics

Euryale ferox seeds are huge, round to slightly oval, and they have a tough outer seed coat, which surrounds the inner embryo and endosperm. The seeds contain starch and other nutrients, which give them stores of energy to release during seed germination and the initial growth of the seedlings (Kapoor et al., 2022; Kumar et al., 2025). The seed germination normally takes place in aquatic sediments, wherein the seeds are covered until they can grow due to favorable environmental conditions. The success of germination and the growth of seedlings can be affected by temperature, water depth, and sediment characteristics (Kadono and Schneider, 1987; Wetzel, 2001). Upon germination, the embryo grows into a small seedling, which forms the first leaves that are submerged above water before the development of the usual floating leaves of plants. This developmental change is a significant phase of the *Euryale ferox* life cycle and a factor in building new plant communities of aquatic environments. The key *Euryale ferox* reproductive structures and their key morphological features are summarized in Table 2.3, which gives a brief description of the structural features of *Euryale ferox* related to flowering, fruit development, and seed formation.

2.6 Growth ecology of *Euryale ferox*

The ecology of *Euryale ferox* Salisb is growth-oriented as it is directly associated with the environment in freshwater wetlands, where the organism is found in nature. Being an aquatic macro-plant and a family member of the *Nymphaea*ceae, the plant has various ecological adaptations that make it the plant that succeeds in shallow water bodies, such as ponds, lakes, and floodplain wetlands. These habitats are characterized by relatively stable water levels, sediments that are rich in nutrients, and sufficient sunlight exposure, which affect the growth, development, and reproductive success of the species (Wetzel, 2001; Thomaz and Cunha, 2010; Bornette and Puijalon, 2011). Similar to most other aquatic plants, *Euryale ferox* is exceedingly ecologically plastic, and it can be adjusted to changes in water depth, temperature, and nutrient availability (Islam et al., 2025). Most of the life cycle of the species goes through one growing season, which involves specific phenological stages, such as seed germination, vegetative growth, flowering, fruit development, and seed maturation (Kadono and Schneider, 1987; Jha and Kargupta, 1991). These ecological processes are significant to the natural ecosystem research and the management of the makhana growing systems.

TABLE 2.3 Morphological characteristics of reproductive structures of *Euryale ferox*.

Reproductive structure	Morphological characteristics
Flower	Solitary, large, purple or violet petals emerging above the water surface
Sepals	Green outer floral parts protecting the developing bud
Stamens	Numerous stamens are arranged around the central gynoecium.
Carpels	Multiple carpels form the ovary and reproductive core.
Fruit	Globose spiny fruit containing multiple seeds
Seeds	Large, spherical seeds with a hard protective seed coat

2.6.1 Habitat characteristics

Euryale ferox is normally found in freshwater environments, such as ponds, oxbow lakes, marshes, and wetlands within floodplains. They are typically shallow water bodies of between 0.5 and 1.5 m deep where the floating leaves of the plants reach the water body, and at the same time, the roots are rooted in the sediment (Jha and Kargupta, 1991; Wetzel, 2001). In these aquatic environments, the sediments are normally enriched with organic matter and nutrients, which make the environment conducive to the growth of plants. *Euryale ferox* rhizosphere associates with the microbial populations that are present in the sediment and that, in turn, contribute to the nutrient cycling and organic matter breaking down (Wetzel, 2001; Chaurasia, 2022; Rather and Sharma, 2023). Such interactions improve access to key nutrients such as nitrogen and phosphorus that are essential in the development of plants. Besides the sediment properties, environmental conditions of water temperature, sunlight accessibility, and hydrological stability are the major factors that affect the distribution of *Euryale ferox*. Its species is mostly adapted to hot climatic conditions, and it is widespread in the tropical and subtropical parts of Asia, such as India, China, and Southeast Asia (Jha and Kargupta, 1991; Wu et al., 2022).

2.6.2 Life cycle and phenological stages

Euryale ferox 's life cycle starts with the germination of the seeds that are buried in the wetland sediments throughout the dormant season. The process of germination tends to take place when the environmental conditions, such as

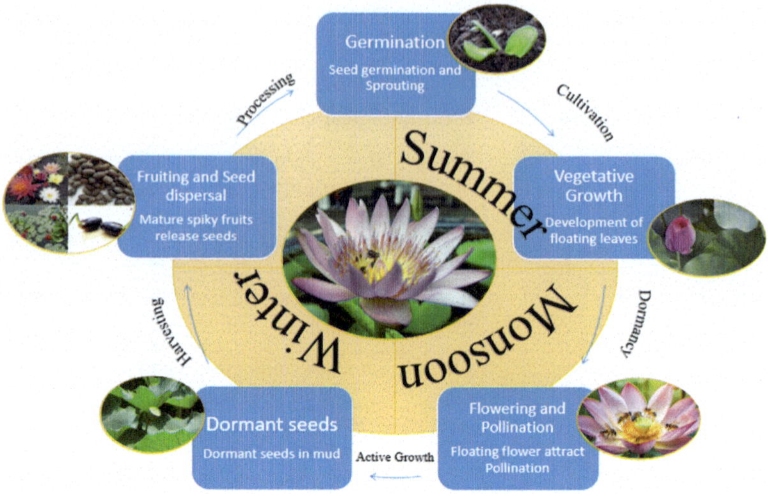

FIGURE 2.4 Life cycle and phenological stages of *Euryale ferox*.

temperature and the depth of water, become conducive to the growth of the seedlings (Kadono and Schneider, 1987; Wetzel, 2001).

After germination, the plant develops juvenile submerged leaves that are under the water surface. The plant will become bigger and larger as it grows, and the floating leaves have their own petioles, which are long. These leaves spread over the water and become the major structure of photosynthesis of the plant. The reproductive period starts when the floral buds grow on long peduncles above the water surface. Pollination and fertilization then occur, followed by fruit and seed development (Schneider and Williamson, 1993). After the seeds have developed, they can be blown into the immediate sediment, and there they lie in an inactive condition until the next season that promises successful growth. Fig. 2.4, summarizing the life cycle and phenological development of the plant *Euryale ferox*, reflects the various stages of development, that is, germination to seed production.

2.6.3 Seasonal growth patterns

The seasonal environmental conditions have a strong impact on the growth of *Euryale ferox*. In other parts of the world, like the Indo-Gangetic plains and East Asia, the development of the plant is significantly linked with monsoon changes and seasonal changes in temperature (Jha and Kargupta, 1991). The germination and early vegetative growth take place in the early growing season, usually at the end of spring or early summer, when the water temperatures start to increase (Islam et al., 2025). With the advancement of the season, the vegetative growth stage of the plant shifts to the rapid stage, which is marked by the growth of floating leaves and the acceleration of photosynthesis. The flowering season normally comes between the mid-to-late growing season, when the environmental parameters, such as the intensity of light and nutrients, are perfect. The maturation of the seed and the growth of the fruit occur shortly. As the growing season ends, the plant gradually decays and the mature seeds fall into the sediment and stay dormant until the start of the next growth cycle (Kadono and Schneider, 1987; Jha and Kargupta, 1991).

2.6.4 Biomass accumulation and productivity

Environmental factors, which include nutrient availability, the depth of water, and the intensity of light, largely determine the productivity of *Euryale ferox*. The plant has the potential to grow massive floating creeper leaves under ideal conditions, which add considerably to the biomass accretion in wetland environments (Wetzel, 2001). *Euryale ferox* has a massive leaf canopy that improves the efficiency of photosynthesis as it maximizes the exposure to the sunlight. This leads to a lot of organic matter production, which promotes the growth of plants and ecology in the wetland environment (Thomaz and Cunha, 2010; Bornette and Puijalon, 2011). Accumulation of biomass in cultivated systems of producing makhana is of great significance, as it affects the seed production and the overall crop production (Jha and Kargupta, 1991). Farmers usually control the degree of water and nutrient concentrations to prepare for the maximum growth of plants and seed production. Besides its agricultural value, *Euryale ferox* biomass also adds to the functioning of wetlands by serving as habitat to aquatic life and its role in nutrient cycling (Wetzel, 2001; Thomaz and Cunha, 2010).

2.7 Environmental factors influencing growth

Environmental conditions in freshwater ecosystems are known to exert a great impact on the growth and development of aquatic macrophytes. In the case of *Euryale ferox* Salisb., the water depth, water temperature, light availability, nutrient concentration, and sediment composition are important factors that influence the establishment of the plants, their productivity, and their reproductive success. Since this species inhabits shallow wetlands and ponds, any minor change that occurs in the environmental parameters can have a significant impact on its physiological functioning and ecological distribution (Wetzel, 2001; Thomaz and Cunha, 2010; Bornette and Puijalon, 2011). Environmental factors affect various aspects of the biology of the plant, such as seed germination, vegetative growth, flowering, and seed production (Yadav et al., 2025). These ecological drivers are especially of interest to natural wetland conservation and to agricultural control of the makhana farming system (Jha and Kargupta, 1991).

2.7.1 Water depth and hydrological conditions

Water depth is also among the most significant ecological factors that regulate the development of the floating-leaved aquatic plants. *Euryale ferox* usually grows in shallow freshwater sources with water levels such that the leaves are covered by the surface, and the roots are attached to the sediment. The best depth of the water is around half a meter to a meter and a half, but the plant can survive moderate changes in the depth of the water based on the environmental conditions in the area (Wetzel, 2001). Hydrological stability is also significant to ensure appropriate conditions of growth. High water content or long-term flooding may disrupt the germination of the seeds and the early establishment of seedlings. On the other hand, deep water can be very shallow, or the wetlands can be temporary, as this can restrict the existence of vegetation. Constant water levels consequently offer good environments for vegetative growth and reproductive development (Thomaz and Cunha, 2010).

2.7.2 Temperature requirements

Temperature is a basic factor in controlling the physiological activities of water plants, such as photosynthesis, respiration, and metabolic rate. *Euryale ferox* is normally related to warm climatic conditions and is best grown in tropical and subtropical areas. The highest growth is evident when the water temperatures are moderate to warm, and thus it encourages a high rate of vegetative growth and flowering (Wetzel, 2001; Bornette and Puijalon, 2011). Changes in the phenological stages of the plant also depend on seasonal changes in temperature. Warmer temperature at spring and early summer promote germination and first vegetative development of seeds, whereas warmer temperatures toward the end of the growing season encourage flowering and fruit set (Kadono and Schneider, 1987).

2.7.3 Light availability and photosynthetic adaptation

Another important issue that influences the growth of *Euryale ferox* is the provision of light. The reason why the plant is well suited to capture sunlight is the fact that it produces large floating leaves, which cover the water surface. The leaves are the main photosynthetic organs, which enable the plant to use solar energy in making biomass by the process of photosynthesizing (Wetzel, 2001; Thomaz and Cunha, 2010). Adequate sunlight is especially significant in the vegetative growth cycle, wherein the leaves are growing fast and amassing biomass. Nevertheless, too much shading caused by other aquatic vegetation or other environmental conditions can cause low efficiency in photosynthesis and slow down the growth of the plants. Floating-leaf is thus one of the most important ecological adaptations that enables *Euryale ferox* to be in direct contact with the sunlight despite being in thick water vegetation (Sculthorpe, 1967).

2.7.4 Nutrient dynamics (nitrogen and phosphorus)

The nutrient content of nitrogen and phosphorus, among others, contributes significantly to the productivity of aquatic macrophytes. Soil sediments in wetlands usually have these types of nutrients, and the roots of *Euryale ferox* take them up. They are also nutrient-rich surroundings, which tend to stimulate intensive vegetative growth and biomass generation (Wetzel, 2001; Chaurasia, 2022; Rather and Sharma, 2023; Yadav et al., 2025). Nevertheless, the over-enrichment of nutrients can cause eutrophication, which can change the composition of the aquatic vegetation and cause water quality changes. It is then crucial to attain balanced nutrient contents in the cultivation systems that are managed with the aim of maximizing the growth of plants without causing adverse ecological effects (Thomaz and Cunha, 2010).

2.7.5 Sediment characteristics and substrate influence

Euryale ferox develops in the sediment, and this sediment is the main substrate for anchoring the root and uptake of nutrients. Nutrient-enriched soft sediments with organic matter are usually deemed friendly to the establishment and growth of the plant. These sediments enable easy penetration of roots and absorption of nutrients (Jha and Kargupta, 1991; Wetzel, 2001; Yadav et al., 2025). The composition of the sediment is capable of affecting microbial activity in the rhizosphere as well. The microorganisms in wetland sediments are involved in significant nutrient mineralization and decomposition of organic matter, thereby increasing nutrient availability to aquatic plants (Chaurasia, 2022; Rather and Sharma, 2023). The significant environmental factors that affect the growth of *Euryale ferox* are presented in Table 2.4, which summarizes the ecological factors that govern the development of the plant in a freshwater environment.

TABLE 2.4 Environmental parameters influencing the growth of *Euryale ferox.*

Environmental factor	Influence on plant growth
Water depth	Determines leaf access to surface and root anchorage
Temperature	Regulates physiological processes and seasonal growth
Light availability	Controls photosynthetic activity and biomass production
Nutrient availability	Influences vegetative growth and productivity
Sediment composition	Affects root development and nutrient uptake

2.8 Ecophysiological adaptations

Aquatic macrophytes are highly adapted both structurally and physiologically so that they can endure the variable nature of the water conditions and other factors that include the changing water levels, low oxygen levels in the sediments, and changes in nutrient levels. Being a floating-leaved aquatic species, *Euryale ferox* Salisb exhibits various ecophysiological mechanisms that enable the species to keep its growth and productivity in freshwater wetlands and ponds. These adaptations include changes in plant morphology, internal gaseous transfer systems, photosynthetic activity, and stress tolerance mechanisms that all help in the survival of the plant in aquatic environments (Sculthorpe, 1967; Wetzel, 2001; Bornette and Puijalon, 2011). These adaptive characteristics have made *Euryale ferox* a successful ecological species in both natural wetlands and cultivated systems, wherein they have increased the ability of the plant to acquire nutrients, to retain physiological activity in submerged environments and to adapt to environmental heterogeneity (Thomaz and Cunha, 2010).

2.8.1 Adaptations to aquatic environments

Aquatic plants have to endure different environmental conditions, which are not similar to those that terrestrial plants have to face. The development of large floating leaves that are in direct contact with the atmosphere is one of the most significant changes that have been experienced in *Euryale ferox*. Such a morphological characteristic enables the plant to receive carbon dioxide and oxygen in the atmosphere and obtain as much light as possible to photosynthesize (Wetzel, 2001; Thomaz and Cunha, 2010). The plant also has long elastic petioles that make an extension of the rhizome to the water column that uphold the floating leaves on the surface. Such structural adjustment helps the plant to position the leaves optimally, even in the event of varying water levels. The floating leaf habit, thus, is one of the major ecological strategies that increases photosynthetic capacity and competitiveness of the plant in the aquatic communities of plants (Sculthorpe, 1967).

2.8.2 Gas exchange and buoyancy mechanisms

The process of gas exchange is one of the most important physiological processes among aquatic plants since there is not always enough oxygen in the submerged sediments. The internal transport of gases of *Euryale ferox* is also achieved with the help of special tissues called aerenchyma, which are complex interconnecting air spaces in the leaves of the plant, in the petioles, and in the stems. Such airways enable transport of oxygen generated by photosynthesis in aerial plant portions to submerged tissues, whereby it facilitates cellular respiration (Sculthorpe, 1967; Wetzel, 2001; Colmer, 2003; Evans,

2004). Aerenchyma tissues also aid in buoyancy, while decreasing the overall density of plant tissues. This structural characteristic assists in keeping the leaves afloat on the surface of water. The effective ventilation of the aerial and submerged tissues, such as this, provides the plant with the ability to remain metabolically active in the oxygen-poor, aquatic sediments (Bornette and Puijalon, 2011).

2.8.3 Stress tolerance mechanisms

The environmental stresses that are commonly associated with aquatic plants include fluctuation in water bodies, temperature, and even imbalances in nutrients. *Euryale ferox* has a number of physiological processes that make the plant withstand such environmental stress conditions (Colmer and Voesenek, 2009). The phytoplankton capacity is increased by the large area of the leaf surface of the plant to facilitate photosynthesis and to effectively absorb carbon in different light conditions. Moreover, the root system of the plant also interacts with microorganisms that live in the sediments and helps in cycling nutrients and decomposing organic matter, and hence the nutrient availability in the rhizosphere (Wetzel, 2001; Chaurasia, 2022; Rather and Sharma, 2023). Defensive spines on vegetative and reproductive parts would also have an indirect impact on stress tolerance through shielding of plant tissues against herbivory. The plant can be able to keep its physiological activity and reproductive capability by reducing tissue damage from aquatic animals.

2.8.4 Adaptation to fluctuating water levels

Rainfall and evaporation in freshwater wetlands and ponds seasonally cause an increase or decrease in the water level in these habitats. *Euryale ferox* has a number of adaptive characteristics that enable it to adapt to these changes in the environment. The petioles of the floating leaves are long and have the ability of adjusting with variation in water depth. An increase in water levels causes the petioles to elongate to keep the leaves on the surface so that they can get the sunlight and other gases in the atmosphere. On the other hand, the plant will have contact with the water surface at the leaf level because its vegetative structures are flexible, in case water levels go down (Thomaz and Cunha, 2010). Such adaptive mechanisms enable *Euryale ferox* to sustain physiological activity in changing hydrological regimes, have been involved in its survival in natural wetland ecosystems, and ways of successfully culturing it in shallow pond systems (Wetzel, 2001).

2.9 Population ecology and distribution

Population ecology is concerned with how species are distributed, how abundant they are, and how they change in their natural habitats. In the case of aquatic macrophytes like *Euryale ferox* Salisb. The environmental factors that have a strong effect on population structure and pattern are the water depth, sediment nature, nutrients, and hydrological stability. The ecological factors establish, maintain, and reproduce plant populations in freshwater ecosystems (Wetzel, 2001; Thomaz and Cunha, 2010; Bornette and Puijalon, 2011).

Euryale ferox is a globally known significant aquatic plant species in Asian wetlands with respect to environmental significance as well as economic characteristics as a food crop. The species occur in groups in shallow water areas where the environment is conducive to the germination of the seed and vegetative development. The ecology and geography distribution of the species is thus significant in the conservation of the wetlands, biodiversity, and future sustainable practices in cultivating the species (Jha and Kargupta, 1991).

2.9.1 Geographic distribution of *Euryale ferox*

Euryale ferox is a species that is mainly found in the temperate and tropical areas of Asia. The species is found in nature in the countries of India, China, Bangladesh, Japan, and some parts of Southeast Asia. In these areas, it is normally related to freshwater wetlands, ponds, floodplain lakes, and slow-moving water bodies where conditions are favorable to its growth and reproduction (Wu et al., 2022). The plant is especially common in the eastern Indo-Gangetic plains, especially in the state of Bihar, where makhana farming has grown to become a major agro-aquatic system of farming. Another important region is China, which is the distributor and grower of *Euryale ferox*, the seeds of which are consumed as food and as traditional medicinal products (Kumar et al., 2025). Climatic factors like temperature and rain pattern are geographical determinants of the distribution of the species, as they affect the hydrological condition of wetlands and ponds. The presence of stable freshwater environments in warm climates offers the best conditions for the setting up and development of the *Euryale ferox* population (Wetzel, 2001).

2.9.2 Habitat preferences and ecological niches

Euryale ferox has a very narrow ecological niche that has a strong relationship with shallow freshwater habitats with nutrient-rich sediments and moderate water depth. The plants are usually found in the ponds, lakes, marshes, and floodplain wetlands where the water is deep enough to keep the floating leaves of the plant at the surface with the roots entrenched in the sediment (Wetzel, 2001; Thomaz and Cunha, 2010). These environments usually have sediments that are rich in organic matter that contain nutrients that are vital in the growth of plants. Hydrological conditions are also relatively stable in the plant since high agitations of water levels resulting in seed germination and premature establishment of seedlings may affect germination and subsequent seedling

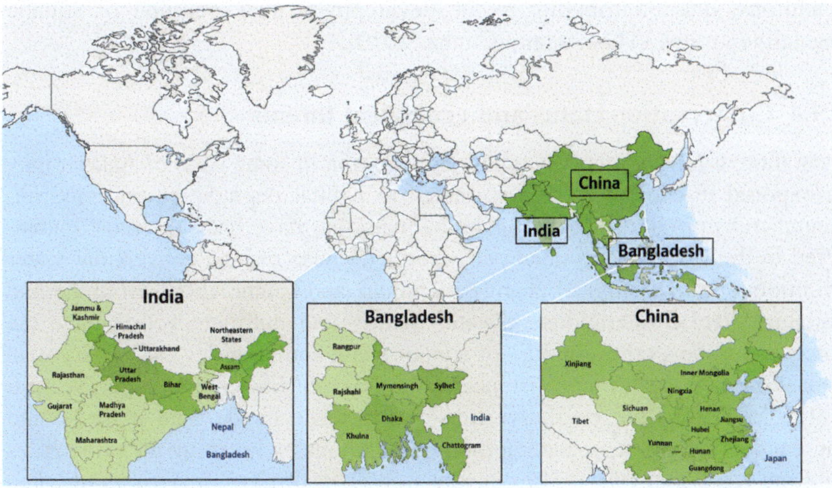

FIGURE 2.5 Geographic distribution of *Euryale ferox* cultivation. Major cultivation regions across South and East Asia, including India, Bangladesh, and China. Key cultivation areas such as Bihar (India), Rajshahi (Bangladesh), and Jiangsu (China). ***Map lines delineate study areas and do not necessarily depict accepted national boundaries.***

growth (Jha and Kargupta, 1991). In algal communities, *Euryale ferox* may be able to take up ecological space, with ecological competition being reduced by submerged vegetation. The huge floating leaves of the plant enable it to cover the water at the surface, which reflects sunlight and forms shady areas that could inhibit the growth of other aquatic organisms (Thomaz and Cunha, 2010; Bornette and Puijalon, 2011). The geographical distribution of *Euryale ferox*, including major cultivation regions across Asia, is presented in Fig. 2.5.

2.9.3 Population dynamics in natural wetlands

Several ecological processes, such as seed production, seed germination, plant survival, and environmental variability, affect the *Euryale ferox* population in natural wetlands. Since the reproduction cycle of the species mainly follows the seed, the quantity and quality of the seed bank available in wetland sediments are important in the stability of the population (Kadono and Schneider, 1987; Wetzel, 2001). The process of seed dispersal is when the seeds fall out of the fruit when the mature fruit has fallen into the water or sediment around it. These seeds can also take quite a long time before they germinate when subjected to the right environment. The recruitment of new plants in a population is therefore affected by seasonal changes in temperature, water level, and availability of nutrients. The population of *Euryale ferox* within various wetland ecosystems shows varying density across time and space with regard to the environmental factors and competition with other aquatic macrophytes. Nonetheless, the species are capable of developing dense stands under good conditions, thus contributing to the development and operation of aquatic vegetative groups (Thomaz and Cunha, 2010).

2.9.4 Conservation status and ecological threats

Even though *Euryale ferox* is relatively common in some parts of Asia, It may be exposed to ecological strains related to habitat degradation and environmental transformation. Some of the habitats that have been the most endangered in the world are wetland ecosystems because of land conversion, water pollution, and hydrological changes (Thomaz and Cunha, 2010; Bornette and Puijalon, 2011). Alterations in water quality and nutrient cycling have the potential to impact the structure of aquatic plant communities and can have an impact on the long-term persistence of the *Euryale ferox* populations (Wetzel, 2001). Moreover, the overharvesting of the seeds in the wild places could cause a decrease in the regeneration of populations by lowering the propagules (Jha and Kargupta, 1991). The conservation plans of *Euryale ferox* would then aim at protecting wetlands, sustainable harvests, and the ecology in which the natural population regeneration can occur. It will be necessary to combine ecological study with conservation planning to sustain the long-term survival of this ecologically and economically worthy aquatic species.

2.10 Cultivation ecology and agronomic aspects

The *Euryale ferox* Salisb, also referred to as makhana or fox nut, cultivation is a significant agro-aquatic system of production in various areas of Asia. Although the species is found in natural wetlands in freshwater bodies, it has also been domesticated and grown in controlled pond ecosystems to obtain its edible seeds. Makhana is especially cultivated in parts of the eastern Indo-Gangetic plains of India and some parts of China, where the crop forms a major source of livelihood for the rural population and economy in the regions (Jha and Kargupta, 1991; Kumar et al., 2025). *Euryale ferox* has ecological needs that render it suitable in shallow waters with high nutrient sediments. High-yielding requires intensive control of environmental variables like depth of water, quality of sediments, and nutrients. It is thus important to learn the ecological and agronomic factors of makhana production to maximize productivity and sustainably manage the aquatic farming systems (Wetzel, 2001; Thomaz and Cunha, 2010).

2.10.1 Traditional cultivation practices

Euryale ferox has long been cultivated traditionally in South and East Asia. Natural ponds, oxbow lakes, and wetlands are the major places where the crop is cultivated in India, where the ponds have a favorable environment to support the growth of the crop. The seeds accumulated in the past harvests are usually gathered by the farmers and aired into very shallow waters at the onset of the growth season (Jha and Kargupta, 1991). Germinated seedlings form submerged leaves and then massive floating leaves which grow over the water body. With the growth of the plants, farmers will control the density of vegetation using their hands in order to ensure the plants are spaced effectively. The cultivation entails many labor-intensive processes like manual weeding and eliminating rival aquatic vegetation. When the fruits are ready, the makhana seeds are harvested by falling into the water. The seeds are harvested at the bottom of the ponds, and then the traditional drying and roasting processes convert the seeds into edible puffed kernels (Ekal and Kumbamoorthy, 2024). Most of the processes in the production of makhana are still based on the use of traditional knowledge and manual labor despite the presence of modern technologies (Kumar et al., 2025).

2.10.2 Environmental requirements for cultivation

The successful growth of *Euryale ferox* requires a favorable environment to enable the plant to grow and reproduce. The depth of water is also one of the major factors since it dictates the ability of the floating leaves to touch the water surface with the roots still attached to the sediment. A depth of water of between 0.5 and 1.5 m has usually been found suitable for growing (Wetzel, 2001; Jha and Kargupta, 1991). The productivity of plants is also influenced by the quality of the sediment. The nutrient-rich sediments with organic matter would also supply the essential nutrients for the vegetative growth and development of seeds. Sediments in most cultivated systems contain organic matter, which increases the activities of microbes and nutrient cycles around the rhizosphere (Wetzel, 2001; Chaurasia, 2022; Rather and Sharma, 2023). Temperature and the availability of sunlight also affect the growth of *Euryale ferox*. The climatic conditions are also favorable to the fast development of vegetation and flowering, including a sufficient amount of sunlight, maintaining high levels of photosynthesis and biomass development (Thomaz and Cunha, 2010).

2.10.3 Growth performance in managed ecosystems

Euryale ferox productivity is affected by the ecological conditions as well as by the practices applied to the managed ecosystems. The plant is capable of creating large floating canopies of leaves in the traditional pond systems, supporting high photosynthetic rates and biomass output. These circumstances assist in the development of large fruits with a high number of seeds, which are treasured because of their nutritional makeup and food uses (Ekal and Kumbamoorthy, 2024). Aquatic plant productivity studies have also indicated that nutrient availability and light interception are major factors that determine the accumulation of biomass in floating-leaved macrophytes (Thomaz and Cunha,

2010; Bornette and Puijalon, 2011). Since *Euryale ferox* has large floating leaves, the plant can trap high quantities of sunlight, and this is favorable to support high levels of photosynthesis. In domesticated ponds, the farmers can control the amount of water and plant density to grow the plants to maximum growth and produce the maximum number of seeds. The practices of such management contribute to the preservation of the good ecological conditions and mitigation of the competition with other aquatic plants (Wetzel, 2001).

2.10.4 Sustainable cultivation strategies

The need to keep up the demand for makhana has made sustainable cultivation practices more significant to sustainability in terms of productivity and conservation of wetland ecosystems. The objectives of sustainable cultivation methods are to stabilize production in agriculture and conserve the environment. One of their strategies is the maintenance of the right depth of water and nutrients in the cultivation ponds. An overload of nutrients may cause eutrophication that might adversely affect the water quality and biodiversity in the water (Wetzel, 2001; Thomaz and Cunha, 2010).

The other significant factor in the sustainable cultivation is the preservation of wetland habitats, which facilitate the occurrence of natural populations of *Euryale ferox*. Ecological information can be used to merge traditional farming systems with ecological knowledge in a bid to ensure that makhana production systems are able to sustain in the end (Jha and Kargupta, 1991). Through encouraging the use of environmentally friendly methods of cultivation, agricultural productivity can be increased whilst maintaining the ecological processes of freshwater wetlands.

2.11 Ecological role in freshwater ecosystems

Aquatic macrophytes are important in ensuring ecological balance and the freshwater ecosystems. The species include *Euryale ferox* Salisb., which are known to be important in increasing the complexity of habitats, cycling of nutrients, and biological productivity in wetlands and shallow waters. *Euryale ferox*, being a floating-leaved macrophyte, is known to interact with the water column and sediment, and therefore affects ecological processes with influence on a large variety of aquatic organisms (Carpenter and Lodge, 1986; Wetzel, 2001; Chambers et al., 2008; Thomaz and Cunha, 2010; Bornette and Puijalon, 2011). *Euryale ferox* has an ecological significance that is not due to its value as an economic crop. In wetlands in the natural habitat, the vegetation plays a role in maintaining stability in the ecosystem by giving home to the animals, facilitating food webs, and nutrient cycles. These ecological roles posit the significance of preserving aquatic macrophyte communities to be able to retain healthy freshwater ecosystems (Carpenter and Lodge, 1986; Wetzel, 2001; Hilt et al., 2006).

2.11.1 Role in aquatic biodiversity

Euryale ferox and other aquatic macrophytes also offer structural complexity in freshwater ecosystems, increasing the habitat diversity while sustaining a variety of aquatic organisms. The plant provides shelter to fish, aquatic invertebrates, and amphibians as the big floating leaves of the plant generate shaded microhabitats on the water surface and as the lower parts of the plant (submerged) provide the shelter (Thomaz and Cunha, 2010; Bornette and Puijalon, 2011). These vegetation formations provide a significant breeding and feeding environment for numerous aquatic species. The invertebrates are usually located on the leaf and stem surface; they either eat organic matter or algae. In their turn, these organisms constitute a significant element of the aquatic food webs because they become prey of fish and other higher trophic levels (Wetzel, 2001). Macrophytes, such as *Euryale ferox*, help enhance the amount of biodiversity in freshwater ecosystems by enhancing the heterogeneity of the habitat (Thomaz and Cunha, 2010).

2.11.2 Interaction with aquatic fauna

The introduction of *Euryale ferox* into the wetlands affects the activity and dispersal of other aquatic animals. Amphibians and insects that live in an aquatic environment may also use the vegetation as their breeding habitat, whereas fish species usually take shelter behind the submerged structures of aquatic flora (Wetzel, 2001; Bornette and Puijalon, 2011). The plant can also be a source of food to some of the herbivorous aquatic animals. Although the existence of defensive spines on the leaves and petioles decreases the number of herbivores, certain organisms might still eat the plant tissues or related microorganisms growing on the surface of the plants. Such interactions reveal the importance of *Euryale ferox* as a good constituent of the aquatic food web and bring about its great importance in the ecology of supporting biological interaction in wetland ecosystems (Thomaz and Cunha, 2010).

2.11.3 Contribution to nutrient cycling

The aquatic macrophytes help in the recycling of nutrients through the absorption of nutrients in the sediments and the surrounding water. Plants, including *Euryale ferox*, absorb such nutrients as nitrogen and phosphorus through root uptake and metabolic processes (Wetzel, 2001; Bornette and Puijalon, 2011). As the plants grow, nutrients are added to the plant biomass. Because of senescence and decomposition of plant tissues, these nutrients are slowly absorbed back into the water body, where other organisms can use them. This is significant in terms of controlling the nutrient resources and the sustainability of the ecosystem (Wetzel, 2001). Aquatic plant presence thus affects the freshwater ecosystem chemistry and stabilizes the nutrient cycles of wetland habitats (Thomaz and Cunha, 2010).

2.11.4 Ecosystem services provided by *Euryale ferox*

Besides the ecological roles it plays in the natural ecosystems, *Euryale ferox* delivers a number of ecosystem services to human societies. It is one of the most influential services, as it helps in sustaining traditional agricultural systems through the use of makhana. The plant seeds are also commonly used as one of the beneficial food items and are a valuable economic tool of societies engaged in wetland agro systems (Jha and Kargupta, 1991; Kumar et al., 2025). Aquatic macrophytes are also involved in improving the quality of water through the absorption of nutrients and trapping suspended particles in sediments. The processes are useful in keeping the ecological integrity of freshwater ecosystems and avoiding excessive accumulation of nutrients that may cause eutrophication (Wetzel, 2001; Thomaz and Cunha, 2010). Moreover, wetlands with macrophytes, such as *Euryale ferox*, promote conservation of biodiversity, climate, and carbon sequestration. These plants are important not only to the natural ecosystems but also to sustainable environmental management due to the ecological services they offer (Bornette and Puijalon, 2011).

2.12 Challenges and future research directions

Although *Euryale ferox* Salisb. has ecological and economic value, the ecological and biological details of its life cycle, as well as rearing, are not well researched. Studies on the species have also been on the rise over the past decades, especially in regard to its nutritional value and agricultural significance. Nonetheless, little extensive research has been done on its taxonomy, ecological relationships, population dynamics, and genetic variation. These gaps in knowledge need to be addressed to enhance scientific knowledge of the species and its sustainable management in both natural and artificial ecosystems (Soltis et al., 2018; Wetzel, 2001; Bornette and Puijalon, 2011). Further studies should thus involve a combination of botanical, ecological, and molecular techniques to have a better insight into the adaptive strategies and evolutionary history of *Euryale ferox*. These will help enhance the conservation practices and sustainable use of this valuable water plant (Borsch et al., 2008).

2.12.1 Knowledge gaps in taxonomy and ecology

Although the taxonomy of *Euryale ferox* has been outlined in the classical literature on botany, more specific taxonomic and ecological research is required to demonstrate the changes in the species distribution in its geographical area. The morphological differences and ecological differences among populations can be attributed to either environmental adaptation or genetic differences. More studies are needed to investigate the intraspecific variation and to identify the existence of different ecological populations or subspecies within the existing taxonomic system. This research would help in gaining a better insight into the evolutionary past and ecological variability of the species (Borsch et al., 2008; Soltis et al., 2018).

2.12.2 Climate change impacts on aquatic macrophytes

Climate change is one of the biggest issues for freshwater ecosystems in the world. Increased temperatures, modified rainfalls, and hydrological regime shifts can be of great significance to the wetland ecosystem and the distribution of aquatic flora. Since *Euryale ferox* is a freshwater-based species within its growth and breeding cycle, the population dynamics and distribution patterns of this species may be influenced by the environmental fluctuations related to climate variability (Thomaz and Cunha, 2010; Bornette and Puijalon, 2011). Alterations in water temperature and hydrological conditions are likely to affect seed germination, growth rates of plants, and reproductive success. Besides this, floods or droughts may hit the wetlands and diminish the accessibility of favorable environments by the species (Wetzel, 2001).

2.12.3 Need for molecular and ecological research

Recent progress in molecular biology and genomics offers novel avenues to research into the evolutionary biology and ecological adaptations of aquatic vegetation. *Euryale ferox*. Recently, genetic studies of its adaptations to aquatic environments have started with some inferences on the evolutionary relationship with other basal angiosperms (Zhang et al., 2020; Wu et al., 2022). Nevertheless, further detailed molecular studies are needed to get to know genetic diversity within and among natural populations. This kind of information would be useful in creating a better variety of crops, protecting genetic materials, and understanding evolutionary adaptations to aquatic life. Long-term ecological studies are also required in addition to the molecular research to keep track of population dynamics and ecosystem interactions with *Euryale ferox*. The combination of molecular data and ecological analysis will be useful in explaining the effects of genetic variation on ecological performance and adaptation (Wetzel, 2001).

2.12.4 Conservation and sustainable utilization

The ecosystems of the wetland that host *Euryale ferox* are under growing pressure due to the human activities that include land conversion, water pollution, and hydrological changes. Conserving these habitats is crucial in the continued survival of the species in the natural population as well as safeguarding the ecological processes of freshwater ecosystems (Wetzel, 2001; Thomaz and Cunha, 2010). The approach to a sustainable use of the *Euryale ferox* is to reduce the imbalance between agricultural production and conservation of the ecology. Better farming techniques, the rehabilitation of the habitats, and genetic variety conservation will be significant toward the long-term viability of the makhana production systems (Jha and Kargupta, 1991). Future research must thus focus on developing integrated conservation and management plans that will ensure the conservation and management of wetland ecosystems, while enabling sustainable farming practices. They will be useful in preserving the ecological and economic value of *Euryale ferox*.

2.13 Conclusions

Euryale ferox Salisb. is a unique aquatic macrophyte belonging to the family *Nymphaea*ceae, and it is still playing an ecological and economic role in freshwater environments in Asia. Being the last survivor of the genus *Euryale*, the plant has the peculiar morphological and physiological features that make it unique among the representatives of the water lily family. The systematic rank and taxonomic position of the species can be seen as having evolutionary importance in the order Nymphaeales, which is one of the earliest branches to divide among flowering plants. Adaptation of *Euryale ferox* is in the morphological structures that ensure the survival of the plant in shallow freshwater environments, such as its massive floating leaf, extensive roots, and distinctive defensive spines. Vegetative and reproductive structures ensure beneficial growth and reproduction of the species in the waters. The solitary nature of the flowers, the spiny nature of the fruits, and the nutritious nature of the seeds in the plant contribute to the reproduction and ecological viability in the wetland ecosystems.

The ecological requirements in the development of *Euryale ferox* are highly related to ecological factors, such as depth of water, nutrients, temperature, and sedimentary features. Those have an effect on important biological processes, such as germination, vegetative growth, flowering, and seed production. In natural wetlands, the species has a role to play in the ecosystem as they increase the complexity of the habitat, provides support to aquatic biodiversity, and is involved in nutrient cycling mechanisms. *Euryale ferox* is of great economic importance, besides having an ecological value, because of its seeds that are popularly eaten as makhana or fox nuts. The use of traditional farming techniques in some areas, such as the Indo-Gangetic plains, shows how the species can be incorporated into agro-aquatic farming systems that will serve the rural people. However, the growing demands on the wetland ecosystems and the environment are making it clear that sustainable management methods are required to safeguard the natural populations and the cultivated systems. The future studies of *Euryale ferox* ought to be aimed at the combination of ecological, molecular, and agronomic studies to comprehend the evolutionary biology, genetic diversity, and adaptation strategies in the species. The interdisciplinary studies that will be made through such research will help in the conservation and sustainable use of this precious aquatic plant. The ecological functions and economic values of *Euryale ferox* can be preserved for future generations using a combination of scientific knowledge and environmentally responsible management.

References

Alahuhta, J., Kosten, S., Akasaka, M., Auderset-Joyce, C., Bolpagni, R., Bove, C.P., Chambers, P. A., Clayton, J., de Winton, M., Ecke, F., et al., 2021. Global patterns in the metacommunity structuring of lake macrophytes. Proc. Natl. Acad. Sci. 118 (3), e2019685118.

Bornette, G., Puijalon, S., 2011. Response of aquatic plants to abiotic factors: a review. Aquat. Sci. 73 (1), 1–14. https://doi.org/10.1007/s00027-010-0162-7.

Borsch, T., Löhne, C., Wiersema, J., 2008. Phylogeny and evolutionary patterns in *Nymphaea*les: integrating genes, genomes and morphology. Taxon 57 (4), 1052–1081. https://www.jstor.org/stable/27756765.

Carpenter, S.R., Lodge, D.M., 1986. Effects of submerged macrophytes on ecosystem processes. Aquat. Bot. 26, 341–370. https://doi.org/10.1016/0304-3770(86)90031-8.

Chambers, P.A., Lacoul, P., Murphy, K.J., Thomaz, S.M., 2008. Global diversity of aquatic macrophytes in freshwater. Hydrobiologia 595, 9–26. https://doi.org/10.1007/s10750-007-9154-6.

Chaurasia, S., 2022. Role of macrophytes: a review. Adv. Zool. Botany 10 (4), 75–81. https://doi.org/10.13189/azb.2022.100401.

Colmer, T.D., 2003. Long-distance transport of gases in plants: a perspective on internal aeration and radial oxygen loss from roots. Plant Cell Environ. 26 (1), 17–36. https://doi.org/10.1046/j.1365-3040.2003.00846.x.

Colmer, T.D., Voesenek, L.A.C.J., 2009. Flooding tolerance: suites of plant traits in variable environments. Funct. Plant Biol. 36 (8), 665–681. https://doi.org/10.1071/FP09144.

Conran, J., 2007. Morphology of Hydatellaceae, an anomalous aquatic family recently recognized as an early-divergent angiosperm lineage. Am. J. Bot. https://doi.org/10.3732/AJB.94.7.1073.

Dudgeon, D., Arthington, A.H., Gessner, M.O., Kawabata, Z.-I., Knowler, D.J., Lévêque, C., Naiman, R.J., Prieur-Richard, A.H., Soto, D., Stiassny, M.L.J., Sullivan, C.A., 2006. Freshwater biodiversity: importance, threats, status, and conservation challenges. Biol. Rev. 81 (2), 163–182. https://doi.org/10.1017/S1464793105006950.

Ekal, V., Kumbamoorthy, S., 2024. Formulation and proximate analysis of makhana-enriched laddoos. Int. J. Nutrit. Pharmacol. Neurol. Dis. 14 (2), 205–212. https://doi.org/10.4103/ijnpnd.ijnpnd_92_23.

Endress, P.K., Doyle, J.A., 2009. Reconstructing the ancestral angiosperm flower and its early diversification. Am. J. Bot. 96 (1), 22–66. https://doi.org/10.3732/ajb.0800047.

Evans, D.E., 2004. Aerenchyma formation. New Phytol. 161 (1), 35–49. https://doi.org/10.1046/j.1469-8137.2003.00907.x.

Heegaard, E., Birks, H.H., Gibson, C.E., Smith, S.J., Wolfe-Murphy, S., 2001. Species–environmental relationships of aquatic macrophytes in Northern Ireland. Aquat. Bot. 70 (3), 175–223. https://doi.org/10.1016/S0304-3770(01)00161-9.

Hilt, S., Gross, E.M., Hupfer, M., et al., 2006. Restoration of submerged vegetation in shallow lakes: a review of ecological requirements. Hydrobiologia 569, 151–164.

Islam, Z., Bashir, K., Singh, S., Siddiqui, S., 2025. Effect of germination on physicochemical, functional, and antinutritional characteristics of makhana seed (*Euryale ferox*) flour. Food Human. 5. https://doi.org/10.1016/j.foohum.2025.100717.

Jha, V., Kargupta, A.N., 1991. Utilization and conservation of *Euryale ferox* Salisbury in Mithila, India. Biol. Conserv. 55 (3), 213–225.

Kadono, Y., Schneider, E.L., 1987. The life history of *Euryale ferox* Salisb. in southwestern Japan with special reference to reproductive ecology. Plant Species Biol. 2 (2), 109–115. https://doi.org/10.1111/j.1442-1984.1987.tb00037.x.

Kapoor, S., Kaur, G., Sharma, A., 2022. *Euryale ferox*: a promising aquatic functional food with nutritional and pharmacological potential. J. Food Biochem. 46 (7), e14435. https://doi.org/10.1111/jfbc.14435.

Kumar, R., Singh, I.S., Jha, V., 2025. Bioactive compounds in *Euryale ferox* Salisb. and their nutritional and therapeutic actions—a review. Int. J. Plant Res. https://doi.org/10.1007/s42535-025-01209-x.

Lesiv, M.S., Polishchuk, A.I., Antonyak, H.L., 2020. Aquatic macrophytes: ecological features and functions. Studia Biol. 14 (2), 79–94. https://doi.org/10.30970/sbi.1402.619.

Liaquat, M., Pasha, I., Ahsin, M., et al., 2022. Roasted fox nuts (*Euryale ferox* L.) contain a higher concentration of phenolics, flavonoids, minerals, and antioxidants, and exhibit a lower Glycemic Index (GI) in human subjects. Food Prod. Proc. Nutr. 4, 1. https://doi.org/10.1186/s43014-021-00081-x.

Löhne, C., Wiersema, J.H., Borsch, T., 2007. Phylogenetic analysis of *Nymphaea*les using chloroplast markers. Bot. J. Linn. Soc. 160 (4), 421–450. https://doi.org/10.1111/j.1095-8339.2007.00659.x.

Rather, Z.A., Sharma, P., 2023. Role of macrophytes in aquatic ecosystem: a case study on Dal Lake, Kashmir. Int. J. Res. Eng. Appl. Manag. 9 (2).

Schneider, E.L., Williamson, P.S., 1993. *Nymphaea*ceae. In: Kubitzki, K., Rohwer, J.G., Bittrich, V. (Eds.), The Families and Genera of Vascular Plants, vol. 2. Springer, pp. 486–493.

Sculthorpe, C.D., 1967. The Biology of Aquatic Vascular Plants. Edward Arnold.

Soltis, D.E., Soltis, P.S., Endress, P.K., 2018. Phylogeny and Evolution of the Angiosperms, second ed. University of Chicago Press. https://doi.org/10.7208/chicago/9780226441757.001.0001.

Thiemer, K., Schneider, S.C., Demars, B.O.L., 2021. Mechanical removal of macrophytes in freshwater ecosystems: implications for ecosystem structure and function. Sci. Total Environ. 782, 146671. https://doi.org/10.1016/j.scitotenv.2021.146671.

Thomaz, S.M., Cunha, E.R., 2010. The role of macrophytes in habitat structuring in aquatic ecosystems: methods of measurement, causes and consequences on animal assemblages' composition and biodiversity. Acta Limnol. Bras. 22 (2), 218–236. https://doi.org/10.4322/actalb.02202011.

Turland, N.J., Wiersema, J.H., Barrie, F.R., Greuter, W., Hawksworth, D.L., Herendeen, P.S., Knapp, S., Kusber, W.H., Li, D.Z., Marhold, K., et al., 2018. International Code of Nomenclature for Algae, Fungi, and Plants (Shenzhen Code). Koeltz Botanical Books.

Wetzel, R.G., 2001. Limnology: Lake and River Ecosystems, third ed. Academic Press.

Wu, P., Zhang, L., Li, Y., et al., 2022. Adaptive evolution of *Euryale ferox* to aquatic environments revealed by genomic analysis. Plant J. 111 (4), 1113–1128. https://doi.org/10.1111/tpj.15717.

Yadav, P.K., Kumar, A., Choudhary, G.L., Gorai, T., Nath, P., Saha, R., Kumar, R., Singh, I.S., 2025. Effect of integrated nutrient management on production performance of makhana (*Euryale ferox* Salisb.) under field conditions. Indian J. Agron. 70 (2), 224–227. https://doi.org/10.59797/ija.v70.i2.5607.

Zedler, J.B., Kercher, S., 2005. Wetland resources: status, trends, ecosystem services, and restorability. Annu. Rev. Environ. Resour. 30, 39–74. https://doi.org/10.1146/annurev.energy.30.050504.144248.

Zhang, L., Chen, F., Zhang, X., Li, Z., Zhao, Y., Lohaus, R., Chang, X., Dong, W., Ho, S.Y.W., Liu, X., et al., 2020. The water lily genome and the early evolution of flowering plants. Nature 577, 79–84. https://doi.org/10.1038/s41586-019-1852-5.

Chapter 3

Agro-climatic requirements and cultivation practices of makhana (*Euryale ferox* Salisb.)

Manoj Kumar and Shailendra M. Raut
ICAR-National Research Centre for Makhana, Darbhanga, Bihar, India

3.1 Introduction

Makhana (*Euryale ferox* Salisb.), also known as Gorgon nut, fox nut, and Euryale seed, is fast emerging as a super food globally owing to its richness in medicinal values and mineral contents (Jha et al., 2018; Kumar et al. 2020a,b). It is an aquatic crop, grown mostly in India, China, Japan, South Korea, Bangladesh, Russia, and a few other countries of the world. Traditionally, Makhana has been grown wild in ponds and natural water bodies with minimal management interventions. However, thanks to a recent rise in global demand, driven by growing awareness about its nutritive and medicinal values, the crop has received unprecedented focus in recent years. In addition, to exploit its economic potential, fox nut is now being increasingly grown in field conditions as well as a high-value cash crop (Kumar et al., 2011; Singh et al., 2020). Reports suggest that field cultivation of Makhana is many folds (usually four to five times) more remunerative than rice and wheat—the two most commonly grown crops in south Asia, including India (Kumar et al. 2020a,b). This transition from traditional water bodies to managed field systems necessitates a comprehensive understanding of the crop's agro-climatic requirements, ecological adaptations, and improved cultivation practices. Scientific management of water, soil, nutrients, and crop operations has therefore become crucial for achieving higher productivity, better quality produce, and enhanced profitability.

3.2 Makhana: A broad-spectrum food

Not many crops are known to have as much cultural, commercial, nutritional, and medicinal significance as Makhana. Thanks to these all-round attributes,

Makhana (Euryale ferox) as a Functional Food. https://doi.org/10.1016/B978-0-443-44421-0.00012-3

Makhana has received unprecedented focus in recent times, nationally and globally. Culturally, Makhana is regarded as "God's Food", as it is used as offering to the god. It is also used as a "Food for Fasting" due to its high carbohydrate content. In Mithila region of Bihar, India, it is customary for a new bride's parents to send Makhana to the grooms as a marital ritual. The custom is grounded on the spermatogenic and aphrodisiac properties of Makhana. It is an integral component of a number of ethnic practices in Mithila and other regions of India (Jha et al., 1991, 2014; Mishra et al., 2003). Apart from its cultural significance, Makhana has, of late, received huge economic importance as well. Hundreds of crores of Makhana is traded every year in India. The total area under Makhana cultivation in India is estimated to be around 45,000 ha, of which Bihar, an eastern state, contributes nearly 85%. India exports Makhana to as many as 20 countries, with USA and UAE being the top two importers of Indian Makhana. Makhana export is further expected to grow at an annual rate of around 25%, with the present value of Makhana exported from India being approximately USD 30 million in 2025. The global Makhana market size is presently around USD 700 million. Domestically, Makhana cultivation, harvesting, popping, and related activities are the mainstay of nearly five lakh families in north Bihar, India. Cultivated scientifically, a farmer can earn up to USD 6000 from 1 hectare of Makhana cultivation. Processing and value addition can further add to the farmers' income.

Health benefits of Makhana are also considered superior to those of dry fruits, such as almonds, walnuts, and many others. It is a good source of energy, protein, carbohydrates, fiber, magnesium, potassium, phosphorus, and micronutrients, including iron and zinc (Jha et al., 2009; Kumar et al., 2016a; Srivastava et al., 2017). Makhana is high in edible fiber and low in calories. It enables weight loss. Its low sodium and cholesterol and high magnesium content makes it beneficial for those suffering from heart diseases, high blood pressure, diabetes, and obesity. It contains many essential amino acids and antiaging biomolecules, including antioxidants (Das et al., 2006; Joseph and Ramesh, 2023; Wang et al., 2025). Glycemic index of Makhana is lower than most high carbohydrate foods, such as rice and bread. Makhana is also known for its spermatogenic and aphrodisiac properties, and has been traditionally used to treat male infertility.

Although the multifaceted significance of Makhana—spanning cultural traditions, economic potential, and nutritional superiority—has positioned it as a broad-spectrum food of global relevance, its successful cultivation remains inherently dependent on specific ecological and environmental conditions. Being an aquatic crop with unique biological and physiological adaptations, *E. ferox* requires a well-defined set of agro-climatic parameters for optimal growth, yield, and quality. With the increasing shift from traditional pond-based systems to scientifically managed field cultivation, a deeper understanding of these agro-climatic requirements has become even more critical. Variations in temperature, water depth, soil characteristics, and hydrological regimes significantly influence plant growth, seed development, and overall productivity. Therefore, a comprehensive examination of the agro-climatic requirements of Makhana is essential not only for sustaining its traditional cultivation but also for expanding its production to new geographical areas and enhancing its commercial viability under changing climatic scenarios.

3.3 Agro-climatic requirements of makhana (*E. ferox* Salisb.)

Makhana is a warm-season aquatic crop that thrives under humid subtropical climatic conditions. Its growth and productivity are governed by a delicate interaction of temperature, rainfall, humidity, and solar radiation. The crop performs optimally within a temperature range of 20–35°C, with warm and humid conditions favoring vigorous vegetative growth and reproductive development. Adequate rainfall, particularly during the monsoon season, ensures sustained water availability in natural and managed ecosystems, while high relative humidity supports physiological processes and reduces evapotranspiration stress (Jha et al., 2018; Kumar et al. 2020a,b).

Being an aquatic species, water regime is the most critical factor influencing Makhana cultivation. The crop requires stagnant or slow-moving freshwater bodies with a stable water depth throughout its growth cycle. Although traditional pond-based systems depend on deeper water levels, modern field-based cultivation practices use shallower water depths with better control and management. Water quality, particularly pH and organic matter content, plays a significant role in nutrient availability and overall plant health. Similarly, soil characteristics—especially clayey texture and high organic content—provide anchorage and act as a reservoir of nutrients (Singh et al., 2020).

In addition to climatic and hydrological factors, the suitability of landforms and ecological conditions also determines the success of Makhana cultivation. Low-lying, waterlogged areas, such as floodplains and wetlands, offer ideal conditions due to their natural water retention capacity. The crop is sensitive to fluctuations in environmental parameters, including temperature extremes and water instability. Therefore, a comprehensive understanding of these agro-climatic requirements is essential for optimizing yield, ensuring quality production, and expanding Makhana cultivation beyond its traditional regions.

Table 3.1 clearly demonstrates that Makhana cultivation is strongly influenced by a combination of climatic and hydrological parameters, with water depth and quality being the most critical determinants. Temperature and humidity regulate physiological growth, while soil characteristics ensure nutrient availability and plant stability. The distinction between pond-based and field-based systems also reflects the increasing role of scientific management in enhancing productivity. Overall, maintaining a stable agro-ecological environment is essential for achieving a higher yield and better-quality Makhana production.

TABLE 3.1 Agro-climatic requirements of makhana.

Parameter	Requirement/Range	Remarks
Temperature	20–35°C (optimum: 25–32°C)	Below 15°C slows growth; above 38°C causes stress
Rainfall	100–250 cm annually	Ensures water availability, especially during monsoon
Relative humidity	50%–90%	High humidity favors growth and reduces water loss
Light	Full sunlight	Essential for photosynthesis and leaf development
Water depth (pond)	1.0–1.8 m	Suitable for traditional systems
Water depth (field)	30–60 cm (optimum: 45 cm)	Suitable for improved field cultivation
Water type	Freshwater (stagnant/slow-moving)	Stable water regime is critical
Water pH	6.5–8.0	Neutral to slightly alkaline preferred
Soil type	Heavy textured soil (clay loam)	High water-holding capacity required
Soil pH	6.5–8.0	Supports nutrient availability
Organic matter	High	Enhances soil fertility and productivity
Soil fertility	Medium to high	Makhana responds highly to secondary and micronutrients
Topography	Low-lying, waterlogged areas	Floodplains and wetlands are ideal

3.4 Makhana cultivation practices

Conventionally, Makhana was cultivated almost exclusively in ponds, oxbow lakes, and other perennial water bodies, primarily by the fisher communities of North Bihar. These traditional systems relied largely on natural ecological conditions with minimal external inputs, resulting in moderate productivity and high labor dependency. However, with the rapid increase in domestic and global demand for Makhana and its recognition as a high-value crop, there has been a growing need to enhance both productivity and the area under cultivation. At the same time, the gradual decline in natural water bodies (Tarate et al., 2025) and increasing pressure on wetland ecosystems have posed serious challenges to the expansion of traditional Makhana cultivation. In response to these constraints, scientific efforts have led to the introduction of Makhana cultivation under field conditions, similar to wetland rice cultivation. This system enables better control over water depth, nutrient management, and crop operations, thereby improving harvesting efficiency and overall productivity.

Field-based cultivation also offers significant advantages in terms of crop diversification and integration into existing cropping systems prevalent in North Bihar and other parts of eastern India. When cultivated scientifically, particularly in areas with assured water availability during the growing season, Makhana can yield substantially higher returns compared with traditional systems. Therefore, a combination of improved traditional pond-based cultivation and scientifically managed field-based systems holds immense potential for enhancing production and meeting the rising global demand for Makhana in regions with suitable edaphoclimatic conditions (Kumar et al., 2011). Based on the ecological conditions, water availability, and level of management, Makhana cultivation can broadly be categorized into three different production systems. These include the traditional pond-based system, the improved pond system, and the field-based cultivation system. Each of these systems has its own set of practices, advantages, and limitations, which are discussed in detail in the following sections.

3.5 Makhana cultivation in ponds

As Makhana is a crop of lentic ecosystem, traditionally, it was being grown in the ponds, natural water bodies, and low land areas, with assured availability of water throughout the Makhana growing season (Figs. 3.1 and 3.2). Seed sowing is not required in pond system as the new crop arises every season from

FIGURE 3.1 Makhana cultivation in pond at ICAR-National Research Center for makhana, Darbhanga, India.

FIGURE 3.2 Cultivation of makhana in the *Chaur* land.

left over seeds of the previous crop. Makhana farming in new ponds, however, may be started by direct sowing or transplanting the plantlets. For direct sown crop, 75–80 kg/ha of Makhana seed is broadcasted in the month of December. In March, plants should be thinned out to leave a line-to-line and plant-to-plant spacing of 1.25 × 1.25 m. Traditionally, little attention has been given to proper nutrient management in large ponds and water bodies, so the yields obtained from this system of cultivation has been low (≈1.4–1.6 t/ha) as compared to newly introduced field system of cultivation (≈2.0–2.5 t/ha) (Kumar et al., 2011).

3.6 Makhana cultivation in *chaur* (seasonal floodplain wetland) areas

In eastern India, particularly in North Bihar, large tracts of low-lying lands locally known as *chaur* (seasonal floodplain wetlands) provide highly suitable ecological conditions for Makhana cultivation. These landscapes are characterized by seasonal inundation during the monsoon, followed by gradual recession of water, creating a unique hydrological regime with relatively stable water availability for a considerable part of the growing season. The presence of fertile alluvial soils and nutrient-rich sediments further enhances their suitability for supporting aquatic crops, such as Makhana.

Traditionally, these *chaur* areas have remained underused or have been used for low-intensity agriculture due to prolonged waterlogging. However, in the context of declining availability of perennial ponds and increasing pressure on conventional water bodies, *chaur* lands offer a promising alternative for expanding Makhana cultivation. Unlike deep pond systems, these areas allow relatively easier crop management and harvesting operations, while still maintaining the essential water conditions required for the crop.

Furthermore, while field-based Makhana cultivation has emerged as a viable option, it is often constrained by high water requirements and management costs. In this context, the utilization of *chaur* lands presents a more resource-efficient and ecologically sustainable approach for scaling up Makhana production (Raut et al., 2024). Therefore, the promotion of Makhana cultivation in *chaur* areas can play a significant role in enhancing production, improving farmers' income, and meeting the rising demand for Makhana both within India and globally.

3.7 Field system of Makhana cultivation

Field cultivation of Makhana (*E. ferox*) has emerged as an important approach to expand the cultivation area and improve productivity (Fig. 3.3). Unlike traditional pond systems, field-based cultivation allows better control over nutrient management, water levels, and plant protection, while also facilitating intercultural operations and improving harvesting efficiency. These advantages collectively enable higher yield potential under field conditions (Kumar et al. 2020a,b). In India, the area under Makhana cultivation has increased nearly threefold over the last 10 years, rising from about 13,000 to around 40,000 ha in Bihar alone, with the majority of the newly brought area being under field-based cultivation. This shift reflects the growing adoption of managed field systems to meet rising domestic and global demand.

3.8 Makhana nursery management

Effective nursery management is a cornerstone of successful field cultivation of Makhana (*E. ferox*), as it determines the establishment, vigor, and uniformity of seedlings that will be transplanted into the main field. Unlike pond systems, where seedlings can emerge naturally from residual seeds, field

FIGURE 3.3 Makhana cultivation in field at ICAR-National Research Center for makhana, Darbhanga, India.

cultivation requires a dedicated nursery area (about 500 m^2 per hectare) to raise healthy plantlets under controlled conditions prior to transplanting. Optimal nursery preparation begins with land preparation, involving two to three deep ploughings to ensure good soil structure and aeration, which enhances root penetration and early seedling growth. The soil should be clayey and rich in organic matter, which supports nutrient availability and water retention, essential for aquatic seed establishment (Kumar et al., 2011; Raut et al., 2024). Nursery beds are enriched with organic manure (such as well-decomposed farmyard manure) to ensure baseline fertility and to promote healthy microbial activity in the soil. In addition to organic inputs, balanced nutrients in terms of nitrogen (N), phosphorus (P), and potassium (K) are applied based on fertilizer recommendations for Makhana cultivation in field conditions, typically following an integrated nutrient management approach. Micronutrients, such as zinc (Zn) and boron (B), may also be included when soil tests indicate deficiencies, as these elements are critical for early seedling development and stress tolerance. Neem cake or neem-based products can be incorporated into the nursery soil to enhance pest suppression, especially against soil-borne insects, without resorting to harmful chemical pesticides.

Seed sowing in the nursery is usually done in December, when the prepared bed is filled with water to about 30-cm depth. Healthy, bold seeds (about 20 kg per hectare equivalent) are uniformly broadcasted to achieve optimum seedling density. Maintaining a stable water level through the nursery period (December through March) is crucial (Fig. 3.4). During this period, regular monitoring for insect pests, such as aphids, is important, and ecofriendly pest management, such as neem oil sprays or introduction of natural predators, is advised to protect seedlings without damaging beneficial organisms or the aquatic environment. By March, the seedlings attain

FIGURE 3.4 Makhana nursery being raised in field at two different sites in Darbhanga, India.

sufficient size and vigor, becoming ready for transplantation into prepared field plots with optimal spacing.

To sum up, meticulous nursery management—with attention to soil preparation, balanced NPK fertilization, organic amendments, water-level management, and ecofriendly pest control—provides a strong foundation for vigorous seedling growth and contributes significantly to achieving high field yields and sustainable Makhana production. This approach aligns with integrated nutrient management principles developed for its cultivation in North Bihar, which emphasize balanced fertilization and soil fertility maintenance for improved crop performance.

3.9 Transplanting

Healthy Makhana plantlets (when leaves are of 15–20 cm diameter) should be uprooted carefully from nursery so as to avoid any damage to root and transplanted in a well prepared field in the month of March/April, maintaining a line-to-line and plant-to-plant distance of 1.25 m each (Fig. 3.5) (Kumar et al. 2011; Kumar et al. 2020a,b). Time interval between uprooting of seedlings and their transplanting should be kept at minimum possible. While pressing the root in soil during transplanting, care should be taken to ensure that an upward facing "light-pink colored finger-like structure" (3–5 cm in length) at the junction of root and shoot should not be pushed inside the mud of the floor. Gap filling, in case few seedlings do not survive, may be done to ensure uniform plant stand in the field.

3.10 Nutrient management in makhana

Nutrient management in Makhana under field conditions has emerged as a critical determinant of productivity, profitability, and seed quality, particularly

FIGURE 3.5 Uprooted makhana plant for transplanting (*left*) and transplanted makhana field (*right*) at ICAR-National Research Center for makhana, Darbhanga, India.

in the context of its transition from traditional pond-based systems to intensified cultivation practices. Unlike pond ecosystems where nutrient cycling is largely self-sustaining through decomposition of plant residues, field-grown Makhana requires deliberate and balanced external nutrient inputs to maintain soil fertility and sustain crop performance (Kumari et al., 2014; Kumar et al. 2020a,b). The available research, particularly the studies of Singh et al. (2020) and Kumar (2023), clearly establishes that Makhana responds not only to primary nutrients (N, P, K) but also significantly to organic matter, secondary nutrients, such as magnesium, and micronutrients, including zinc, boron and copper, all of which play distinct physiological and biochemical roles in influencing crop growth, yield attributes, and seed quality.

Application of primary nutrients forms the foundational component of nutrient management in Makhana. Nitrogen is directly involved in vegetative growth through its role in chlorophyll formation, amino acid synthesis, and overall biomass production. In Makhana, which produces very large floating leaves, adequate nitrogen supply ensures the development of a robust photosynthetic canopy, thereby enhancing assimilate production. Phosphorus plays a crucial role in root development, energy transfer through ATP, and early establishment of transplanted seedlings, which is particularly important in field conditions where root anchorage and nutrient uptake efficiency determine subsequent growth. Potassium regulates osmotic balance, enzyme activation, and translocation of photosynthates from source to sink, thereby directly influencing seed filling and quality (Havlin et al., 2015). Experimental evidence indicates that the application of NPK at 100:60:40 kg ha^{-1} significantly improves seed yield compared to control, with yield enhancement of about 19%–26% reported under similar agro-climatic conditions (Singh et al., 2020; Kumar, 2023). However, increasing NPK levels beyond this threshold does not result in proportional yield gains (Fig. 3.6), suggesting the operation of Liebig's Law of Minimum, wherein the availability of other limiting nutrients restricts further response to primary nutrient application (Singh et al., 2020). This clearly indicates that while NPK is necessary, it is not sufficient alone to achieve the full yield potential of Makhana.

The role of organic matter, particularly farmyard manure (FYM), is equally crucial and often more pronounced than that of chemical fertilizers alone. Organic carbon is a key indicator of soil health and influences a wide range of physical, chemical, and biological properties of soil. Application of FYM at 20 t ha^{-1} has been shown to significantly improve soil structure by reducing bulk density and increasing porosity, thereby facilitating better root penetration and aeration even under submerged conditions typical of Makhana cultivation (Singh et al., 2020). Improved soil physical conditions enhance root proliferation and increase the effective volume of soil explored by roots for nutrient uptake. Additionally, FYM increases water-holding capacity, which is particularly beneficial in maintaining stable water regimes in field systems. Chemically, organic matter acts as a reservoir of nutrients and improves cation

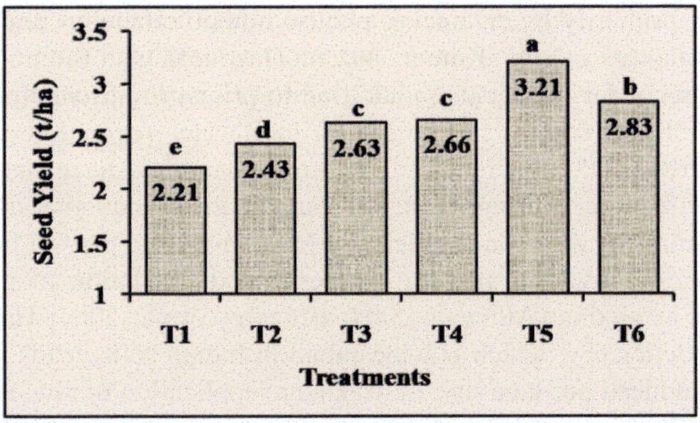

FIGURE 3.6 Effect of various doses and combinations of fertilizers and manures on seed yield of makhana. *Different letter/s at the top of bars indicate significant differences at P* $\leq$*.05. Treatments comprised* T_1 *(control, no fertilizer or manure),* T_2 *(NPK @ 75:45:*30 kg *ha*$^{-1}$), T_3 *(NPK @ 100:60:*40 kg *ha*$^{-1}$), T_4 *(NPK @ 125:75:*50 kg *ha*$^{-1}$), T_5 *(NPK @ 100:60:*40 kg *ha*$^{-1}$ + *FYM @ 20 t ha*$^{-1}$), *and* T_6 *(FYM @ 20 t ha*−1).

exchange capacity, thereby enhancing the retention and gradual release of nutrients, such as nitrogen, phosphorus, and potassium. More importantly, it plays a vital role in increasing the availability of micronutrients, such as zinc and copper, through chelation processes, preventing their fixation in unavailable forms (Kumar et al., 2016b). Biologically, FYM stimulates microbial activity, which accelerates nutrient mineralization and improves nutrient cycling. These combined effects translate into substantial yield improvements, with studies reporting up to 45% higher yield when FYM is integrated with NPK compared with control, and significantly higher than sole application of fertilizers (Fig. 3.6) (Singh et al., 2020). The strong positive association between organic carbon and Makhana productivity observed in pond-based systems further reinforces the importance of organic inputs (Dutta et al., 1986).

Secondary nutrients, particularly magnesium, also play a critical role in Makhana nutrition, although their importance has only recently been recognized. Magnesium is the central atom of the chlorophyll molecule and is therefore directly involved in photosynthesis. Given the large leaf area of Makhana, any improvement in chlorophyll content and photosynthetic efficiency can have a substantial impact on biomass production and subsequent yield (Cakmak, 2013; Gerendas and Fuhrs, 2015). Magnesium also functions as an activator of several enzymes involved in carbohydrate metabolism and energy transfer, thereby influencing overall plant growth and development (Guo et al., 2016; Lukowiak et al., 2016). In many Indian soils, including those of North Bihar, magnesium deficiency or imbalance with calcium is common, which can limit nutrient uptake and plant performance. Application of magnesium @ 5 kg ha^{-1} along with NPK has been shown to improve yield significantly, primarily by enhancing photosynthetic efficiency and correcting nutrient imbalances in soil (Kumar, 2023). This highlights the importance of considering secondary nutrients in addition to primary nutrients for achieving balanced plant nutrition.

Micronutrients, though required in small quantities, have disproportionately large effects on plant growth, particularly in reproductive development and seed formation, which are critical for Makhana. Zinc plays a fundamental role in enzyme activation, protein synthesis, and regulation of growth hormones, such as auxins (Alloway, 2004; Broadley et al., 2007; Hafeez et al., 2013). Zinc deficiency, which is widespread in Indian soils, leads to impaired growth and reduced productivity. In Makhana, application of zinc at 5 kg ha^{-1} along with NPK has been found to significantly improve yield, likely due to its role in enhancing metabolic activities and improving overall plant vigor (Kumar, 2023). Boron is another essential micronutrient that is particularly important for reproductive processes. It is involved in cell wall formation, pollen viability, fertilization, and translocation of sugars from leaves to developing seeds (Marschner, 1995; Szulc and Rutkowska, 2013). Adequate boron supply ensures proper seed setting and filling, which directly influences yield and seed quality. Application of boron at 1.0 kg ha^{-1} along with NPK has been shown to improve yield significantly, reflecting its critical role in reproductive growth (Kumar, 2023).

Among micronutrients, copper has shown the most pronounced effect on Makhana productivity, especially when applied as a foliar spray (Kumar, 2023). Copper is involved in several physiological processes, including photosynthesis, respiration, lignin synthesis, and enzyme regulation (Yruela, 2009). More importantly, it plays a vital role in reproductive development, particularly in pollen formation, fertilization, and seed development (Havlin et al., 2015). Foliar application of copper at 0.1% concentration (in form of $CuSO_4$) around 45 days after transplanting has been found to result in the highest yield improvement among all micronutrient treatments, with yield increases of over 60% compared with control and significantly higher than nitrogen, phosphorus and potassium (NPK) alone (Kumar, 2023). The effectiveness of foliar application in Makhana can be attributed to the unique morphology of the crop, characterized by large, horizontally oriented leaves that provide an extensive surface area for nutrient absorption. Foliar feeding allows for rapid correction of nutrient deficiencies and ensures direct availability of nutrients to the plant during critical growth stages, bypassing soil-related constraints, such as fixation and leaching. Additionally, copper has protective effects against fungal and bacterial diseases, which may further contribute to improved crop performance.

The combined application of primary, secondary, and micronutrients has been shown to produce synergistic effects on the yield and quality of Makhana. When magnesium, zinc, boron, and copper are applied together with NPK,

TABLE 3.2 Effect of nutrients' application on seed yield and yield parameters of fox nut.

Treatments	Number of fruits plant^{-1}	Number of seeds fruit^{-1}	Test weight (100 seed wt.) (g)	Seed yield (t ha^{-1})
T_1: Control	7.33 ^{b}	55.3 ^{b}	73.2 ^{d}	1.81^{e}
T_2: NPK @ 100:60:40 kg ha^{-1}	7.63 ^{ab}	59.1 ^{b}	84.6^{c}	2.28^{d}
T_3: NPK + Mg @ 5 kg ha^{-1}	7.78^{a}	66.3^{a}	93.1 ^{b}	2.65^{c}
T_4: NPK + Zn @ 5 kg ha^{-1}	7.76^{a}	67.2^{a}	92.6 ^{b}	2.71^{c}
T_5: NPK + Cu (0.1% spray)	7.86^{a}	68.3^{a}	103.8^{a}	2.93^{b}
T_6: NPK + B @ 1 kg ha^{-1}	7.68^{a}	67.7^{a}	91.2 ^{b}	2.66^{c}
T_7: NPK + Mg + Zn + Cu + B	7.82^{a}	71.3^{a}	109.5^{a}	3.27^{a}

Different superscript letters within a column indicate significant differences among treatment means ($P \leq 0.05$).

yield improvements of up to 80% over control and more than 40% over NPK alone have been reported by Kumar (2023) (Table 3.2). This demonstrates that balanced nutrition not only enhances individual nutrient use efficiency but also improves overall plant physiological functioning. Yield improvement in Makhana is closely associated with key yield attributes, such as number of fruits per plant, number of seeds per fruit, seed size, and test weight.

Nutrient application, particularly micronutrients, has been found to significantly improve seed size (Fig. 3.7) and test weight (Table 3.2), which are important quality parameters in Makhana (Kumar, 2023). Copper, in particular, had a pronounced effect on size and weight of Makhana seeds. The highest proportion of bold seeds ($\geq$1.0 cm diameter) was recorded under $CuSO_4$-treated plots, indicating a strong role of copper in improving seed development. This can be attributed to the involvement of Cu in reproductive processes, such as pollen formation, fertilization, and the efficient translocation of photosynthates to developing seeds (Yruela, 2009; Havlin et al., 2015; Kumar, 2023). Adequate copper nutrition under balanced fertilization thus enhances seed filling, resulting in larger and more marketable seeds.

An important observation from these studies is that increased yield due to nutrient application does not necessarily lead to the dilution of nutrient concentration in seeds. On the contrary, application of micronutrients and

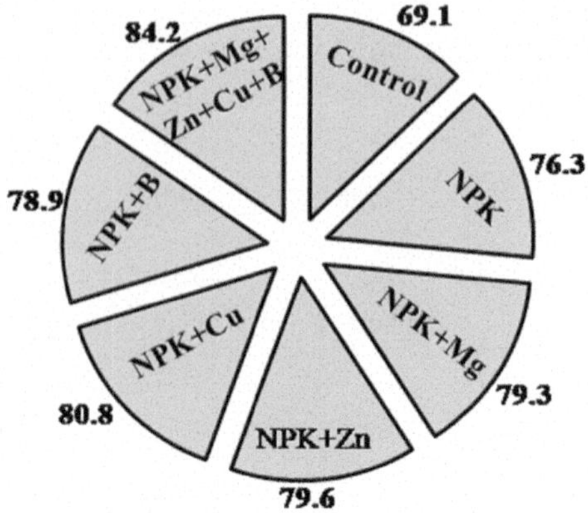

FIGURE 3.7 Effect of nutrients' application on seed size of makhana in terms of percentage of seeds with ≥1.0 cm diameter. Description of the nutrients doses can be seen in Table 3.2.

magnesium has been found to maintain or even improve nutrient concentration and protein content in seeds, indicating a quality-sustaining effect (Kumar, 2023). This is particularly important in the context of Makhana's growing demand as a nutritionally rich food, where quality parameters are as important as yield.

From an economic perspective, integrated nutrient management, involving the combined use of NPK and FYM, has been found to provide the highest benefit-cost ratio, indicating that such practices are not only agronomically effective but also economically viable for farmers (Singh et al., 2020). The improvement in soil health, nutrient availability, and crop productivity under integrated approaches ensures sustainability of Makhana cultivation in the long run.

Based on the synthesis of available research, the most effective nutrient management strategy for Makhana under field conditions involves the application of NPK at 100:60:40 kg ha^{-1}, supplemented with FYM at 20 t ha^{-1} to improve soil health and nutrient availability. In addition, the application of magnesium @5.0 kg ha^{-1}, zinc @5.0 kg ha^{-1}, and boron @1.0 kg ha^{-1} through soil, along with foliar spray of copper sulphate at 0.1% concentration during the vegetative phase, particularly around 45 days after transplanting, is recommended for achieving higher yield and better seed quality (Singh et al., 2020; Kumar, 2023). This integrated approach ensures balanced nutrition, improved nutrient-use efficiency, enhanced photosynthetic activity, better reproductive success, and ultimately, higher productivity and profitability for Makhana under field conditions.

3.11 Water management in makhana

Being an aquatic crop, Makhana requires a continuous water regime for its growth and development, and can be successfully cultivated in a wide range of water depths, from as low as 20–30 cm to as high as 10–12 feet (Kumar et al., 2021a) (Fig. 3.8). However, the efficiency of water use, plant growth behavior, and ultimately, the productivity of the crop are strongly influenced by how water is managed rather than merely how much water is present. Therefore, water management in Makhana is not just about maintaining standing water, but about optimizing depth, timing, and source of water in accordance with crop growth stages and local climatic conditions.

A key principle emerging from recent field experiences is that judicious utilization of rainwater should form the backbone of Makhana cultivation. Since the crop requires substantial quantities of water, aligning peak crop water requirement period with the onset of monsoon becomes critical. Of late, farmers are depending heavily on groundwater irrigation, but such indiscriminate extraction is neither economical nor sustainable, particularly in regions, such as North Bihar, where groundwater depletion is becoming a growing concern (Kumar et al., 2025; Sharma and Kumari, 2024). A key consideration in successful Makhana cultivation is the selection of site with abundant natural availability of the surface water, such as ponds, *chaurs*, low-lying areas, where dependence on groundwater for irrigation is minimum. Due to declining rainfall and increasing surface temperature induced by the

FIGURE 3.8 Makhana crop growing in 30-cm water depth (*left*) and in nearly 10-feet water depth (*right*).

FIGURE 3.9 A complete failure of makhana crop due to water shortage in a farm at Darbhanga, India during July 2025.

changing climate in the major Makhana-growing areas of the north Bihar, India (Tarate et al., 2025), farmers are frequently witnessing crop failures due to water shortage in peak growth period (Fig. 3.9).

Adjusting the cropping calendar, even if it involves delaying transplanting by a few weeks, can significantly reduce dependence on irrigation and ensure that the crop benefits from natural rainfall. This approach not only conserves groundwater but also reduces the cost of cultivation and improves long-term sustainability of Makhana farming systems. There has been a long-standing perception among traditional growers that Makhana requires deep-water conditions, typically 5–6 feet or more, for optimal growth. This belief is rooted in the conventional practice of cultivating Makhana in natural ponds, where such depths are common. However, recent scientific studies and field-level observations have challenged this notion. It has now been clearly demonstrated that Makhana can be grown very successfully under shallow water conditions of about 1.0–2.0 feet, particularly in field-based systems (Jana, 2017). In fact, shallower water depths offer several physiological and agronomic advantages over deep-water conditions (Kumar et al., 2021a).

Under deep-water conditions, Makhana plants tend to invest a significant portion of their energy and biomass in elongating their petioles (leaf stalks) to ensure that leaves can reach and float on the water surface. This is essentially a survival mechanism that enables the plant to maintain photosynthetic activity. However, this excessive vegetative growth comes at the cost of reproductive development. In contrast, when grown under shallow water conditions, the plant does not need to expend energy on the elongation of vegetative structures. Instead, there is a preferential allocation of assimilates toward reproductive organs, such as flowers and seeds, resulting in improved harvest index and higher seed yield. Thus, shallow water cultivation not only saves water but also enhances productivity.

Despite the advantages of shallow water, maintaining an optimum and stable water depth throughout the crop growth period is crucial. The water level should be sufficient to allow leaves to float freely and to ensure that developing fruits remain properly submerged. This is particularly important during the reproductive phase, when fluctuations in water level can adversely affect pollination, fruit development, and seed filling. Sudden reduction in water depth during this stage may expose fruits or restrict their development, while excessive water may again divert plant energy toward vegetative adaptation.

Another important aspect of water management relates to the timing of harvest and water depth at maturity. In late-sown crops, if water depth becomes too shallow toward the end of the season, there is a risk of premature germination of seeds within the field itself. After fruit bursting, seeds settle at the bottom of the field, and under shallow water conditions combined with favorable temperature (particularly during November–December), they may start germinating due to exposure to sunlight and warmth. Such in situ germination leads to significant losses in both yield and quality. Therefore, maintaining adequate water depth until harvest is essential to prevent this problem. At the same time, extremely shallow water during harvesting operations makes it difficult for workers to sweep, collect, and clean the seeds embedded in the muddy field bottom. Thus, a balanced water level is necessary not only for crop growth but also for efficient harvesting.

Water management in Makhana also assumes great importance under conditions of flooding, which is a common occurrence in North Bihar. While Makhana is inherently tolerant to standing water, extreme flood situations can still damage the crop, particularly when water currents become strong or when a sudden rise in water level submerges leaves beyond their adaptive capacity. Based on field experiences and farmer innovations, certain practices can help mitigate flood-related damage. These include strengthening field bunds to prevent the entry of strong water currents, and installing net fencing to make sure that the seeds are not washed away from the farm by flowing water (Kumar, 2020). Selection of appropriate field sites with better water retention but a lower flood vulnerability also plays a role in long-term risk reduction.

Equally important are the strategies aimed at reducing water requirement without compromising productivity. One of the most effective approaches is the adoption of field-based cultivation with controlled water depth (one to two feet), as opposed to traditional deep pond systems. Proper land leveling ensures uniform water distribution and prevents localized water stress or excess. Strengthening bunds and minimizing seepage losses help in conserving water within the field. A recent success story by Kumar et al. (2021a) showed a Makhana farmer using a simple, cost-effective, and easily applicable technique of "polythene lining of the bunds" (Fig. 3.11) to reduce the loss of costly

FIGURE 3.10 Net fencing to prevent the seeds from flowing out of the farm along with running flood water *Courtesy Kumar, M. 2020. Makhana triumphs over flood: a success story from Darbhanga, Bihar. Biotica Res. Today. 2(11), 1113-1116.*

irrigation water by lateral seepage through the bunds, thereby reducing the irrigation frequency required for maintaining the minimum-required water depth in the Makhana field, particularly before the onset of the rainy season (March–May), the water-scarce period in North Bihar, India (Fig. 3.10). Scheduling transplanting in synchrony with monsoon rains reduces the need for supplemental irrigation. Additionally, maintaining a minimum-required layer of water rather than an excessive depth during vegetative stages can significantly reduce the total water requirement while still supporting healthy crop growth. These practices collectively contribute to improving water-use efficiency in Makhana cultivation.

FIGURE 3.11 A Makhana farmer standing on the bund with "polythene lining".

On the whole, effective water management in Makhana involves a combination of traditional knowledge and scientific insights. Although the crop can tolerate a wide range of water depths, maintaining an optimum depth of about 1–2 feet under field conditions is most beneficial for maximizing yield and water-use efficiency. Emphasis should be placed on rainwater utilization, avoidance of excessive groundwater extraction, careful timing of nursery and transplanting, and stage-specific water-level management. Additional strategies for flood protection and water conservation further enhance the sustainability and resilience of Makhana cultivation systems. By adopting these practices, farmers can achieve higher productivity, better resource-use efficiency, and long-term sustainability in Makhana farming.

3.12 Plant protection

Makhana was once regarded as a hardy crop, as not many insect pests and disease occurrences, causing substantial yield losses, were reported. However, with increasing commercialization of Makhana farming and intensive research efforts directed toward improving Makhana yield, many weeds, insect pests, and diseases have been found to cause yield losses ranging from 20%–25% to the total crop failure. Aphids (*Rhopalosiphum nymphaeae*), caseworm (*Elophila crisonalis* W.), root borer (*Donacia delesserti*), leaf midge or rib borer (*Chironomus* spp.), and sucking bug (*Plea liturata* F.) are the major insect pests of Makhana (Kumar et al., 2021b). Aphids' and sucking bugs' attacks are usually observed in tender leaves of young plants, more often in nursery and initial stages of crop growth after transplanting. Caseworm and root borers are more damaging in the later stage of crop growth. Caseworm usually attacks on flower and root borers on root. Gastropod (snails and slugs) infestation is also frequently observed in Makhana these days (Fig. 3.12). They remain attached

FIGURE 3.12 Aphid-infested young leaves of makhana (*left*); gastropod infestation in makhana (*middle* and *right*).

with under surface of the leaves, shoots and fibrous roots, and feed on plant tissue of the host plant, causing fast decay of affected plants.

Seed treatment with imidacloprid 70WS @5 gm/kg seed and root dip treatment with imidacloprid 70 WS @ 5 mL/L of water for 30 min at the time of transplanting can be helpful in controlling the insect pests of Makhana. Foliar spray of imidacloprid (a systemic insecticide) can also be given in standing crop @1–2 mL/lit water (once or twice depending on intensity of insects' infestation) to control the insect pests of Makhana (Kumar et al., 2021b). As Makhana is grown in aquatic environment, using high quantities of chemical pesticides is not advisable due to associated environmental concerns. Organic pesticides are instead better advised for controlling pests in Makhana field. Neem cake @100 kg/ha can be used as basal dose during land preparation to control root borer. Foliar spray of neem oil (3.0 mL/L water) has also been found effective in controlling insect pests, such as aphids and caseworm.

Leaf blight, caused by *Alternaria tenuis,* is the most serious fungal disease of Makhana. Symptoms appear on the upper surface of the leaves, which include leaf spots and blighting. Light brown leaf spots usually have concentric rings, which look like a target board with a yellow halo. As leaf spotting increases, blighting and premature defoliation occur in Makhana. Foliar sprays with copper oxychloride or dithane M-45 @0.3% twice or thrice at fortnightly interval have been found very effective to check the disease. Root rot disease is believed to be caused by Phytophthora species. Apparently, healthy plants suddenly show wilting and yellowing. The roots appear dark brown or black, which become soft and rotten. Plants infected with root rot should be removed from the field and destroyed. Algal boom is also being increasingly reported in intensively managed Makhana fields receiving heavy doses of fertilizers (Fig. 3.13). It can be managed by controlled application of $CuSO_4$, which can also add nutrients and give some plant protection advantages.

FIGURE 3.13 Algal boom in makhana nursery (*left*) and in main field (*right*).

FIGURE 3.14 Weeds usually found in makhana farms of north Bihar, India.

Weeds appear mostly in early stages of Makhana growth, when there remains a lot of open space between the plants (Fig. 3.14). Manual weeding is required once in nursery, and once during the initial 30–40 days after transplanting. Makhana leaves cover the entire water surface thereafter, requiring no further efforts for weed control.

3.13 Harvesting and storage

Makhana transplanted in the month of March and early April starts flowering and fruiting from May (nearly 2 months after transplanting) and continues up to October/November. It takes around 40–45 days (after flowering) for the fruits to become fully mature. Thereafter, the fruits start rupturing and the seeds with pinkish cover remain afloat on the water surface (Fig. 3.15).

FIGURE 3.15 Makhana flower and fruit (*left*); seeds floating on the water surface (*right*).

FIGURE 3.16 Makhana harvesting (*left*) and harvested seed (*right*).

After 2–3 days, they get settled on the bottom of the field. Since flowering, fruiting, and rupturing of the fruits are highly unsynchronized in nature, a practical decision has to be taken before harvesting as to how much time to allow for the fruiting and rupturing to avoid any significant yield loss and to accommodate the crop/s to follow in the cropping sequence on the same field. A crop transplanted in March can be harvested in early August, although fruiting may continue up to November, if left unharvested. Makhana seeds are collected from the floor of the field using a local device called "*Ganja*" (Kumar et al., 2011) (Fig. 3.16). Three rounds of harvesting are usually required to harvest nearly 90% of the seeds. Harvest may further continue if the price of harvested produce is higher than the cost of harvesting. Using scientifically devised cultivation techniques, farmers can get a seed yield of up to 30q/ha from an improved variety. Seed yield even in excess of 35q/ha is also observed at research stations and progressive farmer's field. After proper cleaning of seeds (removing mud, snails and slugs, weeds, and other extraneous materials), they are packed in gunny bags and stored in water tanks or ponds to keep them moist for further processing. Sprinkling of water at regular intervals also serves the purpose. When stored in shallow water depth of ponds or trenches, some live covering material, such as water hyacinth, may be placed over water surface (with Makhana bags there under) to protect the seeds from warming effect of sun light, which may otherwise cause germination of seeds if kept for long.

3.14 Postharvest processing

After harvesting, black seeds of Makhana have to undergo a series of highly skillful procedures, including sun drying, seed grading, preheating, tempering, roasting, popping, and polishing, before we see the consumable white pops of Makhana (Fig. 3.17). Optimum level of "Sun Drying" is extremely important which determines the success of subsequent procedures leading to the desired quality of Makhana pops. Sun drying reduces seeds' moisture content from an

FIGURE 3.17 Makhana seed drying, roasting, and popping (*left* to *right*).

initial 37% to around 30%. Sun-dried seeds are graded in different size groups using sieves of desired diameters. Seven size grades (0.4–1.2 cm) are usually obtained through sieving. However, with increasing commercial interest in Makhana processing, seeds are graded in 16 different size groups by many commercial processers. This is to facilitate uniform striking of the roasted Makhana seeds during popping operations, which leads to better percentage of good-quality well-rounded pops, and lesser proportion of unpopped or semipopped products, locally known as *Murra* and *Thurri* (Kumar et al., 2011).

After sun drying and size grading, seeds are subjected to first heating (preheating). Seeds are heated in earthen pitcher or cast iron pan, with continuous stirring. Preheating is usually done at a temperature of 250–300 °C for 5–6 min, which reduces seed moisture to nearly 20%. Preheated seeds are then stored for 2–3 days at ambient condition, a process known as "Tempering". Tempering reverses the process of thermal expansion of the seeds (differential thermal expansion of different parts) caused during preheating. This alternate swelling and shrinking of the seed loosen the kernel within hard seed coat and helps to equilibrate the moisture within the seed. After 2–3 days of seed tempering, seeds are subjected to final "Roasting and Popping". This is the most skill-demanding activity during the entire post-harvest processing of Makhana. Nearly 250 g of tempered seeds are roasted at a time in a cast iron pan at a temperature of 290–340°C for nearly 90–130 s. Cracking sounds coming from the seeds being roasted are heard to judge the readiness of seeds to be struck for final popping. Ready seeds (8–10 at a time) are scooped from pan, kept on a flat wooden block, and struck with a wooden hammer, subsequent to which the black seed coat breaks and the white kernel pops out in expanded form, known as *Lawa* or Makhana we happen to see in market. Yield of Makhana thus obtained is usually 40% of the raw seed weight, though it may vary with the quality of the raw seeds. Red specks are usually seen adhering to the popped Makhana, which are the remains of the red pericarp, covering the kernel of the raw seeds before popping. Makhana is polished to remove these red specks and to give it more white and shiny appearance. Polishing is done immediately after popping, while Makhana pops are still hot, by rubbing and vigorously shaking them in a bamboo-made basket (Kumar et al., 2011).

Popped Makhana is separated in two grades at producer level, i.e., *lawa* and *thurri*. Lawa is the fully popped white-colored Makhana, while *thurri* is semipopped, hard and reddish in color. For trading purposes, Makhana is categorized into multiple grades depending on their size and quality. *Lawa* top is the highest quality grade followed by *Lawa, Olwa, Murra,* and *Thurri* in order of reducing quality grade. These days, "*Suta*" a traditional unit of length measurement, is frequently used in Makhana trading circle. One *suta* equals 3.175 mm, as eight *sutas* make an inch (25.4 mm). Makhana with diameter of more than six *suta* is considered the best in quality and fetches the best market price. These are hand-picked Makhana pops. Makhana with a diameter of more than five *suta* is also high in demand among consumers. With increasing commercial interests in Makhana processing, machines are also being increasingly used these days for the entire postharvest processing operations (Seed drying–grading–roasting–popping–polishing) of Makhana seeds. Mechanization can substantially reduce the human drudgery involved in Makhana harvesting and processing.

3.15 Value addition

In view of the rapidly growing commercial importance of Makhana consequent to the increasing global awareness about its nutritional and medicinal importance, value addition in Makhana has witnessed a sudden surge in recent couple of years. A range of value-added products are now available in the market to cater to the varying consumer preferences. Roasted Makhana, salted Makhana, flavored Makhana, Makhana kheer mix, Makhana *ladoos*, Makhana barfi, Makhana chocolates, and many other Makhana-based consumables are making their presence felt to the health conscious consumers of India and abroad. Makhana flour can be prepared from low-grade Makhana (*Thurri* and *Murra*) which cannot be profitably sold as Makhana pops. A variety of dishes can be prepared from Makhana flour, using either pure Makhana flour or mixing it with wheat flour and others. Makhana chapatti, *puri*, *paratha, etc.* are some of the easy options that a consumer would be interested in on regular basis. Baby food can also be prepared from Makhana, thanks to easy digestibility and rich mineral contents therein.

3.16 Makhana in cropping system

One major advantage of cultivating Makhana in field condition is the possibility of growing some other crops in the same field after harvesting of Makhana, which increases both the cropping intensity and farmers' income. With meticulous planning, as many as two other crops can be taken in a year on the same piece of land after harvesting of Makhana and before transplanting in the next season. April-transplanted Makhana can be harvested in August, subsequent to which a short duration of rice variety can be grown followed by wheat or some other *rabi* crops that can be harvested by the middle of April, leaving the field available for Makhana transplanting for the next cropping cycle. Apart from Makhana–rice–wheat, Makhana–water chestnut–berseem or other cropping sequence as suited/needed by a particular farmer can be adopted for better farm income (Kumar et al., 2011).

3.17 Income from makhana farming

Makhana has emerged as one of the most profitable crops in eastern India, particularly in regions with relatively better water availability, and is currently cultivated over about 45,000 ha in the country. In recent years, it has gained recognition as a nutrient-rich superfood due to its high mineral content and medicinal value, leading to a steady rise in both domestic and international demand; at present, Makhana is exported from India to more than 20 countries, which has significantly strengthened its market price over the past 5 years. With the adoption of improved scientific cultivation practices, farmers can earn a net income ranging from ₹300,000 to ₹500,000 per hectare, which is approximately USD 3600 to USD 6000 per hectare (depending on exchange rate variations). The income can be further enhanced through processing and value-addition activities, such as grading, roasting, flavoring, and attractive packaging, which fetch higher prices in the market. Traditionally, a major share of the cost of cultivation is incurred in harvesting and postharvest operations (Fig. 3.18), as these processes have largely been manual and labor-intensive; however, the availability of improved machines and mechanized technologies for harvesting and processing is gradually reducing these costs, improving efficiency, and ultimately, maximizing the profitability of Makhana farming and its associated processing activities.

3.18 Makhana *cum* fish culture

In an integrated aquaculture system, Makhana cultivation is combined with fish farming to optimize the use of water resources and improve productivity (Verma et al., 2008; Raut et al., 2024). For this integrated aquaculture system, two models can be applied according to their suitability: *Vacant Space System* and the *Pen System*. These models are designed to provide optimal environments for both Makhana growth and fish rearing, balancing the needs of both crops while maintaining ecological sustainability.

3.18.1 Vacant space system

In this model, about 10%–20% of the total water spread area is kept free from Makhana plants to create a dedicated space for fish rearing (Fig. 3.19). This

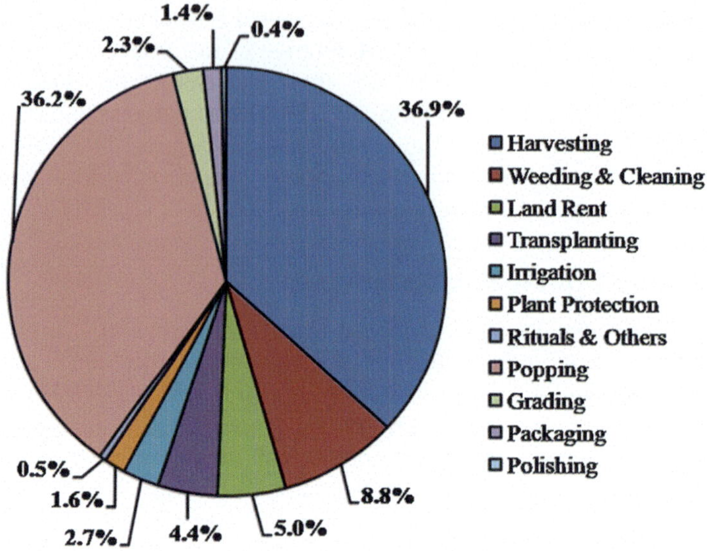

FIGURE 3.18 Activity-wise partitioning of the cost of cultivation and processing of makhana.

FIGURE 3.19 Vacant space system in *Chaur* area for fish rearing with makhana.

vacant space is crucial for several reasons. First, it allows better penetration of solar radiation into the water, which is essential for maintaining water quality and supporting the growth of phytoplankton, which serves as a natural food source for fish. The open space also helps to oxygenate the water, compensating for oxygen depletion that often occurs in areas covered by dense Makhana leaves, where light penetration is low and organic decomposition is high. The vacant space is enclosed using bamboo (*Bambusa* spp.) poles, nylon threads, or netting to prevent the fish from entering areas densely covered by Makhana or spreading Makhana leaves into the vacant space. When selecting the site for this vacant space, the bottom topography of the pond or *Chaur* is a key factor to consider. Areas where the water depth is more than 1 meter are ideal for applying this model, as deeper waters provide better living conditions for the fish, such as maintaining more stable water temperatures and oxygen levels. This model ensures that both fish and Makhana can coexist without competition for space and resources. It allows fish to grow in an oxygen-rich environment while Makhana thrives in the rest of the pond. Regular monitoring of the water quality and fish health is important in this system to maintain a balance between plant decomposition and oxygen supply.

3.18.2 Pen system

The pen system is another approach for integrating Makhana and fish culture. In this model, a portion of the pond of *Chaur* lands, preferably a marginal or peripheral site, is enclosed using bamboo poles or netting. The enclosed area, or "pen," is specifically reserved for fish rearing while Makhana is cultivated in the rest of the water body. The pen system offers several advantages. First, it creates a controlled environment for fish rearing, protecting them from predators while still allowing access to natural food sources, such as detritus, plankton, and small invertebrates. Second, it prevents fish from damaging Makhana plants, which is important since some fish species, particularly herbivorous types, may feed on aquatic vegetation. The pen also facilitates easier fish management, including stocking, feeding, and harvesting, since the enclosed area can be more easily accessed than the larger, open-water body. This model is particularly effective in *Chaur* lands that have irregular topography or where complete separation of fish and Makhana is not feasible. The pen system allows for better utilization of unused marginal sites. Both the vacant space system and pen system are effective models for integrating Makhana and fish farming in *Chaur* lands. The choice between them depends on the specific characteristics of the water body, such as depth, topography, and water quality. Both models, however, focus on optimizing the ecological balance between Makhana plants and fish, ensuring that solar radiation and oxygen levels are maintained for healthy fish growth while maximizing the economic returns from both crops. Proper implementation of these models requires careful planning, regular monitoring, and adaptive management practices to ensure the sustainability of the integrated aquaculture system.

3.19 Compatible fish species with makhana

In the aquaculture system of North Bihar, India, fish farming traditionally follows the polyculture method, which includes Indian Major Carps (IMC), such as Catla (*Labeo catla)*, rohu (*Labeo rohita*), and mrigal (*Cirrhinus mrigal*), along with exotic carps, such as silver carp (*Hypophthalmichthys molitrix*), grass carp (*Ctenopharyngodon idella*), and common carp (*Cyprinus *carpio*). However, in the Makhana *cum* fish culture system, the selection of fish species in this system should prioritize fast-growing fish to enhance productivity and reduce cultivation cycles. Using species with diverse feeding habits—such as surface, column, and bottom feeders—ensures efficient use of the water body's ecological niches. Avoiding macro-vegetation feeder species is crucial, as they may damage Makhana crops. In addition to traditional carp species, detritivores, and air-breathing fish, such as magur (*Clarias magur*), singhi (*Heteropneustes fossilis*), spiny eel (*Macrognathus* spp.), Climbing perch/Kawai (*Anabas testudineus*), and *Channa* species are particularly suitable for cultivation alongside Makhana. These species are highly resilient, capable of thriving in the muddy, anoxic conditions typical of Makhana ponds. Their high survival rates make them an excellent choice for integrated systems. Lastly, market demand should guide fish selection, ensuring the farming system's economic viability. This thoughtful approach maximizes production efficiency while maintaining sustainability in Makhana-fish-integrated farming (Raut et al., 2024).

3.20 Stocking of fish fingerling

Stocking of advance fish fry (size 20–30 mm, 2–4 g) in Makhana-fish integrated systems coincides with the Makhana plant's life cycle, specifically between its vegetative growth (sprawling leaves) and flowering periods (Fig. 3.20). This timing is crucial because the water body's conditions during this phase provide an ideal environment for the stocking of fish. The dense aquatic vegetation from Makhana offers shelter for fish, while the water's nutrient balance supports optimal growth.

For effective fish culture, species such as catla, rohu, mrigal, bata carp (*labeo bata*), climbing perch (*Anabas testudineus*), *channa* spp., magur (*Clarias magur*), and singhi (*Heteropneustes fossilis*) can be stocked. The stocking ratio should be balanced according to their ecological niches, such as catla, rohu, and mrigal, primarily feeding on plankton and column feeders, while magur and singhi are detritivores that use different food resources. A stocking density of one fingerling per 2 square meters is recommended. This density ensures enough space and resources for the fish to grow without overcrowding, which could lead to competition for food and oxygen. Proper stocking management enhances the productivity of both Makhana and fish, making the system sustainable and economically beneficial for farmers.

Feed management is critical in the integrated Makhana-fish farming system due to the lower primary productivity typically found in Makhana ecosystems. Since the aquatic environment under Makhana cultivation has less natural production of phytoplankton and other primary food sources, supplemental feeding becomes essential for ensuring the optimal growth of fish from the fingerling stage to marketable size. Locally available feed materials like rice or wheat bran mixed with mustard oil cake in a 2:1 ratio provide a cost-effective

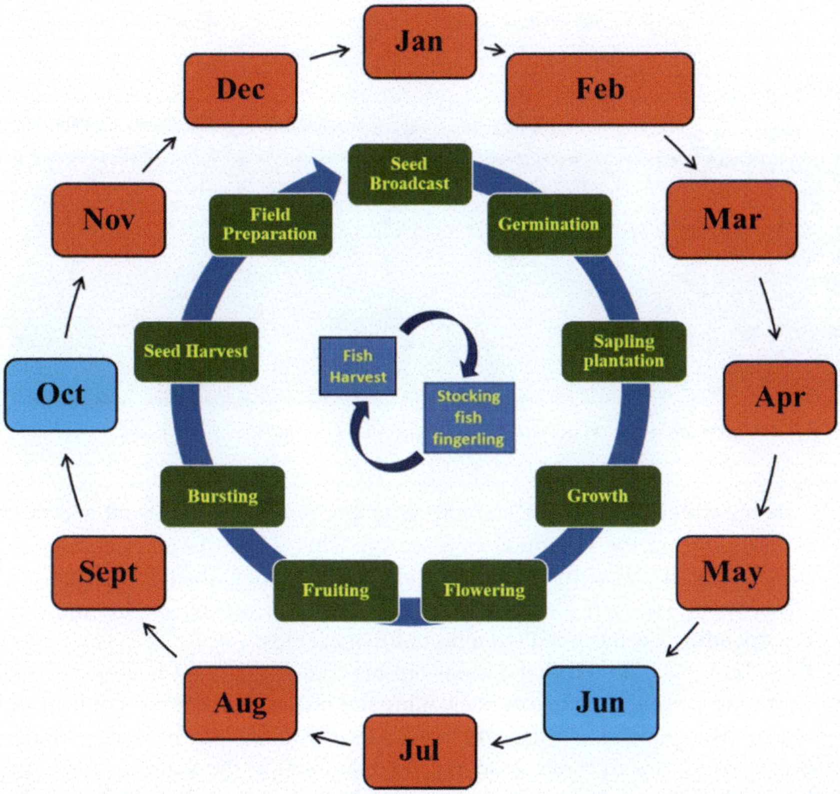

FIGURE 3.20 Life cycle of makhana crop and timing for the fish fingerling stocking and harvesting.

and nutritious feed base. Adding a small quantity of vitamin and mineral premixes further improves fish health and growth rates. Additionally, pelleted, sinking commercial feed available in the local market offers better results and is easy to use. The recommended feeding rate is 2%–5% of the fish's body mass, adjusted according to their growth rate. The daily ration should be divided into two equal portions and applied in the morning and evening.

3.21 Harvesting of fish and makhana seed

After completing the life cycle of Makhana (*E. ferox*) during the month of July–August, its fruit bursts open, scattering seeds at the bottom of the pond. The seeds are then collected manually using a locally crafted bamboo tool known as a *Ganja* (Fig. 3.21). This traditional harvesting method is labor-intensive but effective in gathering Makhana seeds from the pond bed. Simultaneously, fish harvesting is also conducted after the Makhana life cycle

FIGURE 3.21 Harvesting of makhana seed from the *Chaur* land area.

ends, as the water temperature starts to drop and fish growth rates slow down, particularly during the post-monsoon season. This timing is ideal because it allows the fish to grow to marketable size without negatively affecting the Makhana crop. The fish are harvested either through netting or draining the pond, depending on the pond's depth and fish species.

Once both fish and Makhana seeds are harvested, the pond is prepared for the next crop season. This involves cleaning the *chaur* area, drying, eradicating unwanted fish, aquatic weeds, and algae, fertilization, and manuring. Such premanagement practices are essential for maintaining the water quality and ecological balance necessary for successful Makhana and fish cultivation in the following season. This cycle of harvesting and pond management ensures the sustainability and economic viability of the integrated farming system Raut et al. (2024).

3.22 Economic analysis of the makhana *cum* fish culture

Integrating fish culture with Makhana cultivation in *Chaur* lands of North Bihar offers a significant opportunity to boost farmers' income by 50%–60%. In a study by Raut et al. (2024), the net income from Makhana cum fish culture was reported to be around $4500 per hectare as compared to $3000 per hectare from solo Makhana farming. This increase is primarily due to the synergetic relationship between Makhana and fish, which optimizes land and water use with minimal supportive management practices. Makhana, a highly nutritious aquatic plant, grows in shallow water, making it ideal for cultivation in *Chaur* land (i. e., low-lying depressions). Fish, on the other hand, thrive in the same aquatic environment, creating a mutually beneficial ecosystem.

The estimated production from 1–1.5 ha of *Chaur* land can yield 1.5 t of Makhana seeds and 1–1.5 t of fish per cultivation cycle. The presence of fish in

Makhana fields contributes to natural nutrient cycling, enhancing the overall water quality and providing essential nutrients that promote Makhana growth. Species such as catla, rohu, and other minor carp are well-suited for these aquatic systems. Moreover, species such as magur and singhi are well-suited for these aquatic systems, as they are air-breathing and thrive in low-oxygen environments. In fact, the production of more than 1.6 tones has been recorded in a previous study.

Minimal management practices, such as maintaining water levels and quality, fish stocking density and its ratio, Makhana variety, and farming practices, ensure the sustainability of this integrated system. The additional fish yield offers a steady income alongside the Makhana harvest, diversifying revenue streams and reducing economic risks for farmers. This ecofriendly approach enhances productivity, promotes sustainable aquaculture, and provides an economically viable solution for *Chaur* land utilization.

3.23 Comparison of makhana cultivation systems

Makhana (*E. ferox*) can be cultivated under three broad systems—traditional pond, chaur (seasonal floodplain wetland), and field-based systems—each with distinct ecological, agronomic, and economic characteristics. Table 3.3 presents a comparative overview of these systems in terms of water regime, soil type, management requirements, yield potential, and other relevant parameters.

TABLE 3.3 Comparative overview of makhana cultivation systems.

Parameter	Traditional pond system	Chaur (seasonal floodplain wetland) system	Field-based system
Water regime	Deep, perennial water (1–1.8 m)	Seasonal inundation (0.5–1.2 m), water recedes gradually	Controlled shallow water (30–60 cm), requires irrigation
Soil type	Clayey bottom of pond, rich in organic matter	Alluvial, fertile, nutrient-rich sediments	Clayey to clay loam, amended if required
Water management	Minimal control, depends on natural water levels	Moderate control, mainly manages recession timing	High control, continuous irrigation required
Sowing/ Planting	Seedlings transplanted in pond	Seedlings planted after partial recession	Seedlings planted like wetland rice with preprepared beds

Continued

TABLE 3.3 Comparative overview of makhana cultivation systems.—cont'd

Parameter	Traditional pond system	Chaur (seasonal floodplain wetland) system	Field-based system
Harvesting efficiency	Labor-intensive, traditional methods	Moderate, easier than deep pond due to shallower water	High, mechanization possible
Yield potential	Moderate (1–1.5 t/ha)	Moderate to high (1.8–2.5 t/ha)	High (2.5–3.5 t/ha)
Input requirement	Low	Moderate (water and nutrient management during recession)	High (irrigation, fertilization, and mechanization)
Risk factors	Flood, drought, water quality	Flooding, drought if water recedes early	Water scarcity, high input costs, and labor dependency
Economic potential	Moderate	Moderate to high	High
Sustainability	High, ecofriendly	Moderate, better than field	Moderate, requires more resources
Area expansion feasibility	Limited due to declining ponds	High, underutilized low-lying lands	Moderate, depends on irrigation infrastructure

From Table 3.3, it is evident that each system has its unique advantages and limitations. The traditional pond system is culturally established, ecofriendly, and requires minimal management, but it is limited in area due to the gradual decline of natural ponds and comparatively lower yield potential. The *chaur* system, using seasonal floodplain wetlands, strikes a balance between natural resource efficiency and productivity. Its moderate water depth and fertile alluvial soils allow relatively easier crop management and harvesting compared with deep pond systems, while still supporting moderate-to-high yields. This makes chaur-based cultivation a highly promising option for expanding Makhana production in regions, such as North Bihar and beyond. Finally, the field-based system offers the highest yield and economic return, largely because of controlled water management and mechanization, but it is resource-intensive and requires careful planning for irrigation, nutrient management, and labor. Thus, integrating traditional pond cultivation with the expansion of chaur and field-based systems can optimize Makhana production, improve farmers' income, and help meet the growing domestic and global demand, all while maintaining ecological sustainability.

References

Alloway, B.J., 2004. Zinc in Soils and Crop Nutrition. International Zinc Association Publications, Brussels, pp. 1–116.

Broadley, M.R., White, P.J., Hammond, J.P., Zelko, I., Lux, A., 2007. Zinc in plants. New Phytol. 173, 677–702.

Cakmak, I., 2013. Magnesium in crop production, food quality and human health. Plant Soil 368, 1–4.

Das, S., Der, P., Raychaudhary, U., Maulike, N., Das, D.K., 2006. The effect of *Euryale ferox* Salisb. (makhana), a herb of aquatic origin on myocardial ischemic reperfusion injury. Mol. Cell. Biochem. 289 (12), 55–63.

Dutta, R.N., Jha, U.N., Jha, S.N., 1986. Relationship of biomass yield of Makhana (E. *ferox*) with soil properties and water quality. Plant Soil 95 (3), 345–350.

Gerendas, J., Fuhrs, H., 2015. The significance of magnesium for crop quality. Plant Soil 368, 101–128.

Guo, W., Nazim, H., Liang, Z., Yang, D., 2016. Magnesium deficiency in plants: an urgent problem. Crop J. 4, 83–91.

Hafeez, B., Khanif, Y.M., Saleem, M., 2013. Role of zinc in plant nutrition-A review. Am. J. Exp. Agric. 3 (2), 374–391.

Havlin, J.L., Beaton, J.D., Tisdale, S.L., Nelson, W.L. 2015. Soil Fertility and Fertilizers: An Introduction to Nutrient Management. Upper Saddle River, New Jersey.USA.

Jana, B.R., 2017. Impact of water depth on growth of gorgon nut and associated weed under wetland ecosystem of North Bihar province in India. J. Crop Weed 13 (1), 77–82.

Jha, V., Kargupta, A.N., Dutta, R.N., Jha, U.N., Mishra, R.K., Saraswati, K.C., 1991. Utilization and conservation of *Euryale ferox* salisbury in Mithila (North Bihar), India. Aquat. Bot. 39 (3–4), 295–314.

Jha, V., Kargupta, A.N., Prasad, K., 2009. Makhana (*Euryale ferox*): a review of its nutritional composition and therapeutic properties. Indian J. Trad. Knowled. 8 (3), 395–400.

Jha, V., Shalini, R., Kumari, A., Jha, P., Sah, N.K., 2018. Aquacultural, nutritional and therapeutic biology of delicious seeds of *Euryale ferox* Salisb.: a mini review. Curr. Pharm. Biotechnol. 19, 545–555.

Jha, V., Verma, A.B., Jha, P., Jha, M., Kumar, R., 2014. Wetlands in north Bihar provide a basis to its sustainable development. J. Aqua. Biol. Fisher. (2), 843–851.

Joseph, A., Ramesh, G., 2023. Nutrient analysis, phytochemical and antioxidant activity of a food product formulated with fox nuts (*Euryale ferox*). Asian J. Biol. Life Sci. 12 (2), 279–285.

Kumar, L., Gupta, V.K., Jha, B.K., Singh, I.S., Bhatt, B.P., Singh, A.K., 2011. Status of Makhana (*Euryale ferox* Salisb.) Cultivation in India. Technical Bulletin No. R-32/PAT-21. ICAR Research Complex for Eastern Region, Patna.

Kumar, L., Singh, A.K., Bhatt, B.P., 2016a. Nutritional status of recently developed makhana (gorgon nut) variety - Swarna Vaidehi. J. AgriSearch 3 (4), 199–205.

Kumar, M., 2020. Makhana triumphs over flood: a success story from Darbhanga, Bihar. Biotica Res. Today 2 (11), 1113–1116.

Kumar, M., 2023. Effect of secondary and micronutrients on yield and quality of fox nut (*Euryale ferox* Salisb.) in subtropical climate. J. Plant Nutr. 46 (2), 212–218. https://doi.org/10.1080/01904167.2022.2067051.

Kumar, M., Jha, A.K., Hazarika, S., Verma, B.C., Choudhury, B.U., Ramesh, T., Moirangthem, P., Kumar, R., Brajendra, Rajkhowa, D.J., Kumar, A., Devi, M.H., 2016b. Micronutrients (B, Zn, Mo) for improving crop production on acidic soils of northeast India. Natl. Acad. Sci. Lett. 39 (2), 85–89.

Kumar, M., Padala, V.K., Ramya, N., Chakraborty, A., Dey, J.K., 2021b. Plant health issues in Fox Nut/Makhana (*Euryale ferox*): an agronomic perspective. J. Plant Health Iss. 2 (1), 30–35.

Kumar, M., Raut, S.M., Bhatt, B.P., Kumar, L., 2020a. Scientific cultivation of Makhana for improving farmers' livelihood in eastern India. Biotica Res. Today 2 (7), 670–672.

Kumar, M., Shekhar, D., Kumari, A., 2021a. Reducing water requirement for Makhana farming: a case report. Biotica Res. Today 3 (6), 505–507.

Kumar, M., Shekhar, D., Kumari, A., Prasad, R.P., Bhatt, B.P., 2020b. Makhana farming for maximizing farm income: a success story. Biotica Res. Today 2 (9), 962–965.

Kumar, M., Tarate, S.B., Rout, R.K., Padala, V.K., Raut, S.M., 2025. Water crisis in Bihar: causes, impacts and the road to recovery. Biotica Res. Today 7 (9), 258–260.

Kumari, A., Singh, I.S., Kumar, L., Kumar, A., Kumar, R., Gupta, V.K., 2014. Morphological characteristics of Makhana germplasm of Manipur under Darbhanga conditions. J. AgriSearch 1 (3), 157–160.

Lukowiak, R., Grzebisz, W., Barlog, P., 2016. Magnesium management in the soil-crop system – a crop rotation approach. Plant Soil Environ. 62 (9), 395–401.

Marschner, H., 1995. Mineral Nutrition of Higher Plants, second ed. Academic Press, London.

Mishra, R. K., Jha, V., Dehadrai, P.V. 2003. Makhana. ICAR, Krishi Anusandhan Bhavan, Pusa, New Delhi.

Raut, S. M., Jana, B. R., Kumar, M., Singh, I. S., Sarma, K., Padala, V. K., Ahirwal, S. K., Tarate, S. B., Rout, R. K., & Kumar, A. 2024. Integration of Makhana with fish culture in chaur land for improving farmers' livelihood. Technical Bulletin No. 01. ICAR–National Research Centre for Makhana, Darbhanga, Bihar, India. pp. 27.

Sharma, Y., Kumari, M., 2024. A study on groundwater irrigation expansion in Bihar, India. Asian J. Agricult. Extens. Econ. Sociol. 42 (5), 53–60.

Singh, I.S., Kumar, M., Raut, S.M., Thakur, A.K., Singh, S.P., 2020. Integrated nutrient management package for field cultivation of Makhana in North Bihar. J. AgriSearch 7 (3), 138–141.

Srivastava, S., Mishra, H.N., Jha, A., 2017. Nutritional and therapeutic potential of fox nut (*Euryale ferox* Salisb.). J. Food Sci. Technol. 54 (6), 1521–1531.

Szulc, W., Rutkowska, B., 2013. Diagnostics of boron deficiency for plants in reference to boron concentration in the soil solution. Plant Soil Environ. 59 (8), 372–377.

Tarate, S.B., Singh, I.S., Raut, S.M., Madane, D.A., 2025. Waterbody decline and warming trends in aquatic ecosystems of selected districts of North Bihar in India (1995–2024): a remote sensing perspective. Theor. Appl. Climatol. https://doi.org/10.1007/s00704-025-05820-5.

Verma, A.M., Jha, V., Ahmad, S.H., 2008. Fish–Makhana (*Euryale ferox* Salisb.) integration: a case study of sustainable aquafarming system in North Bihar. J. Indian Fish. Assoc. 35, 87–96.

Wang, Y., Wei, H., Wang, S., Yu, F., Zhai, L., Yang, J., Wei, Z., 2025. γ-Aminobutyric acid treatment enhances quality and improves antioxidant activities of *Euryale ferox* during post-harvest storage. Front. Nutr. 12, 1527555.

Yruela, I., 2009. Copper in plants: acquisition, transport and interactions. Funct. Plant Biol. 36 (5), 409–430.

Chapter 4

Genetic diversity, breeding, and crop improvement strategies in makhana (*Euryale ferox* Salisb.)

Lokendra Kumar[1], Charu Lata[1], Naresh Kumar[2], Vijeta Sagwal[1], Priti Duhan[3], Arpita Sharma[1], Tushadri Singh[1] and Vikas Joon[4]
[1]*ICAR-Indian Institute of Wheat and Barley Research, Karnal, Haryana, India;* [2]*Eternal University, Baru Sahib, Himachal Pradesh, India;* [3]*K.L. Mehta Dayanand College for Women, Faridabad, Haryana, India;* [4]*ICAR-Central Plantation Crops Research Institute Kasaragod, Kasaragod, Kerala, India*

4.1 Introduction

Makhana (*Euryale ferox* Salisb.), also known as fox nut, Prickly water lily, semen euryales, Qian shiand gorgon nut, is an ancient aquatic plant of great nutritional and cultural significance. It has different names in various Indian languages, for example, Padambijabh-Sanskrit, Makhana Gujarati, Mallunipdananu-Telgu, Makane-Marathi, Jaivar-Punjabi, and Tagarai-Tamil (Sidh and Sharma, 2019) and is relished as a dry fruit all over the country. Traditionally, makhana is the delicacy of Mithila, Darbhanga, and Madhubani regions of Bihar state. Euryale ferox seeds have been reported earlier in their uses in Ayurveda and Chinese preparations for the treatment of renal disorders, chronic diarrhea, excessive leucorrhea, and hepatic dysfunction. It belongs to the family Nymphaeaceae, which also includes the well-known water lilies, and is the only extant species of the genus *Euryale*. This is the only plant species that is native to Eastern and Southern Asia, covering parts of India, Bangladesh, China, Korea, and Japan (He et al., 2019; Yamada et al., 2001). India is the world's leading producer of makhana, accounting for more than 85% of global output, with the crop predominantly cultivated in the north-eastern plains, particularly in Bihar, eastern Uttar Pradesh, Manipur, West Bengal, and parts of Madhya Pradesh. Within India, the Mithilanchal region of Bihar—covering Darbhanga, Madhubani, and Katihar districts—has emerged

Makhana (Euryale ferox) as a Functional Food. https://doi.org/10.1016/B978-0-443-44421-0.00005-6

as the primary center of makhana cultivation and trade, often referred to as the "*Makhana* Bowl of India" (Jha et al., 1991).

Makhana thrives in stagnant water bodies such as ponds, oxbow lakes, and wetlands in tropical monsoon conditions. Most favorable atmospheric conditions to grow are tropical and subtropical climates, with temperatures ranging from 20–35°C, relative humidity between 50% and 90%, and rainfall between 100 and 250°cm. The plant exhibits a distinctive morphology characterized by large floating circular leaves with spiny undersurfaces, robust rhizomatous stems, and solitary bisexual purple flowers. This is a perineal plant, rooted in the soil with leaves that are round, broad, and blade-shaped, thus remaining afloat on the surface of a water body. Rhizomes lie deep in the bottom of the pond with the help of dense fleshy roots, and the leaf lamina and vascular bundles have many sharp prickles. Its unique floral mechanism involves heterostyly, which promotes cross-pollination and maintains genetic variability within populations (Mandal et al., 2010). The fruits are large and spiny, containing numerous hard-coated black seeds. After harvesting and processing, the seeds yield the edible popped kernels known as "makhana," which form an important component of traditional diets and local economies.

The whole world recognizes the exceptional nutritional composition and therapeutic properties of the edible seeds of *Euryale ferox* (Fig. 4.1). Nutritional composition of makhana seeds makes them an ideal low-calorie food due to low fat level, along with being rich in carbohydrates and proteins (10%–12%). They are also a source of essential amino acids, contributing

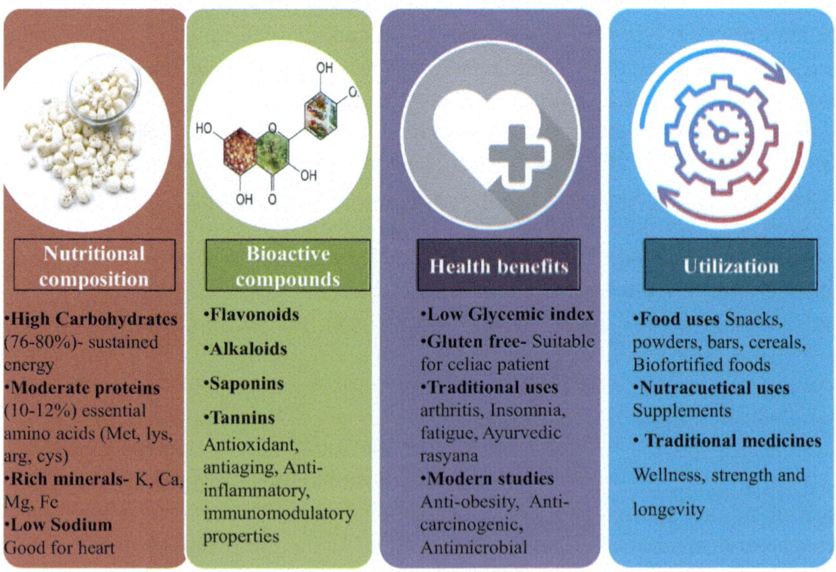

FIGURE 4.1 Nutritional composition and therapeutic properties of *Euryale ferox*.

to their high-quality protein profile. Makhana contains significant quantities of minerals such as potassium, phosphorus, magnesium, calcium, and iron, while being low in sodium. The high potassium-to-sodium ratio supports cardiovascular health and aids in blood pressure regulation. In addition, makhana contains abundant phytochemicals, including flavonoids, alkaloids, saponins, and phenolic compounds, which contribute to its antioxidant, antiinflammatory, and antiaging properties (Dutta et al., 1986).

Makhana seeds are highly suitable for diabetic and celiac patients due to their low glycemic index (GI $\approx$ 55) and gluten-free nature. It has long been utilized in Ayurvedic, Unani, and Chinese traditional medicine systems, where it is regarded as a *rasayana* (rejuvenating) food that enhances strength and longevity. Ayurvedic texts mention its efficacy in managing conditions such as arthritis, insomnia, nephritis, and infertility. Modern pharmacological studies have corroborated these traditional claims, demonstrating the nephroprotective, hepatoprotective, antiobesity, neuroprotective, and anticarcinogenic potential of seed and leaf extracts. The antioxidant potential of makhana has been linked to its ability to scavenge reactive oxygen species (ROS) and prevent oxidative stress-induced cellular damage. Consequently, makhana has gained global attention as a functional food and nutraceutical crop, offering scope for development of health-promoting products and value-added derivatives such as protein powders, snack formulations, and fortified supplements (Mittal et al., 2020).

Makhana holds immense socio-economic importance in rural and wetland-based livelihoods, beyond its nutritional and medicinal attributes. It is a traditional cash crop of the Mithila region, sustaining thousands of farming and fishing families. Its cultivation often occurs in integrated fish–makhana systems, which optimize resource use by combining aquaculture with plant production. The crop also plays an ecological role in wetland conservation, providing habitat for aquatic fauna and maintaining water quality through nutrient absorption. This integrated system contributes significantly to income diversification, employment generation, and food security in economically marginalized regions (Kumar et al., 2016a). The production and value chain of makhana remains basically unorganized and traditional, despite its nutritional importance. Cultivation processes are mostly manual and confined to small ponds or natural wetlands. The labor-intensive nature of pond preparation, seed sowing, and manual harvesting makes cultivation physically demanding and economically risky. Postharvest processing (drying, roasting, and popping) requires specialized skills and involves substantial hard work. Traditional inefficient processing methods are resulting in variable quality and low recovery rates. The average productivity of makhana in India is only 1.2–1.5 t per hectare, far below its potential yield of 2.5–3.0 t achievable under optimized management methods (Raut et al., 2020). Furthermore, postharvest losses, lack of mechanization, limited access to improved seeds, and absence of organized marketing infrastructure continue to hinder the economic potential of this high-value crop.

Scientific research on *Euryale ferox* has historically been limited compared with other aquatic crops such as lotus (*Nelumbo nucifera*) or water chestnut (*Trapa natans*). The early research in India focused mainly on its ecology, agronomy, and pond management, with limited attention to genetics or crop improvement. Consequently, systematic breeding programs are virtually absent, and no standardized high-yielding varieties have been released so far. The available cultivars are essentially landraces maintained by farmers through mass selection and traditional propagation. There is a need to prioritize this crop on a scientific and commercial level, as it has immense nutritional and economic value. In this chapter, the authors focused on different aspects of the makhana crop, such as agronomic and economic importance, genetic diversity, conventional breeding approaches, molecular breeding and omics interventions, crop improvement strategies, and future directions.

4.2 Botanical and agronomic overview of the makhana crop

4.2.1 Taxonomy and classification kingdom: Plantae

Family: Nymphaeaceae

Genus: *Euryale* →

Species: *E. ferox Salisb.*

Makhana (*Euryale ferox* Salisb.) is an economically and nutritionally important aquatic crop belonging to the family **Nymphaeaceae.** Taxonomically, it is classified under the Kingdom *Plantae*, Family Nymphaeaceae, Genus *Euryale*, and Species *E. ferox* Salisb.

4.2.2 Morphological features

This is one of the few domesticated aquatic crops grown for their edible seeds. *Euryale ferox* is a prickly, floating-leaved aquatic plant with huge, round leaves that float on the water's surface and have spiky undersides. The plant produces **solitary, bisexual, purple flowers,** which are typically borne on long peduncles that rise above the water level. The plant develops **round, black seeds** after pollination, enclosed in a spongy fruit; the seeds are covered by a **mucilaginous aril,** which protects the developing embryo and aids in buoyancy. The seeds are an economic part of the plant and, once processed, form the puffed edible kernels commonly known as makhana or fox nuts.

4.2.3 Reproductive biology

E. ferox is **predominantly self-pollinated,** ensuring a degree of genetic uniformity in cultivated populations. However, **occasional cross-pollination** may occur, mediated by insects or through water movement, contributing a little bit of genetic diversity. The flowers open during early morning hours, and fertilization generally takes place within the same flower. Because limited natural outcrossing limits variety within cultivated germplasm, the crop's unusual reproductive strategy—which combines self-pollination and vegetative propagation through seeds—has consequences for its genetic development (Saxena et al., 2025).

4.2.4 Ecological adaptation and production status

Ecologically, *E. ferox* demonstrates significant adaptation to aquatic and semiaquatic environments, thriving in stagnant or slow-moving freshwater bodies such as ponds, lakes, and wetlands. It prefers a tropical to subtropical climate, with optimal growth observed in regions having high humidity and temperatures ranging from 20°C to 35°C. The plant needs clayey, shallow, rich soil that can hold onto water for the duration of its growth. Because the crop is often grown in premonsoon or early monsoon months and harvested in late postmonsoon times when the water level recedes, its life cycle closely corresponds with monsoonal trends. The plant's thorny leaves and stems offer a natural defense against aquatic animals, and the seeds sprout underwater. Because of its plasticity, E. ferox is an ecologically robust species that can maintain productivity in the face of changing hydrological circumstances. More than 90% of the global production of makhana is produced in India and holds substantial economic importance. Within India, the state of Bihar is recognized as the major production hub, accounting for nearly 80% of total output, followed by parts of eastern Uttar Pradesh, Madhya Pradesh, and Manipur (Singh et al., 2023). The crop forms a vital component of the rural economy in these regions, supporting thousands of farming families involved in pond-based cultivation, seed collection, and processing (Raut et al., 2025). Despite its popularity, makhana cultivation is still mostly unstructured and traditional, depending on antiquated methods and missing contemporary agronomic treatments. However, the necessity of systematic improvement initiatives, such as improved pond management, seed selection, and post-harvest technology, has started to be highlighted by current research and policy attention.

4.3 Importance and utilization

The edible seeds of *Euryale ferox* Salisb. are highly valued for their exceptional nutritional profile and diverse therapeutic attributes. These seeds represent a unique aquatic crop characterized by a rich array of bioactive compounds and a balanced composition of macronutrients.

4.3.1 Nutritional and medicinal properties

Nutritionally, makhana is dominated by carbohydrates (76%–80%), which provide sustained energy release, and contains moderate amounts of proteins

(10%–12%) along with very low-fat levels (0.5%–2%). This combination makes it an ideal low-calorie and health-promoting food suitable for people of all ages. The protein fraction is particularly noteworthy due to its enrichment in essential amino acids such as methionine, cysteine, lysine, and arginine, which contribute to enhancing metabolic functioning, immune support, and tissue repair (Jana et al., 2025). Makhana seeds are also notable by its dense mineral composition. It contains substantial quantities of potassium, phosphorus, magnesium, calcium, and iron, while it is naturally low in sodium. This makes a high potassium-to-sodium ratio, which plays a critical role in maintaining cardiovascular health and regulating blood pressure. Furthermore, makhana is a rich source of phytochemicals such as flavonoids, alkaloids, saponins, tannins, and phenolic compounds. These bioactive constituents have been coupled with strong antiaging, antioxidant, antiinflammatory, and immune-modulatory properties, contributing significantly to their therapeutic value (Kapoor et al., 2022).

4.3.2 Industrial and socio-economic importance

The seeds also have a low glycemic index (GI $\approx$ 55), allowing for gradual glucose release and making them suitable for diabetic patients and those suffering from metabolic disorders. Their gluten free nature further makes them appropriate for patients with celiac disease or gluten sensitivity. Owing to these nutritional and medicinal properties, makhana has been deeply embedded in classical traditional medicine systems such as Ayurveda, Unani, and Traditional Chinese Medicine. In Ayurveda, it is classified as *rasayana* (rejuvenating food), believed to enhance stamina, strength, and longevity. Classical texts describe makhana's efficacy in alleviating conditions such as arthritis, insomnia, nephritis, chronic fatigue, and infertility (Kumari et al., 2019). Modern experimental and pharmacological studies have scientifically analyzed and validated several traditional claims. Extracts of *Euryale ferox* seeds and leaves have demonstrated hepatoprotective, nephroprotective, anti-obesity, neuroprotective, anticarcinogenic, and antimicrobial activities in various in vivo and in vitro models. The strong antioxidant capacity of makhana is primarily attributed to its phenolic and flavonoid content, which aids in scavenging reactive oxygen species (ROS) and reducing oxidative stress–induced cellular damage, thereby contributing to disease prevention and healthy aging (Jiang et al., 2023).

With rising global demand for nutrient-rich, plant-based foods, makhana has gained prominence as a promising functional and nutraceutical crop. Its mild flavor, effortless processing, and high nutritional density have facilitated the production of multiple value-added products, for example, energy bars, roasted snacks, protein powders, fortified supplements, and functional beverages. Emerging research and market interest continue to highlight makhana's potential as an internationally recognized superfood, offering opportunities for economic growth, value addition, and innovation in the health-food sector.

4.4 Genetic diversity in makhana

The genetic diversity of *Euryale ferox* (makhana) serves as the foundation of crop improvement, enhancement, and long-term conservation, especially as the plant becomes more popular worldwide as a useful aquatic crop.

4.4.1 Germplasm resources

The majority of germplasm resources are preserved in India, where the ICAR–Research Center for Makhana (ICAR-RCER), Darbhanga, and Bihar Agricultural University (BAU), Sabour, encompass limited diversity representing local ecotypes and have the largest collections, which include farmer-developed selections, elite lines, and traditional landraces. A few regional facilities in Eastern and North-Eastern India, as well as the National Bureau of Plant Genetic Resources (NBPGR), New Delhi, house additional accessions (Kumari et al., 2014). Nevertheless, makhana remains underrepresented in global gene banks due to its limited natural distribution, inadequate international research attention, and the largely unexplored diversity outside India and parts of China. This highlights the vital need for systematic certification and global exchange of germplasm to broaden the diversity available for research and breeding programs.

4.4.2 Morphological and phenotypic diversity

Significant morphological and phenotypic variation exists within makhana populations. Studies have documented wide variability in significant agronomic traits in makhana, for example, kernel recovery, seed size, leaf morphology, yield potential, plant vigor, maturity duration, and the extent of leaf prickliness (Kumar et al., 2016a). Swarna Vaidehi variety of makhana was developed by pureline selection method and showed broad seeds and distinctive characters shown below in Fig. 4.2.

Farmers in Bihar, Manipur, Assam, and certain parts of central China cultivate landraces that are remarkably adaptive to local aquatic and agro-climatic conditions, including water depth, temperature regimes, and nutrient variations. The crop's resilience and production in various ecologies are increased by these adaptations, which provide unique phenotypic expressions. Molecular characterization research efforts have further discovered substantial genetic diversity among makhana ecotypes. Marker-based studies using RAPD, ISSR, and SSR markers have been instrumental in analyzing genetic variability and population structure across accessions from Bihar, Manipur,

FIGURE 4.2 *Swarna Vaidehi*, a makhana variety developed through pure line selection, exhibiting broad seeds and distinctive morphological characteristics. (A) Flower of *Swarna Vaidehi*; (B) crop under field conditions; (C) mature seeds and popped seeds demonstrating seed size and popping quality.

and China (Annual Report 2023–24, ICAR-Regional Center for Eastern Region, Patna (Bihar), 2023).

4.4.3 Need for systematic evaluation

Geographic separation and localized domestication are reflected in the moderate to high genetic diversity that these studies consistently describe. In contrast to other aquatic crops, molecular research on makhana is still limited, despite these discoveries. SNP-based genotyping, whole-genome resequencing, and transcriptome profiling are examples of advanced genomic technologies that are only getting started in this crop. Significant progress has been made in comprehending the complex genetic architecture of the species and identifying important genes linked to yield, stress tolerance, and nutritional quality with the start of preliminary draft genome sequencing projects in China and India (Qian et al., 2022). In light of these gaps, a systemic and comprehensive assessment of makhana germplasm is crucial. Combining morphological, agronomic, and molecular information will allow more accurate characterization, reduce redundancy, and help identify genetically unique and elite lines for breeding. The use of germplasm in advanced genetic research, such as association mapping, marker-assisted selection, and trait discovery, will be improved by the creation of core and mini-core collections that effectively reflect the greatest variety with the least amount of duplication. Strengthening genetic diversity research will ultimately support breeding programs target to develop high-yielding, nutritionally superior, and climate-resilient makhana varieties tailored to evolving consumer and environmental needs.

4.5 Reproductive biology and breeding constraints

Makhana has a distinct reproductive system with heterostyly and protogyny, which ensures cross-pollination but also complicates controlled hybridization. Artificial pollination is technically difficult because flowers are solitary and ephemeral, opening in the morning and closing by evening, making artificial pollination technically challenging.

4.5.1 Reproductive system

Flowers of *E. ferox* are mostly cleistogamous, meaning self-pollination occurs while the flower remains closed and often submerged in water. This behavior assures high rates of self-pollination but leads to a very narrow genetic base, which is a major constraint for traditional breeding programs aimed at improving traits. Chasmogamous (open) flowers do occasionally occur, especially later in the season (around July–September), offering a rare opportunity for potential cross-pollination (Jana, 2018). Seed development occurs underwater within spiny fruits, and the hard seed coat leads to prolonged dormancy.

4.5.2 Constraints in breeding

The length of the reproductive cycle is nearly 6 months from planting to seed maturity, which further limits the number of breeding generations per year. Conventional breeding is difficult and time-consuming due to these biological factors, coupled with the aquatic habitat. The majority of improvement initiatives have focused on clonal propagation and mass selection, which has led to slow genetic gain. The identification of elite lines with higher popping quality, thin shell thickness, and early maturity represents a realistic short-term approach; nevertheless, systematic hybridization programs and the use of molecular tools to track inheritance patterns are necessary for long-term progress. Furthermore, there is still little research on the genetic regulation of seed popping characteristics, which are crucial to the customer perspective. Targeted improvement by biochemical or molecular breeding may be made easier with an understanding of the biochemical and biophysical principles governing popping expansion (Kumari et al., 2014; Lokendra Kumar et al., 2015). The narrow genetic base of cultivated makhana, along with the absence of comprehensive germplasm characterization, has severely restricted genetic improvement efforts. Recent studies employing molecular markers such as RAPD, ISSR, and SSR have revealed moderate polymorphism among accessions, indicating potential for genetic improvement. However, the lack of species-specific markers and molecular maps continues to constrain the identification of quantitative trait loci (QTLs) associated with yield and quality traits. However, researchers are utilizing modern molecular technologies to expedite the breeding methodologies for crop improvement of makhana (Kumar et al., 2014). The application of modern omics technologies offers new opportunities to accelerate makhana research. Transcriptomic and metabolomic analyses have provided insights into genes involved in starch metabolism, antioxidant biosynthesis, and stress response (Liu et al., 2018). Comparative genomic studies with *Nymphaea* and *Nelumbo* species have shed light on the evolutionary divergence of *Euryale ferox* and its adaptive features for aquatic environments. Nevertheless, the absence of a reference genome and limited functional genomics studies remain critical bottlenecks. Establishing a draft genome sequence, developing high-throughput markers, and applying marker-assisted selection (MAS) or genomic selection (GS) could substantially enhance the efficiency of breeding programs (Wu et al., 2021).

4.6 Conventional breeding approaches

Despite the nascent stage of formal breeding programs for Makhana (*Euryale ferox* Salisb.), conventional breeding methods form the cornerstone of its genetic improvement. These approaches, primarily reliant on phenotypic selection and farmer participation, have been instrumental in the initial domestication and continued cultivation of this aquatic crop. The challenges posed by its aquatic habitat and self-pollinating nature make these time-tested methods particularly relevant.

4.6.1 Selection and domestication

The cultivation of Makhana is a classic example of informal, farmer-led domestication. Over centuries, local communities, particularly in the wetlands of Bihar and Manipur in India, have selectively propagated plants from wild populations based on desirable agronomic and quality traits (Kumar et al., 2021). This long-term, unconscious selection pressure has led to the development of distinct landraces adapted to specific local agro-climatic conditions. The primary traits targeted in this process include high popping percentage, seed size and uniformity, yield potential, and reduced prickliness. This farmer-driven domestication has created a valuable, albeit poorly characterized, reservoir of genetic variation that serves as the foundation for all formal breeding efforts (Gaurav et al., 2022).

4.6.2 Mass selection

Mass selection is one of the simplest and most widely practiced conventional methods in Makhana improvement. It involves the identification and bulk harvesting of seeds from superior phenotypes within a heterogeneous population of local landraces. Researchers from institutions like the ICAR-Research Complex for Eastern Region, Darbhanga, have undertaken systematic surveys of farmer fields and natural populations to identify such elite genotypes (Kumar et al., 2020). These selected plants are evaluated for traits such as early maturity, high seed number per pod, and disease tolerance. The seeds from these selected plants are then composited to form an improved population for the next cycle of cultivation and selection. While effective for simply inherited traits, mass selection's efficiency is limited for complex traits with low heritability due to the inability to control pollination fully (Khadatkar et al., 2020).

4.6.3 Participatory varietal selection (PVS)

Recognizing the deep-rooted traditional knowledge of Makhana growers, Participatory Varietal Selection (PVS) has emerged as a highly effective approach. In PVS, farmers are actively involved in the evaluation and selection of promising lines or landraces in their own fields. This ensures that the selected varieties not only have high yield potential but also align with farmers' preferences for traits such as taste, popping ability, and adaptation to local management practices (Lebot and Viquez-Zamora, 2025; Raut et al., 2025). This collaborative model between research institutions, NGOs, and farmers accelerates the adoption of new varieties and enhances their relevance. For instance, on-farm trials of early-maturing lines in North Bihar have shown higher adoption rates when local farmers were involved in the selection process from the beginning (Singh et al., 2014).

4.6.4 Variety development efforts

Formal variety development in Makhana is still in its infancy, but it has gained momentum in recent years. The primary focus has been on the evaluation and purification of landraces, as the development of synthetic varieties or hybrids is hampered by the lack of standardized hybridization techniques. A significant breeding objective has been to develop early-maturing varieties to reduce the crop duration, which typically extends from June to December. This allows for better water management and can enable crop rotation. ICAR-RCER, Darbhanga, has identified and propagated early-maturing genotypes that flower and set seed approximately 2–3 weeks earlier than traditional landraces (Sattar and Khan, 2016; ICAR- Research Complex for Eastern Region, Patna, Bihar, Annual Report, 2022). Preliminary varietal trials are ongoing at various regional agricultural stations across Bihar and Manipur. These trials evaluate advanced breeding lines for consistent yield, high popping recovery, and nutritional quality. Makhana cultivation is vulnerable to fluctuations in water depth and temperature; selection for resilience to such abiotic stresses is a key priority. Selections are being made from populations that show tolerance to transient drought conditions during the seedling stage and adaptability to varying water quality (Jana et al., 2018).

4.7 Molecular breeding and omics interventions

The limitations of conventional breeding in Makhana, driven by its long lifecycle, aquatic nature, and predominantly self-pollinating reproductive system, have intensified the need for advanced molecular approaches. The integration of omics technologies—genomics, transcriptomics, proteomics, and metabolomics—offers unprecedented opportunities to dissect the genetic architecture of complex traits and accelerate the breeding cycle. While still in its early stages, molecular breeding in Makhana is poised for a significant leap forward with the ongoing development of genomic resources.

4.7.1 Marker-assisted selection (MAS)

The foundation of molecular breeding is the establishment of a robust marker-trait association. Initial studies in Makhana have focused on utilizing molecular markers to assess genetic diversity, but their application in Marker-

Assisted Selection (MAS) is now emerging. Early genetic diversity studies relied on Random Amplified Polymorphic DNA (RAPD) and Inter-Simple Sequence Repeat (ISSR) markers due to their low cost and no requirement for prior sequence information (Jana et al., 2025a,b). However, the recent development of Simple Sequence Repeat (SSR) or microsatellite markers from transcriptome sequences has provided more reproducible, co-dominant, and genome-specific markers (Chapman, 2019). Preliminary studies have begun to link these SSR markers with key agronomic traits. For instance, association mapping in a diverse panel of Makhana accessions has identified potential SSR markers correlated with seed size and popping expansion volume (Kumar et al., 2018). Validating these markers in biparental populations is the next critical step toward implementing MAS for these quality traits, which are difficult and time-consuming to phenotype accurately.

4.7.2 Transcriptomics and genomics

Genomic resources are the cornerstone of modern plant breeding, and significant efforts are now underway to bridge this gap in Makhana. RNA-sequencing (RNA-seq) has been a powerful tool to glimpse into the functional genome of *E. ferox* without a reference genome. Several studies have deployed RNA-seq on developing seeds and stress-treated plants. For example, transcriptome analysis of seeds at different developmental stages has uncovered key genes involved in the biosynthesis of starch, flavonoids, and unsaturated fatty acids, which are crucial for nutritional quality and popping ability (Mu et al., 2025; Liu et al., 2018). Furthermore, comparative transcriptomics of genotypes under abiotic stress (e.g., salinity and drought) has identified candidate genes encoding for transcription factors (e.g., MYB, NAC, and WRKY) and protective proteins like Late Embryogenesis Abundant (LEA) proteins, providing targets for breeding climate-resilient varieties (Wang et al., 2025a). The absence of a reference genome has been the single biggest bottleneck for advanced genetic studies in Makhana. The large and complex genome poses assembly challenges, but long-read sequencing technologies are being employed to overcome these. The availability of a chromosome-level genome assembly will revolutionize Makhana research, enabling genome-wide association studies (GWAS), whole-genome sequencing-based breeding, and the precise identification of genes and promoters.

4.7.3 Proteomics and metabolomics

While genomics reveals potential, proteomics and metabolomics provide a direct snapshot of the functional molecules and end-products that define phenotype. Proteomic profiling using mass spectrometry has been used to characterize the seed storage proteins, confirming their high nutritional quality and identifying major allergenic proteins, if any (Guo and Barkla, 2025).

Furthermore, comparative proteomic analysis of genotypes with varying degrees of stress tolerance can reveal key proteins involved in defense mechanisms, offering biomarkers for selection. Metabolite profiling is particularly relevant for Makhana given its high value as a nutraceutical. Gas Chromatography-Mass Spectrometry (GC-MS) and Liquid Chromatography-Mass Spectrometry (LC-MS) have been used to comprehensively profile the seed metabolome, quantifying sugars, organic acids, amino acids, and a diverse range of bioactive compounds like polyphenols and phytosterols (Farag et al., 2025). This data not only validates the medicinal properties of Makhana but also allows breeders to select for genotypes with superior nutritional and health-promoting metabolite profiles.

4.7.4 Bioinformatics and comparative genomics

Given the limited species-specific data, bioinformatics and comparative genomics serve as a bridge to transfer knowledge from model and related plant species. The analysis of omics data heavily relies on bioinformatics pipelines for sequence assembly, annotation, gene prediction, and differential expression analysis. Customized pipelines are being developed to handle the unique features of the Makhana transcriptome and genome (Burks and Azad, 2021). Makhana's phylogenetic placement within the Nymphaeaceae family, which includes the water lily (*Nymphaea* spp.), provides a valuable opportunity for comparative genomics. By comparing the genomic sequences of *E. ferox* with the published genomes of *Nymphaea colorata* and other basal angiosperms, researchers can identify conserved orthologous genes related to flowering time, seed development, and aquatic adaptation (Zhang et al., 2020). This approach helps in annotating the Makhana genome and predicting the function of unknown genes.

4.8 Biotechnological approaches

Biotechnological tools offer a powerful suite of methods to overcome the inherent constraints of conventional breeding in Makhana, such as its long reproductive cycle, difficulties in hybridization, and narrow genetic base. Table 4.1 summarize the biotechnological, molecular, and analytical approaches employed in *Euryale ferox*. These approaches, ranging from in vitro culture to modern genome editing, hold the promise of accelerating crop improvement by creating novel genetic variation, enabling rapid clonal propagation, and allowing for precise genetic modifications.

4.8.1 In vitro propagation and tissue culture

The establishment of efficient in vitro propagation protocols is critical for the conservation, multiplication, and genetic manipulation of elite Makhana

TABLE 4.1 Summary of biotechnological, molecular, and analytical approaches employed in *Euryale ferox*, along with their specific applications and major findings.

Technique used	Application	Key Outcome	References
Embryo Rescue/ Culture	In vitro propagation using zygotic embryos on MS medium supplemented with 2,4-D.	Optimized protocol for high-frequency regeneration; essential for rescuing crosses and rapid multiplication of elite lines.	Kumar et al. (2020)
Gamma Irradiation	Exposing seeds to Cobalt-60 gamma rays (doses: 0–500 Gy) to induce variation.	Doses >100 Gy were lethal. Lower doses induced variation in plant height and antioxidant enzymes (SOD, APX), creating a base for stress-tolerant lines.	Kumar et al. (2019, 2020)
Genetic diversity analysis	To fingerprint germplasm from Bihar, Manipur, and Bengal with use of RAPD, ISSR, and SSR markers.	Identified genetically distinct clusters for hybridization. revealed that 100-seed weight contributed most to genetic divergence (57%).	Kumar et al. (2016b)
Chloroplast sequencing	Sequencing the complete chloroplast genome.	Determined the genome size (159,930 bp) and phylogenetic relationship, placing *E. ferox* close to *Nuphar advena* (yellow pond-lily).	Guo and Barkla (2025)
RNA-sequencing (RNA-seq)	Sequencing seed transcriptomes at different developmental stages (10, 20, 30 days after flowering).	Identified key genes (PAL, CHS, F3H) regulating flavonoid biosynthesis and starch accumulation during seed maturation.	Liu et al. (2018)
GC-MS profiling	Profiling bioactive compounds in raw versus popped seeds.	Identified 36 metabolites including tocopherols, gallic acid, and omega-3 fatty acids, validating its nutraceutical potential.	Sinha and Jha (2025)

genotypes. Successful micropropagation has been achieved using various explants, including shoot tips, nodal segments, and immature embryos. The use of cytokinins like Benzylaminopurine (BAP) in the culture medium has been effective in inducing direct shoot organogenesis, thereby minimizing somaclonal variation and ensuring the production of true-to-type plants (Mohiuddin et al., 2025). This protocol is vital for the rapid multiplication of identified elite lines and endangered germplasm. The induction of callus, an unorganized mass of cells, is the first step toward genetic transformation and somatic embryogenesis. Researchers have standardized callus induction from leaf and petiole explants on Murashige and Skoog (MS) medium supplemented with auxins such as 2,4-Dichlorophenoxyacetic acid (2,4-D) (Wang et al., 2025b). Furthermore, the successful progression of this callus into somatic embryos—embryo-like structures that can develop into whole plants—has been reported. This somatic embryogenesis system is a promising pathway for large-scale clonal propagation and offers a potential target tissue for genetic transformation experiments. In vitro techniques are indispensable for the long-term conservation of Makhana germplasm. Slow-growth storage conditions at low temperatures can extend subculture intervals, reducing maintenance costs and the risk of somaclonal variation. For long-term preservation, cryopreservation protocols, where explants like shoot apices are stored in liquid nitrogen (−196°C), are being optimized. This method allows for the secure conservation of genetic resources for decades (Vollmer et al., 2021).

4.8.2 Mutation breeding

Induced mutation is a proven method to create rapid and novel genetic variation, especially in a crop like Makhana with a limited genetic base. Both physical and chemical mutagens have been employed. Gamma radiation and ethyl methanesulfonate (EMS) are the most commonly used mutagens. Treatments are applied to seeds or in vitro cultured tissues to create mutant populations (M1 generation) (Ahmad et al., 2023; Saeed Awan et al., 2021). Large M2 populations derived from treated materials are screened for desirable mutants. Promising traits targeted include selection for mutants with softer spines on leaves and stems to facilitate harvesting, identification of mutants with altered seed size, improved popping percentage, or enhanced nutritional content, selection for compact plant architecture suitable for high-density cultivation, screening for enhanced tolerance to salinity and drought at the seedling stage (Chen et al., 2020). These mutant lines serve as valuable prebreeding material for incorporation into conventional breeding programs.

An integrated improvement strategy for makhana, combining conventional breeding, molecular approaches, multi-omics technologies, and modern biotechnological tools, is illustrated in Fig. 4.3. Traditional mass selection and participatory varietal trials are supported by marker-assisted selection to enhance yield, seed size, and popping quality. Genomics, transcriptomics, proteomics, and metabolomics provide insights into genes, proteins, and metabolites associated with nutritional value and quality traits in makhana. Bioinformatics enables comparative genome analysis, while tissue culture and

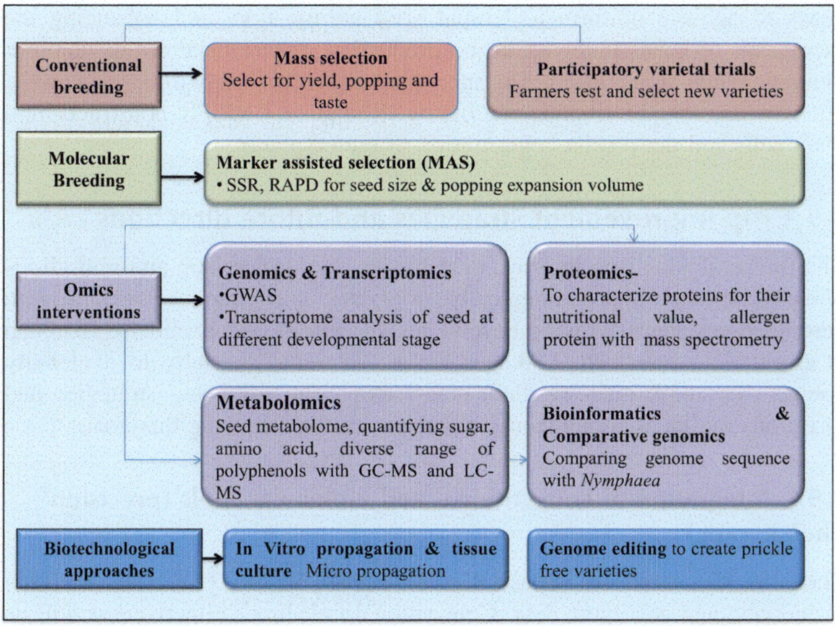

FIGURE 4.3 An integrated strategy combining conventional breeding, molecular tools, multi-omics, and biotechnology for the improvement of yield, quality, and nutritional traits.

genome editing offer new possibilities for rapid propagation and development of improved, farmer- and consumer-preferred makhana varieties.

4.8.3 Genetic transformation and genome editing prospects

While still in the developmental phase, genetic engineering holds the ultimate key to precision breeding in Makhana. The prerequisite for genetic transformation is an efficient and reproducible *Agrobacterium tumefaciens*-mediated or biolistic gene delivery system. While stable genetic transformation has not yet been publicly reported, ongoing research is focused on optimizing the conditions using somatic embryos and callus as target tissues, with reporter genes like β-glucuronidase (GUS) (Tilgam et al., 2024). The major challenges include the selection of transformed tissues and the regeneration of whole plants from them. The Clustered Regularly Interspaced Short Palindromic Repeats (CRISPR/Cas9) system represents a paradigm shift for crop improvement. Its application in Makhana is imminent, pending the availability of a high-quality reference genome and a stable transformation protocol. Future prospects for CRISPR/Cas9 are vast and could target knocking out genes responsible for any identified antinutritional compounds, editing genes involved in spine development to create "prickle-free" varieties, modifying genes to achieve uniform and rapid seed germination, and enhancing the expression of genes involved in the production of beneficial antioxidants or vitamins (Devi et al., 2022). The success of CRISPR/Cas9 will depend on the efficient delivery of ribonucleoproteins (RNPs) or CRISPR constructs into plant cells and the precise regeneration of edited plants.

4.9 Crop improvement strategies and future directions

The future of Makhana improvement hinges on a synergistic and multidisciplinary approach that strategically integrates conventional wisdom with cutting-edge science. Moving beyond isolated efforts, a consolidated roadmap is essential to transform Makhana from a regional specialty to a globally competitive and climate-resilient smart crop. The following strategies and directions outline a comprehensive framework for achieving this goal.

4.9.1 Integration of conventional and molecular tools (precision phenotyping)

The most immediate and impactful strategy is the creation of a bridge between phenotypic and genotypic data. Conventional breeding's effectiveness will be exponentially increased when guided by molecular markers. Establish a "precision phenotyping" pipeline for core germplasm collections. This involves meticulously recording high-throughput phenotypic data for complex traits like yield components, popping quality under standardized conditions, and abiotic stress response indices. This data will then be correlated with genotypic data from SSR and single-nucleotide polymorphism (SNP) markers derived from genome sequencing (Bhat, 2015). This integration will allow breeders to practice Genomic Selection (GS), where a model trained on the genotype and phenotype reference population can predict the breeding value of new lines based on genotype alone, significantly shortening the breeding cycle (Merrick et al., 2022).

4.9.2 Development of efficient hybridization protocols

The predominantly self-pollinating nature and complex floral biology of Makhana have limited genetic recombination. Breaking this barrier is crucial for harnessing hybrid vigor and creating novel genetic combinations. To develop the efficient hybridization protocol requires efficient strategy to solve the challenge of cross-pollination which in include the detailed analysis of flower opening time, anther dehiscence, and stigma receptivity to identify the best window for emasculation and pollination, need of development of nondestructive and efficient emasculation techniques, potentially leveraging thermal or chemical methods to avoid physical damage to the delicate floral structures and exploration strategies to enhance insect-mediated cross-pollination in controlled aquatic environments. These strategies can cause the successful hybridization, which will enable the development of synthetic varieties and, eventually, hybrid Makhana, potentially leading to significant yield breakthroughs (Yahaya et al., 2020).

4.9.3 Genome sequencing and pangenome analysis for trait discovery

While a reference genome is a critical first step, capturing the full spectrum of genetic diversity is the key to unlocking complex traits. Following the completion of a high-quality reference genome for one elite genotype, the next step should be the whole-genome re-sequencing of hundreds of diverse accessions from India, China, and Japan (Tao et al., 2019). This will facilitate the construction of a Makhana pangenome, which represents the entire set of genes within the species, including the "dispensable" genes present only in some individuals. Pangenome analysis is powerful for identifying genes associated with local adaptation, disease resistance, and unique quality traits that are absent from the single reference genome (Tranchant-Dubreuil et al., 2019).

4.9.4 Advanced marker-trait association mapping

Moving beyond preliminary associations to robust, validated gene-trait links is essential for molecular breeding. Strategies include the conduct of large-scale Genome-Wide Association Studies (GWAS) using the re-sequenced germplasm panel. This approach leverages historical recombination events to identify marker-trait associations with high resolution for complex polygenic traits such as yield, popping quality, abiotic stress tolerance, and nutritional quality. GWAS will identify highly diagnostic SNP markers for use in MAS and will pinpoint candidate genes for functional validation via biotechnological approaches (Sonah et al., 2015).

4.9.5 Institutionalizing participatory breeding models

To ensure the rapid adoption and relevance of improved varieties, farmer involvement must be institutionalized throughout the breeding pipeline. Approaches include the formalization of the Participatory Plant Breeding (PPB) network that connects national research institutions (e.g., ICAR-RCER and NBPGR), state agricultural universities, NGOs, and farmer cooperatives. Farmers should be involved not just in the final selection but also in setting breeding priorities and evaluating early-generation breeding lines in their own ponds (Sthapit et al., 1996). This model ensures that developed varieties are locally adapted, meet farmer and consumer preferences, and are disseminated effectively through established trust networks, thereby enhancing livelihood security.

4.9.6 Promotion and seed system development

A superior variety is useless without a robust seed system to deliver it to farmers. So, for that, there is a requirement of a formal Quality Declared Seed (QDS) production system for Makhana, which involves maintaining breeder seed of released varieties under strict genetic purity in controlled environments, potentially leveraging in vitro propagation for rapid scaling (Khadatkar et al., 2020). Empowering progressive farmers with the technical knowledge to produce certified foundation and certified seeds. Establishing community-based seed hubs in key Makhana-growing districts to ensure timely access to quality planting material. A formal seed system will break the current cycle of dependency on farmer-saved seeds, ensuring higher productivity and uniformity across the Makhana cultivation landscape.

4.10 Conservation and sustainable utilization

The long-term survival and agricultural potential of *Euryale ferox* are inextricably linked to the conservation of its genetic diversity and the adoption of sustainable management practices. As a species dependent on fragile wetland ecosystems, a strategic blend of in situ and ex situ conservation, coupled with policies that incentivize sustainable use, is essential to prevent genetic erosion and ensure the crop's continued contribution to food security and rural livelihoods.

4.10.1 In situ conservation: Protecting natural Arenas of diversity

In situ conservation involves the protection of species within their natural habitats and the ecosystems where they have evolved. For Makhana, this is a critical strategy for maintaining evolving gene pools. Many water bodies in Bihar, Manipur, and Assam in India, as well as in regions of China, host wild and semiwild populations of Makhana. These populations are reservoirs of unique alleles and adaptive traits, such as resilience to local pests, diseases, and abiotic stresses (Ahmad, 2020). Establishing designated "Makhana Genetic Sanctuaries" or community-conserved wetlands in these areas is paramount. This requires collaboration between forest departments, local communities, and agricultural agencies to regulate harvesting and protect these habitats from pollution, siltation, and conversion for other uses. Supporting the continued cultivation of diverse local landraces by farmers is a dynamic form of in situ conservation. These landraces are continuously adapting to changing local conditions and farmer selections. Initiatives that link on-farm conservation with market incentives, such as premium prices for unique landracebased products, can encourage farmers to maintain this diversity rather than shifting to a single, potentially genetically uniform, improved variety (Arunachalam, 2000).

4.10.2 Ex situ conservation: Safeguarding in gene banks

Ex situ conservation involves the preservation of genetic resources outside their natural habitat. This provides a secure backup against the loss of diversity in the wild and facilitates access for breeding programs. Conventional seed storage is the most common method. However, Makhana seeds are recalcitrant, meaning they cannot tolerate the low moisture content and cold temperatures used for storing orthodox seeds (like rice or wheat) without losing viability (Pammenter and Berjak, 2014). Research at the National Bureau of Plant Genetic Resources (NBPGR) focuses on identifying the optimal storage conditions—such as specific moisture content levels and temperatures just above freezing—to extend seed longevity for several years. Despite these challenges, maintaining working collections in seed banks remains a vital short-to-medium-term strategy, such as in vitro conservation and cryopreservation. For long-term conservation, in vitro techniques are superior. Using slow-growth storage, where plantlets are maintained on culture media with growth retardants or at reduced temperatures, can extend subculture intervals to 12–24 months, significantly reducing costs and labor (Renau-Morata et al., 2006). This method is ideal for the active conservation of elite and endangered genotypes. Cryopreservation is the most secure long-term ex situ conservation strategy. It involves storing plant tissues, such as shoot apices or embryonic axes, in liquid nitrogen at $-196°C$, where all metabolic activities cease. Protocols for the cryopreservation of Makhana are being optimized, involving techniques like vitrification to prevent lethal ice crystal formation (Kaviani, 2011). Successful implementation will ensure the safe conservation of germplasm for centuries.

4.10.3 Sustainable harvesting and ecosystem management

The commercialization of Makhana must not come at the cost of wetland health. Sustainable practices are essential for the crop's ecological and agricultural perpetuity. Implementing community-based rules that dictate harvesting times and methods can prevent over-exploitation. For instance, allowing a certain proportion of seeds to sink and regenerate naturally ensures the population's sustainability. Promoting Makhana cultivation as part of a polyculture system with compatible fish species can enhance ecosystem productivity and farmer income. This integrated system can improve nutrient cycling, reduce the need for external inputs, and increase overall resilience (Raut et al., 2025). Furthermore, avoiding the use of chemical pesticides and fertilizers protects water quality and aquatic biodiversity.

4.10.4 Policy interventions for mainstreaming conservation

Strategic policy support is required to elevate Makhana conservation and sustainable use to a national priority. Advocating for the explicit inclusion of Makhana in national crop improvement programs, such as those led by the Indian Council of Agricultural Research (ICAR), and in schemes promoting aquaculture and wetland development would channel crucial research funding and institutional support. Strengthening and enforcing the GI tag for regions like "Bihar Makhana" can add market value, which in turn creates economic incentives for farmers to maintain traditional, diverse varieties linked to their region. Developing clear frameworks for Access and Benefit-Sharing (ABS) in accordance with the Nagoya Protocol is also necessary when genetic resources are used for commercial product development. Policies that support the development of efficient processing, branding, and market linkages for Makhana products can increase profitability for farmers, making the cultivation and conservation of diverse genetic resources a more economically viable endeavor.

4.11 Conclusion

Makhana (*Euryale ferox* Salisb.) is a unique and important aquatic crop with enormous potential for sustainable wetland agriculture, livelihood enhancement, and nutritional security. However, limited genetic resources, inadequate breeding infrastructure, and poor integration of contemporary genomic methods have hindered its transformation from a traditional, region-specific crop to a scientifically developed and globally competitive commodity. This chapter emphasizes the significant genetic and phenotypic heterogeneity found in native germplasm, offering opportunities for genetic enhancement if systematically explored and utilized. A comprehensive breeding strategy that combines traditional selection with molecular and omics-based methods, backed by thorough germplasm characterization, genome sequencing, and marker-trait association studies, will be necessary for makhana improvement in the future. Once basic genomic resources are developed, advances in tissue culture, mutant breeding, and new genome-editing tools have the potential to accelerate trait improvement. Conservation initiatives, the sustainable use of wetland ecosystems, and participatory breeding programs that actively involve local stakeholders and farmers are all equally important. Improving production and resilience will fill the research gaps, especially in the areas of genetics, hybridization methods, seed systems, and mechanization. Makhana can emerge as a model aquatic crop that promotes inclusive rural development, nutritional security, and climate-smart agriculture on a national and international scale with concerted research, institutional assistance, and supportive policies.

References

Ahmad, A., 2020. A study on constraints of Makhana cultivation and suitable measures for its better development. J. Pharmacogn. Phytochem. 9 (1), 984–987. https://doi.org/10.22271/phyto.2020.v9.i1q.10580.

Ahmad, J., Jan, S., Javid, W., Bhat, S.A., Tahir, I., 2023. Development and phenotypic screening of gamma rays and ethyl methanesulphonate mutant populations in tartary buckwheat (Fagopyrum tartaricum (L.) Gaertn). J. Botany 163, 646–658. https://doi.org/10.1016/j.sajb.2023.10.012.

Annual Report 2023-24, ICAR-regional Center for Eastern Region, Patna (Bihar). (2023).

Arunachalam, V., 2000. Participatory conservation: a means of encouraging community biodiversity. Plant Genet. Resour. Newsl. 122, 1–6.

Bhat, J.A., 2015. Phenomics: a challenge for crop improvement in genomic era. Mol. Plant Breed. 1923–8266. https://doi.org/10.5376/mpb.2015.06.0022.

Burks, D.J., Azad, R.K., 2021. RNA-seq data analysis pipeline for plants: transcriptome assembly, alignment, and differential expression analysis. In: Plant Metabolic Engineering: Methods and Protocols. Springer US, New York, NY, pp. 47–60.

Chapman, M.A., 2019. Optimizing depth and type of high-throughput sequencing data for microsatellite discovery. Appl. Plant Sci. 7 (11). https://doi.org/10.1002/aps3.11298.

Chen, T., Huang, L., Wang, M., Huang, Y., Zeng, R., Wang, X., Wang, L., Wan, S., Zhang, L., 2020. Ethyl methyl sulfonate-induced mutagenesis and its effects on peanut agronomic, yield and quality traits. Agronomy 10 (5), 655. https://doi.org/10.3390/agronomy10050655.

Devi, R., Chauhan, S., Dhillon, T.S., 2022. Genome editing for vegetable crop improvement: challenges and future prospects. Front. Genet. 13, 16648021. https://doi.org/10.3389/fgene.2022.1037091.

Dutta, R.N., Jha, S.N., Jha, U.N., 1986. Plant contents and quality of makhana (Euryale ferox). Plant Soil 96 (3), 429–432. https://doi.org/10.1007/bf02375148.

Farag, M.A., Khattab, A.R., Farghal, H.H., Ismail, W.M., Fahmy, H.A., 2025. Gas chromatography/mass spectrometry-based metabolite profiling of chia and quinoa seeds in comparison with wheat and oat. Phytochem. Anal. 36 (8), 2301–2320. https://doi.org/10.1002/pca.3398.

Gaurav, A., Nath, D., Barbhuiya, R.I., Pradhan, R.C., 2022. Physical characterization and mass modeling of traditionally popped makhana (euryale ferox Salisbury) variety: Sabour-1 makhana. J. Food Process Eng. 45 (6). https://doi.org/10.1111/jfpe.13927.

Guo, Q., Barkla, B.J., 2025. Proteomic insights into edible nut seeds: nutritional value, allergenicity, stress responses, and processing effects. Agronomy 15 (10), 2353. https://doi.org/10.3390/agronomy15102353.

He, S., Wang, D., Zhang, Y., Yang, S., Xiang, L., Wei, D., Zhang, M., Qin, J., 2019. Chemical components and biological activities of the essential oil from traditional medicinal food, euryale ferox Salisb., Seeds. J. Ess. Oil Bearing Plants 22 (1), 73–81. https://doi.org/10.1080/0972060x.2019.1595165.

ICAR- Research Complex for Eastern Region, Patna, Bihar, Annual Report. (2022).

Jana, B.R., 2018. Flower characteristics and pollination behavior of Euryale ferox (Salisb.). Am. J. Plant Sci. 9 (4), 722–731.

Jana, B.R., Srivastava, A., Idris, M., 2018. Climate change effects and quality makhana (euryale ferox Salisb.) production under wetland ecosystem of northnorth Bihar. Int. J. Curr. Microbiol. Appl. Sci. 7 (12), 578–587. https://doi.org/10.20546/ijcmas.2018.712.072.

Jana, B.R., Anupama, K., Lakshmi Vaishna, R., Das, A., 2025a. Development of two improved water chestnut varieties (trapa spp.) and their characterization using RAPD and ISSR markers. J. Adv. Biol. Biotechnol. 28 (7), 975–986. https://doi.org/10.9734/jabb/2025/v28i72614.

Jana, B.R., Bhatt, B.P., Srivastava, A., 2025b. Biochemical components of makhana (euryale ferox Salisb) and its relevance to anti-obesity effect in human. Asian J. Dairy Food Res 44 (4), 614–619. https://doi.org/10.18805/ajdfr.DR-1871.

Jha, V., Kargupta, A.N., Dutta, R.N., Jha, U.N., Mishra, R.K., Saraswati, K.C., 1991. Utilization and conservation of euryale ferox Salisbury in Mathila (north Bihar), India. Aquatic Bot. 39 (3–4), 295–314. https://doi.org/10.1016/0304-3770(91)90005-.

Jiang, J., Ou, H., Chen, R., Lu, H., Zhou, L., Yang, Z., Ethnopharmacological, T., 2023. Phytochemical, and pharmacological review of euryale ferox Salisb.: a Chinese medicine food homology. Molecules 28 (11), 4399. https://doi.org/10.3390/molecules28114399.

Kapoor, S., Kaur, A., Kaur, R., Kumar, V., Choudhary, M., 2022. Euryale ferox, a prominent superfood: nutritional, pharmaceutical, and its economical importance. J. Food Biochem. 46 (12). https://doi.org/10.1111/jfbc.14435.

Kaviani, B., 2011. Conservation of plant genetic resources by cryopreservation. Australian J. Crop Sci. 5 (6), 778–800.

Khadatkar, A., Mehta, C.R., Gite, L.P., 2020. Makhana (Euryle ferox Salisb.): a high-valued aquatic food crop with emphasis on its agronomic management – a review. Sci. Horticult. 261, 108995. https://doi.org/10.1016/j.scienta.2019.108995.

Kumar, A., Singh, A.K., Kumari, S., Kumar, B., 2019. Socio-economic characteristics of makhana growers in relation to livelihood and earning. Indian J. Ext. Educ. 55 (4), 226–228.

Kumar, S., Singh, I., Nandi, B., Singh, C., 2014. Application and effect of addition of popped makhana flour on the properties and qualities of bun. Int. J. Food Nutr. Sci. 3, 116–120.

Kumar, A., Yadav, P.K., Nat, P., Nath, S., Kumari, U., 2016a. Euryale ferox Salisb. Is an immense potential aquatic cash fruit crop: a review. International Journal of Advances in Agricultural Science and Technology 3 (5), 35–49.

Kumar, L., Singh, A.K., Bhatt, B.P., 2016b. Nutritional status of recently developed makhana (gorgon nut) variety "Swarna Vaidehi". J. AgriSearch 3 (4). https://doi.org/10.21921/jas.v3i4.6701.

Kumar, N., Shikha, D., Kumari, S., Choudhary, B.K., Kumar, L., Singh, I.S., 2018. SSR-based DNA fingerprinting and diversity assessment among Indian germplasm of euryale ferox: an aquatic underutilized and neglected food crop. Appl. Biochem. Biotechnol. 185 (1), 34–41. https://doi.org/10.1007/s12010-017-2643-9.

Kumar, M., Raut, S.M., Bhatt, B.P., Kumar, L., 2020. Scientific cultivation of makhana for improving farmers' livelihood in eastern India. Biotica Res. Today 2 (7), 670–672.

Kumar, A., Singh, A.K., Kumar, B., Kumari, S., 2021. A study of knowledge and traditional wisdom of makhana cultivation in Bihar. Indian J. Trad. Knowl. 20 (2), 532–536.

Kumari, A., Singh, I.S., Kumar, L., Kumar, A.M.I.T., Kumar, R., Gupta, V.K., 2014. Morphological characteristics of makhana germplasm of Manipur under Darbhanga conditions. J. AgriSearch 1 (3), 157–160.

Kumari, R., Jakhar, D.S., Kumar, P., 2019. Nutritional, and medicinal importance of makhana (*Euryale ferox* Salisb.). Marumegh 4, 53–55.

Lebot, V., Viquez-Zamora, C., 2025. Constraints to participatory plant breeding and participatory variety selection of tropical root and tuber crops: a review. Euphytica 221 (7). https://doi.org/10.1007/s10681-025-03561-0.

Liu, X., He, Z., Yin, Y., Xu, X., Wu, W., Li, L., 2018. Transcriptome sequencing and analysis during seed growth and development in Euryale ferox Salisb. BMC Genom. 19 (1), 14712164. https://doi.org/10.1186/s12864-018-4707-9.

Lokendra Kumar, L.K., Choudhary, A.K., Bhatt, B.P., Singh, K.P., 2015. Genetic Divergence in Makhana (*Euryale Ferox* Salisb.).

Mandal, R.N., Saha, G.S., Sarangi, N., 2010. Harvest and processing of Makhana (Euryale ferox Salisb.) - an unique assemblage of traditional knowledge. Indian J. Trad. Knowl. 9 (4), 684–688.

Merrick, L.F., Herr, A.W., Sandhu, K.S., Lozada, D.N., Carter, A.H., 2022. Utilizing genomic selection for wheat population development and improvement. Agronomy 12 (2). https://doi.org/10.3390/agronomy12020522.

Mittal, R., Sharma, S., Mittal, A., 2020. A critical review on ethnobotanical and pharmacological aspects of euryale ferox Salisb. Pharm. J. 12 (6), 1444–1454. https://doi.org/10.5530/pj.2020.12.199.

Mohiuddin, A.K.M., Christine Abdullah, Z., Chowdhury, M.K.U., Harikrishna, K., Napis, S., 2025. Relationship between induction of novel somaclonal variants and types of organogenesis in Muskmelon (Cucumis melo L.). J. Agricult. Marine Sci. 27 (1), 90–98. https://doi.org/10.53541/jams.vol27iss1pp90-98.

Mu, Y., Sun, Y., Wu, Y., Yi, L., Yu, H., Zhang, S., 2025. Transcriptome analysis reveals metabolic pathways and key genes involved in Oleic acid formation of sunflower (Helianthus annuus L.). Int. J. Mol. Sci. 26 (14), 6757. https://doi.org/10.3390/ijms26146757.

Pammenter, N.W., Berjak, P., 2014. physiology of desiccation-sensitive (recalcitrant) seeds and the implications for cryopreservation. Int. J. Plant Sci. 175 (1), 21–28. https://doi.org/10.1086/673302.

Qian, L.H., Wu, J.Y., Wang, Y., Zou, X., Zhou, G.C., Sun, X.Q., 2022. Genome-wide analysis of NBS-LRR genes from an early-diverging angiosperm euryale ferox. Front. Genet. 13. https://doi.org/10.3389/fgene.2022.880071.

Raut, S., Gupta, N., Everard, M., Singh, I.S., 2020. Commercially and medicinally significant aquatic macrophytes: potential for improving livelihood security of indigenous communities in northern Bihar, India. J. Threatened Taxa 12 (13), 16819–16830. https://doi.org/10.11609/jott.5640.12.13.16819-16830.

Raut, S., Sarma, K., Singh, I.S., Singh, J., Padala, V.K., Ahirwal, S.K., Kumar, T., Kumar, R., Jana, B.R., Bhatt, B.P., 2025. Performance and sustainability of the makhana (Euryale ferox Salisbury) cum fish culture system in India. Indian J. Fisher. 72 (2), 84–93. https://doi.org/10.21077/ijf.2025.72.2.147354-10.

Renau-Morata, B., Arrillaga, I., Segura, J., 2006. In vitro storage of cedar shoot cultures under minimal growth conditions. Plant Cell Rep. 25 (7), 636–642. https://doi.org/10.1007/s00299-006-0129-2.

Saeed Awan, F., Sadia, B., Altaf, J., Habib, M., Hameed, K., Hussain, S., 2021. Genetic Variability through Induced Mutation. IntechOpen. https://doi.org/10.5772/intechopen.95027.

Sattar, A., Khan, S.A., 2016. Using agroclimatic approach for crop planning under rainfed condition in Darbhanga district of Bihar. MAUSAM 67 (3), 583–590. https://doi.org/10.54302/mausam.v67i3.1372.

Saxena, A., Srivastava, D.S., Srivastava, A., Garg, S.K., 2025. Suitability of underutilized aquatic cash crop Euryale ferox Salisb. for low-land agriculture and enhancing farmer's income. Vegetos 38 (6), 2333–2340. https://doi.org/10.1007/s42535-024-01030-y.

Sidh, V., Sharma, O., 2019. Medicinal use of makhana (euryale ferox Salisb): a review article. World J. Pharm. Med. Res. 5 (9), 103–104.

Singh, I.S., Kumar, L., Singh, A.K., Bhatt, B.P., Singh, S.P., 2014. Impact of makhana (*Euryale ferox* Salisb.) based cropping systems on nutrient dynamics of inceptisols of North Bihar. Int. J. Agric. Stat. Sci. 10, 125–129.

Singh, H.K., Singh, P., Singh, R., Kumar, A., Kumar, A., 2023. Performance of Makhana (Euryale ferox Salisb.) as a highly profitable venture in waterlogged areas prone to flooding. Int. J. Stat. Appl. Math. 8 (4S), 653–656. https://doi.org/10.22271/maths.2023.v8.i4si.1133.

Sinha, C.B.K., Jha, V., 2025. Identification of nutraceutically important metabolites in raw and popped makhana (*Euryale ferox* Salisb.) seeds using GCMS technique. Int. J. Plant Environ. 11 (3), 514–524.

Sonah, H., O'Donoughue, L., Cober, E., Rajcan, I., Belzile, F., 2015. Identification of loci governing eight agronomic traits using a GBS-GWAS approach and validation by QTL mapping in soya bean. Plant Biotechnol. J. 13 (2), 211–221. https://doi.org/10.1111/pbi.12249Canada.

Sthapit, B.R., Joshi, K.D., Witcombe, J.R., 1996. Farmer participatory crop improvement. III. Participatory plant breeding, a case study for rice in Nepal. Exper. Agricult. 32 (4), 479–496. https://doi.org/10.1017/s001447970000154x.

Tao, Y., Jordan, D.R., Mace, E.S., 2019. Crop genomics goes beyond a single reference genome. Trends in Plant Sci. 24 (12), 1072–1074. https://doi.org/10.1016/j.tplants.2019.10.001.

Tilgam, J., Bhattacharjee, S., Paul, K., Jaiswal, S., Saakre, M., Kumari, P., Vijayan, J., Sreevathsa, R., Pattanayak, D., 2024. Development of an efficient and reproducible embryonic axis-targeted tissue culture-based transformation protocol for Pigeonpea (Cajanus cajan). Legume Res. Int. J. 47 (Of). https://doi.org/10.18805/lr-5249.

Tranchant-Dubreuil, C., Rouard, M., Sabot, F., 2019. Plant pangenome: Impacts on phenotypes and evolution. Ann. Plant Rev. 2 (2), 453–478. https://doi.org/10.1002/9781119312994.apr0664.

Vollmer, R., Espirilla, J., Villagaray, R., Cárdenas, J., Castro, M., Sánchez, J.C., Manrique-Carpintero, N., Ellis, D., Anglin, N.L., 2021. Cryopreservation of potato shoot tips for long-term storage. Methods Mole. Biol. 2354, 21–54. https://doi.org/10.1007/978-1-0716-1609-3_2.

Wang, S., Tang, R., Wang, F., Pan, Y., Duan, Y., Xue, L., Zeng, D., Chen, J., Peng, D., 2025a. Establishment of an in vitro regeneration system and analysis of endogenous hormone dynamics in Melastoma dodecandrum. Horticulturae 11 (8), 875. https://doi.org/10.3390/horticulturae11080875.

Wang, X., Hu, Y., Dong, Y., Zhang, L., Wang, B., 2025b. Abiotic stress-regulated LEA gene mediates the response to drought, salinity, and cold stress in Medicago sativa L. Plant Cell Physiol. 66 (5), 781–796. https://doi.org/10.1093/pcp/pcaf020.

Wu, P., Liu, A., Li, L.J., 2021. Metabolomics and transcriptome analysis of the biosynthesis mechanism of flavonoids in the seeds of Euryale ferox Salisb at different developmental stages. Mol. Genet. Genom. 296 (4), 953–970. https://doi.org/10.1007/s00438-021-01790-1.

Yahaya, M.A., Shimelis, H., Laing, M., Mohammed, M.S., Mathew, I., 2020. Methodologies for hybridization in predominantly self-pollinating crops: a review. J. Crop Improv. 34 (2), 268–289. https://doi.org/10.1080/15427528.2019.1698483.

Yamada, T., Imaichi, R., Kato, M., 2001. Developmental morphology of ovules and seeds of Nymphaeales. Am. J. Bot. 88 (6), 963–974. https://doi.org/10.2307/2657077.

Zhang, L., Chen, F., Zhang, X., Li, Z., Zhao, Y., Lohaus, R., Chang, X., Dong, W., Ho, S.Y.W., Liu, X., Song, A., Chen, J., Guo, W., Wang, Z., Zhuang, Y., Wang, H., Chen, X., Hu, J., Liu, Y., Qin, Y., Wang, K., Dong, S., Liu, Y., Zhang, S., Yu, X., Wu, Q., Wang, L., Yan, X., Jiao, Y., Kong, H., Zhou, X., Yu, C., Chen, Y., Li, F., Wang, J., Chen, W., Chen, X., Jia, Q., Zhang, C., Jiang, Y., Zhang, W., Liu, G., Fu, J., Chen, F., Ma, H., Van de Peer, Y., Tang, H., 2020. The water lily genome and the early evolution of flowering plants. Nature 577 (7788), 79–84. https://doi.org/10.1038/s41586-019-1852-5.

Chapter 5

Postharvest handling, drying, and storage of makhana seeds

Pakkiranna Sivamma[1], Mounika Edde[2], Ganjahalli Vani[3], Ritu Kukde[4] and Kalyani Kumari[1]

[1]*ICAR-National Institute of Seed Science & Technology, Mau, Uttar Pradesh, India;* [2]*College of Agricultural Engineering, PJTAU, Sangareddy, Telangana, India;* [3]*ICAR-Central Institute of Agricultural Engineering, Bhopal, Madhya Pradesh, India;* [4]*ICAR-Central Institute of Post Harvest Engineering and Technology, Ludhiana, Punjab, India*

5.1 Introduction

Makhana (*Euryale ferox*), also known as fox nut or gorgon nut, is an aquatic seed crop belonging to the family Nymphaeaceae. It grows in stagnant or slow moving shallow water bodies like ponds, swamps, ditches, and land depressions. The crop has been cultivated in Asia for over 2000 years and holds considerable agronomic, nutritional, and socioeconomic importance in India (Devi et al., 2020). Currently, makhana production is mainly concentrated in the lower Gangetic plains, particularly in North Bihar and the adjoining regions of West Bengal, Assam, and Odisha. It provides a crucial source of income for rural and marginalized communities (Kumar and Kumar, 2015).

Makhana is a hydrophytic plant valued both as a food and a medicinal resource, with applications in nutrition, traditional medicine, and the rural economy. Makhana is highly nutritious due to its low fat content, high quality protein, and presence of several health promoting bioactive compounds. The high amino acid index (89%–93%) and better arginine + lysine/proline ratio (4.7–7.6) make it superior in comparison to other cereals for achieving the amino acid requirements of the human body (Jha et al., 1991). In traditional Indian and Chinese systems of medicine, makhana is suggested for managing issues related to the respiratory, circulatory, digestive, excretory, and reproductive systems (Jha and Prasad, 1993). These qualities have increased the demand and commercial value for makhana in recent years.

Cultivation, harvesting, and postharvest handling of makhana seeds are based on traditional farming methods, largely managed by the Mallah community, which possesses extensive indigenous knowledge of aquatic resource

Makhana (Euryale ferox) as a Functional Food. https://doi.org/10.1016/B978-0-443-44421-0.00007-X

management (Gaurav et al., 2022). After physiological maturity, makhana fruits rupture naturally, releasing seeds that settle at the pond bottom and are collected manually (Mani, 2019). At harvest, raw makhana seeds contain high moisture and often contain mud, organic matter, and microorganisms, making them prone to mechanical damage, spoilage, and quality loss. Therefore, postharvest handling of raw makhana seeds is crucial for maintaining seed quality, storage, and market value.

Postharvest management of raw makhana seeds involves a sequence of operations including proper initial handling, cleaning and washing, drying to safe moisture levels, size grading for quality differentiation, and storage before marketing or further use. Among these operations, drying is a vital unit process, as excessive moisture retention accelerates microbial growth, seed discoloration, and quality degradation during storage, whereas inappropriate drying conditions may induce surface cracking and internal stress within the seed (Sharma et al., 2023). Therefore, the selection of suitable drying methods and operating conditions is essential for preserving seed integrity and extending storage life.

Storage of raw makhana seeds presents additional constraints due to their hygroscopic nature and sensitivity to ambient temperature and relative humidity. Moisture migration, fungal infestation, oxidative changes, and physical degradation during storage can lead to significant postharvest losses if appropriate storage techniques and quality control measures are not adopted. Despite the economic importance of makhana, postharvest operations remain largely conventional, labor-intensive, and poorly mechanized, with limited scientific standardization (Kumar et al., 2011b).

In view of these considerations, this chapter focuses on postharvest handling, drying, and storage of raw makhana seeds, covering harvesting and initial handling, cleaning and washing, drying methods, size grading, storage techniques, effect of seed properties, and future researchable issues. A systematic understanding of these aspects is essential for reducing postharvest losses, improving seed quality, enhancing economic returns, and supporting sustainable development of the makhana value chain.

5.2 Harvesting and initial handling

Harvesting primarily involves collection of dispersed makhana seeds from bottom of the pond. Makhana cultivation and harvesting are labor-intensive, tedious, and time-consuming processes (Khadatkar et al., 2015). Makhana crop spread over water body, its flower and fruit are illustrated in Figs. 5.1–5.3A. The key factors limiting makhana cultivation are the lack of pond ownership, high drudgery associated with cultivation operation, limited access to scientific knowledge and improved cultivation practices, lack of improved varieties, short lease periods and the labor intensive production process (Kumar et al., 2011a, 2011b). Makhana cultivation is the primary source of

FIGURE 5.1 Makhana crop.

FIGURE 5.2 Makhana flower.

FIGURE 5.3A Makhana flower changing into fruit.

livelihood for the Mallah community; however, its development has received limited attention. Since most of them do not own ponds, they do not receive fair returns. Since skilled workers from the Mallah community typically perform the operation from generation to generation, attention must be paid to improve makhana harvesting practices in order to lessen drudgery and enhance their standard of living. During harvesting, muck goes into the worker's ear, eye, nose, and mouth. Additionally, workers frequently have skin conditions like rashes, itching, etc., as well as injuries from the pond's sharp thrones and prickles, leech bites, etc (Khadatkar et al., 2015, 2020).

5.2.1 Collection

Locals refer to the gathered makhana seeds as giri. Labor intensive harvesting by expert divers from the Mallah community start harvesting at the end of August and continue until October in the morning times, from 6:00 to 11:00 a. m. and sometimes up to 3:00 p.m. According to Khadatkar et al. (2015), harvesting is carried out manually in two or three stages using traditional tools like Kaara and Ganjaa (Fig. 5.3B–D). In the conventional method, "*Kaara*" (bamboo pole) is fixed in the pond, and the worker descends to the bottom, lies down, and uses both palms to drag the mud close to it. Four to five people together collect seeds from the bottom at a time. They cover up to 2–3 m radius around pole during collection. Approximately 1–1.5 h are required for collecting the seeds surrounding a "*Kaara*" pole (Mandal et al., 2010). Near the base, a mound of mud forms, which is subsequently filtered by a "*Ganjaa*" bamboo screen. The gathered black seeds are placed in an earthen container after being passed through water on top (Khadatkar et al., 2020). A single makhana plant produces about 100 seeds and there are about 10,000 plants in 1 ha of pond area (Khatatkar et al., 2014). The yield of raw makhana seed is

FIGURE 5.3B Kaara (Khadatkar et al., 2020).

FIGURE 5.3C Ganjaa.

FIGURE 5.3D Ganjaa.

about 2.5–3 t/ha of pond area (Jha and Prasad, 1994). Raw and mature makhana seeds are depicted in Figs. 5.4 and 5.5. After cleaning, the family's woman and kids put the seeds out on a bamboo or plastic mat to dry in the sun. Additionally, they manually remove the snails and other detritus from the seed. After being sun-dried for six to seven hours, the seeds are graded (Jalaj et al., 2019).

According to Khadatkar et al. (2015), the improved makhana harvesting system (Fig. 5.6) greatly increased efficiency and decreased drudgery in makhana harvesting from ponds. It consists of a floating platform with a compressed air cylinder, hose line with regulator, and a tiny diving equipment. When compared to the conventional approach, the technology increased harvesting production by 33.6%. Additionally, it decreased the body parts discomfort score by 61% and the overall discomfort rate by 50%. The study found that compared to traditional manual equipment and methods, the enhanced system makes harvesting more comfortable, less labor-intensive, and significantly more productive (Khadatkar et al., 2020).

FIGURE 5.4 Raw makhana seeds.

FIGURE 5.5 Mature makhana seeds.

FIGURE 5.6 Improved system for harvesting makhana (Khadatkar et al., 2020).

The labor costs for harvesting and the economics of makhana agriculture in the conventional system account for between 50 and 60% of the total cost of cultivation. In ponds, harvesting often occurs in two to three stages, with harvesting charges starting at ₹ 37.76–50.35/kg for the first harvest, ₹ 50.35–75.52/kg for the second, ₹ 75.52–100.7/kg for the third, and so on, if the harvesting period continues. It was found that when the harvesting period lengthens, fewer seeds are collected, which results in an increase in the amount paid for harvesting (Khadatkar et al., 2015). Additionally, depending on the type of integrated farming system employed, the cost of cultivating makhana in an agricultural field ranged from ₹ 111,426.61 to ₹ 153,818.59 (Khadatkar et al., 2020).

5.3 Cleaning & washing

When seeds are brought ashore, their membrane coverings are removed by the feet rubbing against one another. The seeds are then taken home for additional processing after being properly cleaned once or twice with clean water to get rid of any mud residue, etc. Raw makhana seeds must be cleaned and washed after being removed from the pond in order to get rid of silt, algae, adherent mud, decomposing plant matter, dead fish pieces, snails, mud, stones, and other unnecessary debris. In addition to ensuring consistent heat transfer and effective popping in later processing stages, this procedure shields against microbiological contamination, off-flavors, and quality degradation during drying and storage (Jha and Prasad, 1994).

5.4 Drying of raw makhana seeds

Drying is an essential postharvesting process during the raw makhana seed processing since freshly harvested seeds have high moisture levels and thus they are very perishable and subject to microbial decay, enzymatic degradation and loss of quality unless dried immediately (Mounika et al., 2021). Raw makhana seeds contain approximately 37% (w.b.) moisture content after harvesting from pond (Singh et al., 2021b). Adequate drying lowers the moisture levels to a safe concentration, which ensures that storage is convenient, easy to handle, and can be subjected to other required processing tasks like roasting and popping. The major aims of drying are preservation of quality, shelf life, and conserving of favorable physical and nutritional properties. Monitored moisture reduction inhibits both fungus and nutrient loss and is also found to support the structural integrity of the seeds that is directly related to popping expansion and end product quality. Overall, the drying process contributes to the stability of the storage, marketability, and potential of the makhana seeds in long storage as it suppresses the spoilage reactions (Islam et al., 2025). The different drying methods are compared and tabulated in Table 5.1.

TABLE 5.1 Comparison of drying methods for raw makhana seeds.

Drying method	Energy consumption	Impact on product quality	Scalability and practical applicability
Sun drying	Very low	• uneven moisture removal • risk of discoloration and surface cracking • high contamination and microbial risk	• poor scalability • highly weather dependent • suitable only for small, traditional operations
Shade drying	Very low	• better color retention • reduced shell cracking • minimal thermal damage	• limited scalability • requires large space and long time • applicable for quality-focused traditional processing
Hot-air (tray) drying	Moderate	• uniform drying • improved hygiene • controlled quality retention	• good scalability • suitable for laboratory, pilot, and small commercial units • easy process control
Microwave drying	High (but time-efficient)	• rapid moisture removal • better internal moisture uniformity • risk of overheating if left uncontrolled	• moderate to high scalability • suitable for industrial use with advanced control systems
Fluidized bed drying	Moderate	• uniform heat and mass transfer • reduced drying variability • good physical integrity	• good scalability • suitable for semi-mechanized and industrial processing
Ultrasound-assisted microwave vacuum drying	High initial, reduced overall	• enhanced moisture diffusion • minimal structural damage • superior quality retention	• high scalability potential • suitable for advanced industrial applications • requires high capital investment

5.4.1 Traditional drying practices

In traditional makhana producing regions, drying is commonly carried out using natural methods such as sun drying and shade drying. Sun drying involves spreading the washed seeds in thin layers on clean surfaces and exposing them to direct sunlight for several days (approx 2–3 days) until the moisture reduced to 25%–30% (w.b.) (Devi et al., 2020; Jana et al., 2019; Sharma et al., 2024a). During popping season, a day for processors typically starts at 3:00 a.m. and ends at 11:00 a.m. usually. Periodic turning of seeds is required to ensure uniform moisture removal. Although sun drying is economical and widely practiced, it is highly dependent on weather conditions and may result in uneven drying and contamination.

In shade drying, the cleaned seeds are placed in a single thin layer without coming into direct contact with sunlight. This process is based on natural convective mass transfer where moisture gradually moves from the interior of the seed to its surface, where it evaporates to the outside air. The ambient temperatures are usually maintained between 25 and 35°C and regular rotation of seeds is to be done to achieve a uniform drying in the seeds and also to avoid a localized accumulation of moisture. This slow drying action reduces thermal stress, surface/shell cracking, discoloration and structural damage of the internal structure to the starchy endosperm. This method preserves seed quality compared to sun drying but it is time-consuming (Sharma et al., 2023).

Other research studies reported that cleaned makhana seeds are dispersed for 2–3 h under bright sunlight, even for 7–8 h in case of cloudy weather on a mat or cemented yard with bright sunshine to reduce the moisture for easy transportation and temporary storage by around 31%, appropriate for storage (Mandal et al., 2010). The traditional drying methods has several limitations such as weather dependence, long drying time, uneven moisture removal, high labor requirements, unhygienic, and requires large drying areas. It also increases the risk of microbial contamination and spoilage due to prolonged exposure.

5.4.2 Mechanical drying methods

Mechanical drying methods have been adopted over the past years to surmount the constriction of traditional drying. One of the most widely utilized methods is the hot air drying whereby hot air is introduced into the enclosure of the seeds at the desired temperatures. This approach offers equal drying, increased hygiene, and decreases the duration of drying. Through controlled temperature drying, the thermal damages are also minimized and seed quality is preserved.

5.4.2.1 Tray drying

Tray drying is a controlled hot-air drying method used for drying makhana seeds in laboratory, pilot-scale, and small commercial processing systems. The cleaned seeds are spread evenly in a single seed thickness on food-grade perforated stainless steel trays and placed inside the drying chamber. Heated air is circulated in the chamber in a controlled manner, allowing regulated removal of moisture from the seeds. Drying is normally carried out at 60°C for 24 h until it reaches safe moisture level for storage (Devi and Deka, 2022). Drying temperature and duration directly affect structural integrity. The trays are spaced well to allow even air flow and are rearranged periodically during drying to reduce variation within the dryer. This is repeated until the seeds reach the required level of moisture that can be stored or further processed like roasting and popping. Tray drying is appropriate to obtain a homogeneous drying process of makhana seeds under standardized processing conditions (Devi and Deka, 2022).

In spite of the fact that there has been a limited amount of progressive research conducted on drying techniques for raw makhana seeds; however, novel drying methods such as microwave drying, fluidized bed drying, and ultrasound-assisted microwave vacuum drying may be investigated. These technologies have demonstrated promising results in terms of improving drying efficiency and quality retention in a variety of aquatic crop seeds.

5.4.2.2 Microwave drying

Microwaves hold an area in the electromagnetic spectrum between radio waves and infrared waves, characterized by a frequency that is greater than that of radio waves and lower than that of infrared waves (Dehghannya and Habibi-Ghods, 2025; Kumar and Shukla, 2014). Microwaves are categorized as nonionizing radiation, as microwave energy solely induces the vibration of water molecules in food, resulting in heat generation. In contrast to ionizing radiation, it does not modify chemical bonds, generate toxic chemicals, or cause structural alterations in nutrients beyond those related to standard thermal effects (Bakshi et al., 2023). Microwave heating has widespread applications in food preparation, encompassing drying and preservation (Chandrasekaran et al., 2013). Microwave heating functions via two mechanisms: dipole rotation and ionic polarization (Franco et al., 2015). Polar molecules, such as water, are commonly present in foods. Upon exposure to an electric field, polar molecules within the food orient themselves in accordance with the field's polarity. Consequently, the water molecules experience reorientation due to the ongoing changes in field polarity, leading to friction between the molecules and the surrounding medium, thereby producing heat. The ionic polarization method involves ion collisions that produce heat when food is subjected to an electric field, resulting from the enhanced motion of charged ions (Verma et al., 2020). This heating method provides benefits like an elevated drying rate and more consistent internal drying.

The elevated frequencies in microwaves facilitate fast energy transmission, leading to increased heating rates. Furthermore, while heating initiates at the food's core and radiates outward to the surface when subjected to microwaves, internal heating achieves greater uniformity (Nguyen and Songsermpong, 2022). In microwave drying, parameters including cross-section, shape, composition, volume, and processing equipment of the food significantly affect the energy absorption and distribution within the microwave oven chamber (Bakshi et al., 2023). The wavelength of microwaves spans from 1 m to 1 mm. Frequencies are often categorized into three ranges: ultra-high (0.3–3 GHz), super-high (3–30 GHz), and extremely high (30–300 GHz) frequencies (Orsat et al., 2017).

The primary parameters affecting microwave heating and heat dispersion are the dielectric properties of the food item and the penetration depth (von Horstig et al., 2022). The capacity of a material to transform microwave energy into heat is demonstrated by its dielectric properties (Gao et al., 2023). This can be stated in terms of the dielectric constant/relative permittivity and dielectric loss (Wang et al., 2003), where the dielectric constant refers to a material's ability to polarize and store electrical energy in response to an applied electric field, whereas dielectric loss pertains to energy dissipation as thermal energy (Verma et al., 2020). The dielectric characteristics are affected by microwave frequency, moisture content, food density, and temperature applied. Industrial microwave ovens generally work at a frequency of 915 MHz, whereas household microwave ovens function at 2450 MHz (Sumnu and Sahin, 2005). Microwave power consumption in the food sector has significantly risen due to its advantages over conventional technologies, such as precise control and speedy starting and shutdown capabilities (Kidmose and Martens, 1999).

Microwave drying has been effectively utilized for lotus (*Nelumbo nucifera Gaertn.*) seeds. An increase in microwave output power from 231 to 539 W results in a reduction of drying time from 36 to 13 min. Conversely, an increase in mass load from 100 to 200 g leads to an extension of drying time from 18 to 30 min. The moisture ratio throughout the drying process is effectively represented by the Page model, with predicted values aligning closely with experimental observations. Given the structural and compositional similarities between lotus and makhana seeds, the application of this microwave drying technique merits investigation for makhana to facilitate speedy moisture extraction, minimize processing duration, and enhance drying efficacy. The implementation of kinetic modeling, illustrated by the Page model, can significantly enhance the optimization of microwave parameters, thereby facilitating controlled drying and preserving the quality of makhana seeds (Tian et al., 2013).

5.4.2.3 Fluidized bed drying

Fluidized bed drying can be adopted as an effective hot-air drying technique for raw makhana seeds, especially in cases that involve quick and even moisture evaporation under controlled environment. The cleaned raw makhana seeds are deposited into the drying chamber on a perforated platform. The bottom of the chamber provides heated air which enters the chamber and once the velocity of the air is faster than the minimum fluidization velocity, the seeds become suspended and mixed uniformly. This fluid movement makes the process of mixing seeds and contact with the drying air constant. The airflow rate is controlled to be a little above the minimum fluidization requirement to ensure that the fluidization process is held steady without substantial seed movement or losses. This is carried on until the seeds have reached the required level of moisture to proceed with the next activity that includes handling or storage or other processes. Air temperature, airflow rate and drying duration are the factors that allow fluidized bed drying to be a possible process to standardize drying of raw makhana seeds in research and semi-mechanized processing. The drying temperature is typically regulated based on the initial moisture content of the seeds and the target final moisture level, in order to prevent thermal damage and ensure the preservation of quality. According to the findings presented by Manju et al. (2021), the drying of moringa seeds at an outlet temperature of 50°C for a duration of 3 h does not negatively impact the quality of the seeds. A comparable drying protocol can be implemented for raw makhana seeds, to facilitate a gradual decrease in moisture content while maintaining structural integrity and functional characteristics.

5.4.2.4 Ultrasound-assisted microwave vacuum drying

Numerous drawbacks of traditional hot-air drying technologies have been recognized, including significant energy use, quality degradation of the finished product, and inconsistent drying. Consequently, pretreatment is commonly used prior to drying. Pretreatment before drying is a thoroughly investigated field, with many methods having been established. Thermal pretreatment can eliminate bacteria, soften texture, and enhance the drying rate. Nevertheless, it results in inferior product quality (Huang et al., 2020).

Conversely, nonthermal technologies (e.g., ultrasonic and pulsed electric fields) may serve as superior alternatives to mitigate these limitations. The pretreatment of food products with ultrasound prior to drying has gained significant attention in recent years, demonstrating the ability to substantially reduce drying time. Wiktor et al. (2019) demonstrated that pretreatments resulted in a drying time reduction of up to 40% and enhanced retention of carotenoid content in carrots. The application of ultrasound diminished both the overall drying duration and energy expenditure while simultaneously improving product quality.

Ultrasound is a type of mechanical wave with a frequency range of 20 kHz–1 MHz. Ultrasound pretreatment typically involves the application of ultrasound via an ultrasonic bath or a probe to transmit ultrasound waves into an aqueous medium, such as distilled water. The acoustic waves cause compression and expansion, generate microscopic channels inside the sample, and thus result in the leakage of intracellular fluid into the surrounding environment. When the ultrasound power is sufficiently elevated, the rarefaction cycle can exceed the cohesive forces of the liquid molecules, resulting in the formation of cavitation bubbles. The bubbles progressively enlarge to a critical dimension until they become unstable and violently collapse. The bursting of cavitation bubbles generates rapid and substantial fluctuations in localized pressure and temperature, resulting in the breakdown of water molecules into highly reactive free radicals, including H^+ and OH^-. These free radicals may expedite specific chemical reactions and alter other molecules (Huang et al., 2020).

Ultrasound-assisted microwave vacuum drying has been effectively utilized for the drying of lotus seeds, demonstrating its potential as an advanced technique in this field. In this process, pretreatment is conducted by immersing approximately 70 g of seeds in 1000 mL of distilled water and applying ultrasonic treatment with a probe of approximately 1.9 cm in diameter. The probe functions in a pulsed mode, alternating between 5 s of operation and 5 s of idleness, and is submerged to a depth of approximately 2.5 cm to guarantee consistent energy distribution. Ultrasonic intensity within the range of 6–11 W/cm^2 is given for a duration of approximately 10 min (Lu et al., 2017; Zhao et al., 2021). Subsequent to the pretreatment phase, the seeds undergo drying in a microwave vacuum dryer, operating at a microwave power density of 15 W/g under a reduced pressure of 20 kPa. This process results in complete drying within a time frame of 8–15 min, contingent upon the ultrasonic intensity employed (Chen, 2010; Lu et al., 2017; Tian et al., 2013; Zhao et al., 2021). The ultrasonic treatment can also be done in ultrasonic bath, where the seeds can be kept in distilled water for 10 min with frequency of about 20–80 kHz, with power intensity levels of 0.75 and 1.5 W/g and then followed by microwave vacuum drying (Zhao et al., 2017).

This comprehensive method improves mass transfer, decreases drying duration, and enhances uniformity in the drying process of lotus seeds. Given the structural similarities between lotus and makhana seeds, the application of ultrasound-assisted microwave vacuum drying presents a promising option for investigation as an innovative and time-efficient drying method for makhana. The optimization of ultrasonic intensity and microwave power could facilitate a fast reduction in moisture content while maintaining the functional and structural quality characteristics of makhana seeds.

5.4.3 Effect of drying parameters on makhana

Makhana seed drying is sensitive, and it is necessary to monitor the main process parameters to guarantee the removal of moisture and quality preservation. The temperature in drying must be moderate and not excessive or too low to prevent cracking of seeds, hardening of the surface and damaging the inside structure thus altering the popping performance. Effective heat and mass transfer and uniform pattern of movement of air are necessary as the seeds of makhana do not dry evenly. The drying time should be made optimum so as to reach the desired final moisture content without overdrying and thereby decreasing the popping yield and quality of the product (Chauhan et al., 2023).

The humidity of drying air also has major impact on the rate of drying where low humidity facilitates quicker extraction of moisture in the makhana seeds. Drying efficiency is also influenced by loading density; too much seed loading inhibits the movement of air and leads to incomplete drying whereas optimal loading provides an apt movement of air and even-handed elimination of moisture. These parameters are important to be properly regulated to facilitate the quality, structural integrity, as well as storage stability of makhana seeds (Zhao et al., 2018; Sun et al., 2024).

5.4.4 Effect of drying on quality attributes of makhana seeds

Drying markedly influences the quality attributes of makhana seeds by altering their physical, mechanical, and textural properties. Physical properties such as size, shape, color, and bulk density are directly affected by the degree to which the moisture was removed. Controlled drying contributes toward preserving the color and to reducing overshrinkage, thereby retaining their natural appearance of seeds. Improper drying conditions such as high temperatures or uneven airflow, may cause irregular shrinkage and surface discoloration, which affects the consumer acceptance of product. Moisture reduction during drying also affects bulk density and porosity, which are important for packaging, storage, and subsequent heat transfer during roasting and popping (Zhao et al., 2021; Islam et al., 2025).

Drying also has an impact on mechanical and textural properties of makhana. The lower the moisture content, the seed become harder and it resist to breakage, here it affects their handling and processing behavior. Properly dried seeds exhibit sufficient strength to endure mechanical stresses throughout their storage and transportation without cracking. Excessive drying makes the seed brittle, which leads to breakage and reduces the popping yield. Textural properties such as firmness and internal structure are essential in popping performance since they define the expansion ratio and product crispness. The drying conditions should be optimized to maintain structural integrity and elasticity of the seeds to guarantee desirable mechanical stability and textural quality in the final popped product (Sharma et al., 2024b).

The functional behavior of makhana seeds during popping is directly related with the condition of starch and cellular structure maintained during the drying process. Proper expansion of makhana when exposed to high temperatures is made possible by the preservation of the natural arrangement of starch components by gentle and regulated drying. This structure may be disrupted by harsh drying conditions, which would cause the popped product to expand poorly and become less crisp. In order to preserve desired functional qualities, dependable processing performance, and acceptable quality of makhana based products, careful regulation of drying conditions is crucial (Lien, 2020; Sharma et al., 2024a).

5.4.5 Drying kinetics and mathematical modeling

The drying kinetics of raw makhana seeds could be explained using the thin-layer drying theory, which measures the loss of moisture as a time dependent parameter during drying under given conditions. Drying of makhana seeds falls under falling-rate period. This suggests that the drying process could be controlled by the diffusion of moisture inside the seeds rather than the surface evaporation governs the drying mechanism. This has direct implications for process design, as it necessitates the use of diffusion-based models for accurate prediction and control of moisture removal.

The drying process involves the removal of moisture which is measured using the moisture ratio (MR) which helps to normalize drying data collected at varying temperatures, air velocities, level of microwave power, and mass loading conditions. This normalization is imperative in the comparison of the laboratory scale, pilot scale, and industrial scale drying processes.

$$MR = \frac{M_t - M_e}{M_o - M_e}. \tag{5.1}$$

Where, M_t= Moisture content at drying time t, M_o = initial moisture content, and M_e = equilibrium moisture content. Under practical makhana drying conditions, particularly in high-temperature or microwave-assisted drying, M_e is relatively small and may be neglected, allowing the simplified form to be used for process analysis

$$MR = M_t/M_0. \tag{5.2}$$

The drying behavior of makhana seeds can be theoretically represented using a number of thin-layer drying models. These models can be useful in makhana processing as they generate easy mathematical correlations between the drying duration and process variables, which can be controlled, including temperature, microwave power, airflow, and mass load (Zhao et al., 2021).

Newton (Lewis) model

$$MR = \exp(-kt) \ldots \tag{5.3}$$

Henderson–Pabis model

$$MR = aexp(-kt) \tag{5.4}$$

Page model

$$MR = aexp(-kt^n) \tag{5.5}$$

Midilli–Kucuk model

$$MR = aexp(-kt^n) + bt\ldots \tag{5.6}$$

Where, k—drying rate constant, a & b—empirical coefficients, and n—model exponent. These parameters have physical and practical meaning. Modifications in k are quantitative indicators of the intensity of drying and can be used to compare dissimilar drying techniques and working conditions applied during the processing of makhana.

Performance evaluation parameters like the coefficient of determination (R^2), chi-square (χ^2) and root mean square error (RMSE), can be used to assess the models' fit. In addition to mathematical fitting, such criteria justify the choice of models that could be used to have reliable predictions of drying time and make decisions based on process optimization and scale-up.

The equations of the models such as Page model or Midilli–Kucuk model equations can possibly to achieve better agreement with the experimental results due to their ability to capture nonlinear moisture diffusion behavior commonly observed in industrial drying systems (Tian et al., 2013; Zhao et al., 2017; Lien, 2020).

The effective moisture diffusivity can be estimated using Fick's second law of diffusion to characterize the internal moisture movement during drying. The simplified form can be written as follows, assuming spherical geometry for makhana seeds and a homogeneous initial moisture distribution (Lien, 2020).

$$ln(MR) = \frac{\pi^2 D_{\text{eff}}}{L^2} t \tag{5.7}$$

Where, L is the characteristic length (radius) of the seed, D_{eff} can be estimated by the slope of the linear portion of $ln(MR)$ against time plot and it serves as a critical design parameter, governing drying time, throughput, and energy requirements during large-scale operations.

The effect of temperature on moisture diffusivity can be described using an Arrhenius-type relationship:

$$D_{\text{eff}} = D_o \exp\left(\frac{E_a}{RT}\right) \tag{5.8}$$

where D_o—pre-exponential factor, E_a—activation energy for moisture diffusion, R—universal gas constant, and T—absolute temperature. The data on activation energy makes it possible to extrapolate the drying behavior to industrial temperatures and to aid in choosing the best operating temperatures to compromise drying activity, energy use, and product quality when scaling up.

Factors such as microwave output power and mass load play a significant role in determining the drying time, with high power results in shorter drying periods (Tian et al., 2013). Likewise, ultrasonic pretreatment can be used to improve internal moisture diffusion through forming microstructural routes that result in increased effective diffusivity and reduced drying time in combination with industrial microwave vacuum drying technology. The quantitative interpretation of these improvements is possible in terms of the changes in model parameters and diffusivity values (Zhao et al., 2017).

In general, both drying kinetics and mathematical modeling can give a direct correlation between experimental drying results and the practical makhana processing. These models allow to predict the time of drying, address the differences in drying technologies, optimize the operating parameters, and scale up the experiments conducted under laboratory conditions to pilot and industrial-scale systems and guarantee efficient moisture removal, low energy use, and the stable quality of the products.

5.4.6 Energy and sustainability analysis of drying methods

As processing evolves from traditional to mechanized techniques, energy consumption and environmental sustainability become more and more significant factors in drying of raw makhana seeds. Conventional drying techniques, including sun and shade drying, require minimal external energy input, predominantly depending on solar radiation and natural convection processes (Sivamma et al., 2021a). As a result, the direct carbon footprint is minimal. Nonetheless, these methodologies exhibit a reliance on weather conditions, require significant time investment, and are susceptible to postharvest losses resulting from microbial contamination, uneven drying processes, and unforeseen precipitation. The aforementioned losses contribute to an increased environmental burden by diminishing processing efficiency and product yield.

Conversely, mechanical drying techniques such as hot-air tray drying, fluidized bed drying, and hybrid microwave drying systems provide enhanced control, decreased drying durations, and minimized quantitative and qualitative losses. Nevertheless, substantial electrical or thermal energy input is necessary. Traditional hot-air dryers generally demonstrate a moderate level of energy efficiency, given that a significant amount of the supplied heat is lost through the exhaust air. Microwave and microwave-vacuum systems exhibit accelerated drying rates and may offer reduced specific drying times; however, their energy intensity per unit mass can be considerable if not properly optimized. Consequently, the optimization of processes via temperature regulation, airflow management, and load control is crucial for achieving a balance between product quality and energy efficiency in mechanical drying systems.

Solar-assisted and hybrid drying systems represent a convincing intermediate solution. Solar cabinet dryers, solar tunnel dryers, and solar dryers that incorporate auxiliary electric or biomass heaters can markedly decrease reliance on grid electricity while offering enhanced control compared to traditional open sun drying methods. Hybrid systems that integrate solar predrying with subsequent mechanical finishing drying have the potential to decrease overall energy consumption and operational expenses. The integration of thermal storage materials or photovoltaic-powered fans significantly contributes to the sustainability and reliability of the system in the context of varying climatic conditions (Sivamma et al., 2021c). Aluminum foil and black mulch sheet could serve as cost-effective alternatives for enhancing and maintaining heat retention (Sivamma et al., 2021b).

Choosing a drying technology has significant implications for smallholder farmers in main makhana-producing regions from a sustainability perspective. Although advanced systems like ultrasound-assisted microwave vacuum drying provide enhanced quality retention, the substantial capital investment required restricts their implementation at the farm level. Conversely, low-cost solar-assisted or community-scale mechanical dryers contribute to improved product uniformity, a reduction in postharvest losses, an enhancement of market value, and an increase in income stability. The promotion of energy-efficient, scalable, and locally adaptable drying technologies is essential for fostering climate-resilient processing, mitigating carbon emissions, and enhancing the sustainability of livelihoods within makhana production systems (Rout et al., 2024).

In summary, the integration of energy-efficient drying strategies with sustainable rural processing infrastructure is crucial for achieving a balance among economic viability, environmental responsibility, and product quality within makhana value chains.

5.5 Grading

Grading of Makhana seeds is a crucial step for various postharvest seed handling processes. Grading may occur before or after drying. The grading of makhana seeds ensures equal heating of each nut during roasting, hence enhancing processing efficiency. Dried seed mixtures may consist of various sizes and shapes. Currently, farmers utilize five to seven sieves of varying mesh sizes for the grading of raw seed. These sieves consist of iron sheets encased in wooden frames. Nonetheless, the manual procedure is time-consuming, and the gradations are inadequate (Sundaram et al., 2014).

The makhana seed grader is utilized for grading seeds according to their size. Seeds must be classified into seven sizes. The complete gradation procedure necessitates the use of sieves of seven distinct sizes, labeled No. 1–7. Sieve number 1 possesses the largest pore diameter (1.2 cm), whilst sieve number 7 has the smallest diameter (0.4 cm). Gradation commences with a No. 1 sieve and concludes with a No. 7 sieve. The entire procedure is conducted manually (Verma et al., 2007).

5.5.1 Pedal operated makhana seed grader

A pedal-operated makhana grader was designed and developed at the ICAR-Research Complex for the Eastern Region in Patna for small-scale makhana farmers. It comprises three primary components: the main frame, the sieve unit, and the operator unit. The grading device demonstrated high efficiency in the separating procedure of makhana seeds. The machine graded the seeds into four categories, which were obtained from four distinct outlets. The highest seed grades for grades one, two, three, and four were 99.15%, 98.74%, 95.68%, and 83.65%, respectively, achieved at a crank speed of 45 rpm and a sieve angle of 15 degrees. The capacity is around 700–750 kg per hour. Two individuals are necessary for the efficient functioning of the grader: one to operate the pedals and another to fill the makhana seeds from the sieve outlet (Sundaram et al., 2014).

5.5.2 Effect of grading on hygroscopicity

For instance, a 5 kg makhana seed is categorized into three grades based on average diameter and sphericity: grade 1 (large, round, and fully expanded) with a diameter exceeding 15 mm and sphericity greater than 0.8; grade 2 (irregular, flattened, and overexpanded) with any diameter and sphericity less than or equal to 0.8; and grade 3 (small, round, and partially expanded) with a diameter of 15 mm or less and sphericity greater than 0.8. Graded samples are air dried to achieve reduced moisture content. The samples with moisture contents of 5%, 10%, 15%, and 20% are individually packed in polyethylene bags and stored in separate desiccators at room temperature (Jha, 1999). The grading of makhana by size considerably affects its equilibrium moisture content. The equilibrium moisture content for classes 1, 2, and 3 makhana ranges from 11.5% to 50.6% (d.b.), 12.6%–55.1% (d.b.), and 16%–58.9% (d. b.), respectively, with relative humidities between 11.2% and 88% at 30°C. Larger graded makhana kernels typically demonstrate reduced equilibrium moisture content at a specific relative humidity in comparison to smaller grades. This may result from structural variations across different grades of makhana, which could influence moisture absorption and retention ability. As a result, finer grades generally stabilize at elevated moisture levels, rendering them more vulnerable to moisture absorption during storage. Grading is crucial in assessing hygroscopic behavior and storage stability, while consistent grading facilitates predictable equilibrium moisture content and enhanced shelf life (Jha, 1999).

5.6 Storage techniques

Postharvest storage of makhana seeds is a technologically challenging step due to the high seed moisture requirement and strict functional quality constraints associated with popping. Unlike conventional cereals and oilseeds, makhana seeds are not dried to low moisture contents; instead, they are maintained at approximately 30%–31% moisture, which is essential for generating the internal steam pressure required for endosperm expansion during popping.

Consequently, makhana seeds remain physiologically active after harvest and exhibit limited storability under ambient conditions.

Traditionally, cleaned seeds are sun-dried briefly to remove surface moisture and then stored for only 20–25 days. During storage, small quantities of water are intermittently sprinkled to prevent excessive dehydration and preserve freshness. Traditional methods storage of makhana seeds are commonly carried out in cylindrical bamboo containers coated with cow dung and wrapped in coarse cotton cloth, which provide limited buffering against temperature and humidity fluctuations (Mandal et al., 2010). However, some of the farmers utilize cemented water tanks filled with water for the storage of seeds until they are ready for processing (Singh et al., 2021a). Although the thick and rigid seed coat confers partial resistance to insect infestation (Jha and Prasad,1994), these traditional methods offer minimal control over oxygen availability, relative humidity, microbial proliferation, and biochemical reactions, resulting in rapid physiological aging, off-flavor development, and loss of popping efficiency.

To overcome these limitations, several advanced storage technologies developed in postharvest and seed science may be adapted for makhana seeds. Among these, low-temperature storage combined with controlled relative humidity is one of the most promising strategies. Lowering temperature reduces respiration rate and enzymatic kinetics, while humidity regulation minimizes moisture migration, thereby slowing oxidative and hydrolytic reactions responsible for quality loss (FAO, 2011). Even moderate chilling can significantly extend safe storage duration in high-moisture biological materials by suppressing metabolic activity and microbial proliferation.

Hermetic storage systems represent another potential approach (Sivamma et al., 2022). These systems restrict gas exchange with the external environment, allowing the internal atmosphere to shift toward reduced O_2 and elevated CO_2 levels as a result of seed respiration (Ramachandra et al., 2021). Such atmospheres suppress oxidative reactions and microbial growth and stabilize moisture conditions within the storage unit (Guru et al., 2022). Although hermetic technologies are primarily designed for dry commodities, their ability to buffer environmental fluctuations suggests potential applicability to makhana when combined with temperature and humidity control.

Further technological opportunities lie in modified and controlled atmosphere storage (MAS/CAS). Deliberate reduction of oxygen availability suppresses aerobic respiration and oxidative metabolic pathways, while elevated CO_2 disrupts microbial activity and respiratory enzyme function (Conyers and Bell, 2003; Navarro, 2012). From a food chemistry standpoint, hypoxic and hypercarbic conditions limit lipid peroxidation, pigment degradation, and oxidative enzyme activity, which are central mechanisms of storage-induced quality deterioration (Tutuncu and Emekci, 2019). Although systematic investigations on makhana are scarce, the documented efficacy of MAS/CAS in other high-value seeds and food materials indicates strong potential for maintaining biochemical stability and functional quality during short-term storage.

In addition, vacuum or low-pressure storage may be explored as a supplementary technique, as reduced oxygen partial pressure restricts aerobic metabolism and oxidative reactions (Navarro et al., 2004; Sivamma et al., 2021d). When integrated with cooling and humidity regulation, low-pressure systems could further suppress deterioration kinetics; however, careful optimization is required to avoid excessive moisture loss, which would compromise popping performance.

Overall, modernization of makhana storage requires a transition from purely traditional methods toward integrated systems combining temperature control, humidity regulation, and atmospheric modification. Development of crop-specific storage protocols based on these principles could substantially reduce postharvest losses, stabilize biochemical quality, and improve processing efficiency prior to popping.

5.7 Effect of gamma treatment on raw makhana seed

Gamma irradiation serves as a potent method for inducing variability in makhana. Exposure of seeds to varying doses results in notable changes in germination, growth, physiological, and biochemical characteristics. The percentage of germination exhibits a progressive decline in response to increasing radiation doses. The untreated seeds exhibit a germination rate of 94%, which significantly declines to 10% when exposed to 500 Gy. The survival percentage exhibits a comparable pattern, declining from 92% in untreated seeds to 34% at 100 Gy, with no instances of plant survival observed beyond 100 Gy. The results demonstrate that elevated levels of gamma radiation significantly compromise cellular integrity and the metabolic processes essential for the establishment of seedlings (Kumar et al., 2019).

The parameters of growth are markedly affected by irradiation. The plant height diminishes following exposure to 50 Gy, measuring 64 cm, and further decreases to 62 cm at a 100 Gy dosage. The number of leaves per plant exhibits an increase at 50 Gy, reaching an average of 8.33 leaves, ahead of declining to 6 leaves at 100 Gy. The leaf area decreases from 1820 to 1260 cm^2 at a radiation dose of 50 Gy, followed by a partial recovery to 1686 cm^2 at 100 Gy. At a radiation dose of 50 Gy, there is a significant reduction in root length, measuring 37 cm. However, at 100 Gy, the root length is comparable to that of the untreated seed, recorded at 58 cm. The density of thorns on the upper leaf surface rises to 19 thorns/cm^2 at 50 Gy, up from 13 thorns/cm^2, and subsequently decreases to 12 thorns/cm^2 at 100 Gy (Kumar et al., 2019). The observed morphological modifications suggest that gamma radiation influences meristematic activity, cell division, and tissue differentiation.

From the perspective of photosynthetic pigments, the content of chlorophyll a rises to 110 μg/g fresh weight at 100 Gy, in contrast to 84 μg/g at 50 Gy, indicating an approximate increase of 28% relative to the untreated seeds. The total chlorophyll content attains a level of 240 μg/g fresh weight at a radiation dose of 100 Gy, representing an increase of approximately 20% compared to the untreated seeds. Chlorophyll *b* exhibits minimal variation. The photosynthetic rate exhibits an increase from 7 mmol $CO_2/m^2/s$ to 12 mmol $CO_2/m^2/s$ in irradiated plants, reflecting an enhancement of approximately 42%. Nevertheless, the transpiration rate exhibits a reduction of 18% at 50 Gy and 13% at 100 Gy, whereas stomatal conductance shows a decline of 16% and 8.3%, respectively. The intracellular concentration of CO_2 exhibits a reduction of 61% at a dosage of 50 Gy when compared to untreated seeds (Kumar et al., 2019). The findings indicate enhanced photosynthetic efficiency associated with controlled water loss during moderate irradiation stress.

Gamma radiation induces significant biochemical alterations. The total soluble protein content exhibits a reduction of 20% when subjected to a dose of 100 Gy. The activities of antioxidant enzymes exhibit a notable decline, with superoxide dismutase (SOD) activity diminishing by 50% at 50 Gy and by 75% at 100 Gy. Similarly, ascorbate peroxidase (APX) activity shows a reduction of 50% at 50 Gy and an 89% decrease at 100 Gy. The observed reductions suggest the initiation of oxidative stress at elevated doses. In contrast, osmolytes such as proline and glycine betaine exhibit a progressive accumulation in response to increasing radiation doses, with peak levels observed at 100 Gy (Kumar et al., 2019), indicating the activation of mechanisms associated with stress adaptation.

Gamma radiation additionally modifies the composition of minerals. The content of sodium (Na^+) exhibits a reduction of 46% at a radiation dose of 50 Gy and a decrease of 72% at 100 Gy. The content of potassium (K^+) exhibits a reduction of 42% at a dose of 50 Gy, whereas no significant alteration is noted at 100 Gy. The Na^+/K^+ ratio exhibits variability contingent upon dosage, suggesting alterations in ionic equilibrium. In summary, moderate doses, specifically at 100 Gy (Kumar et al., 2019), contribute positively to chlorophyll content and photosynthetic efficiency while also introducing physiological variability. Conversely, elevated doses have a detrimental impact on germination and survival rates. Meticulous optimization of radiation dosage is crucial to attain advantageous variability while minimizing undue physiological damage.

5.8 Physical, frictional and nutritional properties of raw makhana seed

The physical, frictional, and nutritional characteristics of raw makhana seeds are essential in influencing the efficiency and efficacy of cleaning, drying, grading, and storage processes.

Physical properties including size, shape, bulk density, true density, porosity, and thousand-seed weight are essential for equipment design and process optimization. Size and shape dictate the choice of sieves, graders, and separators in the cleaning and grading processes, whereas bulk density and porosity affect airflow resistance and thermal transfer during drying. Moisture content and seed geometry influence drying kinetics and uniformity, hence affecting processing efficiency and the quality of the final product. Makhana seeds (Swarna Vaidehi) are generally spherical in shape and have a deep black seed coat. The seeds typically weigh about 92–98 g per 100 seeds, with an average diameter ranging from 9.5 to 10.2 mm (Kumar et al., 2016). The physical qualities of makhana seeds (at 10.78% dry basis) are detailed in Table 5.2.

Frictional characteristics, such as the coefficient of friction and angle of repose, are crucial for handling, transportation, and storage. The characteristics of raw makhana seeds dictate their flow behavior on various contact surfaces, affecting hopper design, chute angle, and conveyor efficiency. Thorough evaluation of frictional properties reduces clogging, irregular flow, and damage caused by friction during cleaning and grading, while ensuring secure and effective storage in bins and silos. The frictional characteristics of makhana seeds are presented in Table 5.3.

TABLE 5.2 Physical properties of makhana seeds (at 10.78% d.b.) (Sundaram et al., 2014).

Properties	Range of values
Length, mm	7.95–13.61
Width, mm	7.93–12.72
Thickness, mm	7.74–13.03
Geometric mean diameter, mm	7.87–13.06
Sphericity	95.25–99.03
Bulk density, g/cm^3	0.49–0.54
True density, g/cm^3	0.84–0.90
Angle of repose, degree	26.40–28.18

TABLE 5.3 Frictional properties of makhana seeds (at 10.78% d.b.) (Sundaram et al., 2014).

Material	Coefficient of friction
GI sheet	0.32–0.35
Plywood	0.4–0.44
Plastic laminates	0.37–0.4

Nutritional properties, including moisture, fat, protein, and carbohydrate content, are essential for postharvest stability and quality preservation. Elevated initial moisture and nutrient-dense composition enhance vulnerability to microbial proliferation, lipid oxidation, and enzymatic deterioration. Understanding these features aids in choosing suitable drying conditions to diminish moisture to safe storage levels while preserving nutritional quality. Maintaining appropriate temperature and humidity during storage is crucial for nutrient preservation, quality retention, and shelf life extension. Raw makhana (Swarna Vaidehi) seeds have a commendable nutritional profile, with carbohydrates comprising the majority (57%), followed by amylose (19%), and protein (7.2%). The raw seeds exhibit comparatively low levels of fat, total ash, and crude fibre, measured at 0.3%, 0.3%, and 0.5%, respectively. The nutritional and energy value of makhana seeds is enhanced by their carbohydrate content along with moderate levels of protein. Raw makhana seeds have a calorific content of about 259 kcal/100 g, which emphasises their significance as a potentially high energy dietary source (Kumar et al., 2016). The makhana seeds also provide a variety of essential macro and micronutrients that enhance their nutritional quality. Among the minerals, phosphorus is available at the highest level (66.1 mg/100 g), followed by sodium (48.2 mg/100 g) and potassium (35.6 mg/100 g). The seeds are composed of significant levels of magnesium (11.3 mg/100 g) and calcium (9.5 mg/100 g), in addition to minor amounts of iron (0.8 mg/100 g), manganese (0.9 mg/100 g), zinc (0.9 mg/100 g), and copper (0.3 mg/100 g). The presence of these minerals increases the nutritional value of makhana seeds and reinforces their function as a nutrient-dense food component (Kumar et al., 2016).

A comprehensive understanding of the physical, frictional, and nutritional characteristics of raw makhana seeds facilitates the development of effective cleaning, drying, grading, and storage methods, thereby minimizing losses, enhancing processing efficiency, and improving the quality of the final product.

5.9 Microstructure of raw seed

The makhana kernel, in its raw form, displays a dense and heterogeneous microstructure. Devi et al. (2020) demonstrate through scanning electron micrography the existence of well-defined crystalline starch granules that are loosely arranged within the endosperm matrix. The crystalline particles are predominantly made up of regions rich in amylopectin, organized in a structured lattice, which is a defining feature of native starch. The crystallinity is a result of the double-helical configuration of amylopectin chains, which plays a significant role in imparting rigidity and structural stability to the raw kernel. The presence of void spaces in the raw makhana seed is significant during subsequent thermal processing, as these voids serve as initial sites for moisture migration and vapor pressure buildup. During the initial roasting process, the swelling and gelatinization of starch took place, resulting in the filling of these voids. Upon roasting, the conversion of the crystalline amylopectin structure into an amorphous state takes place resulting in paste-like consistency in the surface of the roasted kernel. Alongside starch granules, the raw kernel encompasses embedded protein bodies. The intact seed coat encasing the kernel serves to further limit moisture loss and uphold internal moisture balance (Devi et al., 2020). The moisture content present in raw makhana is generally adequate to enable starch gelatinization when subjected to heat; however, in its raw form, the water molecules are securely bound within the crystalline structures of starch. The limited mobility of water plays a significant role in determining the firmness and hardness of the raw kernel.

The raw makhana kernel can be characterized as a semicrystalline structure, consisting of loosely organized starch granules, interspersed with voids and containing embedded protein bodies. The existence of crystalline starch regions, air spaces, and bound moisture forms the structural basis that subsequently influences the swelling, gelatinization, and expansion characteristics noted during roasting and popping.

5.10 Future prospects

The future of raw makhana seed processing is promising, driven by growing demand for makhana as a functional, gluten-free, and low-fat food product. Improvements in mechanized and automated cleaning, grading, and drying systems are anticipated to diminish labor reliance, enhance throughput, and guarantee uniform quality. The implementation of sensor-based graders, optical sorters, and automated size categorization can improve precision and reduce postharvest losses.

Innovative drying and pretreatment technologies, including fluidized bed drying, microwave-assisted and ultrasound drying, have been explored in several studies, revealing benefits such as decreased drying duration, enhanced moisture uniformity, and superior quality preservation of makhana seeds. Based on these findings, additional investigation into infrared drying, hybrid drying systems (microwave–hot air or infrared–fluidized bed), and controlled or modified atmosphere drying may be conducted to improve energy efficiency and maintain nutritional and functional properties. These methods can facilitate rapid and uniform moisture removal, reduce quality deterioration, and enhance storage stability by ensuring safe and consistent moisture levels across various grades of raw makhana seeds.

The use of digital monitoring and intelligent storage systems, featuring real-time sensors for moisture, temperature, and humidity, will enhance store management and minimize spoilage. Enhanced packaging methods, including moisture-barrier and modified atmosphere packaging, can significantly prolong the shelf life of raw makhana seeds.

From a value-addition aspect, the processing of raw makhana seeds into various products, including popped makhana, flours, protein-enriched components, and ready-to-eat formulations, creates new market potential. The incorporation of by-products and damaged seeds in nutraceuticals and functional foods enhances both sustainability and profitability.

Future prospects depend on the modernization of processing chains, the adoption of energy-efficient technologies, the standardization of quality, and the integration of value-added product development, collectively enhancing the economic viability and global competitiveness of the makhana processing sector.

5.11 Conclusions

The efficient processing of raw makhana seeds is essential for guaranteeing quality, safety, and storage stability before popping and value enhancement. Thorough cleaning, drying, and grading, influenced by knowledge of the physical, frictional, and hygroscopic characteristics of the seeds, markedly diminish postharvest losses and enhance process uniformity. Among these operations, optimizing drying process through controlled temperature, airflow, or hybrid energy systems is essential for achieving uniform moisture reduction, while also maintaining structural integrity and minimizing energy consumption. The implementation of advanced grading technologies, such as mechanized and sensor-driven systems, has the potential to enhance consistency, reduce reliance on manual labor, and elevate market standards. Achieving and maintaining optimal moisture levels is essential to prevent microbiological spoilage and deterioration of quality during storage. Enhanced storage methodologies incorporating moisture and humidity regulation significantly contribute to the mitigation of spoilage and the preservation of quality integrity. The standardized and meticulously regulated processing of raw makhana seeds establishes the basis for prolonged shelf life, increased processing efficiency, and uniform quality of subsequent makhana products.

References

Bakshi, N., Jain, S., Raman, A., Pant, T., 2023. Microwave: An Overview. Elsevier BV, pp. 19–59. https://doi.org/10.1016/b978-0-323-95991-9.00014-x.

Chandrasekaran, S., Ramanathan, S., Basak, T., 2013. Microwave food processing—a review. Food Res. Int. 52 (1), 243–261. https://doi.org/10.1016/j.foodres.2013.02.033.

Chauhan, D.S., Lal, A.B., Charles, A.P.R., Singh, A.P., Khare, A., Vashisht, P., 2023. Preparation of gluten free cookie using chestnut and foxnut flour blend: composition optimization through response surface methodology. Asian J. Dairy Food Res. 44 (1), 84–91. https://doi.org/10.18805/ajdfr.dr-2178.

Chen, F., 2010. Studies on the Microwave Vacuum Drying of Lotus-Seed. MS Thesis. Fujian Agriculture and Forestry University.

Conyers, S.T., Bell, C.H., 2003. The effect of modified atmospheres on the survival of the eggs of four storage mite species. Exp. Appl. Acarol. 31 (1–2), 115–130. https://doi.org/10.1023/B:APPA.0000005110.19969.23.

Dehghannya, J., Habibi-Ghods, M., 2025. Microwave drying of food materials: principles, hybrid techniques, modeling approaches, and emerging innovations. Compr. Rev. Food Sci. Food Saf. 24 (6). https://doi.org/10.1111/1541-4337.70312, 1541-4337.

Devi, M.B., Deka, S.C., 2022. Physicochemical properties and structure of starches of foxnut (Euryale ferox Salisb.) from India and its application. J. Food Process. Preserv. 46 (2). https://doi.org/10.1111/jfpp.16262.

Devi, M., Sharma, K., Jha, S., Arora, S., Patel, S., Kumar, Y., Vishwakarma, R., 2020. Effect of popping on physicochemical, technological, antioxidant, and microstructural properties of makhana seed. J. Food Process. Preserv. 44 (10). https://doi.org/10.1111/jfpp.14787, 0145-8892.

FAO, 2011. Guide to Seed Handling in Genebanks. Food and Agriculture Organization of the United Nations, Rome, Italy.

Franco, A.P., Yamamoto, L.Y., Tadini, C.C., Gut, J.A.W., 2015. Dielectric properties of green coconut water relevant to microwave processing: effect of temperature and field frequency. J. Food Eng. 155, 69–78. https://doi.org/10.1016/j.jfoodeng.2015.01.011.

Gao, J., Wu, M., Du, S., Zhang, H., Wang, S., Ling, B., 2023. Recent advances in food processing by radio frequency heating techniques: a review of equipment aspects. J. Food Eng. 357, 111609. https://doi.org/10.1016/j.jfoodeng.2023.111609.

Gaurav, A., Nath, D., Barbhuiya, R.I., Pradhan, R.C., 2022. Physical characterization and mass modeling of traditionally popped makhana (*Euryale ferox Salisbury*) variety: Sabour-1 makhana. J. Food Process. Eng. 45 (6). https://doi.org/10.1111/jfpe.13927.

Guru, P.N., Mridula, D., Dukare, A.S., Ghodki, B.M., Paschapur, A.U., Samal, I., Nikhil Raj, M., Padala, V.K., Rajashekhar, M., Subbanna, A.R.N.S., 2022. A comprehensive review on advances in storage pest management: Current scenario and future prospects. Front. Sustain. Food Syst. 6. https://doi.org/10.3389/fsufs.2022.993341.

Huang, D., Men, K., Li, D., Wen, T., Gong, Z., Sunden, B., Wu, Z., 2020. Application of ultrasound technology in the drying of food products. Ultrason. Sonochem. 63, 104950. https://doi.org/10.1016/j.ultsonch.2019.104950.

Islam, Z., Bashir, K., Singh, S., Rastogi, M., Siddiqui, S., Punia Bangar, S., 2025. Structural and functional properties of starch from popped seeds of makhana (*Euryale ferox*). Starch - Stärke 77 (7). https://doi.org/10.1002/star.70030, 0038–9056.

Jalaj, R.K., Jalaj, V.K., Kumar, P., SMS KVK, Araria, M.S.c. (PB & G). CAU, Imphal, & SMS (Agromet) KVK, Araria, 2019. Traditional methods of harvesting and processing of makhana (*Euryale ferox* Salisb.) adopted in Araria District of Bihar. Int. J. Curr. Microbiol. Appl. Sci. 321–326. www.ijcmas.com/special/9/R.%20K.%20Jalaj,%20et%20al.pdf.

Jana, B.R., Srivastava, A., Idris, M., 2019. New makhana (*Euryale ferox*Salisb.) processed products for health benefit. J. Food Process. Technol. 10 (4), 789. https://doi.org/10.4172/2157-7110.1000789.

Jha, S.N., 1999. Physical and hygroscopic properties of makhana. J. Agric. Eng. Res. 72 (2), 145–150. https://doi.org/10.1006/jaer.1998.0356.

Jha, S.N., Prasad, S., 1993. Physical and thermal properties of gorgon nut. J. Food Process. Eng. 16 (3), 237–245. https://doi.org/10.1111/j.1745-4530.1993.tb00319.x.

Jha, S.N., Prasad, S., 1994. Gorgon fruit or makhana its cultivation and processing. Indian Hortic. 39 (2), 18–20.

Jha, V., Barat, G.K., Jha, U.N., 1991. Nutritional evaluation of euryale ferox salisb. makhana. J. Food Sci. Technol. 28 (5), 326–328.

Khadatkar, A., Gite, L.P., Gupta, V.K., 2015. Interventions to reduce drudgery of workers in the traditional method of harvesting Makhana (*Euryale ferox* salisb.) seeds from ponds. Curr. Sci. 109 (7), 1332–1337.

Khadatkar, A., Mehta, C.R., Gite, L.P., 2020. Makhana (*Euryle ferox* Salisb.): a high-valued aquatic food crop with emphasis on its agronomic management – a review. Sci. Hortic. 261, 108995. https://doi.org/10.1016/j.scienta.2019.108995, 03044238.

Khatatkar, A., Gupta, V.K., Bakudkar, H., Narwaria, B.S., Sharma, S., 2014. Drudery involved in traditional way of harvesting makhana seeds (*Euryle ferox* salisb) from ponds. In: Bhartiya Krishi Anusandhan Patrika, 29, pp. 25–30.

Kidmose, U., Martens, H.J., 1999. Changes in texture, microstructure and nutritional quality of carrot slices during blanching and freezing. J. Sci. Food Agric. 79 (12), 1747–1753. https://doi.org/10.1002/(SICI)1097-0010(199909)79:12ESCAPED_CHARACTER_LESS_THAN1747::AID-JSFA429>3.0.CO;2-B.

Kumar, S., Prakash Kumar, S., 2015. Effect of different grades on physical and mechanical properties of popped makhana. Int. J. Process. Post Harvest Technol. 6 (1), 101–108. https://doi.org/10.15740/HAS/IJPPHT/6.1/101-108, 09765638.

Kumar, S., Shukla, S., 2014. Concepts and applications of microwave engineering. PHI Learning Pvt. Ltd.

Kumar, U., Kumar, A., Singh, K.M., 2011. Constraints and drudgery in makhana cultivation. SSRN Electron. J. https://doi.org/10.2139/ssrn.2061652.

Kumar, L., Gupta, V.K., Jha, B.K., Singh, I.S., Bhatt, B.P., Singh, A.K., 2011. Status of Makhana Cultivation in India. Technical Bulletin no.: R-32/PAT-21, ICAR Research Complex for Eastern Region.

Kumar, L., Singh, A.K., Bhatt, B.P., 2016. Nutritional status of recently developed makhana (gorgon nut) variety "Swarna Vaidehi". J. AgriSearch 3 (4). https://doi.org/10.21921/jas.v3i4.6701, 2348-8867.

Kumar, N., Rani, S., Kuamr, G., Kumari, S., Singh, I.S., Gautam, S., Choudhary, B.K., 2019. Physiological and biochemical responses of Makhana (*Euryale ferox*) to gamma irradiation, 15730689 J. Biol. Phys. 45 (1), 1–12. https://doi.org/10.1007/s10867-018-9511-x.

Lien, D.T.P., 2020. Influence of different drying techniques on drying characteristics and quality aspects of pink Lotus (*Nelumbo nucifera*) flowers. J. Adv. Res. Food Sci. Nutr. 03 (01), 16–23. https://doi.org/10.24321/2582.3892.202003.

Lu, X., Zheng, Z., Huang, L., Cao, R., Zheng, Y., Yu, H., Xiao, J., Miao, S., Zheng, B., 2017. Optimization of ultrasonic-microwave assisted extraction of oligosaccharides from lotus (*Nelumbo nucifera* Gaertn.) seeds. Ind. Crop. Prod. 107, 546–557. https://doi.org/10.1016/j.indcrop.2017.05.060.

Mandal, R.N., Saha, G.S., Sarangi, N., 2010. Harvest and processing of Makhana (*Euryale ferox* Salisb.) - an unique assemblage of traditional knowledge, 09725938 Indian J. Tradit. Knowl. 9 (4), 684–688.

Mani, A., 2019. Post harvest technology, processing and value addition of makhana. Agrobios Online 18 (2), 55–59.

Manju, K.M., Rekha, P., Kumar, N., 2021. Effect of fluidized-bed and freeze-drying techniques on physicochemical, nutritional, thermal, and structural properties of Moringa oleifera flowers, leaves, and seeds. J. Food Process. Preserv. 45 (9). https://doi.org/10.1111/jfpp.15719.

Mounika, E., Sivamma, P., Snehitha, R., 2021. Atomization techniques in spray drying: a Review. Pharm. Innov. 10 (5), 454–461.

Navarro, S., 2012. The use of modified and controlled atmospheres for the disinfestation of stored products. J. Pest. Sci. 85 (3), 301–322. https://doi.org/10.1007/s10340-012-0424-3.

Navarro, S., Adler, C., Riudavets, J., Stejskal, V., 2004. Working group "integrated protection of stored products.". In: Proceedings of the Meeting CompteRendu de la Réunion, IOBC WPRS BULLETIN, vol 27 (Czechia).

Nguyen, T.P., Songsermpong, S., 2022. Microwave processing technology for food safety and quality: a review. Agric. Nat. Resour. 56 (1), 57–72 retrieved from. https://li01.tci-thaijo.org/index.php/anres/article/view/253973.

Orsat, V., Raghavan, G.S.V., Krishnaswamy, K., 2017. The microwave processing of foods, microwave technology for food processing: an Overview of current and future applications. In: The Microwave Processing of Foods, second ed. Woodhead Publishing, pp. 100–116.

Ramachandra, C.T., Ashoka, H.G., Mahesh Kumar, G., Shivanna, B., Ray, B.R.M., Sivamma, P., 2021. Design and development of environmental friendly sub-baric storage bin. Int. J. Environ. Clim. Change 19–28. https://doi.org/10.9734/ijecc/2021/v11i1030488.

Rout, R.K., Kumar, A., Sivamma, P., Prakash, R., Kukde, R.B., Misra, S., Rao, P.S., Naik, R., Ravindra, R.M., Murthy, G.R.K., Avinashilingam, N.A.V., Yashavanth, B.S., Srinivasa Rao, C., 2024. Role of agro-processing towards climate resilience in agriculture. In: Research and Technological Advances for Resilient Agriculture. ICAR-NAARM, pp. 205–244.

Sharma, K., Vishwakarma, R.K., Patel, S., Jha, S.N., Devi, M., 2023. Organic Production, Post-harvest Processing, and Value-Added Intervention of Fox Nuts (*Euryale Ferox*) for Export Transforming Organic Agri-Produce into Processed Food Products: Post-COVID-19 Challenges and Opportunities. Apple Academic Press, India, pp. 363–393. Available from: https://www.appleacademicpress.com/transforming-organic-agri-produce-into-processed-food-products-post-covid-19-challenges-and-opportunities/9781774911921.

Sharma, K., Patel, S., Jha, S.N., Devi, M., Nandan, S., Vishwakarma, R.K., 2024a. Aging-induced changes in popping, structural, functional, thermal, and crystalline properties of roasted makhana seeds. Food Bioprocess Technol. 17 (12), 5023–5037. https://doi.org/10.1007/s11947-024-03428-7.

Sharma, K., Patel, S., Jha, S.N., Devi, M., Nandan, S., Vishwakarma, R.K., 2024b. Change in popping, structural, functional, and crystalline properties of roasted makhana seeds in prolong storage: an FTIR, GC-MS, SEM, XRD study. Research Square. https://doi.org/10.21203/rs.3.rs-3910663/v1.

Singh, D.K., Kumar, A., Kumar, U., Chandra, N., Singh, I.S., 2021b. Value addition in makhana through post-harvest operations. Food Sci. Rep. 2 (2), 29–32.

Singh, R., Viswakarma, R.K., Jha, S.N., 2021b. Development and commissioning of pilot-plant for makhana processing. Int. J. Food Sci. Nutr. 6 (4), 91–94.

Sivamma, P., Mounika, E., Carolin Rathinakumari, K., Senthil Kumaran, G., Bindu, B., 2021a. Comparative study between modified solar dryer and open sun drying for drying of onion slices. Curr. J. Appl. Sci. Technol. 23–29. https://doi.org/10.9734/cjast/2021/v40i731324.

Sivamma, P., Mounika, E., Carolin Rathinakumari, K., Senthil Kumaran, G., Bindu, B., 2021b. Development of heat and heat retention in the solar green house dryer. Curr. J. Appl. Sci. Technol. 51–62. https://doi.org/10.9734/cjast/2021/v40i731329.

Sivamma, P., Mounika, E., Carolin Rathina Kumari, K., Senthil Kumaran, G., Bindu, B., 2021c. Drying white onion slices in a modified solar greenhouse dryer using aluminum foil and a black mulch sheet. Pharm. Innov. 10 (4), 1156–1160.

Sivamma, P., Mounika, E., Naga Hari Sairam, N., Jagannadha Rao, P.V.K., 2021d. Applications of vacuum technology in food processing. Pharm. Innov. 10, 914–918.

Sivamma, P., Kumar, G.M., Ashoka, H.G., Ramachandra, C.T., Shivanna, B., Ray, B.R., 2022. Fuzzy logic for sensory evaluation of rice stored in sub-baric storage Bin. Indian J. Ecol. 49 (6), 2377–2383. https://doi.org/10.55362/IJE/2022/3836.

Sumnu, G., Sahin, S., 2005. Recent Developments in Microwave Heating. Elsevier BV, pp. 419–444. https://doi.org/10.1016/b978-012676757-5/50018-9.

Sun, Z., Chen, J., Dai, T., Lv, C., Liang, R., Liu, W., Liu, C., Deng, L., 2024. Effect of maturity on the drying characteristics of lotus seed and molecular structure, gelation and digestive properties of its starch. Carbohydr. Polym. 345, 122589. https://doi.org/10.1016/j.carbpol.2024.122589.

Sundaram, P.K., Sarkar, B., Mondal, S., 2014. Design and performance evaluation of pedal operated Makhana (*Euryale ferox* Salisb) seed grader. Res. J. Agricul. Sci. 5 (3), 428–443.

Tian, Y., Liang, J., Zeng, H., Zheng, B., 2013. Microwave drying characteristics and kinetics of lotus (*Nelumbo nucifera* Gaertn.) seeds. Int. J. Food Eng. 9 (1), 91–98. https://doi.org/10.1515/ijfe-2012-0161.

Tutuncu, S., Emekci, M., 2019. Comparative efficacy of modified atmospheres enriched with carbon dioxide against Cadra (= Ephestia) cautella. J. Sci. Food Agric. 99 (13), 5962–5968. https://doi.org/10.1002/jsfa.9871.

Verma, A.M., Pgdif, R.N., Datta, S., 2007. Integration of Carp Culture with Makhana (*Euryale Ferox* Salisb.)–A Route to Crop Diversification. Central Institute of Fisheries Education, Kolkata-91, India.

Verma, D.K., Kumar Mahanti, N., Thakur, M., Kumar Chakraborty, S., Prakash Srivastav, P., Heating, M., 2020. Alternative Thermal Process Technology for Food Application. Informa UK Limited, pp. 25–67. https://doi.org/10.1201/9780429297335-2.

von Horstig, M.W., Schoo, A., Loellhoeffel, T., Mayer, J.K., Kwade, A., 2022. A perspective on innovative drying methods for energy-efficient solvent-based production of lithium-ion battery electrodes, 21944296 Energy Technol. 10 (12). https://doi.org/10.1002/ente.202200689.

Wang, Y., Wig, T.D., Tang, J., Hallberg, L.M., 2003. Dielectric properties of foods relevant to RF and microwave pasteurization and sterilization. J. Food Eng. 57 (3), 257–268. https://doi.org/10.1016/S0260-8774(02)00306-0.

Wiktor, A., Dadan, M., Nowacka, M., Rybak, K., Witrowa-Rajchert, D., 2019. The impact of combination of pulsed electric field and ultrasound treatment on air drying kinetics and quality of carrot tissue. LWT 110, 71–79. https://doi.org/10.1016/j.lwt.2019.04.060.

Zhao, Y., Wang, W., Zheng, B., Miao, S., Tian, Y., 2017. Mathematical modeling and influence of ultrasonic pretreatment on microwave vacuum drying kinetics of lotus (*Nelumbo nucifera* Gaertn.) seeds. Dry. Technol. 35 (5), 553–563. https://doi.org/10.1080/07373937.2016.1193512.

Zhao, Y., Jiang, Y., Zheng, Y., Li, Z., Zhang, Y., Zheng, B., Lo, Y.M., Miao, S., Tian, Y., 2018. Effects of microwave vacuum drying on macroscopic properties and microstructure of Lotus (*Nelumbo nucifera* Gaertn.) seeds. Int. J. Food Eng. 14 (3). https://doi.org/10.1515/ijfe-2017-0313.

Zhao, Y., Zhu, H., Xu, J., Zhuang, W., Zheng, B., Martin Lo, Y., Huang, Z., Tian, Y., 2021. Microwave vacuum drying of lotus (*Nelumbo nucifera* Gaertn.) seeds: effects of ultrasonic pretreatment on color, antioxidant activity, and rehydration capacity. LWT 149, 111603. https://doi.org/10.1016/j.lwt.2021.111603.

Chapter 6

Traditional and modern processing techniques for *Makhana* popping

Mathangi Raja Sekhar[1], Pakkiranna Sivamma[2], Naveen Kumar Mahanti[3], Mounika Edde[4], Bogala Pravallika[5], Om Prakash[1] and H.L. Kushwaha[1]

[1]*ICAR-Central Arid Zone Research Institute, Jodhpur, Rajasthan, India;* [2]*ICAR-National Institute of Seed Science & Technology, Mau, Uttar Pradesh, India;* [3]*Post-harvest Technology Research Station, Dr. YSR Horticultural University, Venkataramannagudem, Andhra Pradesh, India;* [4]*Sri Krishnadevaraya College of Agricultural Sciences, Anantapur, Andhra Pradesh, India;* [5]*Department of Processing and Food Engineering, Dr. NTR College of Agricultural Engineering, Bapatla, Andhra Pradesh, India*

6.1 Introduction

Makhana, Phool Makhana, Flower *Makhana,* and Popped *Makhana* are popularly named in hindi language of India for the fox nut (Scientific name: *Euryale ferox*; Family: Nymphaeaceae) is also known as Gorgon nut. Fox nut is an aquatic plant, native of Southern and Eastern Asia. India, especially northern part of Bihar is the largest producer of World. Other Indian states viz. West Bengal, Assam, Manipur Tripura, Odissa also produces Fox nuts. Followed by India, China, Nepal, Bangladesh, Thailand, Japan, Korea and Russia are the producers of fox nut in the world. Fox nut is considered as *sattvic* food (nourishing, balancing and pure food) in India. Apart from food, it has been preferred during festivals as diet of religious fast, offering for purity and devoid of neither over stimulating nor under nutritious properties. Fox nut cultivation and processing has been the new employment and livelihoods opportunity for thousands of rural households in its producing states of India. The tedious process is associated with harvesting and processing involving various unit operations such as harvesting, fruit breaking, washing, drying, grading, conditioning, roasting, popping, winnowing, cooling, polishing, grading, packaging and storage. Besides the long chain of unit operations, fox nut has emerged as a high value agro-industrial crop due to its strong

Makhana (Euryale ferox) as a Functional Food. https://doi.org/10.1016/B978-0-443-44421-0.00005-6

nutritional profile, increased consumer preference and high scope for mechanization and value addition especially in India. Fox nut is traditionally limited to small-scale and household processing, which is now evolving into a commercially viable, export-oriented industry, presenting significant opportunities in cultivation, processing, product development, and related sectors. Around 60 to 63 thousand Metric Tonnes of *Makhana* produced in 35 to 40-thousand-hectare pond area as per Production Estimates of APEDA during 2025 (APEDA, 2025). Whereas the global market for Makhana has been increasing from 150.2 USD in 2024 to 265.88 USD by 2031 with a CAGR (Compound Annual Growth Rate) of 8.50% (Cognitive Market Research, 2025). The value chain or unit operations of makhana processing demands intensive labor and skill. Roasting and Popping are crucial stages of processing that determines ultimate quality of popped makhana. The time and temperature of roasting and prompt force application on roasted seed is key to maximize the quality and quantity of the popped makhana. This chapter discusses various aspects of makhana popping and its value chain including physicochemical properties, traditional and modern methods of popping, advancements in popping technologies, pre and post popping operations, recent mechanization, food safety and quality standards and energy and its impact on environment. It will bring out all the processing aspects emerging in makhana production under limelight.

Need for scientific understanding of processing techniques
Processing of makhana (*Euryale ferox*), especially roasting and popping, has traditionally been a manual, labor-intensive, and skill-dependent task. The conventional method of processing, which entails roasting in open pans and manually pounding the seeds with a wooden mallet to induce popping, is both labor-intensive and time-consuming. These processes are susceptible to inconsistent outcomes due to their dependence on human expertise in managing temperature and timing, which directly influences product quality (e.g., size, expansion, and surface properties). The process additionally subjects workers to safety hazards and significant toil resulting from extended manipulation of heated materials. This has been recognized as a limitation of traditional practices, emphasizing the need for controlled scientific approaches to improve worker safety and ensure product consistency.

The production and quality of popped kernels are greatly influenced by systemic process design, which includes pre-processing steps including drying, roasting, and tempering. For example, a series of drying, roasting, tempering, and second roasting produced 97% grade-1 makhana with shorter processing times than conventional procedures, according to Jha and Prasad (1996) controlled experimentation on processing combinations. Comprehending the physical and hygroscopic characteristics of makhana seeds is crucial for the development of mechanised and optimized processing machinery. Physical attributes such as moisture content, density, heat transfer properties, and expansion behavior directly affect roasting and popping efficacy. Research within this domain establishes the scientific foundation for machine design and process parameter determination, which cannot be solely inferred from empirical traditional knowledge (Jha, 1999). Even after popping, makhana milling and conditioning behavior has scientific implications for product development and value addition (powders, flours). Studies on how moisture content and feed rate affect particle size distribution during milling reveal that a scientific understanding of material behavior is essential for decreasing product waste, improving milling efficiency, and assuring product quality (Jha and Sharma, 2010).

6.2 Raw material characteristics

Makhana (*Euryale ferox*) seed moisture content, size, density, physical shape, and engineering features affect popping performance, including expansion, popping efficiency, and product quality. These material qualities must be scientifically understood to optimize traditional and mechanised popping procedures.

6.2.1 Physicochemical properties of makhana

6.2.1.1 Moisture content

The popping process is driven by seed moisture. Heating rapidly evaporates interior water, causing internal pressure to open up the hard shell and expand the starchy kernel with popping. The expansion ratio, popped kernel recovery, and popping efficiency depend on raw seed moisture. Research on the physical properties of *makhana* indicates that variations in moisture content substantially influence engineering and physical properties such as density, size, and flow behavior, which are indirectly related to the seeds' responses during bursting and heating processes. For instance, an increase in moisture content elevates bulk and true densities as well as the angle of repose, while decreasing porosity, thereby demonstrating that moisture influences seed structure and heat transfer characteristics during bursting. This indicates that managing seed moisture content prior to roasting and bursting is essential for attaining optimal expansion and maintaining consistent quality (Jha, 1999). Furthermore, storage studies indicate that maintaining optimal moisture content in roasted seeds is essential for preserving popping efficiency and recovery, as seeds stored at approximately 15.5% dry basis moisture exhibited increased popped makhana recovery and expansion throughout the storage period (Sharma et al., 2024a, 2024b, 2024c, 2024d).

6.2.1.2 Kernel size & shell hardness

Seed size and grade directly impact popping behavior, as seed dimensions determine heat absorption and expansion capacity. Larger, uniform seeds are generally more likely to produce larger, completely expanded makhana, whereas smaller or irregular seeds may remain partially popped or result in reduced expansion. Engineering analyses of physical properties indicate that various grades (classified according to seed dimensions) exhibit significant variations in physical behaviors, including bulk density, true density, and frictional characteristics (Vikash Kumar et al., 2013). These variations suggest that size grade affects the mechanical interactions with processing equipment and, consequently, the behavior during heat transfer and pressure accumulation in popping.

6.2.1.3 Physical and hygroscopic properties

Fundamental physical properties such as bulk density, particle density, porosity, angle of repose, and friction coefficients govern the behavior of seeds within processing lines and influence the transmission of heat and moisture through the raw material. These properties also affect the rate and uniformity of heat penetration during roasting, prior to bursting. A study by Jha (1999) measured these properties across different moisture levels and found that bulk density varied approximately from 56.5 to 84.6 kg/m^3, whereas particle density ranged from approximately 105–240.6 kg/m^3. Porosity reduced and friction increased with increasing moisture content, suggesting that material behavior is altered by moisture conditioning, which subsequently influences heat transfer and popping behavior. These physical characteristics data are crucial for the design of roasting and popping equipment capable of producing homogeneous thermal conditions while controlling material flow effectively without segregation or clogging.

The hygroscopic characteristic of the gorgon nut plays a crucial role in both traditional and mechanised processing of its popped form, Makhana, as it directly influences the effective transformation of the dense kernel into a light, edible puff. This process fundamentally depends on regulated internal steam pressure (Jha et al., 1991; Jha and Prasad, 1996). Popping occurs when a sufficient quantity of moisture within the starchy endosperm is swiftly heated to generate superheated water vapor, whose pressure overcomes the structural integrity of the hard seed coat, resulting in a rapid expansion (Jha and Prasad, 1996). The raw, harvested gorgon nut exhibits a high initial moisture content, which must be meticulously reduced through drying and first-stage roasting to achieve the optimal conditions for popping. Studies has demonstrated that the ideal moisture content for effective popping generally ranges between 11% and 16% (dry basis), as moisture levels below this range result in insufficient steam generation, while higher levels restrict the necessary pressure buildup (Sharma et al., 2024a, 2024b, 2024c, 2024d; Jha and Prasad, 1996). Crucially, the nut's hygroscopic nature requires the tempering process, a storage period (typically 48–72 h) following its first roasting, during which the retained moisture redistributes evenly throughout the kernel and shell. This step ensures that the kernel is properly conditioned for simultaneous gelatinization and explosive rupture during the final high-temperature roasting and subsequent mechanical impact (Jha and Prasad, 1993). Therefore, the capacity of the gorgon nut to absorb, retain, and redistribute moisture is utilized specifically to control the internal thermodynamic conditions, rendering the hygroscopic property the fundamental mechanism determining the ultimate quality, expansion ratio, and commercial recovery rate of Makhana.

6.2.1.4 Role of washing, drying, grading

Gorgon nut popping and quality depend on washing, drying, and grading. Washing eliminates adherent silt, organic impurities, and microbial contamination, ensuring consistent heat transfer during roasting and minimizing contamination problems. Controlled drying is essential for optimal seed moisture content because moisture affects internal steam pressure needed for shell rupture and kernel expansion. Poor drying produces poor popping performance or shell breaking. Grading by seed size and uniformity ensures constant heating and pressure development since larger seeds pop and expand more than smaller or mixed grades. Standardised pre-processing is essential for effective popping since proper dehydrating and grading combinations increase makhana yield and quality (Jha and Prasad, 1996; Jha, 1999).

6.2.1.5 Chemical composition

The proximate composition of the gorgon nut kernel critically influences the popping efficiency and ultimate quality of Makhana, a process inherently dependent on the kernel's capacity to function as a high-pressure steam vessel. The carbohydrate (starch) content, accounting for over 75% of the kernel, is the principal structural component that experiences gelatinization and volumetric expansion (Jha and Prasad, 1996). However, the most crucial proximate factor is moisture content, which must be carefully conditioned within a narrow optimal range (typically 11%–16% dry basis) following drying and tempering; this precise moisture level is vital for producing the required superheated steam pressure necessary to rupture the hard shell and attain maximum expansion (Jha and Prasad, 1993; Sharma et al., 2024a, 2024b, 2024c, 2024d). Furthermore, the gorgon nut's elevated protein content (ranging from 9% to 16%) and extremely low fat content (approximately 0.1%–0.5%) are beneficial for popping. The minimal fat content is particularly desirable as fat can prematurely plasticize and impair the strength of the starch-protein matrix. Conversely, the low fat level contributes in maintaining the structural integrity required for high-pressure resistance and consequent high expansion ratios (Jha et al., 1991). Consequently, the interaction of high starch content, precisely regulated moisture levels, and minimal fat content governs the thermodynamic and structural efficacy of the Makhana bursting process.

The pop properties of gorgon nut (Makhana) are profoundly affected by its starch composition, especially the ratio of amylose to amylopectin, a principle that works true for different popping cereals, though with structural differences. In cereals such as popcorn, a greater proportion of amylose (generally approximately 25% of total starch) is directly associated with a dense, vitreous endosperm structure. This dense matrix is essential as it withstands the substantial internal vapor pressure generated during rapid heating, ultimately resulting in an explosive rupture and significant expansion volume (Jha and Prasad, 1993, 1996). Similarly, rice and millet varieties optimized for puffing or bursting typically possess an intermediate amylose content, which maintains a balance between structural integrity (provided by amylose) and elasticity and swelling capacity (provided by amylopectin) (Sharma et al., 2024a, 2024b, 2024c, 2024d). The gorgon nut kernel, which exhibits highly variable amylose content depending on maturity (ranging from 7.63% to 23.73%), depends on this same structural basis; the complex, multi-stage popping process including the essential tempering step is deliberately designed to condition the starch matrix, thereby ensuring the appropriate molecular structures for optimal gelatinization and subsequent expansion (Sharma et al., 2024a, 2024b, 2024c, 2024d). Although the mechanical popping of *Makhana* is distinctive, necessitating the release from a hard shell following heat treatment, the fundamental mechanism of volumetric expansion is generally dictated by starch components: amylose imparts the structural integrity required to form the pressure vessel, while amylopectin supplies the essential polymer elasticity that results in the final puff's large, porous structure.

6.3 Traditional makhana processing & popping techniques

Popping is the process of generating superheated water vapor within the conditioned nut by heating its internal moisture, followed by an abrupt pressure release that results in a volumetric expansion of the kernel. The expanded interior of the nut produced through this process is referred to as *makhana* in India (Jha and Prasad, 1998). Nearly half of all makhana production still relies on traditional methods. This is because high-quality output depends heavily on human judgment; a skilled operator ensures superior results, whereas a lack of expertise leads to rapid quality deterioration.

6.3.1 Historical development of traditional processing

The traditional methods of processing *makhana*, especially roasting and popping, have developed over centuries through indigenous knowledge and practices, primarily in the wetland areas of eastern India. The process is predominantly manual, requiring significant skill and following a sequential order, involving sun or solar drying of raw seeds, followed by roasting in open iron pans or earthen furnaces, and concluding with hand popping using wooden mallets. Effective popping depends significantly on the operator's capacity to assess temperature, roasting time, and seed moisture content during roasting, as these factors influence internal pressure accumulation and shell rupture. Although conventional techniques yield *makhana* with favorable sensory qualities, they are marked by significant manual labor, low energy efficiency, variable product quality, and limited scalability. Studies have documented that the lack of controlled heating and standardised operating parameters leads to inconsistent popping recovery and expansion ratio, highlighting the importance of scientifically understanding and enhancing these traditional techniques while maintaining their cultural significance (Jha and Prasad, 1996; Jha, 1999).

6.3.2 Harvesting makhana

Makhana is traditionally collected using a labor-intensive manual method conducted in shallow ponds, reservoirs, oxbow lakes, and wetlands, primarily in Bihar and neighboring eastern Indian states. The harvesting process is not a single operation but is conducted in successive phases owing to the uneven flowering, varying fruit maturity, and different underwater disintegration behavior. Mature fruits form beneath the water surface, and upon senescence, the spiny pericarp progressively softens and decomposes, facilitating seed detachment and sinking to the pond bottom. However, entirely relying on natural disintegration is avoided, as extended submersion results in seed loss, deep burial within sediment, microbial degradation, and reduced popping quality. Therefore, the harvesting process is carried out in three intended phases. In the initial stage, fully matured yet intact fruits are tediously harvested by skilled fishermen working in waters ranging from 1 to 2 m in depth. This stage is the most strenuous and costly but produces seeds of superior quality with optimal popping and kernel whiteness. During the second stage, partially disintegrated fruits releases disintegrated seeds that are gathered from soft bottom sediment through manual collection or shallow diving. Seeds acquired at this stage exhibit moderate moisture content and maintain acceptable commercial quality. The third stage entails the retrieval of fully submerged seeds imbedded deeply within pond sediment following the complete disintegration of the fruit. Although this stage involves lower skill and cost, the seeds are subjected to extended exposure to water and sediment, leading to inadequate popping performance and lower market value. Overall, manual harvesting remains the most common approach in makhana harvesting owing to the absence of appropriate mechanization. However, it substantially increases production costs, intensifies laborious efforts, and introduces variations in final quality, thereby harvesting is the most crucial and challenging phase of the makhana value chain (Jha et al., 1991; ICAR-NRCM, 2019; Singh et al., 2020) (Tables 6.1 and 6.2).

TABLE 6.1 Stages of *makhana* harvesting and associated seed quality.

Harvest stage	Material harvested	Duration of water exposure	Popping quality	Market grade
Stage I	Intact mature fruits	Minimum	Excellent	Premium
Stage II	Loose seeds	Moderate	Good–Fair	Commercial
Stage III	Sunken seeds	Maximum	Poor	Low-grade

TABLE 6.2 Labor cost and difficulty level in manual *makhana* harvesting.

Harvest stage	Labor skill required	Approx. Labor cost (₹/ha)	Difficulty level	Physical risk
Stage I—Fruit harvest	Highly skilled	18,000–25,000	Very high	Very high
Stage II—Loose seed harvest	Semi-skilled	10,000–15,000	Moderate	Medium
Stage III—Sunken seed harvest	Low–semi skilled	5000–8000	Low-moderate	Low

6.3.3 Sun drying

In conventional makhana processing, sun drying is an essential post-harvest operation carried out right after seed extraction to decrease the higher initial moisture content of raw seeds (approximately 35%–45% wb) to a safe level suitable for storage and subsequent roasting and popping (ICAR-NRCM, 2019; Saha and Jha, 2018). The seeds are distributed in thin layers on bamboo mats, jute cloth, or earthen floors and are subjected to direct sunlight for 4–6 h daily over a period of 5–7 days, with regular manual rotation to promote uniform moisture removal (Jha et al., 1991; Singh et al., 2020). During the night, the partially desiccated seeds are piled or sheltered to prevent moisture absorption from condensation and to reduce thermal stress. This gradual and intermittent drying process promotes uniform hardening of the seed coat and stabilization of the internal starch–protein matrix, which are crucial for acquiring higher popping expansion, improved whiteness, and reduced breakage during roasting (Saha and Jha, 2018; Mishra et al., 2021). Inadequate or rapid drying often results in seed coat cracking, fungal infestation, and reduced popping quality, indicating sun drying as one of the most critical stages in traditional makhana processing in terms of quality assurance (Mujumdar, 2014).

6.3.4 Grading

Irregularly shaped makhana seeds are graded into different size fractions using a set of sieves with varying screen opening sizes. Grading promotes uniform heat transfer and a simultaneous response of crackling noise during roasting. Therefore, size grading is a prerequisite to achieve uniform roasting and to enable removal of all grains at the same time.

6.3.5 Roasting and popping

Traditional roasting, commonly referred to as the *kadhai* method, is the most critical phase in makhana processing, as it directly influences popping efficiency, its size, whiteness, and market grade. In this procedure, sun-dried raw makhana seeds (approximately 10%–12% moisture content, wet basis) are roasted in heavy gauge iron pans (*kadhai*) placed over a biomass-fueled or LPG stove. Iron pans are preferred due to their high thermal capacity and consistent heat distribution, which facilitate steady and controlled heat transfer to the seed (Jha et al., 1991; ICAR-NRCM, 2019). A small quantity of seeds (1–2 kg) is consistently agitated with a wooden spatula to avoid localized overheating and burning. Roasting is performed at elevated pan surface temperatures ranging from approximately 180–220°C for a brief period, until audible cracking noises signal internal pressure accumulation and shell loosening (Saha and Jha, 2018; Mishra et al., 2021). As per Jha and Prasad (1998), a roasting time ranging from 1.5 to 2.20 min is critical. Roasting beyond this duration causes rupture of makhana in the hot pan itself, which may pose a risk of injury to operators in close vicinity.

Following roasting, the hot seeds are promptly placed on a wooden block and manually broken open with a wooden mallet (Fig. 6.1). This effect induces a rapid release of superheated steam confined within the kernel, leading to abrupt expansion and rupture of the starch structure. In recent time mallet popping method replaced with impact popping machines as shown in Fig. 6.1. Proper roasting ensures consistent heat distribution, optimal moisture–temperature interaction, and sufficient hardening of the seed coat, all of which are crucial for achieving a high expansion ratio and excellent white popped makhana. Under-roasting results in insufficient popping and firm spots, while over-roasting leads to charring, excessive browning, and increased fracturing (Singh et al., 2020; Saha and Jha, 2018). Despite its huge labor requirements and dependence on operator expertise, the kadhai method continues to be used which is due to its low capital investment and capacity to yield high-quality popped makhana when executed with proper management.

6.4 Modern mechanical makhana popping

Makhana (*Euryale ferox*) popping is a critical post-harvest operation that transforms hard, inedible seeds into expanded, edible kernels through controlled roasting and sudden pressure release. Recognizing the severe

FIGURE 6.1 Traditional roasting by mud pot with mallet-based popping, open pan roasting, and recently adopted popping machine.

drudgery, low productivity, and safety concerns associated with traditional manual popping, ICAR-CIPHET, Ludhiana, developed a continuous mechanized makhana popping system that significantly modernized the process (Anonymous, 2015). The ICAR-CIPHET machine operates on the principle of controlled thermal conditioning followed by mechanical impact–induced popping. Raw wet seeds are fed through a hopper into a screw-conveyor-based roaster comprising two continuous barrels (each 1 m in length). One barrel is heated using a thermic oil system (3 kW), while the second employs electrical band heaters (3 × 3 kW). The roasting temperature is precisely maintained between 200 and 340°C using temperature controller-cum-indicators, and residence time is regulated by varying the screw speed between 10 and 50 RPM. After roasting, seeds undergo impact in a closed casing, resulting in shell rupture and kernel expansion. The system processes 35–40 kg h^{-1} of raw wet seeds, achieving about 95% decortication efficiency and 90%–94% popping efficiency. Constructed mainly from mild steel, the unit weighs

~250 kg, has dimensions of 2 × 1.5 × 2.5 m, and costs approximately ₹3.5 lakhs, making it suitable for small- and medium-scale enterprises (Fig. 6.2).

Complementary to the ICAR-CIPHET development, Kumar et al. (2019) designed a makhana popping machine to mechanize grading, roasting, and popping operations. Their system consists of a pre-heater-cum-roaster unit followed by an impeller-based popping chamber, where popping is induced by high-speed mechanical impact rather than manual beating. The roasting chamber operates at temperatures ranging from 230 to 310°C, using multiple electrical heaters (13 heaters of 500 W each), and the impeller speed is varied between 1800 and 2800 rpm. The optimum performance was achieved at an impeller speed of approximately 2600 rpm and a feed rate of about 10 kg h^{-1}, yielding popping efficiencies of 80%–85%. The impeller speed had a greater influence on popping efficiency than feed rate, while excessive speeds increased kernel breakage. The economic advantage of mechanization, reporting a profit margin of about ₹9.25 per kg of processed makhana, underscoring the socio-economic relevance of such machines in traditional makhana-growing regions.

Several commercial makhana popping and shell-removal machines are currently available in India, indicating steady progress toward mechanization of this traditionally manual operation. Blacknut Agrifood Machinery Private Limited, Ambala, Haryana, manufactures makhana popping machines with capacities in the range of 25–100 kg h^{-1} (Fig. 6.3), aimed at small- and

FIGURE 6.2 Makhana popping machine developed by ICAR-CIPHET, Ludhiana.

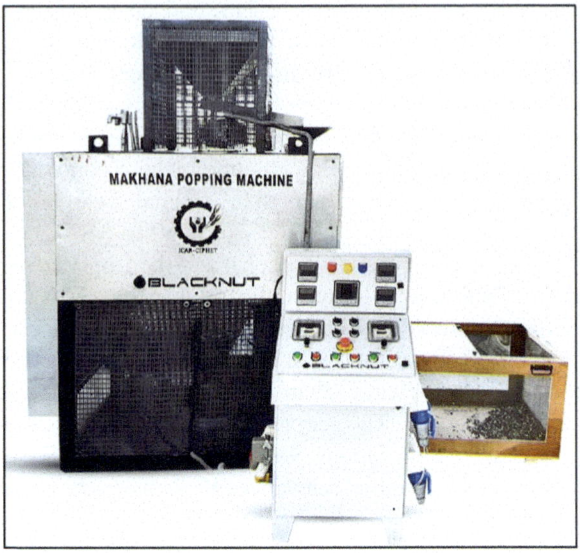

FIGURE 6.3 Makhana popping machine developed by Blacknut Agrifood Machinery Private Limited, Haryana, India.

medium-scale processors seeking higher throughput and uniform popping performance. These systems focus on improved thermal control and reduced labor involvement compared to indigenous roasting methods. In parallel, DIVY-OM Process Machinery & Equipment, New Delhi, offers makhana shell-removing and popping systems with capacities ranging from 40 to 200 kg h^{-1} made up of SS 304, employing an air and vibration assisted mechanism. The combined action of controlled airflow and vibration promotes efficient shell detachment while maintaining low kernel breakage (<5%) and high shell separation efficiency.

Additionally, Evershine Engineers and ARF Industries, Noida, India, has developed versatile makhana popping machines that emphasize flexibility in heating and process control (Fig. 6.4). These machines utilize LPG, natural gas, or electric heating systems to maintain consistent roasting temperatures required for optimal popping. A rotating drum or chamber design ensures uniform heat distribution and minimizes localized overheating or burning of seeds. The systems are equipped with adjustable temperature and time controls, allowing processors to tailor operating conditions for different seed sizes and moisture levels. They offers models covering a wide capacity range, from small-scale units (5–10 kg h^{-1}) to industrial-grade machines exceeding 100 kg h^{-1}. Fabrication using food-grade stainless steel enhances durability, hygiene, and ease of cleaning, while design features focusing on low noise and energy efficiency reduce operational costs. Collectively, these commercial

FIGURE 6.4 Makhana popping machine developed by AJW Industry, Evershine Industries and ARF Industries, Noida, India.

technologies highlight the diversification of heating mechanisms and mechanical designs in makhana popping, offering scalable, hygienic, and quality-oriented solutions for modern processing enterprises.

From a technological standpoint, all commercially available popping machines are attached with electrical resistance heating for roasting, followed by impact-based popping mechanisms. While these systems represent a major leap over indigenous manual methods, electrical coil or band-heater-based roasting is inherently limited by relatively slow heating rates, high energy consumption, and potential non-uniform temperature distribution within the seed mass. The future research should move toward rapid and advanced heating technologies, such as infrared (IR), microwave, and hybrid microwave-infrared systems, to further enhance makhana popping efficiency. Microwave heating can provide volumetric and rapid internal heating, facilitating uniform moisture vapourization and pressure build-up within the kernel, while infrared heating offers high surface heat flux and precise control over roasting intensity. A hybrid approach combining microwave pre-conditioning with infrared surface finishing could significantly reduce roasting time, improve energy efficiency, minimize kernel breakage, and enhance popping yield and product uniformity. Integrating such advanced heating systems with controlled impact mechanisms represents a promising direction for next-generation makhana processing technologies aimed at sustainability, scalability, and superior product quality.

6.5 Advanced processing technologies

6.5.1 Microwave-assisted popping

To overcome the constraints of traditional popping, electromagnetic waves, specifically microwaves, are employed, facilitating enhanced energy efficiency over a short duration (Vukadinović et al., 2024). Makhana seeds possess a dense structure that complicates their gelatinization during heating, hence influencing the release of nutrients and functional substances. The structural modifications of starch during thermal processing and its interactions with other macronutrients, including lipids and proteins, are the primary factors influencing the flavor and texture of processed foods. Furthermore, it influences the breakdown of starch in processed foods. The microwave offers several advantages over conventional methods, including minimum heating or cooking at elevated temperatures for reduced durations and the elimination of antinutritional components, leading to reduced nutritional degradation and superior quality final products (Li et al., 2022).

Microwaves are electromagnetic waves with frequencies ranging from 300 MHz to 300 GHz and wavelengths from 1 mm to 1 m, wherein heating occurs due to the absorption of microwaves by dielectric materials. Typically, industrial microwave ovens utilize a frequency of 915 MHz, while domestic microwave ovens operate at 2450 MHz (Sumnu and Sahin, 2005). Nonetheless, microwave power consumption has rapidly increased in the food sector due to its advantages over traditional methods, including precise control and quick startup and shutdown capabilities (Kidmose and Martens, 1999).

The primary components of microwave popping consist of a microwave source, a waveguide, and an applicator. The core component of the microwave oven is a magnetron, which comprises a vacuum tube including a central electron emitting cathode surrounded by an anode (Regier et al., 2010). When high voltage current flows through a magnetron, it is transformed into microwave energy, emitting high-frequency radiant energy. The polarity of radiant energy transitions from positive to negative at elevated frequencies. Magnetron power typically ranges from 300 to 3000 W; however, other magnetrons are employed in industrial microwaves to enhance power output. The heating chamber retains the product for heating using microwaves directed by the microwave guide (Verma et al., 2020).

6.5.1.1 Microwave heating mechanism

In microwave processing, the dielectric material produces heat by dipole rotation and ionic polarisation. Typically, polar molecules such as water are found in food substances. When subjected to an electric field, polar molecules in the food align according to the field's polarity. Subsequently, the water molecules undergo reorientation as a result of the continuous alterations in field polarity, resulting in friction between the molecules and the surrounding medium, therefore generating heat. The ionic polarisation approach involves collisions between ions, generating heat when food is exposed to an electric field due to the increased movement of charged ions (Verma et al., 2020).

Makhana seeds are typically subjected to treatment at 900 W for 90 s at a frequency of 2450 MHz during microwave processing (Li et al., 2022). In this method, heat is generated at the center of the product and transferred to the surface, a phenomenon that is uncommon in conventional heating methods. Microwave heating is associated with ease of cleaning, volumetric heating (the interaction of electromagnetic fields with food), and rapid processing that preserves quality, whereas traditional techniques experience prolonged processing times (due to slow thermal diffusion), which diminishes product quality. Microwave heating typically necessitates one-fourth the time to achieve processing temperature compared to conventional methods (Tewari, 2007).

Heat energy is generated during microwave application from electromagnetic energy by selective absorption and dissipation (Verma et al., 2020). The potential for microwave conversion to heat is influenced by the dielectric characteristics of food, which are the primary determinants of electromagnetic energy coupling and distribution. This can be expressed in terms of dielectric constant/relative permittivity and dielectric loss (Wang et al., 2003), where the dielectric constant pertains to a material's capacity to polarize and store electrical energy in reaction to an applied electric field, while dielectric loss concerns energy dissipation as heat (Verma et al., 2020). The dielectric properties are influenced by microwave frequency, moisture content, food density, and applied temperature. Li et al. (2022) reported following property alterations in makhana seed and flour upon microwave application.

6.5.1.1.1 Effects on physical and structural properties

Microwave treatment substantially alters the physical and structural characteristics of makhana seeds. The 1000 grain weight reduces from 309.48 g (raw) to 270.24 g postmicrowave heating, alongside a fall in bulk density (from 761.93 g/L to 752.93 g/L), signifying partial expansion and enhanced porosity. Furthermore, seed hardness diminishes markedly from 57.96 to 49.07 N, indicating the degradation of the compact endosperm structure. Microwave treated makhana seed has cracks, fissures, and a porous interior structure with partially fragmented starch granules, while raw makhana features a compact and smooth matrix. Microwave treatment preserves the A-type crystalline structure, accompanied by a minor decrease in peak intensity, signifying a partial drop in starch crystallinity. Microwave treatment alters the dense structure of makhana into a lighter, porous matrix, enhancing its processability and suitability for value-added food applications.

6.5.1.1.2 Influence on functional properties

Microwave treatment significantly enhances the functional characteristics of makhana seed flour by structural disruption and partial starch gelatinization. The water absorption capacity rises to 2.55 g/g from 1.43 g/g (raw), while the oil absorption capacity increases from 0.87 g/g to 0.93 g/g, signifying improved hydration and fat-binding capabilities. Moreover, solubility rises (from 2.82% to 6.10%), although swelling power reduces (from 9.16% to 7.44%), indicating a partial degradation of starch granules and reduced crystalline structure. Microwave treatment improves the functional efficacy of makhana seeds, rendering them more appropriate for inclusion in value-added and functional food items.

6.5.1.1.3 Changes in color

Microwave treatment induces noticeable color alterations, marked by a reduction in lightness (L*) and a substantial increase in yellowness (b*), probably attributable to minor Maillard reactions and pigment synthesis.

6.5.1.1.4 Effect on pasting properties

Microwave treatment significantly modifies the pasting characteristics of makhana. The maximum viscosity reduces from 763 cP (raw) to 309 cP, signifying partial fragmentation of starch granules and less swelling ability. Also, the final viscosity decreases from 1584 cP to 486.33 cP. The reduction in the viscosity profile are attributed to rigid, unfractured swelling granules, granule dimensions, and a decline in the proportion of amylose leached due to the intensity of heat treatment (Huang et al., 2016). These alterations indicate that microwave heating facilitates partial starch gelatinization and structural modification, enhancing functional properties in flour and dough applications.

6.5.1.1.5 Enhancement of bioactive compounds and antioxidant activity

Microwave heating markedly improves the bioactive profile of makhana seeds. The overall phenolic content rises by approximately 55.87%, whilst the total flavonoid content increases by about 40.34% compared to raw seeds. The DPPH (2,2-diphenyl-1-picrylhydrazyl) radical scavenging activity increases significantly from 24.99% to 73.86%, indicating enhanced antioxidant capacity. The enhancements arise from cell wall rupture and the liberation of bound phenolic compounds, alongside the generation of antioxidant Maillard reaction products during microwave treatment.

Overall, microwave treatment is an effective and fast thermal processing technique for makhana seeds, markedly enhancing their physicochemical, functional, and antioxidant qualities while preserving starch crystallinity. In comparison to traditional methods, microwave heating significantly improves water absorption, bioactive chemical release, and flour solubility, underscoring its considerable potential for both domestic and industrial processing of makhana-based functional meals.

6.5.2 Infrared (IR) popping techniques

Infrared radiation constitutes a segment of the electromagnetic spectrum positioned between the visible spectrum and microwaves. In 1800, Herschel wrote the first scientific paper on infrared heating. The report indicated that exposure of the lenses to sunlight resulted in burning of the paper underneath. In the study, a prism was used to separate the light into different colors and a thermometer was subsequently placed in the different colors of the solar energy spectrum. Researchers observed significant heating in the region beyond the red wavelength, commencing at the end of the visible spectrum. The invisible portion of the spectrum is referred to as infrared. Sixty-six percent of solar radiation is infrared (Rosenthal, 2012). Figs. 6.5 and 6.6 depict the electromagnetic spectrum of several radiation types, including infrared.

Infrared radiation is divided into three types (Rosenthal, 2012):

1. Near infrared (NIR) with wavelength ranging from 0.75 to 1.4 μm.
2. Mid infrared (MIR) with a wavelength between 1.4 and 3 μm.
3. Far infrared (FIR) with wavelength between 3 and 1000 μm.

Therefore, infrared radiation is defined as part of an electromagnetic spectrum whose wavelength ranges from 0.75 to 1000 μm.

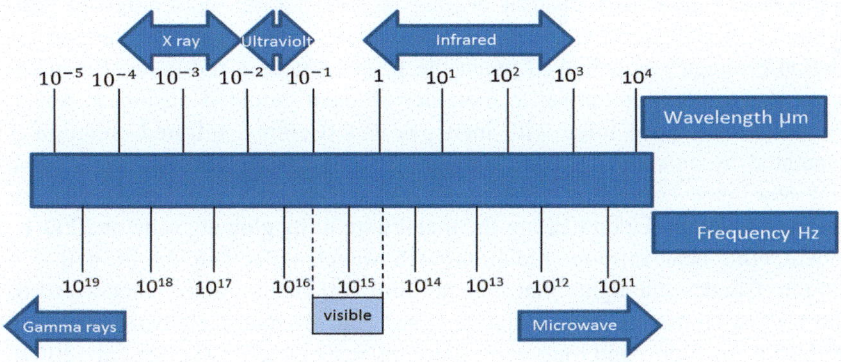

FIGURE 6.5 Electromagnetic radiation spectrum (Aboud et al., 2019).

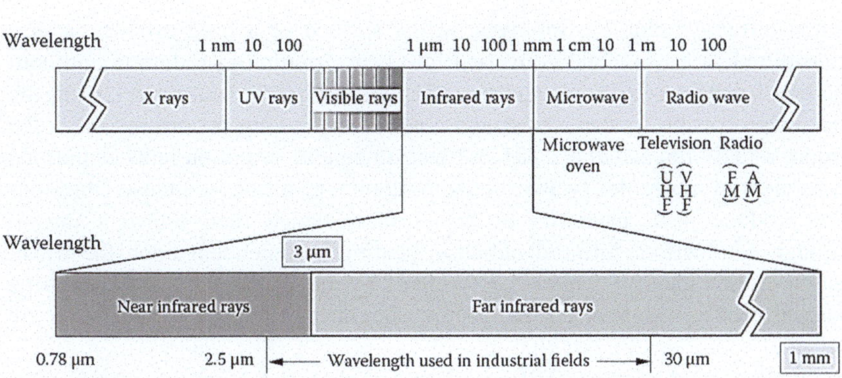

FIGURE 6.6 Spectrum of electromagnetic waves and infrared ranges (Aboud et al., 2019).

6.5.2.1 Infrared sources

Two conventional sources of infrared generation utilized for heating are electric heaters and gas-fired heaters. These techniques generate infrared radiation within a temperature range of 343–1100°C for gas heaters and 1100–2200°C for electric heaters (Aboud et al., 2019). To prevent products from burning, the standard infrared temperature utilized ranges from 650 to 1200°C. The cost of gas heaters is significantly higher; however, their operational costs are comparatively lower than those of systems featuring electric infrared radiation (electric heaters). Infrared electric heaters are more prevalent than gas heaters due to their controllability, rapid heating capabilities, and clean energy output. Infrared emitters have greater versatility in generating the specific wavelength necessary for a given application. Infrared radiation can emit wavelengths that vary from short to long, depending upon the voltage provided to the emitters. The operational efficiency of electric infrared heaters varies between 40% and 70%, whereas infrared gas heaters yield 30%–50% efficiency (Aboud et al., 2019) and emit medium to long wavelengths. The appropriate range for industrial heating is 1.17–5.4 μm, correlating to temperatures of 260–2200°C (Sheridan and Shilton, 1999). Optimal thermal radiation efficiency is achieved when the heater attains a red color. Electrical infrared emitters comprise a metallic filament situated inside a sealed container, which are filled with inert gas or left empty. Infrared radiation is generated by an electric heater through the passage of electric current via a high-resistance conductor, such as nichrome wire, iron-chromium wire, or tungsten filament. Upon heating the metal wire to its glowing state at 2200°K, it will emit near infrared radiation with wavelengths ranging from 0.7 to 1.4 μm. Candescent lights that generate infrared radiation are categorized as short-wave emitters, whereas quartz tubes and resistance elements are categorized as medium- and long-wave emitters, respectively (Aboud et al., 2019).

6.5.2.2 Infrared heating mechanism

When infrared radiation impinges on the surface of a food product, it gets transmitted as a wave and converted into heat. It employs radiation and conduction for heat generation and transfer, with radiative heating occurring on the surface of the food material and conductive heating occurring within the products (Trivittayasil et al., 2011). Electromagnetic radiation induces thermal movement in molecules (water, lipids, proteins), resulting in intense vibrations that produce heat, primarily at the surface, which then transfers inward through conduction. Infrared radiation heating is dependent upon the wavelength of the radiation, which is determined by the temperature of the emitting body, exhibiting an inverse relationship. Starches, proteins, fats, and water are essential food constituents that absorb far infrared (FIR) energy more effectively than near infrared (NIR) radiation (Sakai and Hanzawa, 1994). The penetration depth of infrared (IR) radiation is significantly affected by the food's composition and structure, in addition to the radiation wavelengths, with longer wavelengths facilitating deeper penetration. Consequently, FIR heating is commonly employed in food processing because to its extensive spectral wavelength range and the absorption characteristics of numerous food components (Fakayode et al., 2025). Ramaswamy et al. (2012) indicated that the majority of incident radiation energy in the FIR range is absorbed at the surface, making FIR particularly effective for the surface heating of meals, as evidenced by experimental results.

6.5.2.3 *Infrared heating-assisted fluidized bed (IHFH) treatment for makhana*

The infrared heating-assisted fluidized bed (IHFH) treatment is conducted using an infrared-assisted fluidized bed heating apparatus. Clean makhana seeds are placed in the fluidized bed chamber and are fluidized by a constant stream of hot air to ensure consistent particle motion. Infrared radiation is concurrently delivered from emitters located above the bed, facilitating rapid and uniform heat transfer to the seeds.

During IHFH treatment, the processing temperature is maintained at approximately 150°C, and the seeds are subjected to heating for 20 min under continuous fluidization with a superficial airflow velocity of 6.0 m/s. The synergistic effect of infrared radiation and fluidized hot air facilitates effective moisture extraction, regulated surface heating, and consistent thermal exposure, while preventing localized overheating. Subsequent to treatment, the IHFH-processed makhana seeds are promptly cooled to ambient temperature and preserved in airtight containers (Li et al., 2022).

6.5.2.4 *Effects on physical and structural properties*

The IHFH treatment markedly diminishes the physical density and hardness of makhana seeds. Bulk density decreases from 761.93 g/L (raw) to 618.43 g/L, while the 1000 grain weight reduces from 309.48 to 264.56 g. The hardness decreases substantially from 57.90 to 46.44 N, signifying enhanced friability. The puffing index markedly rises to 1.23, indicating improved expansion characteristics relative to microwave treatment. The treatment facilitates the preservation of the A-type crystalline structure; nonetheless, there is a notable reduction in peak intensity, signifying a partial decline in relative crystallinity due to the greater thermal impact relative to microwave treatments.

6.5.2.5 *Surface morphology and microstructure*

IHFH treatment results in significant surface cracks and fissures, and highly porous structures among the pericarp, episperm, and endosperm layers. This treatment can significantly disrupt the cellular matrix, hence enhancing grinding efficiency and moisture penetration.

6.5.2.6 Color characteristics

IHFH processing markedly modifies the color characteristics of makhana seeds. Lightness (L*) declines from 88.18 (raw) to 81.70, while redness (a*) escalates to 3.35 from 0.79, and yellowness (b*) rises to 9.91 from 7.12. The total color difference (ΔE) attains 7.51, signifying significant browning resulting from intensified Maillard and caramelization reactions during uniform infrared heating.

6.5.2.7 Functional properties

The water absorption capacity rises markedly from 1.43 g/g (raw) to 2.77 g/g, and the oil absorption capacity increases from 0.87 g/g to 0.93 g/g. The swelling power diminishes from 9.16% to 7.02%, while solubility rises from 2.82% to 5.72%, indicating starch alteration and enhanced makhana flour functioning after IHFH treatment.

6.5.2.8 Pasting and structural properties

The IHFH treatment significantly alters pasting properties, leading to a substantial decrease in viscosity profiles. IHFH treatment also substantially alters the pasting properties, although the result varies slightly from that of microwave treatment. The peak viscosity drops to 139 cP from 763 cP (raw), signifying the fragmentation of starch granules resulting from the combined effects of infrared heating and fluidization. Also, the final viscosity falls to 268.67 cP from 1854 cP, indicating improved heat stability and less retrogradation. The findings indicate that IHFH treatment facilitates regulated starch alteration and enhances paste stability for functional food applications.

6.5.2.9 Bioactive compounds and antioxidant activity

IHFH treatment yields superior enhancement of bioactive chemicals relative to microwave treatment. The overall phenolic content rises by 86.63%, while the total flavonoid content increases by 76.90% compared to the raw samples. The DPPH radical scavenging activity rises from 25.00% to 76.87%, and the reducing power escalates from 0.22 to 0.57, demonstrating enhanced antioxidant capability relative to alternative thermal techniques.

Overall, IHFH treatment elicits the most significant physicochemical, structural, and functional alterations in makhana seed and flour. The integration of rapid infrared heating and fluidization improves expansion, water absorption, and the release of bioactive components, rendering IHFH an exceptionally efficient industrial-scale processing technique for the production of value-added makhana-based functional meals.

6.5.3 Comparison between microwave and IHFH treatments

Microwave and infrared heating-assisted fluidized bed treatments both enhance the physical, functional, and bioactive features of makhana seeds, although they demonstrate divergent effects owing to variations in heating methods. Microwave treatment principally facilitates quick volumetric heating, resulting in internal steam generation and partial starch gelatinization, which results in substantial expansion, enhanced water absorption, and elevated antioxidant activity. The IHFH treatment integrates infrared surface heating with fluidized bed air circulation, resulting in a more consistent temperature distribution and regulated moisture removal. This yields an essentially porous structure, improved puffing, and higher preservation of starch crystallinity, while simultaneously augmenting functional and bioactive characteristics. Microwave treatment yields rapid and severe internal modifications, while IHFH provides milder, consistent structural alterations, rendering it especially appropriate for preserving grain integrity while improving functional performance. The comparative effects of microwave and IHFH treatments on makhana seeds and flour are summarized and depicted in Table 6.3.

TABLE 6.3 Effects of microwave and IHFH treatments on *makhana* seeds and flour (Li et al., 2022).

Property	Raw	Microwave treatment	IHFH treatment
Physical properties			
Bulk density (g/L)	761.93	752.93	618.43
Thousand-grain weight (g)	309.48	270.24	264.56
Hardness (N)	57.90	49.07	46.44
Puffing index	1.00	1.01	1.23
Functional properties			
Water absorption capacity (g/g)	1.43	2.55	2.77
Oil absorption capacity (g/g)	0.87	0.93	0.93
Swelling power (%)	9.16	7.44	7.02
Solubility (%)	2.82	6.10	5.72
Color propeties			
Lightness (L*)	88.18	83.75	81.70
Redness (a*)	0.79	1.71	3.35
Yellowness (b*)	7.12	15.11	9.91

Continued

TABLE 6.3 Effects of microwave and IHFH treatments on *makhana* seeds and flour (Li et al., 2022).—cont'd

Property	Raw	Microwave treatment	IHFH treatment
Pasting properties			
Peak viscosity (cP)	763	309	139
Final viscosity (cP)	1854	486.33	268.67
Bioactive compounds and antioxidant activity			
Improvement in total phenolic content (%)	–	55.87	86.63
Improvement in total flavonoid content (%)	–	40.34	76.90
DPPH scavenging activity (%)	25.00	73.86	76.87
Reducing power (at Abs 700 nm)	0.21–0.22	0.56	0.57

6.5.4 Researchable emerging technologies

Emerging technologies employing induction heating, pressure, ultrasound, ohmic heating, and radio frequency can improve the popping and processing efficiency of makhana. Induction heating produces heat using electromagnetic induction, enabling fine temperature regulation for consistent roasting. Pressure-assisted popping pretreats seeds under regulated pressure, thereafter, releasing it rapidly, resulting in internal steam expansion and enhanced puffing. Ultrasound pretreatment induces microstructural disturbances, hence improving heat and mass transmission and raising popping efficiency. Moreover, ohmic heating employs electric current for consistent volumetric heating, while radio-frequency (RF) heating delivers deep, quick heating, facilitating scalable and energy-efficient production for functional makhana products. These sophisticated and developing technologies combined provide promising solutions to enhance product quality, yield, and functional attributes of makhana-based snacks.

6.6 Pre- and post-popping operations

Makhana processing involves series of unit operations before and after the popping operation, that is, pre and post popping operations.

6.6.1 Prepopping operations of makhana

Pre-popping operations are generally used to prepare raw makhana seeds for effective popping and which helps to ensure size uniformity, even moisture distribution and to improve the popping yield.

6.6.1.1 Cleaning

The foremost step in making of makhana is cleaning, which helps to remove the foreign matter such as mud particles, plant residues and other impurities to prevent contamination during the subsequent processing. The cleaning may be done manually or by mechanical means depending on the condition of the raw material. Clean seeds facilitate uniform heat transfer and prevent scorching during roasting (Kumari et al., 2023a, 2023b).

6.6.1.2 Drying

Drying is an essential post harvest operation to the high initial moisture content (Approx 37 ± 1 w.b) of freshly harvested foxnut seeds to a suitable level of moisture content (Approx $30 \pm 1\%$ w.b) for safe storage and further processing steps (Kumar et al., 2023; Kumari et al., 2023b; Sharma et al., 2024a, 2024b, 2024c, 2024d). This method helps in reducing enzymatic activity, microbial growth and quality deterioration during storage. These seeds require uniform drying to avoid internal stress and surface hardening due to the hard outer shell enclosing the kernel as well as slow diffusion from the interior to the surface of seed (Jalaj et al., 2019; Islam et al., 2025).

Drying can be performed using traditional methods such as sun/shade drying or by mechanical methods lie tray drying, cabinet drying and advanced techniques such as fluidized bed drying, microwave vacuum method, ultrasound assisted microwave method ensures even moisture removal, which improves popping efficiency and expansion characteristics whereas improper drying adversely affects expansion and its texture. After drying, the seeds are stored for 20–25 days before processing (Tian et al., 2013; Devi et al., 2020; Manju et al., 2021; Zhao et al., 2021; Devi and Deka, 2022).

6.6.1.3 Size grading

The cleaned and dried makhana seeds are undergo grading based on the size, the seeds are classified into various groups using manual sorting or mechanical sieving systems. Uniform size seeds exhibit similar heat transfer and moisture loss during the roasting and popping operations. Proper grading helps the pressure build up evenly during popping, which makes the kernels expand better and leaves fewer unpopped or partially popped kernels. This process also makes processing faster, products more consistent, and increases market by allowing seeds to be categorized into several commercial grades (Devi et al., 2020).

6.6.1.4 Preheating and tempering

Preheating and tempering are essential intermediate operations carried out prior to popping of makhana (*Euryale ferox*) seeds to enhance heat utilization and ensure uniform expansion. The seeds are pre heated in an iron karahi or other suitable equipment at a temperature range of 200–300°C for a period of 1–3 min

(Mani, 2019), allowing gradual elevation of kernel temperature without inducing thermal damage. This operation helps in improving heat penetration, reducing water activity and decrease pericarp hardness. After being heated, the seeds are tempered for 48–72 h, which means that the seeds are kept still and covered or insulated for specific period (Sharma et al., 2021). This process helps the seeds to reach moisture equilibrium, which lowers stress inside the seeds, prevents cracking on the surface and promotes even vapor generation during the popping operation. The combined impact of these two processes yields higher popping efficiency and increased expansion volume, resulting in superior porous texture, whiteness, and overall product quality (Singh et al., 2021).

6.6.1.5 Popping

The tempered seeds were roasted in a cast iron pan at a temperature range of 225–300°C for a duration of about 2–3 min until its moisture content reaches to 12% w.b. Continuous stirring or agitation is essential during roasting to ensure uniform heat distribution and to prevent localized overheating or scorching of the seeds (Bhatt et al., 2017). During the roasting, the moisture inside the seeds quickly changes to vapor, which leads to a sharp rise in pressure. When the vapor pressure is greater than the mechanical strength of the pericarp, a sudden rupture happens, leading to popping and expansion of the endosperm, kernel, or seed. The effective popping depends on its temperature, exposure period and the moisture achieved during the operations of preheating and tempering (Verma et al., 2010; Sharma et al., 2023).

6.6.2 Post-popping operations of makhana (foxnut)

After the popping process, it undergoes a set of operations to stabilize the popped kernels and to increase its quality, safety and shelf life. These processes are important in ensuring greater uniformity, marketability, and the quality of the end product.

6.6.2.1 Cooling

Immediately after the popping process, the popped kernels were cooled by spreading it in thin layers under ambient conditions or by using mechanical means of air to remove the residual surface heat, preventing the moisture condensation and stabilizing the popped kernel structure (Kumari et al., 2023a, 2023b).

6.6.2.2 Cleaning and removal of un-popped seeds

After the cooling operation, the makhana undergoes cleaning operation which involves the removal of any shell or pericarp fragments adhered and also to separate the unpopped, partially popped, broken and burnt seeds. This step improves its hygiene, appearance and consumer acceptability (Sharma et al., 2023).

6.6.2.3 *Polishing*

This operation is generally used to improve the surface uniformity and appearance of the popped kernels. It involves gently rubbing or tumbling the kernels to remove surface particles, shell fragments, and small rough spots without harming the expanded structure. This procedure makes makhana whiter, smoother, and more visually appealing, which raises its market value (Mahawar, 2016; Mani, 2019).

6.6.2.4 *Grading and size sorting*

The process of grading is done to categorize the popped makhana based on the factors of size, expansion volume, shape, and appearance. The size grading offers consistency in the packaging and price as the bigger and better expanded kernels are considered to be better in quality. The sorting can be performed either by hand or by use of screens and graders (Devi et al., 2020).

6.6.2.5 *Final drying or moisture adjustment*

Popped makhana may hold small amounts of either surface or internal moisture which may damage shelf stability. A mild drying or conditioning process is therefore adopted to reduce the amount of moisture to levels considered safe during storage. This step helps to prevent microbial growth, caking, and loss of crispness during storage (Mahawar, 2016).

6.6.2.6 *Secondary roasting (optional)*

A secondary roasting process may also use in order to increase the crispness and flavor of makhana. This is done at low temperatures over a short period and stringently controlled to avoid the unwanted browning or loss of texture (Jalaj et al., 2019).

6.6.2.7 *Packaging*

The makhana packaging is done with the aim of protecting the popped kernels against moisture absorption, oxidation as well as physical damage during their storage and distribution cycle. Laminated pouches or food-grade polymer films are materials with strong moisture and oxygen barrier properties, which are mostly used in preserving the crispness, whiteness, and general quality of the product. Good quality packaging therefore prolongs shelf life and preserves market value of makhana with ambient storage conditions (Singh et al., 2021).

6.6.2.8 *Storage*

Storage of makhana is carried out to maintain the physical and sensory quality of the popped kernels after post-processing. The product is stored under cool, dry, and hygienic conditions with low relative humidity to prevent moisture absorption and loss of crispness. Proper storage also helps minimize oxidative deterioration, discoloration, and quality degradation, thereby ensuring product stability and acceptable shelf life (Jalaj et al., 2019).

6.7 Quality evaluation and standardization

In makhana production, the quality should be maintained at each and every step. When the makhana seed is popped, a lot of changes occurred in the physical, chemical nutritional properties. It is mandatory to evaluate and standardize the quality of the popped makhana to for ensuring safety, increased market price and processing opportunities. The popped makhana can be used as a functional ingredient in preparation of various food products viz. makhana snacks, makhana powder, ready-to-use breakfast cereals, baby foods, barfi, kalakhand, bread and instant makhana kheer mix. The quality evaluation of makhana mainly includes physical, chemical and nutritional and storage.

6.7.1 Physical quality parameters

The physical quality parameters are important in grading of makhana based on size, texture and residue. Kumar et al. (2015) graded makhana into 4 to 5 grades according to size and light weight. The physical properties such as size, shape, arithmetic and geometric mean diameter, sphericity, aspect ratio, surface area, volume, 1000 kernel weight, bulk density and porosity, frictional properties such as friction coefficient and angle of repose and textural property that is, hardness were measured. Devi et al. (2020) classified the decorticated gorgon nut seed into seven grades based on size such as <7, 7, 8, 9, 10, 11, and 12 mm diameter. Makhana can be graded into 2–3 grades namely Lawa or Rasgulla as grade 1, Murra oa Samundha as grade 2 and Thuri as grade 3 with diameters ≥ 15 mm fully expanded, ≤ 14 mm fully expanded and flattened or irregular of any diameter respectively and color of makhana are creamy white, creamy white with reddish spots and more light reddish spots respectively (Handbook of Makhana, 2022). The other physical quality properties which are essential for popped makhana are given below:

6.7.1.1 Popping percentage

The recovery percentage of popped makhana is estimated by the following formula (Sharma et al., 2024):

$$\text{Popping } \% = \frac{W_{\text{fpm}}}{W_{\text{fpm}} + W_{\text{spm}} + W_{\text{upm}}} \times 100$$

where as

W_{fpm} is weight of fully popped makhana
W_{spm} is weight of semi popped makhana
W_{upm} is weight of unpopped makhana

The popping % recovery was $48.53 \pm 0.50\%$ before storage and $7.28 \pm 0.60\%$ after storing roasted makhana seeds for 150 days (Sharma et al., 2024).

6.7.1.2 Expansion ratio

The expansion ratio is defined as the ratio of geometric mean diameter (GMD) of popped makhana and the geometric mean diameter of makhana seed (Sharma et al., 2024). The equation can be written as

$$\text{Expansion ratio, ER} = \frac{\text{GMD}_{\text{popped makhana}}}{\text{GMD}_{\text{machana seed}}}$$

The expansion ratio depends on the storage periods of roasted seeds before popping. Sharma et al. (2024) stated that the expansion ratio was >2 up to 135 days and thereafter decreased significantly.

6.7.1.3 Bulk density

The bulk density of popped makhana can be determined by ratio of weight of popped makhana filled in a known volume of tarred measuring cylinder (Kumar et al., 2015). The bulk density of fully expanded popped makhana of width >15 mm and thickness >14 mm with 4 and 12% moisture content were 68.18 ± 1.02 and 72.81 ± 1.29 kg/m^3 respectively (Kumar et al., 2015).

6.7.1.4 Chemical & nutritional aspects

The proximate composition of makhana seeds is given in Table 6.4 (Kumar et al., 2023). The protein content of popped makhana increases than that of raw gorgon nut due to the conversion of complex molecules in to individual amino acids and thus popping improves the protein quality and digestibility. The fat content and ash content decrease slightly because of bran removal during decortications of gorgon nut seeds before popping. Popping can increase the total fiber content because the structural change of starch results in availability of more analytically accessible fiber. The carbohydrates decrease on popping due to the increase in the protein and fiber contents. The mineral profile of raw seed and popped makhana was similar except for Ca and Fe contents. Popping of makhana causes thermal degradation of phenolics and flavonoid compounds and decomposition of phytochemicals and thus reduces the total phenolics, flavonoid contents and % inhibition of DPPH antioxidant activity. The mineral profile of makhana contains Calcium, Sodium, Potassium, Magnesium, Phosphorous, Iron, Manganese, Molebdenum, Zinc, Copper are 20.94, 4.06, 48.39, 12.71, 124.01, 2.67, 1.24, 0.04, 1.04, 0.76 mg/100 g dry matter basis (Devi et al., 2020) (Table 6.5).

6.7.2 Contamination risk & food safety

Makhana is an aquatic crop and needs integrated insect and pest management integrated nutrient management. Makhana is highly prone to severe disease

TABLE 6.4 The proximate analysis of various varieties of makhana

	Popped kernel (Swarna vaidehi)		Raw seed		
Proximate composition, (% by weight)	Devi et al. (2020)	Kumar et al. (2016)	Swarna Vaidehi Jana et al. (2025)	Superior Selection-1 Jana et al. (2024)	Swarna Vaidehi Devi et al. (2020)
Moisture content	–	10.4	14.75 ± 0.83	11.15 ± 0.43	30 ± 0.1
Protein	11.03 ± 0.45	8.7	7.24 ± 0.37	9.70 ± 0.62	9.15 ± 0.10
Carbohydrates	84.87 ± 0.60	79.8	77.33 ± 0.97	75.70 ± 1.54	87.52 ± 0.15
Ash content	0.38 ± 0.03	0.4	0.52 ± 0.09	0.52 ± 0.02	0.45 ± 0.03
Crude fiber	3.26 ± 0.06	0.2	0.35 ± 0.05	0.7 ± 0.03	2.55 ± 0.04[a]
Total fat	0.33 ± 0.12	0.5	0.35 ± 0.43	0.31 ± 0.01	0.46 ± 0.04

[a]*Total dietary fiber.*

TABLE 6.5 The anti-oxidant composition of popped makhana and raw gorgon nut (Swarna Vaidehi) (Devi et al., 2020).

Anti-oxidant composition, (% by weight)	Popped makhana	Raw seed
Total phenolic content (mg GAE/g)	1.12 ± 0.06	2.22 ± 0.06
Total flavonoid content (mg RE/100 g)	1.26 ± 0.18	3.23 ± 0.28
Antioxidant activity (% inhibition of DPPH)	17.48 ± 1.2	27.57 ± 2.13

named Alter aria tenuis leaf blight and the attaking primary insects are Aphids *(R. nymphaeae)*, case worm (*E. crisonalis*) and root borer (*D. delesserti*). The contamination risk can be minimized by the use of organic pestiicdes like neem oil spray and providing adequate nutrients (Choudhary and Rawat, 2022). Utilization of carbon derived from discarded Euryale ferox seed shells to eliminate Pb^{2+} ions from a solution (Goswami et al., 2024).

The permitted range of usage for adults per day (given in terms of raw herb/material) 20–30 g (as seed); 3–5 g (as powder) (FSSAI, 2016). The quality standards for market readiness includes the 90% kernel size uniformity, moisture content below 12% dry basis, free of impurities and residue, no husk or dust particles, nutritional accuracy at milligram level, tamper proof and food safe packaging and certification. According to Packaging & Labeling Laws—FSSAI, the labeling should contain following information such as FSSAI License/Registration number, name of the food, list of the ingredients, nutritional information, net weight/quantity, lot/code/batch identification number, date/month/of manufacturer, best before or use by date or date of expiry, declaration regarding vegetarian & non vegetarian, instruction of use, declaration regarding food additives, name & complete address of the manufacturer, importer details and code number.

6.7.3 Shelf-life considerations

Popped makhana is hygroscopic and loses its crunchiness when exposed to environment. Modified atmospheric packaging of makhana can improve the shelf life and maintains its crispness and prevent dust accumulation and contamination. This can be achieved by removing oxygen or air present within the package and filling standard nitrogen. Different sizes from 200 g to 20 kg of makhana packets are available based on the market requirement. However, the roasted makhana seeds can also be stored in poly propylene packages and popping is done instantly by using a popping machine. Sharma et al. (2024) stored roasted seeds of makhana for 6 months in airtight polypropylene packages. There was no change in the popped makhana recovery up to 150 days. Aseptic packaging is one of recent packaging technique Properly lacquered tin cans, glass bottles & jars, lined carton or standalone paper pack, PET or PVC, flexible films/pouches/laminates and jute/plastic/cloth/hdpe wholesale bags are being used for packaging of makhana.

6.8 Energy use & environmental impact

Although Makhana processing is considered eco-friendly, it encounters obstacles in water use, energy efficiency with carbon footprint. It involves sun drying, storage, grading, pre-heating, tempering, roasting and popping followed by polishing, grading, roasting/frying, seasoning/coating and packaging (Directorate of Horticulture, Government of Bihar, India). However, sustainable practices comprising solar drying, innovative roasting techniques and ecologically sound packaging are the ways and means for significant reduction in impact of the environment. Makhana growing supports the biodiversity, while requiring less chemicals as compared to the crops grown in land. The cultivation contributes to generating income of rural communities aligning the Sustainable Development Goals (SDGs) of UN. Makhana based snacks are now promoted as a sustainable substitute of other snacks available in market such as fried foods or chips. Considering the facts Priyadarshi and Ahmad (2025) mentioned its alignment with no poverty (SDG 1), zero hunger (SDG 2), health and well-being (SDG 3), decent work and economic growth (SDG 8) and responsible consumption and production (SDG 12). Kashyap and Shukla (2025) highlighted the importance of sustainability as food wastage has reached 33% when around 9% population is malnourished.

6.8.1 Issues with traditional fuel-intensive methods

Makhana is made edible through high energy inputs in the stages of drying and roasting the seeds using coal or wood. Modern methods may be based on solar dryers and energy-saving roasters, while cutting carbon footprint through electrification from renewable sources. Fossil fuel and Biomass cause emission of greenhouse gases in traditional processing. Although, makhana cultivation is practiced in ponds, its freshwater requirement is lesser for irrigation relative to wheat or rice. Plastics used for their packaging contributes to pollution in the present practice. In recent years, export growth has also drawn attention toward transport emissions.

Fuel-intensive process makes it unsustainable, whereas dependence on non-renewable fuels increases processing cost and thereby reduces profitability to processors at small scale. Firewood requirement leads to deforestation and fuel burning releases particulate matter and greenhouse gases resulted into deterioration in air quality. Exposure to ash, heat and smoke causes issues in eye and respiration process. The effect of hazards is multiplied due to unavailability of equipment inevitable for protection. Traditionally, roasting-popping operations are manual and needs skilled laborer for long periods, failing which it compromises the productivity and quality. Moreover, non-uniform heating resulted as burnt or un-popped seeds in large quantities fetching lesser competitiveness and market value. High cost and little efficiency of fuel make makhana processing unsustainable and limited mechanization additionally creates hurdles in scaling up the production level to meet the demand. Thus, reliability on coal or firewood is linked to inefficient, labor intensive and costly operation, poor quality of output, ecological risk and health hazard.

6.8.2 Energy-efficient modern technologies

1. *Cleaning and grading operation:* Automated washer removes soil with hollow and rotten seeds, whereas mechanized graders distribute makhana among 5–7 size-based seed categories (Kumari et al., 2023a, 2023b). The mechanized and automated operation minimized dependency upon laborers with high throughput.
2. *Roasting operation*: After heating and tempering, single layer of 300g nuts is placed in a pan of cast iron for roasting at 290–340°C in traditional method (Kumari et al., 2023a, 2023b). LPG or electricity-based roasters reduces smoke exposure with uniform heating and improved popping efficiency and saves up to 40% of the energy. Alternatively, their solar energy version uses renewable sources of energy instead of conventional fuels. It offers ease of heat control and uniformity in roasting with reduced wastage. It is associated to lower running costs, while supporting the sustainable increase in units.
3. *De-shelling and polishing and packaging operation*: Mechanized operation of de-shelling efficiently cracks the stiff outer shell, whereas polishing is desired for enhanced appearance and thereby market value. It reduces the labor with improved hygiene. Shelf life can be extended through limiting the absorption of moisture, while maintaining the crispness through practicing mechanized and efficient packaging.

Future lies in mechanization with solar dryers, electric roasters and popping machines working with LPG. Their level of incorporation can significantly reduce dependency on non-renewable fuel sources and environmental degradation resulted as sustainability with reduction in carbon footprint. The worker safety can be ensured, while maintaining the product quality and enhancing production level profitability.

6.8.3 Waste utilization

Utilization of by-products (seed coat and biomass) can reduce waste with lowered environmental repercussions and generate additional income. Their potential uses are in for agriculture, energy and value addition. Production of briquettes or pellets or biogas from seed coat may be useful for reducing the carbon emission. Reduction in landfill wastes, lowered carbon footprint and keeping away from open burns are beneficial to environmental. Composting of seed coats can be practiced to improve fertility of the soil. Seed-coats can also be used in handicraft and decoration upon polishing and treating and can engender employment to rural population in crafting, composting and briquetting. Discarded as low-grade kernels or broken seeds are alternate to poultry and cattle feeds as low-cost supplement for carbohydrate and protein. Filtration and water purification needs activated carbon, which can be prepared from seed shells. Thus, waste utilization is beneficial in economic, environmental, social as well as agricultural aspects.

Considering the facts, it is recommended to promote briquetting or composting units near makhana clusters. Linkages may be established to industries associated to biofuel or activated carbon. Rural artisans may also be trained for creative application of makhana shells (Wardhan et al., 2025).

6.8.4 Carbon footprint considerations

Makhana processing involves fuel, its efficiency and transportation as the factors affecting the carbon footprint. It can be reduced through adapting renewable energy or switching to LPG or electricity for processing operations. The goal focuses on improving transportation, establishment of localized processing hubs, minimizing chemicals and water requirement. Mechanized grading and packaging are also covered for the cause to reduce carbon footprint.

6.8.4.1 Calculation of carbon footprint for makhana processing

It requires the inventory of Makhana comprising reviewable framework for transport, materials and energy to be maintained. The rough approximation of the values related to Makhana processing in north Bihar region of India is considered as case study and the calculations of carbon footprints were presented below. As food supply chain and associated activities contributes around 20% of the total emission of greenhouse gas (Borowski et al., 2020), it is necessary to estimate carbon footprint for makhana processing to cut down energy intensities at possible operations.

6.8.4.2 Calculation

1) Grading, polishing and sealing operation

$$\begin{aligned}\text{Electricity consumption} &= 0.05 - 0.15\ \text{kWh}\\ &= 0.1\ \text{kWh}/\ \textit{kg popped makhana}\ (\textit{Say})\end{aligned}$$

Emission factor for electricity consumption

$$= 0.8 \text{ kg } CO_2 \text{ per kWh electricity}/\textit{kg popped makhana}$$

$$\text{Carbon footprint} = (0.10) \times (0.8) \text{ kg } CO_2 = 0.08 \text{ kg } CO_2/\text{ kg popped makhana}$$

2) Roasting operation

$$\text{LPG consumption} = 0.2 - 0.25 \text{ kg}$$

$$\text{LPG} = 0.25 \text{ kg LPG}/\textit{kg poppeed makhana (Say)}$$

Emission factor for LPG consumption

$$= 3.0 \text{ kg } CO_2 \text{ per kg LPG consumption}$$

$$\text{Carbon footprint} = (0.25) \times (3.0) \text{ kg } CO_2$$

$$= 0.75 \text{ kg } CO_2/\text{ kg popped makhana}$$

3) Packaging operation

$$\text{Multilayer pouch requirement} = 0.015 - 0.025 \text{ kg}$$

$$\approx 0.02 \text{ kg}/\textit{kg popped makhana (Say)}$$

$$\text{Emission factor for pouch consumption} = 6.0 - 7.0 \text{ kg } CO_2$$

$$\approx 6.5 \text{ kg } CO_2/\text{ kg pouch (Say)}$$

$$\text{Carbon footprint} = (0.02) \times (6.5) \text{ kg } CO_2$$

$$= 0.13 \text{ kg } CO_2 \text{ /kg popped makhana}$$

4) Transportation (for each tonne of popped makhana)

$$\text{Transport requirement (farm to processing unit)} = 0 - 40 \text{ km} = 20\text{km } (\textit{Say})$$

Transport requirement (processing unit to market within Bihar)

$$= 50\text{-}300 \text{ km} = 200 \text{ km } (\textit{Say})$$

Transport requirement (processing unit to market for Delhi)

$$= 1,100 - 1,200 \text{ km} = 1,150 \text{ km } (\textit{Say})$$

$$\text{Total transport requirement (within Bihar)} \approx 220 \text{ km}$$

$$\text{Total transport requirement (market for Delhi)} \approx 1,170 \text{ km}$$

$$\text{Emission factor for transport (within Bihar via road)} = 0.08 - 0.12 \text{ kg } CO_2$$

$$= 0.1 \text{ kg } CO_2 \text{ per km / tonne } \textit{popped makhana (Say)}$$

$$\text{Emission factor for transport (market for Delhi via train)} = 0.02 - 0.03 \text{ kg } CO_2$$

$$\approx 0.025 \text{ kg } CO_2 \text{ per tonne per km}$$

$$\text{Carbon footprint (within Bihar via road)} = (20) \times \left(\frac{0.1}{1000}\right)$$

$$+ (200) \times \left(\frac{0.1}{1000}\right) \text{ kg } CO_2 = 0.022 \text{ kg } CO_2/ \text{ kg popped makhana}$$

$$\text{Carbon footprint (market for Delhi via train)} = (20) \times \left(\frac{0.1}{1000}\right)$$

$$+ (1150) \times \left(\frac{0.025}{1000}\right) \text{ kg } CO_2 = 0.03075 \text{ kg } CO_2/ \text{ kg popped makhana}$$

$$\text{Carbon footprint for transportation} \approx 0.022 \ (\textit{for Bihar})$$
$$\textit{and } 0.03075 \ (\textit{for Delhi}) \text{ kg } CO_2/ \text{ kg popped makhana}$$

Hence,

$$\text{Total carbon footprint for makhana processing}$$

$$= \text{Carbon footprint [(Grading, polishing, sealing)}$$
$$+ \text{(Roasting)} + \text{(Packaging)} + \text{Transport]}$$

$$= 0.08 + 0.75 + 0.13 + 0.022 = 0.982 \text{ kg}$$
$$CO_2/ \text{ kg popped makhana } (\textit{for Bihar})$$
$$= 0.08 + 0.75 + 0.13 + 0.03075$$
$$= 0.991 \text{ kg } CO_2/ \text{ kg popped makhana } (\textit{for Delhi})$$

$$\approx 1.0 \text{ kg } CO_2 \ / \text{ kg popped makhana}$$

Carbon footprint of makhana processing can subsequently be reduced through improvement in energy efficiency, fuel switching and optimizing the packaging and logistics.

6.9 Conclusion

The processing of Makhana (*Euryale ferox*) is currently going through an important transition from a traditional, labor-intensive skill to a scientifically driven, mechanised industry. Traditionally, the quality of popped Makhana depended significantly on the indigenous skills of workers directing complicated steps such as sun-drying, tempering, and manual roasting. Although these methods yield high-quality products, they are hindered by low efficiency, variable quality, and extensive laborious processes. Recent developments in post-harvest technology have led to the introduction of continuous mechanised systems, including those developed by ICAR-CIPHET and several commercial producers. These innovations substitute manual labor with regulated thermal conditioning/roasting and mechanical impact mechanisms, significantly improving safety, scalability, and economic feasibility. Moreover, scientific studies into the physicochemical characteristics of the seed, particularly the crucial roles of starch gelatinization and moisture-dependent makhana internal steam pressure, have established the basis for enhancing the performance of these instruments. Looking forward, the industry has the potential to implement advanced thermal technologies such as microwave and infrared systems. These provide volumetric heating and enhanced energy efficiency, with the potential to further improve expansion ratios and nutrient retention. This chapter on makhana popping, highlights wide scope for mechanizing the operations to scale up production processes.

Self-declaration

Artificial Intelligence (AI) tools were used only for English language editing in the preparation of this manuscript. No AI tools were used to generate any kind of content, including text, tables, or figures.

References

Aboud, S.A., Altemimi, A.B., Al-Hilphy, A.R.S., Yi-Chen, L., Cacciola, F., 2019. A comprehensive review on infrared heating applications in food processing. Molecules 24 (22), 4125.

Anonymous, 2015. Technologies machinery and gadgets developed by icar institutes for post -harvest processing preservation and value addition of agricultural produce. In: Agricultural Engineering Division Indian Council of Agricultural Research New Delhi.

Agricultural and Processed Food Products Export Development Authority (APEDA), 2025. Monthly Dashboard – Makhana (Fox-Nut) Sep-2025. https://apeda.gov.in/sites/default/files/2025-10/MIC_Monthly_dashboard_Makhana_30102025.pdf.

Bhatt, B., Gupta, V., Kumar, L., Singh, I.S., Sarkar, B., 2017. *Euryale ferox* (Salisb): promising aquatic food crop of Eastern indo gangetic plains. Int. J. Curr. Microbiol. Appl. Sci. 6 (6), 1914–1921.

Borowski, P.F., 2020. Nexus between water, energy, food and climate change as challenges facing the modern global, European and Polish economy. AIMS Geosci. 6 (4), 397–421.

Choudhary, M., Rawat, A., 2022. Makhana cultivation and management of disease, insect and pest. In: Recent Innovative Approaches in Agricultural Science, II, 978-93-91768-85-0.

Cognitive Market Research, 2025. Fox Nuts Makhana Market Report. https://www.cognitivemarketresearch.com/fox-nuts-makhana-market-report.

Devi, M.B., Deka, S.C., 2022. Physicochemical properties and structure of starches of foxnut (*Euryale ferox Salisb.*) from India and its application. J. Food Process. Preserv. 46 (2), e16262.

Devi, M., Sharma, K., Narayan Jha, S., Arora, S., Patel, S., Kumar, Y., Kumar Vishwakarma, R., 2020. Effect of popping on physicochemical, technological, antioxidant, and microstructural properties of makhana seed. J. Food Process. Preserv. 44 (10), e14787.

Fakayode, O.A., Ojoawo, O.O., Zhou, M., Wahia, H., Ogunlade, C.A., 2025. Revolutionizing food processing with infrared heating: new approaches to quality and efficiency. Food Phys. 2, 100046.

FSSAI, 2016. Food Safety, Standards (Health Supplements, Nutraceuticals,Food for Special Dietary Use, Food for Special Medical Purpose, Functional Food and Novel Food) Regulations.

Goswami, M., Devi, B., Baruah, N.P., Rabha, S., Bharatwaj, A., Sarma, H.P., Devi, A., 2024. Waste to wealth strategies for removal of Pb^{2+} ions from aqueous solution. Inorg. Chem. Commun. 161, 112097.

Handbook of Makhana, 2022. Pradhan Mantri Formalization of Micro Food Processing Enterprises (PMFME) Scheme, Aatmanirbhar Bharat, National Institute of Food Technology Entrepreneurship and Management. An Autonomous Institution under Ministry of Food Processing Industries, Government of India, Sonepat, Haryana, India.

Huang, T.T., Zhou, D.N., Jin, Z.Y., Xu, X.M., Chen, H.Q., 2016. Effect of repeated heat-moisture treatments on digestibility, physicochemical and structural properties of sweet potato starch. Food Hydrocoll. 54, 202–210.

ICAR-National Research Centre for Makhana (ICAR-NRCM), 2019. Makhana Production Technology. ICAR, Darbhanga, Bihar.

Islam, Z., Bashir, K., Singh, S., Rastogi, M., Siddiqui, S., Bangar, S.P., 2025. Structural and functional properties of starch from popped seeds of Makhana (*Euryale ferox*). Starch-Stärke 77 (7), e70030.

Jalaj, R.K., Jalaj, V.K., Kumar, P., 2019. Traditional methods of harvesting and processing of makhana *Euryale ferox* Salisb. Int.J.Curr.Microbiol.App.Sci. 321–326.

Jana, B.R., Bhatt, B.P., Srivastava, A., 2025. Biochemical components of Makhana (*Euryale ferox Salisb*) and its relevance to anti-obesity effect in human. Asian J. Dairy Food Res. 44 (4), 614–619.

Jana, B.R., Kumar, M., Raut, S.M., 2024. Antidiabetic biomolecules and nutrient elements in makhana (*Euryale ferox Salisb.*). Ann. Phytomed. 13 (2), 521–528.

Jha, S.N., 1999. Physical and hygroscopic properties of makhana. J. Agric. Eng. Res. 72 (2), 145–150.

Jha, S.N., Prasad, S., 1993. Physical and thermal properties of gorgon nut. J. Food Process. Eng. 16 (3), 237–245.

Jha, S.N., Prasad, S., 1996. Determination of processing conditions for gorgon nut (Euryale ferox). J. Agric. Eng. Res. 63 (2), 103–111.

Jha, S.N., Sharma, R., 2010. Physical, gravimetric and functional characterization of various milling fractions of popped gorgon nut (*Euryale ferox*). J. Food Sci. Technol. 47 (5), 564–570.

Jha, S.N., Prasad, S., 1998. Mathematical modelling for thermal expansion of gorgon nut during roasting. J. Inst. Eng. India Agric. Eng. Div. 79 (1), 14–18.

Jha, V., Sah, R.S., Prasad, K., 1991. Traditional practices of makhana (*Euryale ferox* Salisb.) cultivation, harvesting and processing in north Bihar. J. Res. Birsa Agric. Univ. 3 (1), 45–52.

Kashyap, A., Shukla, O.J., 2025. Sustainable food supply chain: exploration, identification, and analysis of the critical drivers for the foxnut (Makhana) industry. J. Glob. Oper. Strateg. Sourc. 18 (3), 559–582.

Kidmose, U., Martens, H.J., 1999. Changes in texture, microstructure and nutritional quality of carrot slices during blanching and freezing. J. Sci. Food Agric. 79 (12), 1747–1753.

Kumar, V., Kumar, R., Rajak, D., Kumar, V., Kumar, A., 2013. Influence of Moisture Content and Grade on Engineering Properties of Makhana.

Kumar, S., Sehrawat, R., Kumar, A., Basir, K., Prakash, K.S., 2015. Effect of different grades on physical and mechanical properties of popped Makhana. Trend Biosci. 8 (12), 3194–3201.

Kumar, L., Singh, A.K., Bhatt, B.P., 2016. Nutritional status of recently developed makhana (gorgon nut) variety-Swarna Vaidehi. J. AgriSearch 3 (4), 199–205.

Kumar, J., Kumar, S., Kumar, A., Kumar, S., 2019. Field performance of popping machine for makhana seeds. Int. J. Chem. Stud. 7 (6), 773–777.

Kumar, M.M.M., Goudar, G., Sharma, P., Vishwakarma, R.K., Gogoi, P., Mahajan, A., Chandragiri, A., Kalpuri, S., Geddam, J.J.B., Suresh, C., Chary, M., Venketrajireddy, G., Radhika, M., Sreenu, P., Tulja, B., 2023. Differentiating the nutrient composition, in-vitro starch digestibility, individual polyphenols and anti-oxidant properties of raw and popped makhana (*Euryale ferox*). J. Food Meas. Char. 17, 5828–5844. https://doi.org/10.1007/s11694-023-02074-5.

Kumari, P., Noel, A.S., Rai, A., 2023a. Analysis of the processing cost and constraint faced in foxnut (Makhana) processing. Int. J. Eng. Res. 12 (4), 75–80.

Kumari, P., Noel, A.S., Rai, A., 2023b. Analysis of the processing cost and constraint faced in foxnut (Makhana) processing. TIJER-Int. Res. J. 10 (4), 585–589.

Li, Q., Li, H.T., Bai, Y.P., Zhu, K.R., Huang, P.H., 2022. Effect of thermal treatment on the physicochemical, ultrastructural, and antioxidant characteristics of Euryale ferox seeds and flour. Foods 11 (16), 2404.

Mahawar, H.K., 2016. A report on Makhana (Fox nut) production, processing and supply chain IIT (pp. 3–27. Tech. Rep. https://www.researchgate.net/publication/31896-0974.

Mani, A., 2019. Post harvest technology, processing and value addition of makhana. Agrobios 18 (2), 55–58.

Manju, K.M., Rekha, P., Kumar, N., 2021. Effect of fluidized-bed and freeze-drying techniques on physicochemical, nutritional, thermal, and structural properties of *Moringa oleifera* flowers, leaves, and seeds. J. Food Process. Preserv. 45 (9), e15719.

Mishra, R.K., Kumar, S., Singh, D., 2021. Influence of drying and roasting conditions on physical and popping characteristics of makhana seeds. J. Food Process. Preserv. 45 (3), e15218.

Mujumdar, A.S., 2014. Handbook of Industrial Drying, fourth ed. CRC Press, Boca Raton, FL.

Priyadarshi, A., Ahmad, H., 2025. Makhana and the sustainable development goals: a pathway to nutrition, livelihoods, and environmental sustainability. Int. J. Adv. Multidiscip. Res. Stud. 5, 1481–1483.

Ramaswamy, R., Krishnamurthy, K., Jun, S., 2012. Microbial decontamination of food by infrared (IR) heating. In: Microbial Decontamination in the Food Industry, pp. 450–471.

Regier, M., Rother, M., Schuchmann, H.P., 2010. Alternative heating technologies. In: Processing Effects on Safety and Quality of Foods, pp. 187–229.

Rosenthal, I., 2012. Electromagnetic Radiations in Food Science. Springer-Verlag, Berlin, Germany.

Saha, S., Jha, V., 2018. Effect of harvesting and post-harvest operations on popping quality of makhana (*Euryale ferox*). J. Food Sci. Technol. 55 (9), 3562–3569.

Sakai, N., Hanzawa, T., 1994. Applications and advances in far-infrared heating in Japan. Trends Food Sci. Technol. 5 (11), 357–362.

Sharma, K., Patel, S., Jha, S.N., Devi, M., Vishwakarma, R.K., 2021. Rotating orifice feeding system for continuous and uniform discharge of makhana seeds (*Euryale ferox*). J. Food Eng.

Sharma, K., Vishwakarma, R.K., Patel, S., Jha, S.N., Devi, M., 2023. Organic production, post-harvest processing, and value-added intervention of fox nuts (*Euryale ferox*) for export. In: Transforming Organic Agri-Produce into Processed Food Products. Apple Academic Press, pp. 363–393.

Sharma, K., Patel, S., Devi, M., Vishwakarma, R.K., 2024a. Effect of packaging material and storage period on popping behaviour of roasted makhana seeds. J. Agric. Eng. 61 (4), 525–538.

Sharma, K., Patel, S., Jha, S.N., Devi, M., Nandan, S., Vishwakarma, R.K., 2024b. Change in popping, structural, functional, and crystalline properties of roasted makhana seeds in prolong storage: an FTIR, GC-MS, SEM, XRD study. Research Square 1–18.

Sharma, K., Patel, S., Jha, S.N., Devi, M., Nandan, S., Vishwakarma, R.K., 2024c. Aging-induced changes in popping, structural, functional, thermal, and crystalline properties of roasted makhana seeds. Food Bioprocess Technol. 17 (12), 5023–5037.

Sharma, K., Patel, S., Jha, S.N., Devi, M., Nandan, S., Vishwakarma, R.K., 2024d. Change in Popping, Structural, Functional, and Crystalline Properties of Roasted Makhana Seeds in Prolong Storage: An FTIR, GC-MS, SEM, XRD Study.

Sheridan, P., Shilton, N., 1999. Application of far infra-red radiation to cooking of meat products. J. Food Eng. 41 (3–4), 203–208.

Singh, A.K., Mishra, J.S., Kumar, R., 2020. Constraints and prospects of makhana cultivation and processing in India. Indian J. Agric. Sci. 90 (6), 1085–1090.

Singh, D.K., Kumar, A., Kumar, U., Chandra, N., Singh, I.S., 2021. Value addition in makhana through post-harvest operations. Food Sci. Rep. 2, 29–32.

Sumnu, G., Sahin, S., 2005. Recent developments in microwave heating. In: Emerging Technologies for Food Processing, pp. 419–444.

Tewari, G., 2007. Microwave and radio-frequency heating. In: Advances in Thermal and Non-Thermal Food Preservation, pp. 91–98.

Tian, Y., Liang, J., Zeng, H., Zheng, B., 2013. Microwave drying characteristics and kinetics of lotus (*Nelumbo nucifera Gaertn.*) seeds. Int. J. Food Eng. 9 (1), 91–98.

Trivittayasil, V., Tanaka, F., Uchino, T., 2011. Investigation of deactivation of mold conidia by infrared heating in a model-based approach. J. Food Eng. 104 (4), 565–570.

Verma, A.K., Banerji, B.K., Chakrabarty, D., Datta, S.K., 2010. Studies on makhana (*Euryale ferox* Salisbury). Curr. Sci. 795–800.

Verma, D.K., Mahanti, N.K., Thakur, M., Chakraborty, S.K., Srivastav, P.P., 2020. Microwave heating: alternative thermal process technology for food application. In: Emerging Thermal and Nonthermal Technologies in Food Processing. Apple Academic Press, pp. 25–67.

Vukadinović, J., Srdić, J., Kravić, N., Mladenović Drinić, S., Simić, M., Brankov, M., Dragičević, V., 2024. Assessment of popcorn's bioactive status in response to popping. Molecules 29 (4), 807.

Wang, Y., Wig, T.D., Tang, J., Hallberg, L.M., 2003. Dielectric properties of foods relevant to RF and microwave pasteurization and sterilization. J. Food Eng. 57 (3), 257–268.

Wardhan, H., Suntwal, T., Gulati, A., 2025. Strategies to Boost India's Makhana Exports (No. 25-r-13). Indian Council for Research on International Economic Relations (ICRIER), New Delhi, India.

Zhao, Y., Zhu, H., Xu, J., Zhuang, W., Zheng, B., Lo, Y.M., Tian, Y., 2021. Microwave vacuum drying of lotus (*Nelumbo nucifera Gaertn.*) seeds: effects of ultrasonic pretreatment on color, antioxidant activity, and rehydration capacity. Lwt 149, 111603.

Chapter 7

Physicochemical properties of Makhana seeds and popped kernels

Reza Karimi[1], Ayda Kargar-Ghavidel[1], Rezvane Nategh-Khoshnoud-Roohi[1] and Aliakbar Nategh-Khoshnoud-Roohi[2]
[1]*Department of Food Science and Technology, Faculty of Agricultural Sciences, University of Guilan, Rasht, Iran;* [2]*Department of Cellular and Molecular Biology, Faculty of Basic Sciences, Rasht Branch, Islamic Azad University, Guilan, Iran*

7.1 Introduction

Euryale ferox is a well-known cash crop of the family Nymphaeaceae that is grown in aquatic environments and has acquired international names, including gorgon nut and prickly water lily (Gaurav et al., 2022). Historical and ethnobotanical literature emphasizes that the entire plant has tonic, astringent, and nonobstructive properties, which also highlights the multifaceted nature of the medicinal importance of the plant (Sodi and Kumar, 2019). When the fruits are fully matured, they release a notable number of seeds, and those seeds that are not mature, such as those that are whitish in color, stay on the surface of the pond for several days and then finally descend to the substrate. Generally, the reproductive phase of the plant starts in April (EFSA, 2022). The morphology of the individual plants consists of about eight to nine leaves, containing flavonoids, phenolic acids, and alkaloids, which exhibit their traditional and medicinal use for antiinflammatory and diuretic purposes. Flowers, which begin to emerge in an alternate pattern that fits perfectly between the leaves, are purple in color, but slowly change to any other as they proceed to produce fruits (Khadatkar et al., 2020).

The roots of *E. ferox* are characterized as elongated and fibrous in nature, and most roots usually develop two to three clusters, which have many air pockets that enable them to be stable in water (Khadatkar et al., 2020). Substantial elements, such as iron, copper, and magnesium, exist abundantly in the crop, while essential macronutrients, such as nitrogen, potassium, and phosphorus, are distributed in the vegetative tissues of the plant (Bisht et al., 2025).

Makhana (Euryale ferox) as a Functional Food. https://doi.org/10.1016/B978-0-443-44421-0.00016-0

The seeds are known by their strong outer coating, black color, and spherical shape, and diameters between 4.5 and 14.5 mm. The kernel of the nut is surrounded in the outer cover, which is the edible part. The extraction of kernels from raw seeds is not easy, as it demands traditional procedures of manual popping to make it edible (Arjun et al., 2017).

Findings exhibited that the protein composition of dry endosperm of *E. ferox* seeds contained notable number of amino acids, including arginine and histidine. This seed source had better DPPH and FRAP activities than other counterparts in antioxidant potential assessment. The endosperm is a remarkable source of vital vitamins, such as B_1 and C, and a major source of flavonoids, including quercetin and kaempferol, which are significant in effective glucose absorption in cellular functions (Fig. 7.1) (Jana et al., 2024).

So far, the traditional techniques are the most common mechanisms used for popping *E. ferox* seeds. This can be done by immersing the nuts in water before heating so as to make the extraction of the starchy edible kernel easier. During the process of popping, the moisture that is present inside the seed is superheated, creating vapor, which when suddenly released through pressure, results in a notable amount of expansion to the kernel. The expanded kernel, which is also called popped kernel, is generally known as Makhana in India. Fox nuts (popped Makhana) are a result of *E. ferox* seeds exposed to heat treatment, and this treatment renders them edible and palatable (Liaquat et al., 2022; Sahithi et al., 2022).

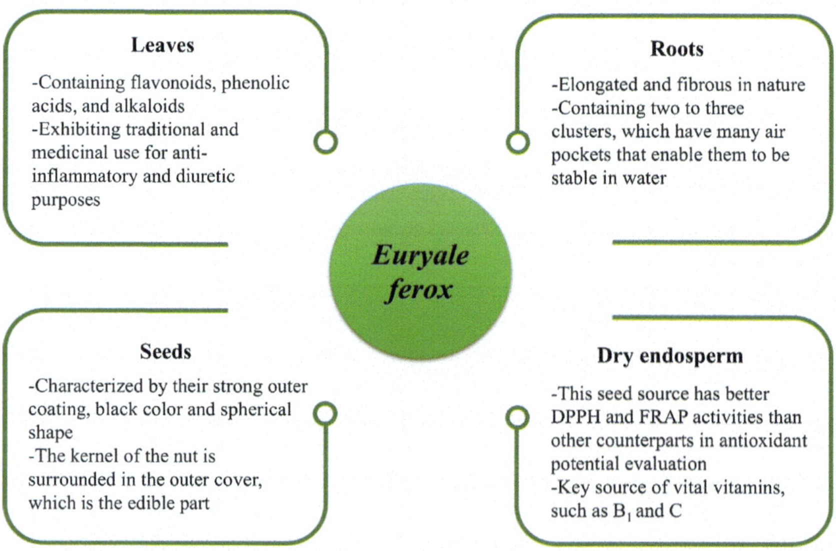

FIGURE 7.1 Bioactive components and functional characteristics of makhana plant parts.

Along with holding a significant promise of being processed in large-scale manufacturing of snack foods, the *E. ferox* seeds can be consumed as a notable ingredient of vegetarian diets after frying (Arjun et al., 2017). As a strong recommendation for preventing cardiovascular diseases, popped *E. ferox* is also a great component used for low-fat diets and the Dietary Approaches to Stop Hypertension (DASH) plan (Haleema et al., 2018).

Makhana pops are substantial sources of energy needed by the body to support optimal physiological functions, since they are remarkable sources of carbohydrates and micronutrients, also including low fat content (Khadatkar et al., 2020; Sodi and Kumar, 2019). In addition to high fiber content, they encompass antiaging enzymes and antioxidant kaempferol, which aid in preventing constipation, repairing damaged proteins, and mitigating inflammatory processes (Sodi and Kumar, 2019).

Negligible amount of cholesterol and sodium with notable gluten-free amino acid composition and high concentration of fibers render Makhana seed pops as an effective food item in the diets of people with cardiovascular diseases, high blood pressure, diabetes, kidney problems, and obesity (Sodi and Kumar, 2019). Research also exhibits that the intake of these seeds also assists in repairing the elasticity of arteries and veins, contributes in maintaining healthy tissues, and consequently may have antiaging effects (Joseph and Ramesh, 2023). The nutrition analysis shows that the fiber composition of *E. ferox* seeds consists approximately 61% of insoluble fiber and 39% of soluble fiber, while starch is considered as the major carbohydrate fraction (Haleema et al., 2018). Along with carbohydrates, the seeds are supplemented with proteins and essential amino acids, along with flavonoids, cyclic dipeptides, vitamin E, sterols, and lipids. Their calories are also noteworthy, with the raw seeds containing about 362 kcal per 100 g and the popped form providing 328 kcal per 100 g (Mittal et al., 2020).

Strong scientific foundation to *E. ferox* seeds and other parts of the plant application is available as a part of traditional medicine to treat urinary tract infections, since studies conducted on nine clinical bacterial strains showed that *E. ferox* seeds and various plant parts methanolic extracts, revealed a strong antibacterial effect (Mittal et al., 2020). In addition to their pharmacological usage, these nuts are considered to be very nutritious as they often take the top position in the list of dried fruits, due to their impressive essential amino acid index ranging between 89% and 93% (Liaquat et al., 2022).

Makhana pops have potent medicinal properties and are prescribed in Indian and Chinese traditions to treat digestive, excretory, and respiratory system disorders (Sodi and Kumar, 2019). Popped Makhana, also sold as Makhana lava, is a valuable cash crop, and the economic importance of its seeds cannot be underestimated (Rathod et al., 2023). The grading of the popped kernels also emphasizes their market value. First-grade Makhana, also called rasgulla accounts for around 54% of the total production, followed by second-grade murra at 35%, and the rest 11% of the total is thurri (Sahithi et al., 2022).

Vacuum-packed Makhana packs are now found in the shelves of most supermarkets in a variety of flavors, including sour cream, onion, chocolate, etc. due to their rise in popularity in the food market. Popped Makhana is applied in various culinary applications, it is added to dishes such as dal Makhana with vegetables, can be eaten raw or roasted with tea or coffee after adding salt, ghee, or other oils to it (Rathod et al., 2023). In the traditional cuisines, seeds of *E. ferox* are often used in curries and puddings, and salted and fried Makhana are one of the popular snacks usually accompanied by tea or coffee. Makhana has cultural and spiritual value beyond its food applications as it is also used in ritual practices. Besides, the seeds are also an excellent source of starch, which is useful in a wide array of industrial manufactures (Sahithi et al., 2022).

7.2 Chemical composition of makhana seeds

The seeds of *E. ferox* fall under the category of a functional food as they contain a dense and diversified nutrient profile. The key part of their composition is carbohydrate, specifically starch, with a notable percentage of high-quality protein. Seeds also include essential dietary minerals, such as magnesium, potassium, iron, and zinc, in addition to substantial levels of essential micronutrients, including vitamins C and E (Kapoor et al., 2022).

7.2.1 Macronutrient profile

People are no longer content with the traditional carbohydrate-based diets, and due to this issue, the transition to protein-based diets is very noticeable in the current dietary trends, where consumers are interested in approaches that improve their health and eating habits (Jana et al., 2019); in this context, Makhana has become a potential functional food, which represents remarkable health-promoting effects. The perisperm is the major edible part of the Makhana with high levels of starch and quality protein content with almost negligible amounts of fat. This type of composition has received a lot of demand in the developed countries, where the issue of obesity has become a major social health issue (Sinha and Jha, 2025). These attributes support the importance of Makhana as a nutrient-rich food with the capability to enhance the health of the diet worldwide.

Fox nuts offer an excellent source of essential amino acids (Fig. 7.2), including seven out of nine needed by the human body, but they do not contain phenylalanine and tryptophan (Joseph and Ramesh, 2023). The endosperm of the raw seed contains specifically high levels of glutamic acid, aspartic acid, arginine, and alanine (Jana et al., 2024). Among them, arginine has a critical role in increasing the insulin sensitivity, and concurrently, glucose metabolism to manage diabetes in the body is effectively achieved. It also helps in the decrease of accumulation of white adipose tissue, which is

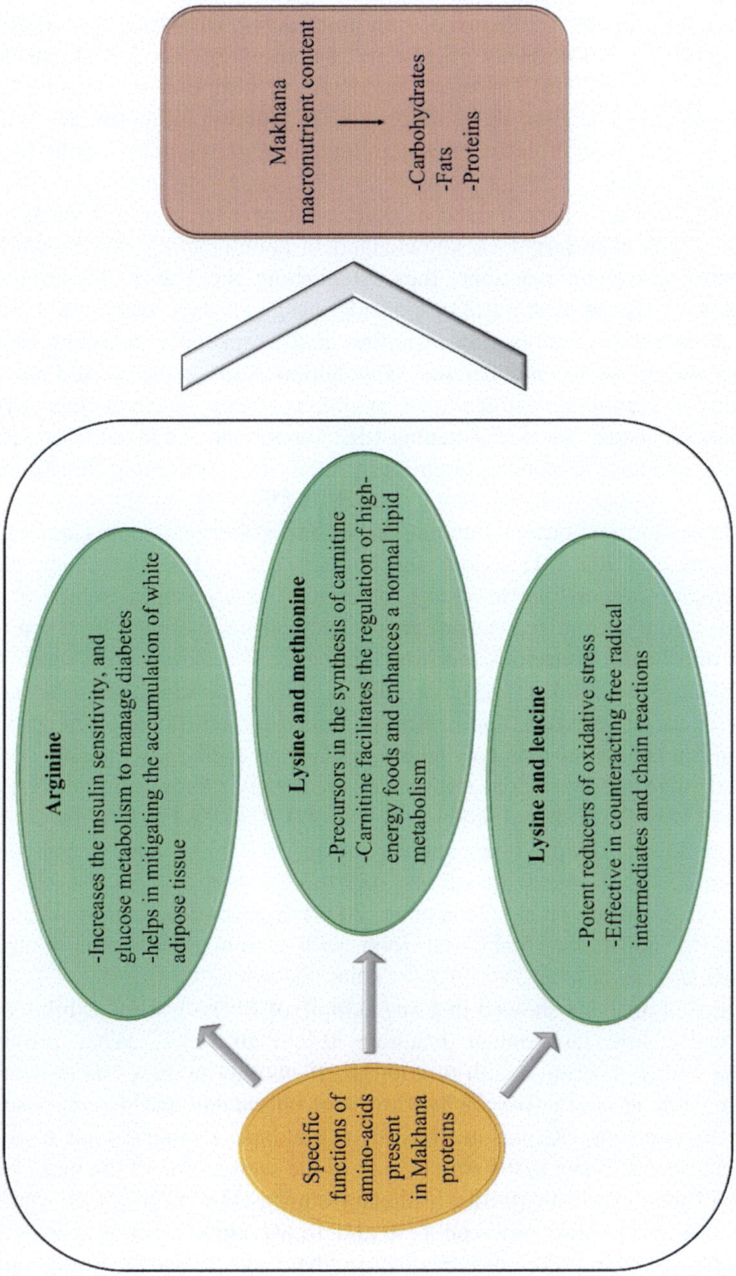

FIGURE 7.2 Functional effects of key amino acids present in makhana protein content.

vital in alleviating the effects of obesity (Jana et al., 2022). Besides, Makhana seeds also contain lysine and methionine, which are essential amino acids and precursors in the synthesis of carnitine. Also, known as a natural fat burner, carnitine facilitates the regulation of high-energy foods and enhances normal lipid metabolism. Therefore, makhana consumption as a regular food supplement could trigger the endogenous production of carnitine, which supports its role as an effective food supplement to maintain metabolic health (Jana et al., 2022).

Amino acids are characterized as potent reducers of oxidative stress, and their usage has been largely acknowledged in counteracting free radical intermediates and chain reactions, thus establishing them as highly effective antioxidants (Rathod et al., 2023). Considering given data, makhana demonstrates an exclusive composition of amino acids, especially including highly concentrated in lysine and leucine. The amino acids serve as critical determinants in reducing intracellular oxidative stress and reducing glyco-oxidation processes in strengthening the functional value of the seeds. Moreover, makhana contains gluten-free protein, thus becoming nutritionally favorable to people with disorders relating to digestion in the presence of gluten; however, it can cause immune responses in allergic people (Jana et al., 2022; Sinha and Jha, 2025; Sodi and Kumar, 2019). Taken together, these characteristics suggest that Makhana serves both as a potential source of antioxidants and a specialized dietary product; therefore, it should be taken into account in a health-conscious nutrition strategy.

Makhana seeds, however, have varying levels of crude protein based on differences in varieties, application of fertilizers, and location of growth. It is important to note that cotyledons and whole seeds have greater levels of protein than the seed coat (Kumar et al., 2016). Compositional analysis of the seed kernel has shown protein content of $9.70 \pm 0.62\%$, total fat content of $0.31 \pm 0.01\%$, crude fiber content of $0.7 \pm 0.03\%$, and carbohydrate content of $75.7 \pm 1.74\%$ with ash and moisture content of $0.52 \pm 0.02\%$ and $11.15 \pm 0.43\%$, respectively (Jana et al., 2024). Studies designate makhana as a source of nutritious, functional food with a remarkable macronutrient composition.

Consistent analysis showed that any sample of *E. ferox* seeds exhibited an exceptionally low fat content reaching less than 0.5%. What provides Makhana with significant health-promoting effects and increases its utilization as a nutraceutical alternative are low levels of fat content and its composition of nutrient elements (Kumar et al., 2016). Makhana's contribution to functional food sector is due to the starchy part of the seed as one of the main keys, which will also create flexibility in dietary choices (Devi, 2022). It contains amylose content ranging between 17% and 18.6% among different varieties, which acts as a fructooligosaccharide in the large intestine. Fructooligosaccharides enhance gut functions due to development of positive microbial activities and digestive processes (Jana et al., 2022).

7.2.2 Micronutrient distribution

E. ferox seeds' mineral content is remarkable, since they contain ash content ranging around 3%–5% (Islam et al., 2025). These seeds are categorized as a valuable vitamin containing resource, including notable levels of vitamin A (0.18%) and vitamin C (0.20%) (Bisht et al., 2025). These micronutrients help increase the nutritional value of the seeds and strengthen the chances of their utilization as a functional food. Nevertheless, there is a significant decrease in key micronutrients in processing, and this is mainly due to the elimination of the bran layer. This section serves as the main source of bioactive compounds, phenols, flavonoids, and various antioxidants. Its removal contributes greatly to the reduction of nutritional content of the final product, hence lowering the health-promoting advantages of processed Makhana (Kapoor et al., 2022). Preserving nutritional features and the bioactive constituents of the product through optimized processing methods seems significant.

Makhana is a source of most of the essential nutrients, such as a moderate amount of magnesium, which has been recognized to be scarce in the human body, which is strongly linked to a higher possibility of obesity (Jana et al., 2022). In addition to magnesium, makhana seeds also include a wide range of trace elements, such as calcium, copper, and sodium, but all the concentrations are likely to drop during the purification of starch (Sodi and Kumar, 2019). It is also important to note that makhana has a high mineral profile of potassium and phosphorus (Jana et al., 2024). All these minerals operate together to provide the seeds with their functional values and help to maintain metabolism, the cardiovascular system, and nutritional balance.

The sodium content of *E. ferox* is 180–240 mg per kilogram, which means that makhana is a sodium-accumulating crop. This is due to high sodium uptake capacity, even when using low-sodium soils, and the reason Makhana is not found in the ponds within salt-affected calcareous areas with high amounts of sodium in the soil (Dutta et al., 1986). Although raw kernels include elevated concentration of phosphorus, comparative studies demonstrate higher amount of minerals such as potassium, iron, and calcium in popped seeds (Kumar et al., 2016). Makhana seeds also have sulfur levels ranging between 47 and 60 mg per 100 g and are also highly abundant in total divalent cations (Jana et al., 2022).

It has been observed that during the harvesting season, seeds tend to accumulate elevated quantities of toxic metals, such as lead, chromium, copper, and cadmium. This contamination is mainly caused by sewage release into the environment by domestic and urban sources, where serious risks to the natural habitats of the plant are known to root from. The amount of toxic elements that were present in the surrounding water and sediments were almost on par with the quantities that were reported in the seeds (Rai et al., 2002). On the other hand, significant concentrations of selenium and chromium, also classified as heavy metals, were reported to exist in the analyzed dry kernel powder sample. Although these heavy elements can beneficially act as antioxidants, they also play a significant role in enhancing insulin sensitivity among diabetic patients. This dual nature of the mineral composition of Makhana as a potential risk factor under polluted conditions and a source of bioactive compounds with therapeutic properties is complicated (Jana et al., 2024).

7.2.3 Phytochemical constituents and antimicrobial activity

Flavonoids are one of the large groups of secondary metabolites and phyto-components present in fruits and vegetables and nuts, and in this case, play an important role in nutritional and functional features (Jana et al., 2024). Flavonoid presence is confirmed by phytochemical assessment. A relative evaluation of the processing stages shows that the raw seeds produce more of these compounds. Thermal popping not only lowers this total variety but also forms a number of novel and distinct constituents that cannot be observed in the raw material (Sinha and Jha, 2025). The European food safety authority has reported that *E. ferox* like any other aquatic plant tends to take up trace elements, heavy metals, cyanotoxins, such as microcystins, and even pesticide residues of the surrounding aquatic environment. In turn, the type of cultivation activities has a direct effect on the concentrations of these contaminants in the plant, emphasizing controlled and sustainable agricultural management (EFSA, 2022).

Antinutritional factors are naturally occurring food substances that inhibit the normal nutrient intake by interfering with their assimilation into the body via various mechanisms, which include deactivation of nutrients, poor digestion, and changes in the metabolism of seeds, thus limiting nutrient absorption. Some of the processing methods, such as soaking, germination, and genetic modification, have been shown to mitigate most of these undesirable plant components (Bisht et al., 2025). Even though the seeds of *E. ferox* have a great variety of health-promoting effects, such as other plant seeds, it is important to note that they also contain antinutritional constituents, such as tannins, phytic acid, and total phenolics. These substances affect the nutritional quality of the seeds by inhibiting the enzymatic activity and lowering mineral bioavailability. Therefore, it is necessary to develop optimal processing protocols that can improve the nutritional quality of Makhana and reduce the negative impact of those antinutritional factors at the same time (Islam et al., 2025).

It has been demonstrated that germination of *E. ferox* seeds has a significant positive effect on its nutritional profile. The protein content measured in the germinated seed flour exceeded 15% compared to 12.31% in the raw flour. This can be linked to the enzyme activities, which are more efficient in breaking down and synthesizing the proteins in the flour. What renders makhana seeds suitable for daily consumption is the germination progress, which reduces the antinutritional factors through depletion of total phenolic and tannin levels. Germinated makhana flour, with enhanced nutritional value and functionality, emerges as a perfect source of developing nutritionally enhanced functional foods (Islam et al., 2025).

Additional research reveals that there is a significant decrease in the total phenolic content in germinated Makhana flour compared to the raw flour. The enzymatic breakdown that is triggered by the polyphenol oxidases and peroxidases is the primary cause of this reduction, and these enzymes oxidize or structurally alter the phenolic compounds and hence accelerate their degradation. Even though the reduced levels of phenolics can mitigate some of the antinutritional effects as seen in cases of poor mineral absorption, it should be noted that the phenolics have their own share of positive antioxidant effects (Islam et al., 2025). Therefore, despite the improvement of nutritional properties by reducing the number of antinutrient components, the resulting decrease in the level of phenolics could negatively affect the properties of antioxidants, which is a possible nutritional substitution (Islam et al., 2025).

Studies on the phytochemical constituents of *E. ferox* seed coat extracts have determined considerable concentrations of bioactive constituents. The total phenolic content and total flavonoid content were quantified at approximately 50.96 mg of gallic acid equivalents per gram and approximately 16.27 mg of quercetin equivalents per gram, respectively, indicating the elevated content of antioxidant compounds in the extract. In other studies, a wide variety of phytochemicals, such as flavonoids, terpenoids, tannins, glycosides, etc., were identified in ethanol and water seed coat extracts (Joseph and Ramesh, 2023).

Phytochemicals known as alkaloids, which are commonly found in the seeds of plants, are examples of compounds of significant pharmacological interest; common representatives, such as caffeine and atropine, have long been examined due to their physiological activity. These alkaloids, together with other phytochemicals that are identified in the fox nut seed coats possess high bioactivity, functioning as central nervous system stimulants, anesthetics, analgesics, hypoglycemic agents, and antioxidants. These results indicate the potential medicinal properties of the *E. ferox* seed coat extracts and justify their use in the traditional and contemporary treatment methods (Joseph and Ramesh, 2023).

When the methanolic seed extract of *E. ferox* was fractionated into multiple solvent components, the ethyl acetate fraction was found to be the strongest inhibitor of endogenous enzymes, which was involved in the decrease of oxidative stress (Mittal et al., 2020). Additional studies have widely investigated the antimicrobial effects of *E. ferox*. Extracts of the seed coat prepared in three different solvents showed significant antimicrobial effects on pathogenic strains, such as *Staphylococcus aureus* and *Escherichia coli*. In addition to the antibacterial activity, extracts of the leaf and seed of *E. ferox* showed antifungal activity, which indeed targets *Candida albicans* and *Penicillium notatum* (Devi, 2022).

7.3 Physical properties of makhana seeds

7.3.1 Physical attributes

Makhana has specific physical properties with moisture content ranging from 5% to 20% on a dry weight (DW) basis, 1000 seed weight reaching 286–384 g, and bulk density ranging from 56.5 to 84.6 kg per cubic meter (Jha, 1999). Kernels in a commercial trade are graded into four categories: lava, murha, turi, and mix, in terms of the size of the popped kernels, as a practice of standardized quality assessment (Kapoor et al., 2022). In addition to physical and commercial properties, the isolated starch granules from Makhana seeds exhibit a specific polyhedral shape, where the diameter of the grains is between 1 and 3 mm. Structural characterization indicates that the amylose content of the starch is significantly high, and crystallinity is comparatively high, as it has been established through X-ray diffraction analysis. The diffraction pattern is that of an A-type crystal structure, which is indicative of a well-organized molecular structure and large crystalline areas. Thermal analysis also indicates that Makhana starch has a high gelatinization temperature and substantial enthalpy and this is why it has a high thermal stability (Kapoor et al., 2022).

7.3.2 Hydration, solubility, and swelling behavior

E. ferox seed processing is a series of steps that are sequential and focused on maximizing the quality of the kernels. The primary stage of the process is drying of seeds, as this stage helps to separate the inner kernel and the outer seed coat. They are dried and then divided into grades by means of a series of sieves. These graded seeds are then heated using pans with a temperature of about 200–350 °C and duration of around 6 min. After the heating process, the seeds are placed in open baskets for 40–50 h, which is a very crucial step that allows the seed coat to completely loosen and detach from the kernel (Kapoor et al., 2022). In addition, functional characteristics of the material increase notably with an elevation in the capacity to absorb water and oil, swelling power, and solubility (Kapoor et al., 2022). It can be concluded that controlled thermal processing not only makes the separation of kernels easier but also enhances the functional and physicochemical characteristics of Makhana seeds, which expands their possible uses in food preparations.

High drying temperatures also altered the functional, physical, and thermal properties of the flour greatly. Microscopic observations showed that the structural integrity of starch granules was changed under high temperatures, which confirmed the development of gelatinization. The enthalpy of gelatinization decreased alongside the mentioned alteration, which is an indicator of the changes in thermal behavior of starch (Bhat and Arya, 2021). Another applied heat treatment was heat moisture treatment (HMT), which used temperatures more than 90 °C, moisture levels below 35%, and durations under

24 h, which proved effective in controlling intrachain interaction and mobility of functional groups in starch. HMT is a highly applicable physical modification method that is environmentally friendly, nontoxic, and cost-effective, which has been used widely to enhance the functional limitations of native starch-based foods (Bhat and Arya, 2020).

HMT is a physical process that rearranges the inner structure of the starch granules and results in a modified starch with a distinct functional profile. Some major effects of HMT are low solubility, decreased granule swelling, less amylose leaching, higher gelatinization, and pasting temperatures, thus leading to better thermal stability and better physicochemical performance in food matrices. Though the variety of effects on starch are dependent on the exact treatment parameters used and the botanical source of the starch. During the treatment of *E. ferox* flour, HMT modified the granular structure, which caused the agglomerated morphology (Bhat and Arya, 2020). The treatment continued to decrease the number of functional characteristics, such as water absorption capacity, solubility, and swelling power. The swelling capacity of HMT-modified flour was significantly less than that of native flour and more pronounced decreases were recorded at harsher treatment conditions. This reduction in swelling can be explained by the high pasting temperature caused by HMT that stimulates the rearrangement of the molecules and can lead to the partial degradation of amylopectin chains in granules of starch (Bhat and Arya, 2020). These data demonstrate such duality of HMT in altering starch structure and functionality with both increased stability and controlled functionality in food usage.

The recent realization of the botanical classification of Makhana as a superfood has greatly enhanced its proper recognition as an aquatic crop. Despite the fact that Makhana and lotus plant (*Nelumbo nucifera*) belong to the same family, Nymphaeaceae, Makhana is mistakenly taken to be the product of the lotus plant (Sinha and Jha, 2025). The lotus seed starch was commercially produced in China, where it has been used for many years as a part of confectionery or a food additive. Annual production is also increasing in line with the increase in demand in the market. However, the rate of using native lotus seed is not very high, mainly because this seed has a strong retrogradation tendency and lacks storage stability. This distinction underlines the significance of differentiating between Makhana and lotus, not only in scientific studies but also in commerce, to clearly understand the full nutritional and functional potential of Makhana (Guo et al., 2019).

Results have shown that lotus seed starch, like most other botanical starches, is a convenient agent to obtain viscosity, stability, gel formation, filler, and moisture retention in the food and industrial sector. This starch is characterized by a high content of amylose, which makes up a significant percentage of about 40% (weight/weight) of its composition and the total starch content of about 500 g per kilogram DW (Guo et al., 2015). A similar trend was recorded in a study involving the examination of functional characteristics of *E. ferox* starch, which has proved to be significantly useful in food applications. Its modified variants have been useful in formulating pickering emulsions while also functioning as a stabilizing agent. This application is associated with wider culinary applications of starches as fat replacements in baked products that adds value to the ingredient in making stable food structures and enhances the textural characteristics (Kapoor et al., 2022), thus, it can be concluded that lotus and Makhana starches have a similar functional potential that demonstrates their relevance in various food and nonfood formulations.

Along with the elevation in the moisture content, the physical behavior of *E. ferox* seeds varies significantly. An increase in moisture content escalates the angle of repose, which is mainly due to an increase in cohesion between particles by increased humidity that increases the frictional force experienced during movement (Arjun et al., 2017). It is also observed that high moisture is associated with a higher value of the static coefficient of friction, which exhibits the same effect of stronger particle cohesion. The high friction alters Makhana seeds material flow features, thus characterizing the moisture content as a determinant variable in seeds operating behavior in the handling, storage, and processing phases (Arjun et al., 2017).

The equilibrium moisture of Makhana seeds enhances with the rise of relative humidity and decreases with an improvement in seed grade, perhaps due to structural variations affecting absorption and retention of moisture (Jha, 1999). These properties are further modified by germination, which increases the water retention potential of fox nut seeds, but at the same time, decreases the ability of seeds to hold oils. Both of these implications are significant to food formulation and innovative product development in that they allow the functional versatility of the seeds to be broadened and a change in the lipid-related properties of the seeds (Bisht et al., 2025).

7.4 Physicochemical changes during popping and heat treatment

7.4.1 Moisture and textural changes

The traditional procedure of *E. ferox* seeds processing involves an initial heating of sun-dried seeds in an earthen pitcher or cast-iron pan over an open flame, with constant stirring of the pan contents between 5 and 10 min at temperatures between 220 and 300°C (Fig. 7.3). This process reduces the moisture content of the seeds to about 20%. Subsequently, the preheated seeds are tempered between 18 and 72 h under ambient conditions, which helps in loosening the kernels in the hard shell. Tempered seeds are then roasted in a single layer at 280 to 340°C for 1.5–3 min with constant stirring until cracking occurs. Cracked seeds are gathered in a short time, spread on a solid surface,

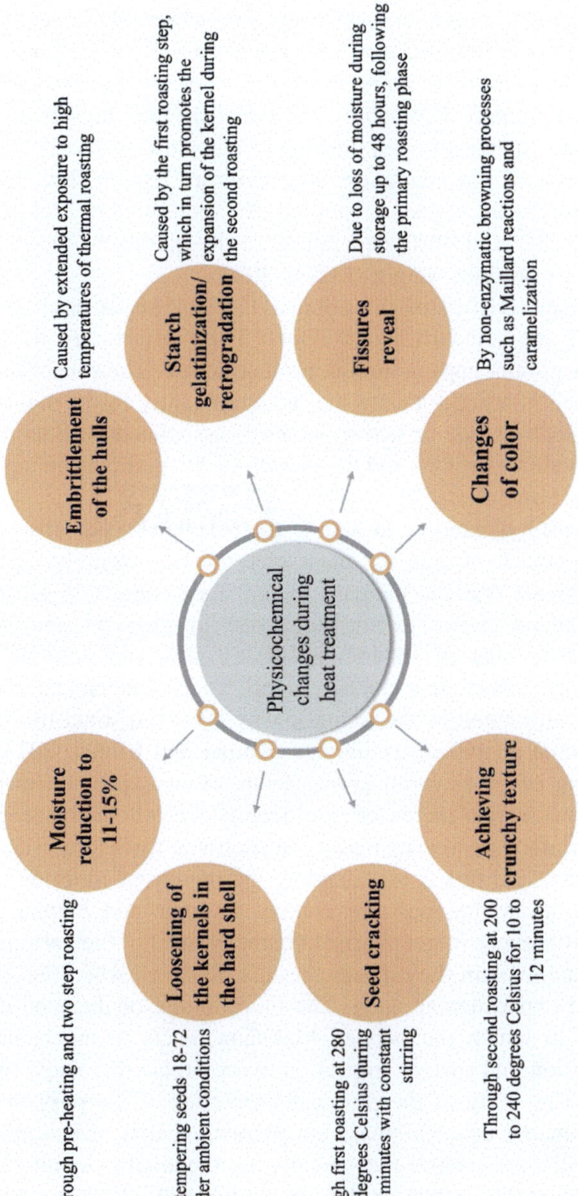

FIGURE 7.3 Operational steps and physicochemical transformations in Makhana thermal processing.

and beaten with a wooden hammer to release the swollen kernel, which is also known as Makhana (EFSA, 2022).

The popped kernels are then roasted again to achieve the crunchy texture in special equipment at 200 to 240 °C for approximately 10–12 min, normally with refined olive oil or olive pomace oil. The step also mitigates the moisture content to approximately 11%–15%. The pops can then be salted or flavored and then packaged under nitrogen flushing to maintain the level of oxygen to a minimum level in closed containers and therefore have a shelf life of 6–12 months as per the standard documentation. The total yield of Makhana usually varies between 35% and 40%, depending on the quality of the raw seeds in comparison to its original weight (EFSA, 2022).

Physical characteristics of the popped *E. ferox* seeds have a distinctive difference from the standard cereal flours. The particles are likely to stick loosely, leaving greater gaps in their general structure. The consequence of this structural change is a bulk density that is significantly lower than that of the normal flours, such as rice or wheat, which is associated with the increase in the size of individual particles and the growth of the particles owing to the rise in porosity (Haleema et al., 2018).

Since they are hygroscopic in nature, the Makhana seeds absorb moisture whenever subjected to it, and the mass increases accordingly. Experimental analyses also reveal that the increase in moisture content in all the studied grades of Makhana shows a consistent rise in physical and gravimetric properties, such as that of sphericity, surface area and volume, but some geometrical properties show a contrary trend. These mechanical and physical properties were evaluated in a systematic way by two distinct moisture levels using conventional analytical techniques (Kumar and Kumar, 2015).

The swelling of seeds in all grades leads to an expansion in their three linear dimensions, which increases significantly with the increase in moisture content. Higher moisture also increases the frictional forces experienced by the seeds on surfaces, and this is explained by the increased cohesion that occurs with an elevation in the moisture content (Kumar and Kumar, 2015). In addition, the differences in geometrical characteristics of the various grades of popped Makhana explain the different results observed when the samples are evaluated in an experimental study. The morphology of the seed directly reflects the way in which the popped Makhana reacts to mechanical forces, including vibration and surface impact, at every processing stage (Kumar and Kumar, 2015).The study of the physical properties of the popped Makhana, including chromatic profiles, mass properties, spatial measurements, and mechanical resilience, contributes greatly to the quality control after harvesting and allows the correct grading procedures (Table 7.1) (Gaurav et al., 2022).

Processors usually keep the temperature of the initial and final roasting consistent to make the operations easier. A constant roasting temperature would therefore be better, with any adjustment; it would require a change in

TABLE 7.1 Effect of thermal processing on physical properties of makhana from selected varieties.

Variety	Geographical origin	Plant part	Thermal processing	Physical profile changes compared to raw makhana	References
–	Market of Saharsa and Purnea	Seed	Roasted stage	Increase in hardness Decreased in moisture content and angle of repose	Kumar et al. (2019)
–	Grocery store at Faisalabad, Pakistan			Increase in redness and calorie density Decreased in bulk density and hardness	Liaquat et al. (2022)
Swarn Vaidehi	Krishi Vigyan Kendra, Dhamtari, Chhattisgarh, India	Kernel	Popped stage	Increase in water absorption capacity, oil absorption capacity, redness, yellowness, hue, and chroma	Devi et al. (2020)

the optimal parameters, such as the roasting and tempering time. A rise in roasting temperatures and durations results in the force required to break the Makhana seed hull significantly declining. This phenomenon is arguably connected to the embrittlement of the hulls caused by extended exposure to high temperatures of thermal roasting (Arjun et al., 2017; Jha and Prasad, 1996). Kernel gelatinization takes place during early roasting in high temperatures, and in this process, the shell is more desiccated than the kernel. Subsequent tempering causes a gradual difference in shrinkage rate between shell and kernel that occurs as they cool to ambient temperature, thus breaking their interface and causing kernels to be detached more easily. The gorgon nuts are subjected to a second roasting that finishes the basic processing of the nuts (Jha and Prasad, 1996). Makhana texture is expected to be soft, chewy, resembling small puffed balls, such as popcorn, with an added crunchiness achieved through the ultimate popping roast (EFSA, 2022).

7.4.2 Surface morphology and color alterations

Evenness in weight and stability in shape are key characteristics that make popped *E. ferox* seeds have a high degree of consumer acceptance, since they increase the appeal of the product. Researchers have used mass-based modeling methods with the purpose of operating successful sorting, which directly relate significant physical features with product weight that results in enhancement in the accuracy of grading and quality control (Gaurav et al., 2022).

Optimization of postharvest handling system is closely associated with aerodynamic characteristics of popped Makhana. Specific characteristics of variations in projected area and density are the important parameters in determining the seeds movement in the air, while also imparting substantial roles in effective separation and grading processes. Considering these properties can be beneficial in designing better conveyance systems where air resistance plays a significant role; this helps eliminate adhered particles, which ultimately improves process efficiency and promotes a better product (Gaurav et al., 2022).

The intact hull-enclosed seeds are first roasted followed by mechanical popping, which involves sequential steps in which each step has a distinct effect on the physical and chemical characteristics of the material. Starch gelatinization and retrogradation are initiated during the first stage of the roasting process, which in turn promotes the subsequent expansion of the kernel during the second roasting (Devi et al., 2020).

To help retaining moisture content in the seed, the seed coat plays an important part since it does not allow immediate loss of moisture at high roasting temperatures. This retained moisture causes high vapor pressure, which causes internal voids. The retrograded starch expands rapidly once the seed coat ruptures to form the typical popped kernel structure of Makhana (Devi et al., 2020).

Surface cracks revealed fissures due to loss of moisture during storage of Makhana kernels following the primary roasting phase up to 48 h. The availability of gelatinized starch and denatured proteins contributed to kernel hardness. The compact kernels were then broken due to an abrupt rupture of the seed coat to become a porous matrix with big cavities (Devi et al., 2020).

In subsequent research, it was established that heat treatment of Makhana kernels triggered the volumetric expansion, which improved the capacity of Makhana kernels to absorb milk. Investigations on popped Makhana also found that there was an increment in the protein contents, with these proteins having hydrophilic properties. Such hydrophilic proteins also helped to deteriorate the structure of the kernels when they were in milk, which is important to note the interplay between compositional changes and functional behavior (Gaurav et al., 2022).

Better processing of *E. ferox* seeds requires sorting and grading of the seeds on the basis of physical qualities, including size and shape. The improper classification can cause the smaller seeds to go through deshelling equipment without breaking, and larger seeds to break into smaller bits instead of being shelled (Kumar and Kumar, 2015).

Makhana kernels are affected by chemical reactions, such as the Maillard reaction and nonenzymatic browning during the process of roasting, which causes browning changes. These reactions are what cause the kernels to obtain their final color, texture, and their unique phenolic and antioxidant properties. To ensure the best health-promoting benefits of *E. ferox*, the protection or improvement of phytochemical contents in processing requires the control of operational conditions to achieve the best results (Liaquat et al., 2022).

The changes of color during high-temperature roasting and popping of grains and seeds are mainly explained by nonenzymatic browning processes, such as the degradation of sugars, oxidation process, Maillard reactions, and caramelization. Although these are natural reactions of the thermal treatment, they usually have a negative influence on the visualization of the final product. As color is one of the main factors of consumer selection, the roasting parameters should be carefully controlled to provide nutritional value and sensory properties (Devi et al., 2020).

7.5 Comparative analysis of makhana seed and popped kernel

7.5.1 Proximate composition comparison

The carbohydrates are the main component of raw seeds, with high levels of moisture and large quantities of protein and fat as their secondary components.

Conversely, popped Makhana exhibits an amplified carbohydrate content with the presence of a lowered moisture, protein, and fat content that leads to drier and more dense nutritional content (Rathod et al., 2023).

Further increase in the availability of plant phenolics in the seed matrix is achieved through heat application during processing. This is because intracellular water is evaporated, and it causes a series of chemical reactions which enhance protein denaturation and structural changes. Therefore, it is crucial to control the thermal process, which positively influences the nutritional value and functionality of Makhana, thus increasing its health-promoting effect (Table 7.2) (Liaquat et al., 2022).

Raw Makhana has a significantly higher essential amino acid composition of approximately 93.63%, whereas the popping form contains an approximation of 89.95%, both surpassing common values seen in many diets, which are plant-based (Sodi and Kumar, 2019). Essential and nonessential amino acids declined compared to raw kernels, except lysine, which increased significantly from 2.19 to 6.34 mg per gram. Histidine was the highest in both forms, then threonine, leucine, and phenylalanine, but the levels were lower in

TABLE 7.2 Impact of thermal processing on the chemical composition of makhana from selected varieties.

Variety	Seed part	Thermal processing	Chemical profile changes compared to raw makhana	References
–	Seed	Roasted stage	Increase in crude ash, crude protein, crude fiber, crude fat, nitrogen free extract, K, Mg, P, Ca, Fe, Na, and Zn	Liaquat et al. (2022)
Swarn Vaidehi	Kernel	Popped stage	Increase in protein and total dietary fiber Decreased in carbohydrates, starch, amylose, amylopectin, essential, and nonessential amino acids (except lysine)	Devi et al. (2020)
–			Decreased in protein, total starch, Fe, Zn, Cu, and phytic acid	Kumar et al. (2023)
Swarn Vaidehi	Seed		Increase in total ash, protein, fat, carbohydrate, K, Mg, and Cu Decreased in crude fiber, amylose, and P	Kumar et al. (2016)

K, Potassium; *Mg*, Magnesium; *P*, Phosphorus; *Ca*, Calcium; *Fe*, Iron; *Na*, Sodium; *Zn*, Zinc; *Cu*, Copper.

the case of the popped seeds. According to this profile, Makhana is a good ingredient that can be added to the infant foods industry to achieve better growth results (Devi et al., 2020).

The protein profile of popped Makhana shows that the amino acids exhibited high contents of arginine, alanine, and tyrosine (expressed per 16-gram basis) relative to eggs. Popping process, though it decreased the level of tyrosine, also elevated the levels of lysine, arginine, threonine, etc., thus improving the nutritional content of the seeds (Sodi and Kumar, 2019).

The fiber content of Makhana is also affected by the thermal processing. Popping adds to the total and soluble dietary fiber, decreases the insoluble fiber content, and mitigates the carbohydrate content compared to the raw kernels, besides the fact that the raw seeds had more moisture holding capacity than that of the popped seeds. These modifications are probably reflecting a simultaneous increase in protein and fiber fractions. Along with this, roasting increases the ash content of fox nuts by about 14%, which means that there is a higher concentration of minerals (Devi et al., 2020; Liaquat et al., 2022). In another research, the raw Makhana total mineral content was estimated slightly less than the popped one. The difference in crude ash content across the samples investigated is due to a gap in the differences in varieties, field conditions, and processing methods (Kumar et al., 2016, 2023).

From the dietary aspect, the fox nuts are unique in terms of having the lowest fat content and a relatively higher proportion of carbohydrates than other nuts. This nutritional content particularly makes them an ideal option in terms of weight loss, and at the same time, provides an increased source of protein and fiber (Liaquat et al., 2022).

In unroasted *E. ferox* seeds, the level of fat is significantly low, around 0.5%. Conversely, the roasted specimens have an elevated fat content, and this elevation is due to enhanced oil absorption during roasting as well as biochemical changes in the oil used in the roasting process, rather than in the kernel itself (EFSA, 2022). Quantifiable fat level of the roasted specimen increases due to two significant biochemical changes. First, high temperature inactivates the native enzymes, such as lipases and esterases, which inhibits the process of lipid catabolism, resulting in its overall accumulation. Second, the heat disrupts complexes between lipids, proteins, and polysaccharides, which were first formed by the action of enzymes. The degradation of these structural components breaks the bound lipids, resulting in a free, extractible state, thus increasing the observed fat content; however, a different study result showed that popping lowers Makhana fat content, which could be due to thermal decomposition of fat molecules into smaller glycerides or fatty acids (Sinha and Jha, 2025). Another reason that explains the fat reduction is the presence of bran, the major source of fat, which gets eliminated during popping (Kumar et al., 2023).

On top of the low intrinsic fat content, the population of hydrophilic side chains produced by amino acids, proteins, and starch is increased in the popped Makhana seeds. The structural change increases their ability to absorb water and they are thus specifically preferred to be included in recipes, such as soups and baked foods. Their high moisture absorption improves texture and functionality, and they can be used in a variety of culinary applications (Haleema et al., 2018). The popping process causes the levels of starch as the major chemical compound of the kernels, to be much reduced, probably due to its transformation into simpler sugars, such as maltose and glucose. This conversion also increases the total amount of sugar in the popped kernels (Devi et al., 2020).

The amount of resistant starch (RS) can be developed through high temperatures used in seed treatment, which is further enhanced by dry heat in depolymerizing and altering glycosidic bonds in oligosaccharides and polysaccharides. In vivo research has established many health advantages of long-term RS intake, such as maintenance of blood glucose responses, inducing satiety effects, amplification of calcium intake, and prebiotic properties (Haleema et al., 2018).

Stability of individual vitamins during the popping process was found to differ considerably, whereby vitamin B_3 showed the highest concentration among B vitamins tested in the raw Makhana, but its levels reduced after popping, indicating heat sensitivity. Vitamin B_2 remained intact in both the raw and popped forms, showing significant resistance to the popping process (Kumar et al., 2023). Vitamin B_6 also decreased significantly postprocess. It is interesting to note that popping at high temperatures could cause a minor, insignificant increase in the measurable α-tocopherol (vitamin E) content, which may have been present in the germ layer of the grain, as observed in some studies. In contrast, the research shows that the total phenolic and flavonoid content of the popped fox nuts decreases relative to the raw seeds and this is mostly because of the breakdown of compounds that are heat-sensitive during popping. Although this decrease can cause an attenuation of antioxidant properties, studies found a decrease in total carotenoid content following popping process and could be potentially attributed to both the inherent characteristics of each grain and the distribution of carotenoids in their tissues (Joseph and Ramesh, 2023; Kumar et al., 2023).

The trace metal analysis of *E. ferox* starch indicated that the level of iron and magnesium with whole and puffed starch was less than 100 mg per kilogram and 2000 mg per kilogram, respectively, in seed meal and whole starch, and about 80 mg per kilogram and 100 mg per kilogram in purified starch (Rathod et al., 2023).

Phenolic compounds are known for their antioxidant effects, due to their ability to chelate metals and to scavenge radicals, which are important in the nutritional value of Makhana. Nonetheless, the process of popping results in a close to half degradation in the total amount of phenolics because of the influence of heat, thus reducing antioxidant activity (Devi et al., 2020).

On the other hand, the antioxidant activity of fox nuts has been observed to be increased by roasting. Research shows that the level of DPPH and FRAP in roasted kernels is much higher than that of their unroasted counterparts. It has been shown that roasting has a positive effect on ferric-reducing capacity, which increases the total antioxidant properties despite the loss of phenolic compounds. Consequently, this twofold finding shows how complicated the interaction is between the thermal processing conditions and the stability and bioactivity of the resulting phytochemicals in Makhana (Joseph and Ramesh, 2023).

The decreased iron levels in the kernels of the *E. ferox* that have been popped indicate that there is certainly an abundant amount of iron retained in the bran fraction, which is lost during the roasting process. On the contrary, the concentration of calcium is observed to be increased in the popped kernels, which means that this mineral is evenly distributed in the raw material. In general, there are no significant differences in the mineral profiles of raw and popped Makhana kernels, and all the differences are insignificant in terms of calcium and iron content (Devi et al., 2020).

Nutritionally, popped Makhana seeds are quite rich in terms of their levels of phosphorus, magnesium, and potassium, and have only low levels of cholesterol, sodium, and saturated fats (Sodi and Kumar, 2019). Atomic absorption spectrometry studies of phosphorus and other minerals as complementary studies confirmed phosphorus as the most abundant mineral, after which there were potassium, calcium, magnesium, and sodium in that order. A small amount of iron and copper was also found, which once again highlighted the predominance of microminerals in the *E. ferox* seed that was subjected to popping (Haleema et al., 2018).

7.5.2 Antioxidant and functional property variation

Phytochemical data of *E. ferox* seed indicate that the seeds have a complex antioxidant system, with a high concentration of tocopherol acetate as one of the important phytochemicals that helps the plant to defend itself. More profiling has revealed a wide range of 129 flavonoid-associated metabolites, including various subclasses, such as flavonols, chalcones, and anthocyanins. Besides this large antioxidant collection, studies also show the occurrence of bioactive compounds, such as gallic acid, corilagin, and ethyl gallate whose hepatoprotective activity is also significant (Kapoor et al., 2022). These substances have a significant inhibitory effect on two significant pathological mechanisms of nonalcoholic fatty liver disease (NAFLD): they lower the levels of abnormal lipid accumulation and relieve oxidative stress. They have the ability to reduce oxidative stress and thus protect the integrity of vital cellular components, including DNA, proteins, lipids and membranes, while preventing the development of NAFLD. These findings indicate the immense nutritional and therapeutic prospects of *E. ferox*, which justifies its classification as an attractive functional food to promote both health and clinical nutrition diet (Kapoor et al., 2022). Values of IC50 ranging from 50 to 100 μg per milliliter is considered a good indicator of strong antioxidant activity, and a value of below 50 μg per milliliter indicates very strong activity. An ethanol–water extract of *E. ferox* seed coats displayed an IC50 of 1.62 μg per milliliter, which was better than that of ascorbic acid 2.28 μg per milliliter and thus supported its extremely powerful antioxidant (Joseph and Ramesh, 2023). Complementary studies found that crude polysaccharide extracts of *E. ferox* have a significant antioxidant activity, both in terms of free radical scavenging activity and reducing power. These polysaccharides, which are obtained by isolation of the seed shell, also enhance cellular defense by both augmenting the actions of endogenous antioxidant enzymes and reducing lipid peroxidation. Their reducing power is equally that of typical antioxidants, including vitamin C and TBHQ, and is supported by strong radical scavenging ability and very low IC50 values. These are bioactive polysaccharides that have been obtained using water extraction techniques, which indicate the functional importance of *E. ferox* in alleviating oxidative stress and enhancing cellular resilience (Kapoor et al., 2022).

In addition to studies based on extraction methods, processing conditions also affect antioxidant potential (Table 7.3). It was demonstrated that the roasting has positive effects on the antioxidant activity of Makhana, and the total phenolic content increased by 36% in comparison to the control (Rathod et al., 2023), Thus, Makhana appears beneficial in the functional food industry as it has natural antioxidant properties, which could be enhanced by controlled processing techniques.

TABLE 7.3 **Effect of thermal processing on bioactive compounds and antioxidant properties in selected varieties of makhana.**

Variety/ Genotype	Seed part	Chemical biomolecules in raw makhana	References
–	Kernel	Vitamin B_2, B_3, B_6, antioxidant activity, total flavonoid content, total phenolic content, and polyphenol, such as gallic acid	Kumar et al. (2023)
Swarn Vaidehi		Antioxidant activity, total flavonoid content, and total phenolic content	Jana et al. (2024)
Superior selection-1	Seeds endosperm	Vitamin B_1, vitamin C, antioxidant activity, and flavonoids, such as quercetin	Devi et al. (2020)

7.6 Future perspectives

It has been found that gamma irradiation has the potential to produce new cultivars of *E. ferox* with improved commercial traits, providing new opportunities for better crop production and developing the market (Khadatkar et al., 2020). Simultaneously, the changing nature of consumer preferences that favor vegetarian, vegan, gluten-free, and nongenetically modified food diets supports the high potential of Makhana. The seeds of this water crop have an outstanding nutritional balance, featuring an intrinsically low-caloric content and low lipids, thus making them a logical ingredient for dietary homeostasis (Bisht et al., 2025). Makhana can both become the candidate of the further breeding approaches to increase its commercial feasibility and be presented as the naturally oriented superfood that could support the preferences of contemporary health-aware consumers.

The industrialists and the Makhana processors are also keen to extract value-added products out of these seeds, and they are pushing the processing further, resulting in the creation of compact and improved products that have less volume. Production of proper products and mechanical systems involves a thorough understanding of the seed's physicochemical properties (Sahithi et al., 2022). Furthermore, the volumetric expansion of Makhana is notable, increasing the cost of transportation, which makes it expensive in the market in most cases (Jha, 1999). Processing is thus important in the generation of smaller derivatives, which is more economical. Mechanized production is highly relevant to physical qualities of the crop; ensuring the preservation of quality and supporting the seasonal, locale-specific cultivation is taken care of by the storage policies at the processing plant (Jha, 1999).

Makhana is currently mostly commercialized as a popped product, and there is a small amount of effort being put into developing the wider value enhancement. Although Makhana is a noncereal, organic, and healthy food, it has not been explored and exploited by most of the world yet (Kumar and Kumar, 2015).

The industrial uses of *E. ferox* are much further than the fact that it can be used only as a food ingredient (Fig. 7.4). The plant is also used in nonfood industries, such as part of cosmetic formulations; examples include the use of extracts of the plant together with other oils extracted from other botanical plants to formulate preparations that make the skin less sensitive. Due to its antimelanogenic effects, it is also used in skin-whitening preparations, mainly by suppressing the activity of tyrosinase and production of melanin (Kapoor et al., 2022).

In medicine, *E. ferox* has a long-established history in the Indian and Chinese traditional medicine, and has been used to treat disorders of the digestive, respiratory, excretory, reproductive, and circulatory systems. The plant has a lot of potential as a natural source of pharmaceutical development due to its accessibility and excellent antioxidant profile (Kapoor et al., 2022).

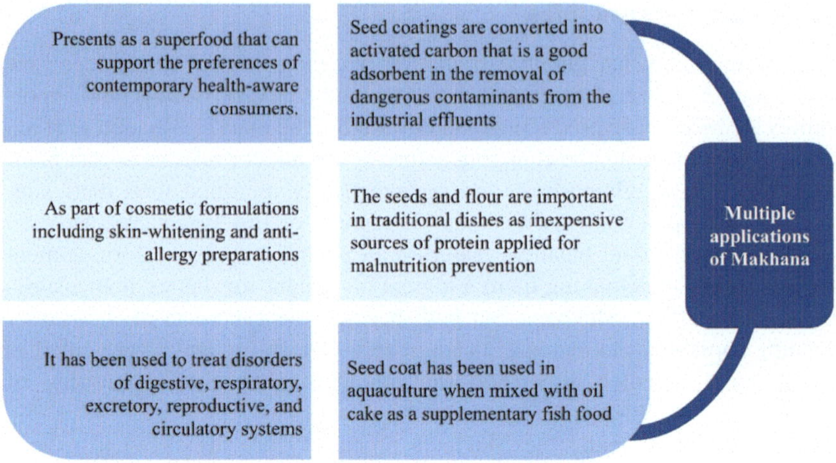

FIGURE 7.4 Multiple uses of makhana across food, cosmetics, medicine, and industry.

Another environmental application includes seed coatings that are converted into activated carbon, which is a good adsorbent in the removal of dangerous contaminants of the industrial effluents (Kapoor et al., 2022).

In addition to these industrial and medicinal applications, roasted fox nut seeds have been shown to have a low glycemic index in the human body due to their resistant starch content. The *E. ferox* resistant starches have shown excellent thermal stability and reduced glycemic response. Consequently, these characteristics make fox nut seeds a promising raw ingredient to develop functional foods that help to reduce metabolic diseases, such as obesity and diabetes (Devi, 2022). These data place *E. ferox* as a versatile crop of significant value, and the applications of the product are restricted not only to health and nutrition but also to cosmetics, industrial, and environmental issues.

The functional food formulation potential of *E. ferox* seeds is mostly based on the starchy constituent that has a flexible base in developing products. Besides containing starchy content, fox nuts also contain key minerals, such as magnesium and potassium, and rank as good sources of foods for hypertension management since potassium is vital in regulating blood pressure (Devi, 2022). Although these seeds have these unique nutritional properties, consumer awareness of the health benefits of this product is quite low, which highlights the necessity for greater dissemination of knowledge and targeted promotion to release its potential in the context of the modern diet (Devi, 2022).

Makhana pops have significant potential in the creation of a wide variety of products, which means that the crop will be able to compete successfully with more popular snacks, including corn, oats, and soybeans (Kumar and Kumar, 2015). In addition to being a snack, the Makhana seeds and flour are important in traditional dishes as inexpensive and gluten-free sources of protein that can be applied to malnutrition prevention. Their future is as a potential source of nutraceuticals and pharmaceuticals, so that Makhana can be useful not only in the cultural food but also in the modern health-conscious food chains (Sahithi et al., 2022).

The consumption of fox nuts in different shapes and forms, such as popped ones or prepared in composite meals, can improve the quality of the diet. This solution not only promotes proper eating habits but also helps in the prevention or delaying lifestyle-related diseases (Joseph and Ramesh, 2023). Other than being a source of nutrition, the Makhana processing also produces byproducts that have new industrial applications. The nonedible seed coat, which has been traditionally discarded as waste and burnt as a source of fuel, has recently been put to a viable use in aquaculture. The seed coat when mixed with oil cake is a supplementary fish food, which is said to enhance yield. The given development presents an excellent opportunity to transform agricultural byproducts into a valuable asset, which supports the larger opportunities of Makhana both as food and as an environmental resource (Sinha and Jha, 2025).

7.7 Conclusions

Makhana has good but underexplored potential and this is due to the unique physicochemical characteristics of its seeds, which are not common in other plant species. The end product, which is the popular popped kernel, is subjected to substantial changes due to heat treatments on the raw seeds. The physical changes that occur in this processing include desirable textural and chemical changes, including alterations in lipid and carbohydrate profiles. These changes ensure that popped Makhana is a product with a wide variety of uses. However, comprehensive research on the physicochemical characteristics of the raw and popped Makhana is required. The future studies should focus more on exploring different processing treatments as a way of improving the functional properties of the seeds and the products obtained. Due to lack of comprehensive studies, it is difficult to identify which treatments are most appropriate and impeding deeper comparisons. To help exploiting this valuable crop more efficiently, processing technologies and functions need to be improved to get a maximum yield and quality. The raw and popped Makhana seeds have a lot of potential in the food industry for the general population and in the diet of specific populations, such as diabetic patients, alongside substantial nonfood industry applications.

Self-declaration

Artificial Intelligence (AI) tools were used only for English language editing in the preparation of this manuscript. No AI tools were used to generate any kind of the content, including text, tables, or figures.

References

Arjun, S., Girish, G., Nagaraju, V.D., Sridhar, B.S., 2017. Physico-fracture characteristics of Makhana (*Euryale ferox*) seeds. Int. J. Food Prop. 20 (Suppl. 2), 1204–1209. https://doi.org/10.1080/10942912.2016.1214843.

Bhat, M.S., Arya, S.S., 2020. Physico-functional, pasting and structural properties of gorgon nut (*Euryale ferox*) flour as affected by heat-moisture and acid treatment. J. Food Meas. Char. 14, 1656–1664. https://doi.org/10.1007/s11694-020-00413-4.

Bhat, M.S., Arya, S.S., 2021. Technofunctional, rheological, thermal and structural properties of Gorgon nut (Eurayle ferox) as affected by drying temperature. J. Food Process. Eng. 44 (7), e13713. https://doi.org/10.1111/jfpe.13713.

Bisht, H., Singh, K., & Islam, Z. (2025). Evaluation of the physicochemical, functional, structural, and anti-nutritional composition of germinated underutilized Fox nut seed flour 8th AMIFOST 2025, BIO Web of Conferences.

Devi, K.R., 2022. A comprehensive review on nutraceutical properties of *Euryale ferox* Salisb. J. Plant Sci. Res. 9 (2), 222.

Devi, M., Sharma, K., Jha, S.N., Arora, S., Patel, S., Kumar, Y., Kumar Vishwakarma, R., 2020. Effect of popping on physicochemical, technological, antioxidant, and microstructural properties of Makhana seed. J. Food Process. Preserv. 44 (10), e14787. https://doi.org/10.1111/jfpp.14787.

Dutta, R.N., Jha, S.N., Jha, U.N., 1986. Plant contents and quality of Makhana (*Euryale ferox*). Plant Soil 96, 429–432. https://doi.org/10.1007/BF02375148.

EFSA, E. F. S. A. (2022). Technical Report on the notification of roasted and popped kernels from the seeds of Euryale ferox Salisb. As a Traditional Food from a Third Country Pursuant to Article 14 of Regulation (EU) 2015/2283 (2022:EN-7538). 9.

Gaurav, A., Nath, D., Barbhuiya, R.I., Pradhan, R.C., 2022. Physical characterization and mass modeling of traditionally popped Makhana (*Euryale ferox* Salisbury) variety: Sabour-1 makhana. J. Food Process. Eng. 45 (6), e13927. https://doi.org/10.1111/jfpe.13927.

Guo, Z., Zeng, S., Lu, X., Zhou, M., Zheng, M., Zheng, B., 2015. Structural and physicochemical properties of Lotus seed starch treated with ultra-high pressure. Food Chem. 186, 223–230. https://doi.org/10.1016/j.foodchem.2015.03.069.

Guo, Z., Zhao, B., Chen, L., Zheng, B., 2019. Physicochemical properties and digestion of Lotus seed starch under high-pressure homogenization. Nutrients 11 (2), 371. https://doi.org/10.3390/nu11020371.

Haleema, K., Srigiripura, C.V., Urooj, A., 2018. Evaluation of popped *Euryale ferox* seeds as a potent functional food based on its nutritional profile, functional attributes and macronutrient digestibility. J. Adv. Food Sci. Technol. 5 (3), 81–87.

Islam, Z., Bashir, K., Singh, S., Siddiqui, S., 2025. Effect of germination on physicochemical, functional and antinutritional characteristics of Makhana seed (*Euryale ferox*) flour. Food Human, 5, 100717. https://doi.org/10.1016/j.foohum.2025.100717.

Jana, B.R., Bhatt, B.P., Srivastava, A., 2022. Biochemical components of Makhana (*Euryale ferox* Salisb) and its relevance to anti-obesity effect in human. Asian J. Dairy Food Res. 44 (4), 614–619. https://doi.org/10.18805/ajdfr.DR-1871.

Jana, B.R., Kumar, M., Raut, S.M., 2024. Antidiabetic biomolecules and nutrient elements in Makhana (*Euryale ferox* Salisb.). Ann. Phytomed.: Int. J. 13 (2), 521–528. https://doi.org/10.54085/ap.2024.13.2.52.

Jana, B.R., Srivastava, A., Idris,, M., 2019. New Makhana (*Euryale ferox* Salisb.) processed products for health benefit. J. Pharmacogn. Phytochem. 8 (2), 1662–1666.

Jha, S.N., 1999. Physical and hygroscopic properties of Makhana. J. Agric. Eng. Res. 72 (2), 145–150. https://doi.org/10.1006/jaer.1998.0356.

Jha, S.N., Prasad, S., 1996. Determination of processing conditions for gorgon nut (*Euryale ferox*). J. Agric. Eng. Res. 63 (2), 103–111. https://doi.org/10.1006/jaer.1996.0012.

Joseph, A., Ramesh, G., 2023. Nutrient analysis, phytochemical and antioxidant activity of a food product formulated with fox nuts (*Euryale ferox*). Asian J. Biol. Life Sci. 12 (2), 279–285. https://doi.org/10.5530/ajbls.2023.12.38.

Kapoor, S., Kaur, A., Kaur, R., Kumar, V., Choudhary, M., 2022. *Euryale ferox*, a prominent superfood: nutritional, pharmaceutical, and its economical importance. J. Food Biochem. 46 (12), e14435. https://doi.org/10.1111/jfbc.14435.

Khadatkar, A., Mehta, C., Gite, L., 2020. Makhana (*Euryale ferox* Salisb.): a high-valued aquatic food crop with emphasis on its agronomic management–A review. Sci. Hortic. 261, 108995. https://doi.org/10.1016/j.scienta.2019.108995.

Kumar, J., Kumar, S., Kumar, A., Kumar, S., 2019. Field performance of popping machine for Makhana seeds. Int. J. Chem. Stud. 7 (SP6), 773–777.

Kumar, L., Singh, A.K., Bhatt, B., 2016. Nutritional status of recently developed Makhana (Gorgon nut) variety-Swarna Vaidehi. J. AgriSearch 3 (4), 199–205. https://doi.org/10.21921/jas.v3i4.6701.

Kumar, M.M., Goudar, G., Sharma, P., Vishwakarma, R., Gogoi, P., Mahajan, A., Chandragiri, A., Kalpuri, S., Babu Geddam, J.J., Suresh, C., Chary, M., Venketrajireddy, G., Radhika, M.S., Sreenu, P., Tulja, B., 2023. Differentiating the nutrient composition, in-vitro starch digestibility, individual polyphenols and anti-oxidant properties of raw and popped Makhana (*Euryale ferox*). J. Food Meas. Char. 17, 5828–5844. https://doi.org/10.1007/s11694-023-02074-5.

Kumar, S., Kumar, S.P., 2015. Effect of different grades on physical and mechanical properties of popped Makhana. Int. J. Process. Post Harvest Technol. 6 (1), 101–108. https://doi.org/10.15740/HAS/IJPPHT/6.1/101-108.

Liaquat, M., Pasha, I., Ahsin, M., Salik, A., 2022. Roasted fox nuts (*Euryale Ferox* L.) contain higher concentration of phenolics, flavonoids, minerals and antioxidants, and exhibit lower Glycemic Index (GI) in human subjects. Food Prod. Proc. Nutr. 4 (1), 1–12. https://doi.org/10.1186/s43014-021-00081-x.

Mittal, R., Sharma, S., Mittal, A., 2020. A critical review on ethnobotanical and pharmacological aspects of *Euryale Ferox* Salisb. Pharmacogn. J. 12 (6), 1444–1454. https://doi.org/10.5530/pj.2020.12.199.

Rai, U.N., Tripathi, R.D., Vajpayee, P., Jha, V., Ali, M.B., 2002. Bioaccumulation of toxic metals (Cr, Cd, Pb and Cu) by seeds of *Euryale ferox* Salisb.(Makhana). Chemosphere 46 (2), 267–272. https://doi.org/10.1016/S0045-6535(01)00087-X.

Rathod, R.V., Neve, G., Jain, A., Giri, P., 2023. A comprehensive review on health benefits and nutritional aspects of foxnut (Makhana). Pharm. Innov. 12 (6), 4432–4438.

Sahithi, G., Thabassum, Z.A.F., kumara, C.S.S., Reddy, M.M., Satish, A., 2022. Estimation of nutritional composition and sensory evaluation of fero nutri bars developed using *Euryale Ferox* (Makhana seeds). Int. J. Food Nutr. Sci. 11 (7), 776–785.

Sinha, C.B.K., Jha, V., 2025. Identification of nutraceutically important metabolites in raw and popped Makhana (*Euryale ferox* Salisb.) seeds using GCMS technique. Int. J. Plant Environ. 11 (03), 514–524. https://doi.org/10.18811/ijpen.v11i03.09.

Sodi, M., Kumar, S., 2019. Integrated Makhana - fish Aquaculture: a sustainable approach for poverty alleviation and livelihood in the rural areas of North-Eastern Bihar. Scholars J. Agricult. Veterinary Sci. 6 (4), 119–127. https://doi.org/10.36347/sjavs.2019.v06i04.001.

Chapter 8

Nutritional composition and health benefits of makhana

Parnasree Sadhukhan[1], Simran Choudhury[1] and Anindita Ray Chakravarti[2]

[1]*Department of Food Science and Nutrition Management, J.D. Birla Institute, Kolkata, West Bengal, India;* [2]*Department of Food and Nutrition, Maharani Kasiswari College, University of Calcutta, Kolkata, West Bengal, India*

8.1 Introduction

Euryale ferox Salisb. (commonly called foxnut, makhana, or gorgon nut) is the edible seed of a spiny aquatic plant in the family Nymphaeaceae. It is primarily cultivated across East, South, and South-East Asia which includes India, China, and Korea (Kumar and Jha, 2024). Its medicinal and nutritional value has been reported in Ayurvedic studies since a long time.

The seeds of foxnut are large, round, and slightly astringent in taste (Jiang et al., 2023). They are most commonly consumed in the popped form (which is locally known as *Makhana* in India). Popping of the seeds involves drying them followed by roasting at high temperature (Shukla et al., 2024). This transforms the seeds into a light and crisp snack with enhanced palatability. Foxnuts reportedly exhibit a number of health benefits. This includes antioxidant activity, hepatoprotective and nephroprotective effects, lipid-lowering action, and potential antidiabetic effects mediated by enzyme inhibition and insulin Pathway modulation (Zhang et al., 2019; Singh et al., 2021). These benefits are attributed to the composition of foxnuts. However, most of the evidence are pre-clinical, with limited human trials.

As per a number of studies, the characteristic composition of makhana is dominated by carbohydrates. It has a moderate protein content, considerably low-fat content and significant amount of dietary fiber. 100 g of un-popped seeds of foxnuts roughly contain 60–75 g carbohydrate, 8–12 g protein and 0.5–5 g fat. The carbohydrate is majorly composed of slowly gelatinizing starch which makes it similar to cereal/pulse in terms of macronutrient composition. The protein content of foxnuts is moderate. However, the amino acid profile includes all the essential amino acids. This is particularly

Makhana (Euryale ferox) as a Functional Food. https://doi.org/10.1016/B978-0-443-44421-0.00008-1

dominated by high amounts of arginine, glutamic acid, leucine, and isoleucine. The common minerals in foxnuts are phosphorus, potassium, magnesium, calcium, and iron (Liaquat et al., 2022). According to most studies, the vitamin content in foxnuts is limited, mostly dominated by Vitamin B1 and Vitamin C (Jana et al., 2022). Presence of saponins, tannins, and phytic acid has also been identified in *E.ferox* as per phytochemical screenings (Zhang et al., 2019).

However, high-temperature processing of the seeds affects the physicochemical properties of foxnuts influencing the digestibility and the bioavailability of the nutrients (Liaquat et al., 2022).

8.2 Nutritional composition of foxnut seeds

8.2.1 Proximate composition

Proximate composition studies of raw and popped foxnut seeds consistently show that the seed is predominantly carbohydrate-rich, with moderate protein and very low-fat content. Factors like cultivar, growing conditions, post-harvest processing (e.g., drying, popping), and analytical methods affect the reported proximate values of foxnuts.; however, typical ranges for the raw seed and the commonly consumed popped product can be summarized as shown in (Table 8.1).

Moisture content is generally low in dried or popped seeds (typically 12.8 % in popped makhana) and 37% in raw harvested seeds as shown in (Table 8.1). Carbohydrate content for popped makhana has been reported in the range of 0–85% on a dry-weight basis, depending on whether starch or available carbohydrate is reported. Analysis of raw seed reports carbohydrates around 65%–75% (Rathod et al., 2023). Protein is commonly reported between ~7 and 11% for popped makhana; some studies report raw seed protein around 8%–12% (dry weight) (Liaquat et al., 2022). Fat content is reported to be <2% in many studies, making foxnut an inherently low-fat snack (Rathod et al., 2023). Total mineral content is generally low-to-moderate, which reflects the mineral content, and it is dependent on soil and cultivar. The calorie contributed by raw foxnuts is 362 kcal/100 g and by popped foxnuts it is 328 kcal/100 g (Sahithi et al., 2012).

8.2.1.1 Effects of processing on the nutritional composition and functional properties

Processing conditions can significantly alter the proximate composition and functional properties. Popping, which is the conventional method involving thermal processing to produce the puffed snack, reduces moisture and thus can concentrate macronutrients per 100 g (dry weight). Changes in starch properties (increased digestibility), enhanced palatability and reduction in some anti-nutrients have been reported post popping of foxnuts. Conversely, germination has been reported to slightly increase measured protein content

TABLE 8.1 Proximate principles of makhana seeds.

Forms of makhana	Moisture content (%)	Total ash (%)	Carbohydrates (%)	Protein (%)	Fat (%)	Total dietary fiber (%)	References
Raw(Unpopped) makhana	37	0.45 ± 0.03	87.52 ± 0.15	9.15 ± 0.10	0.46 ± 0.04	2.55 ± 0.04	Devi (2022)
Popped makhana	12.8	0.38 ± 0.03	84.87 ± 0.60	11.03 ± 0.45	0.33 ± 0.12	3.26 ± 0.06	Devi (2022)
Popped and roasted makhana	4.17 ± 0.15	0.66 ± 0.03	Not Available	14.57 ± 0.29	0.68 ± 0.02	0.51 ± 0.02	Liaquat *et al.* (2022)
Germinated makhana flour	7.4 ± 0.36	2.78 ± 0.40	Not Available	11.5 ± 0.32	0.37 ± 0.04	0.69 ± 0.12	Bisht *et al.* (2025)

and improve bioavailability of some nutrients, while causing the breakdown of some anti-nutritional factors (a common pattern seen in legume and seed germination).

8.2.2 Protein quality and amino acid profile

The composition of foxnuts is dominated by starch followed protein. The protein content although not very significant. But in comparison to other staples, it has many essential amino acids in the right proportions as shown in (Table 8.2).

8.2.2.1 Total protein and digestibility

Foxnut seeds contain moderate amounts of protein and are not considered a very good source of protein in terms of its content. However, the seeds have been noted to possess a favorable essential amino acid profile for a seed. The protein digestibility and overall protein quality have been examined in a few studies, often using indices such as the Essential Amino Acid Index (EAAI) and comparisons with reference amino acid patterns. The protein in the popped makhana is easily digestible. This is due to the reduction in heat-sensitive anti-nutritional factors and structural changes (deamidation, glycation, and oxidation) of protein due to high heat. Germination also significantly affects the protein quality (Bagarinao et al., 2024). Germination affects the protein value in three ways. First, it increases amino acid availability and digestibility by activating enzymes like proteases that break down the protein into peptides and amino acids; second, it reduces the anti-nutritional factors like phytic acid and tannins. This is by the activation of certain enzymes like phytase which can hydrolyze phytic acid into a more digestible form. Tannin is water-soluble in nature which can leach into water during soaking the seeds before or during germination (Sokrab et al., 2012). Lastly, it increases the protein solubility by breaking down proteins into smaller chain peptides (del Rosario Moguel Concha et al., 2022). The impact of a 72-hour germination period on the quality of makhana seed flour was determined by Islam et al., and they reported significant reductions in anti-nutritional factors. Specifically, germination led to a marked decline in phytic acid levels, decreasing from 168.64 mg/100 g to 82.94 mg/100 g, as well as a reduction in the amount of tannin from 3.69 mg/100 g to 2.65 mg/100 g. These reports also indicate that *E. ferox* scores favorably for essential amino acid completeness compared with many cereal staples (Bisht et al., 2025; Islam et al., 2025; Liaquat et al., 2022; Devi, 2022).

8.2.2.2 Amino acid composition

Comprehensive amino acid analysis indicates the presence of essential as well as non-essential amino acids. Among these, relatively high proportions of

TABLE 8.2 Comparative amino acid composition of makhana (*Euryale ferox*) and selected staple foods (wheat, rice, and maize).

Amino acid	Histidine	Isoleucine	Leucine	Lysine	Methionine (S-AA)	Phenylalanine	Threonine	Tryptophan	Valine	Arginine	References
Wheat (*Triticum aestivum*) (g/100 g)	2.6.	4	8.5	2.6	2.2	7	2.8	1	4	5	Shewry and Hey (2015)
Rice (*Oryza sativa*) (g/100 g)	1.19–3.49	2.69–5.18	5.30–9.51	2.20–6.24	0.65–3.49	3.5–6.30	2.09–5.06	Not Available	3.78–6.80	5.30–984	ElShamey *et al.* (2025)
Maize/Corn (*Zea mays*) (g/100 g)	2.7 (typical maize range ~2.5–3.0)	3.4–4.0	7.0–8.5	~1.8–2.5	2.2–2.8	~6.0–7.5	2.8–3.3	1.0–1.3	3.8–4.5	~4.0–5.0	Rouf Shah *et al.* (2016)
Raw makhana (*Euryale ferox*) (g/100 g)	2.7 ± 0.25	3.3–3.5	5.8–6.5	3.8–4.7[a]	3	5.9	3.3–3.5[a]	0.8–1.0	5.2–5.5[a]	10–16[a]	Kapoor *et al.* (2022)

[a]*These amino acids are present at higher levels in makhana but are limited in other commonly consumed staple cereals.*

glutamic acid, aspartic acid, lysine, leucine, and arginine are found in the free or hydrolyzed amino acid fractions. Lysine, often a limiting amino acid in cereal-based diets, is present at levels that make foxnut a useful complementary protein source when combined with cereals or other staples (Jayaprakash et al., 2022). Arginine + lysine/proline ratio is high. Some studies have suggested that the EAAI for raw makhana protein is 93.63%, indicating a reasonable balance of essential amino acids for human nutrition. This can reduce to 89.97% on popping of foxnuts. This reduction in EAAI can be attributed to loss of tyrosine (which is a heat labile amino acid) and there is also a shift in the ratio of leucine to isoleucine (Kapoor et al., 2022; Kumar et al., 2016, 2024a, 2024b). As per Devi et al. (2024) popping of makhana was found to reduce the levels of essential as well as non-essential amino acids, with the notable exception of lysine. The lysine content exhibited a significant increase, rising from 2.19 mg/g in the raw kernels to 6.34 mg/g after popping. In the unprocessed kernels, histidine was the most abundant amino acid (301 mg/g), followed by threonine (122 mg/g), leucine (49.2 mg/g), and phenylalanine (31 mg/g). Although the overall amino acid profile in the popped makhana remained comparable to that of the raw form, the concentrations were significantly reduced post-processing. However, absolute protein content remains lower than that of pulses, so formulations that rely on makhana for protein should account for total protein concentration as well as amino acid balance.

8.2.2.3 Functional qualities and applications

Protein extracted from foxnut seeds has been evaluated for functional properties (solubility, water-holding capacity. and emulsification). While not a major commercial protein isolate source, the functional properties of makhana proteins support its use in composite flours, texturized products, and certain gluten-free formulations. Processing (e.g., heat treatment, germination) can alter protein solubility and digestibility; for example, germination activates certain proteases in the seed that can access the proteins breaking them down into smaller peptide chains increasing their digestibility and can slightly enhance measured protein content due to synthesis of storage and metabolic proteins during sprouting (Bisht et al., 2025).

8.2.3 Carbohydrates, resistant starch, and dietary fiber

8.2.3.1 Starch as the main carbohydrate fraction

In the carbohydrate fraction, starch is the predominant component. The endosperm of the seed is rich in starch. It contributes to around 89% of the total carbohydrate content. The content of amylose is 25% and the rest is amylopectin. Reported starch composition and properties differ across studies, but *E. ferox* starch has been characterized as having a moderate amylose content and distinct physicochemical properties that can be modified by enzymatic or thermal treatments (Kapoor et al., 2022). On popping, the starch content significantly decreases by approximately 38%. The heat treatment breaks the starch polymers into intermediate molecules and free sugars like glucose and maltose. The breakdown of amylose and amylopectin is also essential for the expansion of foxnuts on popping. This results in reduction in the amount of amylose and amylopectin. Analysis under scanning electron microscope revealed that the starch matrix transforms from a highly crystalline structure to a porous matrix. This contributes to the crunchy texture of popped foxnuts or makhana. Popping also increases the degree of gelatinization by 1117% and improves starch digestibility by increasing the rate of starch hydrolysis (Kumar and Jha, 2024).

Kumar et al. conducted a study to check the difference in nutritional composition between popped and raw makhana showed that on popping, the amount of total sugar increased by 154.40% and available carbohydrates increased by 27.45%.

8.2.3.2 Resistant starch

Resistant starch (RS) is the fraction of starch that cannot be digested in the small intestine and accessible to colonic fermentation. It is beneficial for colonic health and because of its reduced glycemic index. RS content in native *E. ferox* starch has been reported in the low-to-moderate range. In raw makhana, it is reported to be between 17% and 18.6% of total starch. This is particularly RS 2. RS can be substantially increased through enzymatic modification (e.g., amylopullulanase treatment) or through retrogradation induced by processing and cooling. Hence, on popping followed by cooling, RS3 starch forms. One notable study documented the production of resistant starch from Indian popped makhana using enzyme-mediated modification and reported increases in RS and altered digestibility profiles, suggesting technological pathways for producing lower-glycemic makhana products. The potential to increase RS is promising for developing functional foods aimed at glycemic management (Biswas et al., 2020). Popping also increases the in-vitro digestibility of the starch which can be attributed to gelatinization of starch (Islam et al., 2024). However, popping has also been reported to increase the glycemic index (GI) of the fox nuts (Mahesh Kumar et al., 2023). The estimated glycemic index (eGI) of raw makhana is 35.63 whereas for popped it is 61.52.

8.2.3.3 Dietary fiber

Total dietary fiber values reported for foxnut vary across studies. Direct fiber determinations and recent compositional studies identify appreciable non-starch polysaccharide fractions, particularly in hulls or fibrous residues. As per USDA, the total dietary fiber content is between 1.1 and 2.5 g per 00 g of raw makhana. This can reduce on popping due to the removal of the outer hull or bran. Fiber contributes to lowered glycemic response, satiety, and colonic fermentation; combined with RS, the fiber profile of makhana may help moderate post-prandial glycemia when consumed as part of composite meals. The degree to which popped makhana retains or loses fiber depends on processing like mechanical abrasion and heat may reduce some insoluble fiber fractions while concentrating other fractions of the fiber (Dega and Barbhai, 2023).

8.2.4 Minerals and trace elements

The mineral composition of makhana includes calcium, magnesium, potassium, sodium, phosphorus, iron, manganese, molybdenum, zinc, copper, and cobalt. Out of the different components of foxnuts, the bran contains significantly higher levels of these minerals. The overall mineral content of raw and popped kernels is very similar. The major difference is in the amounts of calcium and iron. The calcium content observed in the popped samples are higher because of more uniform distribution of calcium in the popped kernels. On the other hand, the reduced iron content in popped makhana is suggestive that iron is predominantly localized in the bran layer, which may be lost or altered during popping (Devi et al., 2024). The process of roasting increases the concentration of these minerals as moisture evaporates (Liaquat et al., 2022). Roasted fox nuts was reported to contain significantly higher quantity of potassium (472 $\mu g\ g^{-1}$), magnesium (263 $\mu g\ g^{-1}$), phosphorus (564 $\mu g\ g^{-1}$), calcium (192 $\mu g\ g^{-1}$), iron (127 $\mu g\ g^{-1}$), sodium (238 $\mu g\ g^{-1}$) and zinc (18 $\mu g\ g^{-1}$) as compared to control (unroasted fox nuts) (Liaquat et al., 2022) as shown in (Table 8.3). Additionally, foxnuts also contain phenols, flavonoids, sterols, and glucosides contributing to its antioxidant profile (Joseph and Ramesh, 2023; Kapoor et al., 2022).

8.2.5 Anti-nutrients

Foxnuts have been reported to contain certain anti-nutrients that can interfere with nutrient absorption and their bioavailability. These anti-nutrients are phytates, oxalates, tannins and saponins. Their content is summarized in (Table 8.3). They are concerned with reducing mineral uptake and protein digestibility. Majority of these are concentrated in the seed coat of the seeds (Wu et al., 2013).

Phytates (myo-inositol hexakisphosphate) are phenolic compounds making them natural antioxidants. This property is beneficial as they provide protection against oxidative damage when their concentrations are moderate.The amount of phytate in the kernel is 87.4 mg per 100 g and 665 mg per 100 g in the bran. As an anti-nutrient, they exert their effects by forming a chelate with major macro-minerals like calcium, iron, and zinc. These chelates are

TABLE 8.3 Micronutrient and anti-nutrient composition of makhana seeds.

	Phosphorus (mg/100g)	Potassium (mg/100g)	Magnesium (mg/100g)	Calcium (mg/100g)	Zinc (mg/100g)	Iron (mg/100g)	Sodium (mg/100g)	Phytic acid (mg/100g)	Tannins (mg CE/g)	References
Raw kernel	132.53	50.25	13.83	14.5	1.5	5.73	2.92	87.36 ± 7.7	0.88 ± 0.12	Devi (2022)
Popped kernel	124.01	48.39	12.71	20.94	1.04	2.67	4.06	Not Detected	Not Detected	Devi (2022)
Roasted	56.4 ± 3.1	47.2 ± 1.8	26.3 ± 2.3	19.2 ± 1.2	1.8 ± 0.3	12.7 ± 1.0	23.8 ± 1.4	Not Available	Not Available	Liaquat *et al.* (2022)
Germinated flour	44.40 ± 0.97	38.43 ± 0.41	22.20 ± 0.37	13.53 ± 0.28	1.33 ± 0.12	9.26 ± 0.46	19.40 ± 0.32	87.14 ± 0.28	2.85 ± 0.24	Bisht *et al.* (2025)

insoluble in nature and are not absorbed in the intestine which reduces their bioavailability (Samtiya et al., 2020). In the same way, oxalates specifically form chelates with calcium ions. But the amount of oxalate in foxnuts is low.

Tannins are also structurally phenols. They are also known to exhibit antimicrobial and anti-oxidant activities. Their concentration is 35.8 mg per 100 g in the bran (Devi, 2022). As an anti-nutrient, they can bind with peptides forming stable complexes with the carbonyl groups on protein molecules, they can also bind with major protein digesting enzymes that are trypsin and chymotrypsin. Both of these result in reduced digestibility and bioavailability of proteins (Wang et al., 2022).

Saponins, which are glycosidic in nature, have also been identified in extracts of foxnuts seeds (quantity not given). Though, the quantity has not been determined. Saponins, as an anti-nutrient, can bind to cholesterol and bile acids, reducing the absorption of fats. However, they also have cholesterol lowering effects (Singh et al., 2017).

Roasting or popping and germinating significantly reduce the amounts of these anti-nutrients in foxnuts. In one study, where the nutritional differences in non-germinated and germinated foxnut flour were tested, it was found that the amount of phytates and tannins reduced from 177.63 to 3.85 mg–87.14 and 2.85 mg respectively per 100 g of germinated flour (Bisht et al., 2025). This change was attributed to the activation of certain polyphenol-degrading enzymes, like polyphenol oxidase, on germination. Since both tannins and phytates are phenols in nature, this enzyme can cause their breakdown. In another study, it was reported that both of these anti-nutrients were undetected on popping of makhana. The reason cited was that popping can remove the bran or seed coat which contains most of the anti-nutrients (Devi, 2022).

8.3 Phytochemicals and bioactive constituents of foxnuts

Foxnuts are also considered a good source of phytochemicals that possess many health-promoting benefits. The details of these compounds are summarized in (Table 8.4). These phytochemicals are primarily found in the seed coat, but some are also found in the kernel.

8.3.1 Phenols, flavonoids, and tannins

According to studies, methanol and aqueous extracts of foxnuts contain a significant quantity of phenols and flavonoids. The total phenolic content (TPC) of raw makhana ranges between 346.02 and 470.62 mg gallic acid equivalent (GAE) per 100 g and the total flavonoid content (TFC) ranges between 4.43 mg catechin equivalents per g, as per reports (Liaquat et al., 2022). These compounds have notable antioxidant capacity and the ability to reduce oxidative stress. Flavonoids, which are a type of secondary metabolite of plants, are polyphenolic compounds with different structures. They are anti-

TABLE 8.4 Identified phytochemicals and bioactive compounds in makhana (*Euryale ferox*).

Sl. No.	Name of the compound	Compound Class	Reported plant part/ fraction	Molecular formula	Structure	Mode of action	References
1	Gallic acid	Phenolic acid	Seed coat/ phenolic extract	$C_7H_6O_5$	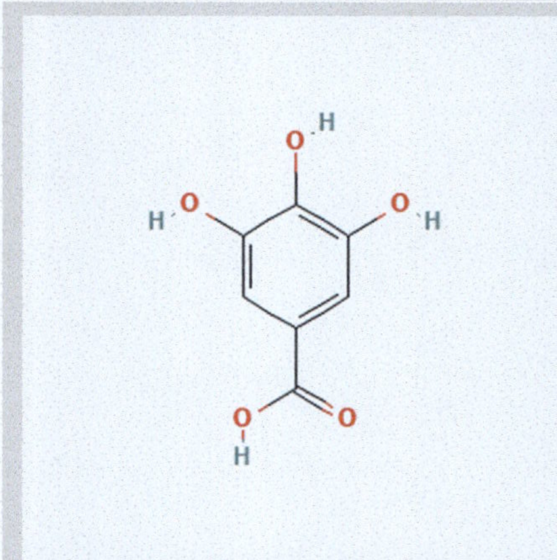	Scavenges ROS and modulates NF-κB–mediated inflammation	Wu et al. (2013), Feng et al. (2018)

Continued

TABLE 8.4 Identified phytochemicals and bioactive compounds in makhana (*Euryale ferox*).—cont'd

Sl. No.	Name of the compound	Compound Class	Reported plant part/ fraction	Molecular formula	Structure	Mode of action	References
2	Protocatechuic acid	Phenolic acid	Seeds (ethanol extract; isolated)	$C_7H_6O_4$	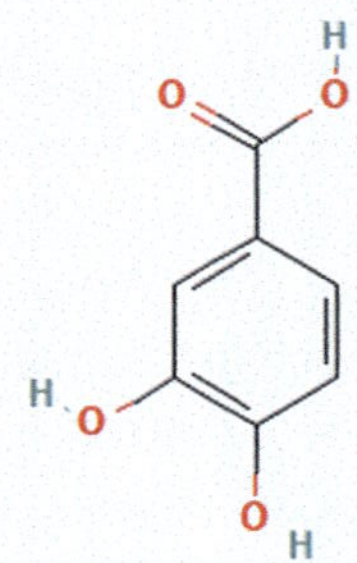	Inhibits lipid peroxidation and suppresses inflammatory signaling	Sun et al. (2014), Kakkar and Bais (2014)

3	Gallic acid ethyl ester	Phenolic derivative (ester)	Seeds (ethanol extract; isolated)	$C_9H_{10}O_5$	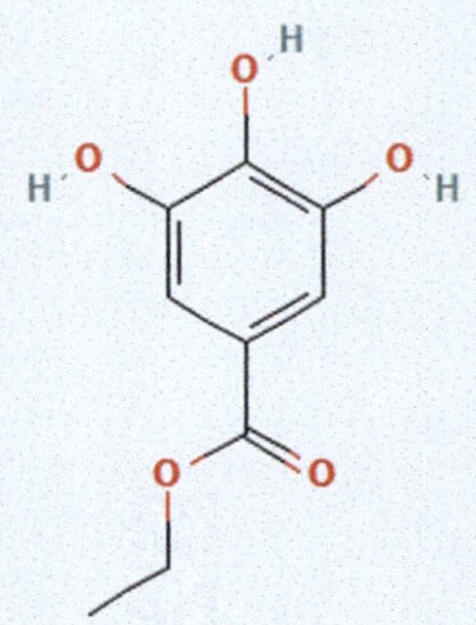	Enhances antioxidant capacity and induces apoptosis in cancer cells	Sun et al. (2014), Choubey et al. (2015)
4	Chlorogenic acid	Phenolic acid (caffeoylquinic)	Seed-coat extract	$C_{16}H_{18}O_9$	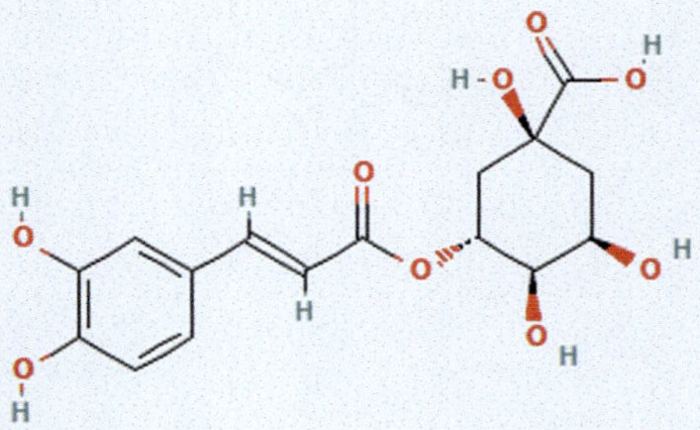	Improves glucose metabolism by inhibiting glucose-6-phosphatase	Wu et al. (2013), Zhang et al. (2023)

Continued

TABLE 8.4 Identified phytochemicals and bioactive compounds in makhana (*Euryale ferox*).—cont'd

Sl. No.	Name of the compound	Compound Class	Reported plant part/ fraction	Molecular formula	Structure	Mode of action	References
5	Catechin	Flavan-3-ol (flavonoid)	Seed-coat/seed extracts	$C_{15}H_{14}O_6$	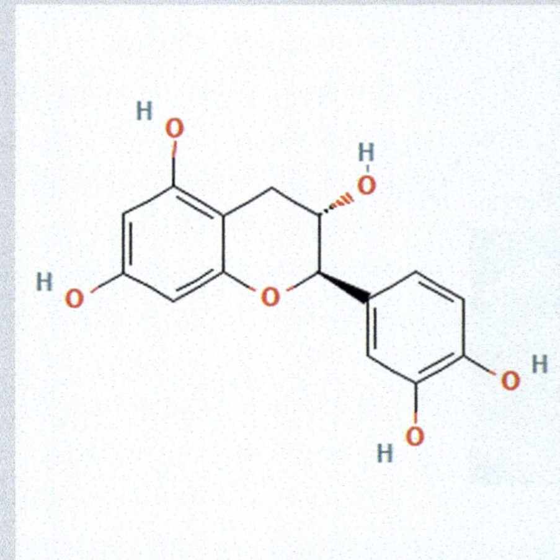	Prevents LDL oxidation and improves endothelial nitric oxide signaling	Zhang et al. (2023),Wu et al. (2013)

6	(−)-Epicatechin	Flavan-3-ol (flavonoid)	Seed/seed-coat extracts	$C_{15}H_{14}O_6$	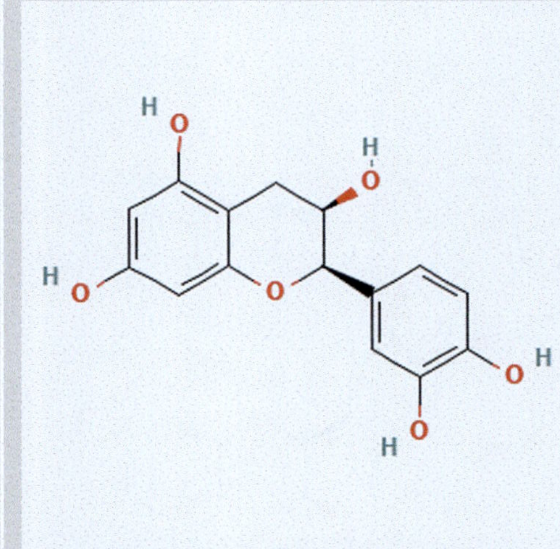	Activates endothelial NO synthase and improves vascular function	Zhang et al. (2023), Wu et al. (2013)

Continued

TABLE 8.4 Identified phytochemicals and bioactive compounds in makhana (*Euryale ferox*).—cont'd

Sl. No.	Name of the compound	Compound Class	Reported plant part/ fraction	Molecular formula	Structure	Mode of action	References
7	Epicatechin-gallate	Galloylated flavanol	Seed-coat extract	$C_{22}H_{18}O_{10}$	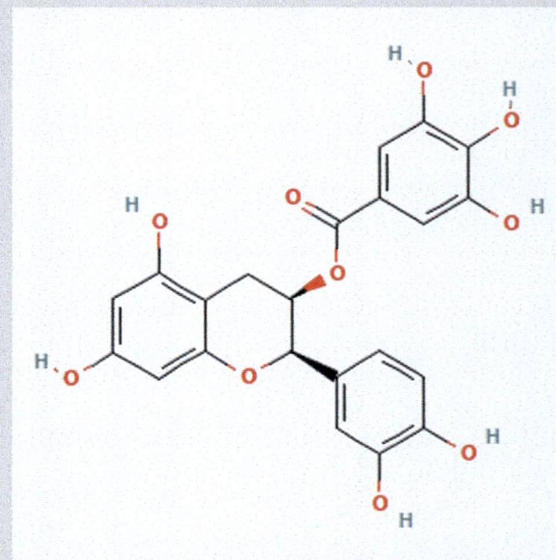	Induces cell-cycle arrest and suppresses tumor cell proliferation	Wu et al. (2013)

8	Procyanidin B2	Proanthocyanidin dimer	Seed extracts	$C_{30}H_{26}O_{12}$	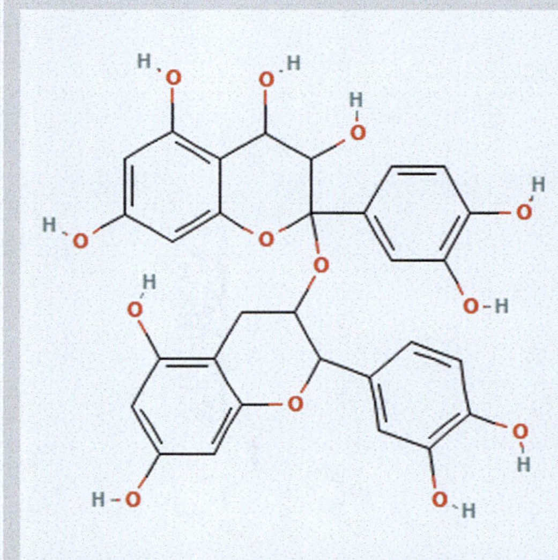	Inhibits platelet aggregation and oxidative stress pathways	Zhang et al. (2023)

Continued

TABLE 8.4 Identified phytochemicals and bioactive compounds in makhana (*Euryale ferox*).—cont'd

Sl. No.	Name of the compound	Compound Class	Reported plant part/ fraction	Molecular formula	Structure	Mode of action	References
9	Digalloyl-hexoside	Galloylated phenolic glycoside	Seed-coat phenolic fraction	$C_{20}H_{20}O_{14}$	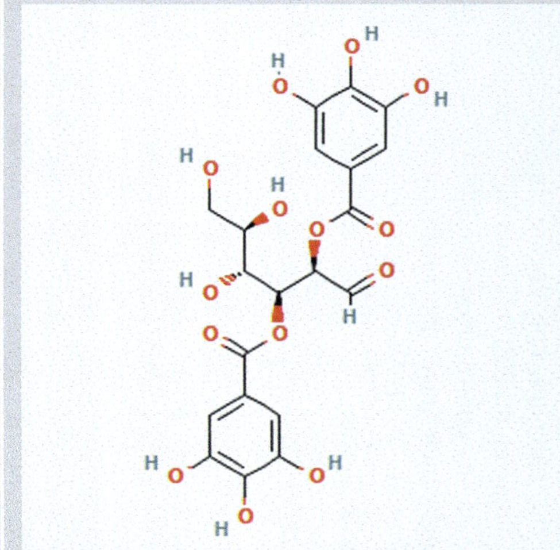	Acts as a polyphenolic antioxidant via radical scavenging	Zhang et al. (2023)

10	Rutin (quercetin-3-O-rutinoside)	Flavonol glycoside	Seed-coat	$C_{27}H_{30}O_{16}$	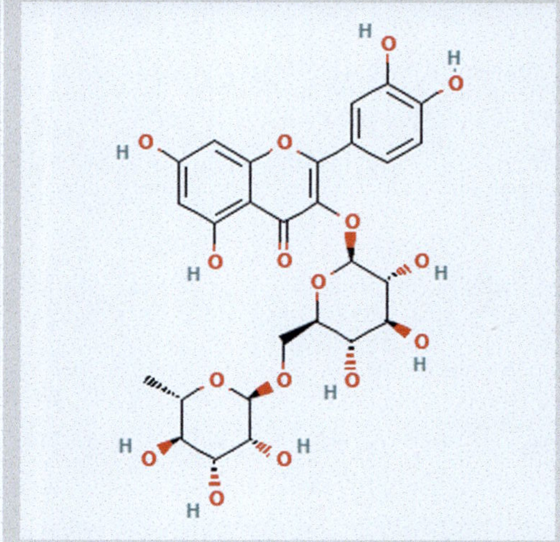	Stabilizes capillaries and downregulates inflammatory enzymes	Wu et al. (2013).
11	5,7,4-Trihydroxyflavanone	Flavanone	Seed-coat phenolic extract	$C_{15}H_{12}O_5$	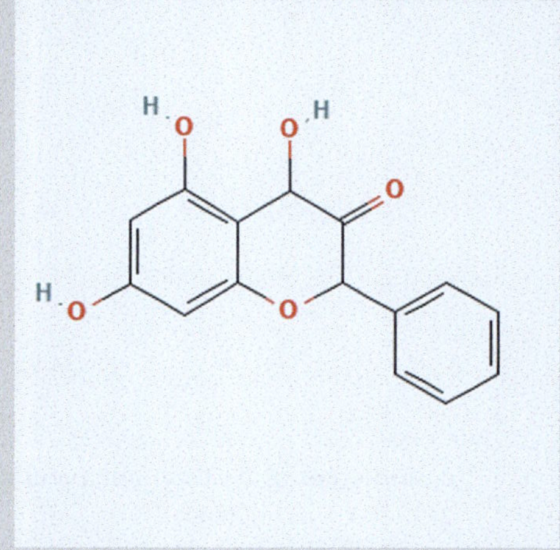	Inhibits microbial enzymes and oxidative stress	Wu et al. (2013)

Continued

TABLE 8.4 Identified phytochemicals and bioactive compounds in makhana (*Euryale ferox*).—cont'd

Sl. No.	Name of the compound	Compound Class	Reported plant part/ fraction	Molecular formula	Structure	Mode of action	References
12	5,7-Dihydroxy-2-(3,4, 5-trihydroxyphenyl)-chroman-4-one	Flavan-type chromanone	Seed-coat phenolic extract	$C_{15}H_{12}O_7$	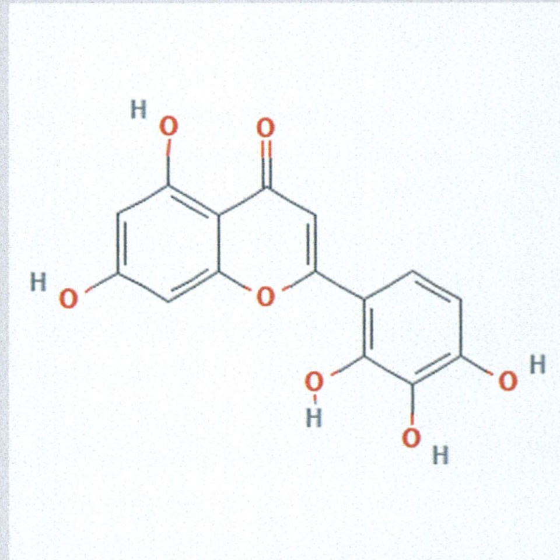	Neutralizes free radicals and modulates inflammatory mediators	Wu et al. (2013)

13	Buddlenol E	Biphenyl lignan-type compound	Seed-coat phenolic fraction	$C_{34}H_{30}O_{11}$	Not Available	Suppresses oxidative stress–induced neuronal damage	Wu et al. (2013)
14	Corilagin (1,2,3-trigalloylhexahydroxydiphenoyl glucose)	Ellagitannin/ hydrolyzable tannin	Shell (seed coat/ shell)	$C_{27}H_{22}O_{18}$	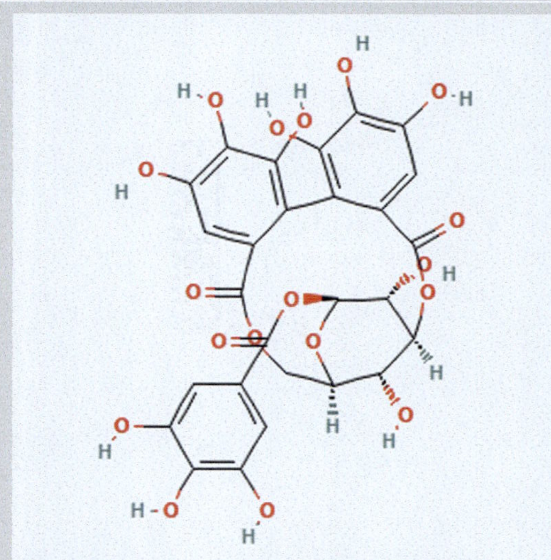	Inhibits TNF-α and protects hepatic cells from oxidative injury	Wu et al. (2023)
15	Euryalin A (new sesquineolignan A)	Sesquineolignan (newly described)	Seeds	$C_{27}H_{30}O_9$	Not Available	Exhibits antioxidant activity through phenolic radical quenching	Song et al. (2011)
16	Euryalin B (new sesquineolignan B)	Sesquineolignan	Seeds	$C_{27}H_{30}O_9$	Not Available	Protects cells against oxidative stress–induced apoptosis	Song et al. (2011)

Continued

TABLE 8.4 Identified phytochemicals and bioactive compounds in makhana (*Euryale ferox*).—cont'd

Sl. No.	Name of the compound	Compound Class	Reported plant part/ fraction	Molecular formula	Structure	Mode of action	References
17	Euryalin C (new sesquineolignan C)	Sesquineolignan	Seeds	$C_{27}H_{30}O_9$	Not Available	Modulates redox balance and inflammatory signaling	Song et al. (2011)
18	rel-(2α,3β)-7-O-methylcedrusin	Biphenyl/lignan derivative	Seeds	$C_{21}H_{24}O_7$	Not Available	Inhibits inflammatory mediators and lipid oxidation	Song et al. (2011)
19	Syringylglycerol-8-O-4-(sinapyl alcohol) ether	Lignin-related dimeric phenylpropanoid	Seeds	$C_{23}H_{30}O_9$	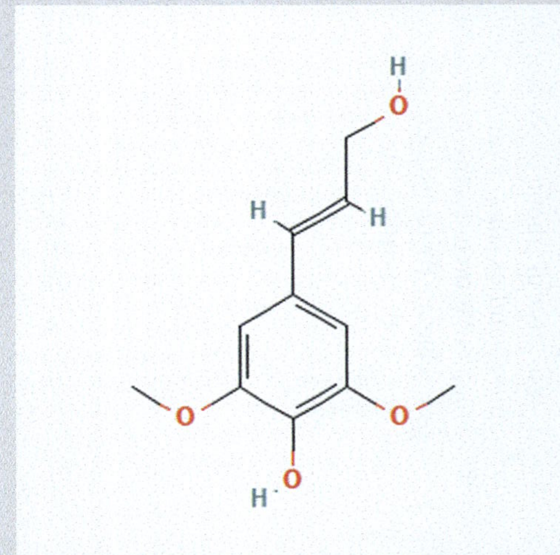	Acts as a lignan antioxidant reducing oxidative damage	Song et al. (2011)

20	(+)-Syringaresinol	Lignan	Seeds	$C_{22}H_{26}O_8$	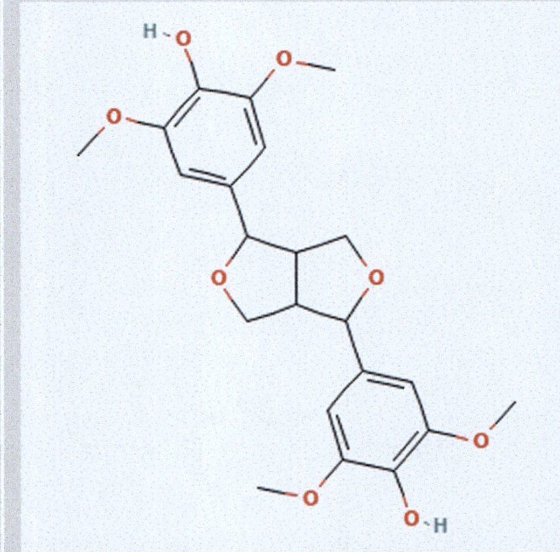	Exhibits phytoestrogenic and antioxidant activity	Song et al. (2011)
21	(1,2,5R,6S)-2-(3,4-dimethoxyphenyl)-6-(3,4-dihydroxyphenyl)-3,7-dioxabicyclo[3.3.0]octane	Bicyclic lignan-type compound	Seeds	$C_{18}H_{20}O_6$	Not Available	Provides cellular protection via antioxidant mechanisms	Song et al. (2011)
22	5,7-Dihydroxy-6,4′-dimethoxyflavone	Flavone (methoxylated)	Seeds (ethanol extract)	$C_{17}H_{14}O_6$	Not Available	Inhibits inflammatory enzymes and tumor growth	Sun et al. (2014)

Continued

TABLE 8.4 Identified phytochemicals and bioactive compounds in makhana (*Euryale ferox*).—cont'd

Sl. No.	Name of the compound	Compound Class	Reported plant part/ fraction	Molecular formula	Structure	Mode of action	References
23	5,7-Dihydroxychromone	Chromone derivative	Seeds (ethanol extract fraction)	$C_9H_6O_4$	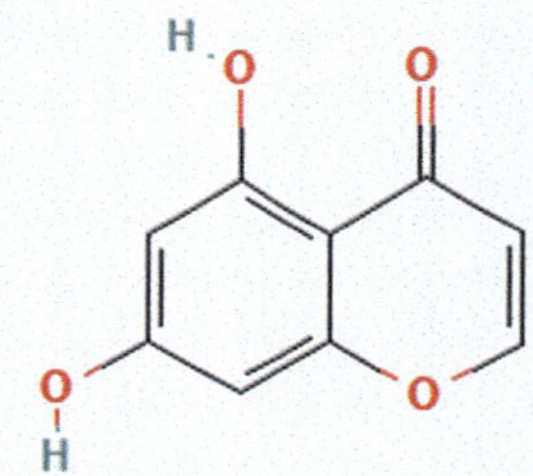	Acts as a flavonoid-like antioxidant	Sun et al. (2014)

24	β-Sitosterol	Plant sterol	Seeds — detected (lipid fraction)	$C_{29}H_{50}O$	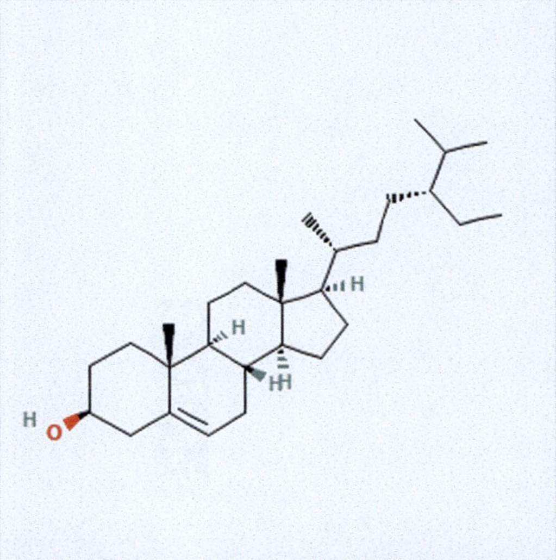	Reduces intestinal cholesterol absorption	Sun et al. (2014)
25	Daucosterol (β-sitosterol-3-O-β-D-glucoside)	Sterol glycoside	Seeds — isolated	$C_{35}H_{60}O_6$	Not Available	Modulates immune response and induces apoptosis	Sun et al. (2014)
26	Protocatechuic aldehyde	Phenolic aldehyde	Seeds - Comparative metabolomics	$C_7H_6O_3$	Not Available	Prevents vascular remodeling via antioxidant action	Liu et al. (2013), Xie et al. (2025)
27	Various cyclic dipeptides/ diketopiperazines (specific individual DKP structures reported in metabolomics)	Cyclic peptides/ small peptides	Seed metabolome (LC-HRMS profiling studies; multiple specific DKPs listed in supplements)	Not Applicable	Not Applicable	Exhibit antimicrobial and antioxidant bioactivity	Xie et al. (2025), Liu et al. (2013)

Continued

TABLE 8.4 Identified phytochemicals and bioactive compounds in makhana (*Euryale ferox*).—cont'd

Sl. No.	Name of the compound	Compound Class	Reported plant part/ fraction	Molecular formula	Structure	Mode of action	References
28	Glycosylated sterols/ cerebrosides/glycolipids (e.g., glucosyl-sterols and identified glycolipid species in lipid profiling)	Glycolipids/ cerebrosides	Seed lipidome	Not Applicable	Not Applicable	Support membrane integrity and immune signaling	Jiang et al. (2023)
29	Resistant starch fractions/ specific oligosaccharide motifs	Polysaccharide/ starch components	Seed endosperm/ polysaccharide isolates	Not Applicable	Not Applicable	Fermented to SCFAs improving gut and metabolic health	Jiang et al. (2022), Zhang et al. (2023)

inflammatory, anti-diabetic and antioxidant in nature. Multiple phenols and flavonoids interact synergistically to neutralize reactive oxygen species (ROS).

In one study, three phenolic compounds were identified in an extract of the seed coat of foxnuts. They are 5,7-dihydroxy-2-(3,4,5-trihydroxyphenyl) chroman-4-one, 5,7,4-trihydroxyflavanone and buddlenol E. Presence of the derivatives of quercetin and kaempferol have also been reported (Wu et al., 2013). Some polyphenols that have been identified in the seed coat are gallic acid, pyrogallol, chlorogenic acid, naringenin, dihydrotricetin and rutin. Resveratrol, 4-O-methyl gallic acid, protocatechuic acid and pectolinarigenin have been qualitatively determined in the seed kernel (Mittal et al., 2020). Tannins, as discussed earlier are also phenols in nature and are present in the seed coat of foxnuts.

However, roasting may affect the concentration of these phytochemicals. In a study, on roasting, the TPC decreased from 2.22 to 1.12 mg GAE per g in the raw kernel and the TFC decreased from 3.23 to 1.26 mg RE per 100 g. This change is primarily attributed to the degradation of heat-sensitive phytochemicals (Devi, 2022).

8.3.2 Alkaloids and saponins

Alkaloids are heterocyclic compounds with at least one nitrogen in their structure. Saponins are glycosidic compounds containing hydrophilic sugar units (the glycone) and one hydrophobic group (aglycone). While the quantitative data on these compounds are scarce, they have been identified in extracts of foxnut seeds qualitatively. The reported saponins have aglycone groups that are steroids or triterpenoids in nature (Kumar et al., 2024a,b).

Roasting may reduce the saponin and alkaloids content of foxnuts, as they are predominantly found in the seed shells which are removed on roasting.

8.3.3 Trace bioactive metabolites

Some trace bioactive metabolites that have been qualitatively analyzed are phytosterols and triterpenoids. Among the phytosterols, β-sitosterol and campesterol have been identified in foxnut seed extracts. These have cholesterol-lowering and anti-inflammatory properties. The quantity of triterpenoids are minute, but they can exhibit strong antioxidative and hepatoprotective effects by regulating enzyme activity involved in detoxification pathways.

As per one study, in a 70% ethanolic extract of foxnut seed shell, the concentration of triterpenoids was 36.7% (Yuan et al., 2013). In another study, two major triterpenoids identified were 2β-hydroxybetulinic acid 3β-oleiate and 2β-hydroxybetulinic acid 3β-caprylate in an ethyl acetate extract of foxnut seeds (Ahmed et al., 2018).

Analytical studies have isolated cerebrosides and novel lipidic constituents from the extracts of foxnuts. These trace metabolites, although present in low concentration, enhance the nutritional benefits of foxnuts (Mittal et al., 2020).

The phytochemical composition of foxnuts adds to its overall nutritional profile and therapeutic potential. Although some phytochemicals like tannins and saponins are anti-nutrients in nature, their presence in moderate quantities promotes foxnut's health benefits.

8.4 Impact of processing on nutritional composition and functional properties

8.4.1 Effects on starch digestibility

The primary effect on starch digestibility is due to a change in the structure of the starch molecules. Processing methods like popping and roasting have been reported to increase starch digestibility. This has been attributed to gelatinization of starch molecules on roasting.

For foxnuts, popping is the most common cooking method used to make the seeds edible. Starch gelatinization is an endothermic process that takes place in the presence of moisture and high heat. As depicted in (Fig. 8.1), when heat is applied, the amylose particles from the starch molecules absorb moisture (from the soaked seeds) and swell, leading to a disorientation in the structure. This makes the starch molecules more accessible to digestive enzymes like alpha and beta amylase (Mahesh Kumar et al., 2023; Bhat and Arya, 2023). Another reason for the increased digestibility is the expansion of the seed kernel on popping. Popping involves a chemical and a physical

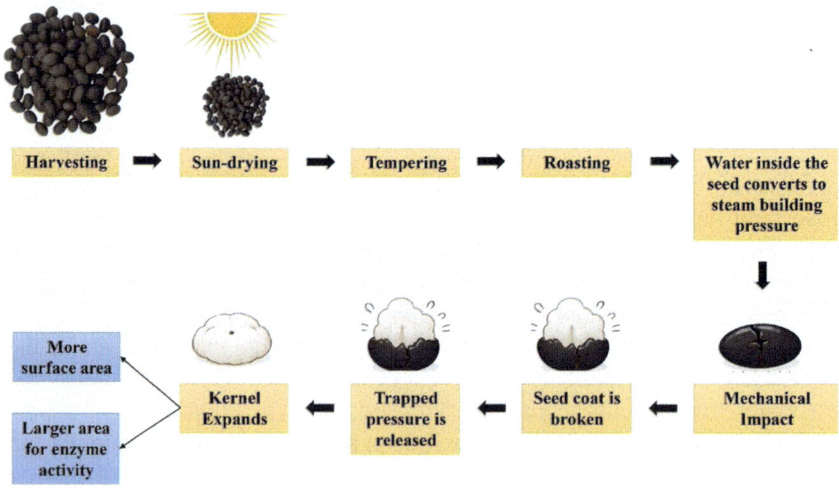

FIGURE 8.1 Process of popping of raw makhana seeds.

transformation in the seed's starchy kernel. The seed coat of foxnuts, which is also known as the pericarp, is a tough, non-porous layer that acts as a pressure vessel. After harvesting the seeds, they are sun-dried to reduce the moisture content. Then there is a step of tempering where the moisture redistributes in the kernel. This also contributes to making the seed coat loose. Then the seeds are heated at a very high temperature that causes the starch to gelatinize. Simultaneously, the water inside the kernel converts to steam, which builds high pressure. To make the seeds pop, they are taken off the heat and struck with a heavy force that breaks the outer seed coat and makes the seed pop. This involves rapid expansion in the gelatinized starch because of the trapped pressure which increases the starch's surface area. A bigger surface area allows for easier enzyme digestibility (Devi, 2022) & Indian Council of Agricultural Research). In one study by Mummadi et al., it has been reported that popping resulted in an increase in starch gelatinization by 1117%. The available starch increased by 27.45% and the total sugar content by 154.40%. However, it has also been noted that retrogradation, which is the realignment of amylose and amylopectin molecules on cooling of starch may reduce starch digestibility, but the effect is considered to be negligible.

There have been studies on reducing starch digestibility of foxnuts with enzymatic modification. One study also reports using enzymatic treatment like amylopullulanase on makhana flour. These enzymes break down the linear chains of starch which reorganizes into a more crystalline structure. This reduces the quantity of rapidly digestible starch and makes these reorganized structures resistant to digestion (Biswas et al., 2020). Some conditions, like heating with moisture, can cause recrystallization of starch. The formation of certain crystalline structures, like the B-type polymorph, can make the starch more resistant to enzymatic digestion.

Foxnuts are often converted to flour with the help of milling. Milling or grinding disrupts the granules of starch breaking them down into smaller particles. This also increases the surface area of the starch molecules which results in better hydrolysis of starch (Yu et al., 2015).

8.4.2 Effect on glycemic index

Foxnuts, as per literature, are often considered a low-GI food in their minimally processed, popped state, making them a popular dietary choice. However, the diverse methods of industrial and domestic preparation, ranging from simple popping to extensive grinding, cooking, and incorporation into composite products can significantly modify the inherent physiochemical structure of its starch and, consequently, its GI.

Various processing conditions have been reported to affect the glycemic index of foxnuts in different ways. As discussed in the previous section, popping increases starch digestibility and the available free sugars which has a significant effect on the glycemic index of starch. Increase in free sugars which can readily get absorbed into the bloodstream increases the glycemic index of foxnuts. A study that compared the eGI of raw foxnuts seeds with popped foxnuts reported that the eGI had increased from 35.63 to 61.52 (Mahesh Kumar et al., 2023).

On the other hand, roasting popped or puffed foxnuts before consumption can decrease its glycemic index. In a study by (Liaquat et al., 2022), 10 healthy subjects aged between 18 and 30 were recruited to check the glycemic response post consuming foxnuts. It was noted that the eGI of foxnuts was 37.05 (with a reference to glucose powder). Based on glycemic index, they are classified into low GI, high GI and moderate GI with values low ($\leq$55), moderate (56–69), or high ($\geq$70).

Further processing often involves frying or seasoning the popped makhana. The addition of fat (oil) like clarified butter or spices can counter the glycemic effect. The presence of fat slows the movement of food and also movement of chyme into the small intestine, which reduces glucose absorption. This can reduce the eGI of seasoned or fried foxnuts (Murillo et al., 2022).

Makhana flour is prepared by milling popped foxnuts into flour followed by sieving to remove the coarse outer layer/seed coat. This reduces the particle size of starch molecules also increases the surface area for the activity of salivary and pancreatic amylase allowing for faster digestion of starch and absorption into the bloodstream. This may increase the eGI of makhana flour (Pruett et al., 2023).

8.4.3 Effect on resistant starch

Starch that can resist digestion in the stomach are known as resistant starch. These are then fermented by the gut microorganisms.The process of gelatinization and retrogradation has an effect on quantity of resistant starch in foxnuts also. As discussed earlier under Section 4.1, the starch granules undergo gelatinization on popping followed by retrogradation when the starch cools down. This increases the amount of resistant starch type 3 (RS3) because the recrystallized sections of starch become less digestible by human digestive enzymes. This also helps in reducing the glycemic index of foxnuts (Gupta and Dabade, 2021).

Studies have shown that the resistant starch content increases significantly in popped makhana compared to raw makhana. One study found that the resistant starch content increased from 21.06% in raw makhana flour to 35.94% in enzymatically modified makhana flour (Biswas et al., 2020). One more study reported increased RS3 content in popped makhana as compared to raw makhana (Mahesh Kumar et al., 2023) (amount not given). RS3 is a type of resistant starch that forms only on recrystallization of gelatinized starch. It is heat-resistant in nature, whereas RS2 starch is naturally found in food (also in foxnuts), but it is heat-labile in nature and becomes easily digestible when food is cooked (Birt et al., 2013).

8.4.4 Effect on phytochemical content and antioxidant activity

Popping, roasting, and germination are techniques that can affect the quantity of phytochemicalas well as the antioxidant activity of the seeds. Roasting of foxnuts has been reported to increase the total phenolic content (TPC) and total flavonoid content (TFC). The reason behind this is the heat from roasting, which helps release the bound phenolic and flavonoid compounds from the seed coat. This increases their availability for extraction in solvents and absorption in the human body. This heat also causes the evaporation of intracellular water, which concentrates the phenolic compounds (Liaquat et al., 2022). In one study, Foxnuts' TPC and TFC were greatly impacted by roasting, which in turn altered the antioxidant potential the nuts displayed after being roasted. TPC was 36% greater in roasted nuts (470.62 mg GAE/100 g) than in the control group (346.02 mg GAE/100 g). When a comparison is made with the control, with a value of 4.15 mg CE/g), roasted nuts had 7% higher TFC value.

Maillard browning is one very common phenomenon that occurs under high temperatures. When amino acids and reducing sugars react to form compounds called melanoidins, it leads to brown coloration in the product. Melanoidins are mostly phenols and flavonoids structurally that contribute to the total phytochemical content (Chandrasekara and Shahidi, 2011). It has been reported that the melanoidins produced improve the stability against oxidation, aroma, and flavor of oils that can be extracted from the nuts (Suri et al., 2019). However, some of the phytochemicals are also heat-sensitive. So, processes like roasting that involve high heat can lead to thermal degradation of these compounds. In a study by Kumar et al., it was established that raw makhana kernels were dominated by catechin and gallic acid equivalents that had a reduction by 58% and 49% respectively when popping. The TPC of raw and popped makhana was determined to be 2.65 mg per 100 g and 1.81 mg per 100 g, respectively (Mahesh Kumar et al., 2023).

Roasting has been shown to significantly enhance antioxidant activity, as measured by common assays like DPPH (2,2-diphenyl-1-picrylhydrazyl) and FRAP (Ferric Reducing Antioxidant Power). This is attributed to the release of bound phytochemicals and the formation of new antioxidant compounds via the Maillard reaction (Liaquat et al., 2022), reports a 30% increase in antioxidant activity as determined by DPPH activity upon roasting of foxnuts. Studies that majorly focus on the effect of popping alone report a decrease in antioxidant activity. As discussed earlier, this is due to the thermal degradation of the heat-sensitive phenolic compounds (Rathod et al., 2023).

The germination process can also affect the seed's antioxidant capacity. A study comparing the antioxidant capacity between non-germinated foxnut seed flour (NGFP) and germinated foxnut seed flour (GFP), reported that the DPPH radical scavenging activity improved from 43.35 $\pm$ 0.48b % in NGFP to 54.37$\pm$ 0.36a % in GFP. This demonstrates a considerable increase in antioxidant activity post-germination. This can be a result of increased enzymatic activity during germination that produces antioxidant-rich peptide molecules (Bisht et al., 2025).

8.5 Health benefits of foxnuts & associated mechanisms of action

Foxnuts exhibit many health benefits supported by their excellent antioxidant content, mineral composition and amino acid profile. These include cardio-protective activity, reduction of oxidation and inflammation activity, hepato-protective activity, cytotoxic and anti-cancer activity, neurocognitive and anti-fatigue activity. The mechanisms of action for these are discussed below.

8.5.1 Anti-diabetic activity

Foxnuts have significant anti-diabetic qualities. This is primarily due to the presence of many antioxidants, minerals and resistant starch. Antioxidants have the ability to enhance insulin sensitivity and control blood glucose levels. This is due to their influence on various pathways. Mainly, they are known to fight oxidative stress. Generation of ROS is known to damage the insulin-secreting beta cells of the pancreas. This can be protected with the help of antioxidants (Ahmed et al., 2018). If the nutritional quality of foxnuts is considered, they exhibit a moderate GI. This refers to a slow and steady absorption of glucose into the bloodstream, which can have low impact on the blood glucose profile of people with hyperglycemia. They also contain ample amounts of fiber. Fiber is known to slow down digestion and absorption of glucose into the bloodstream which improves glycemic control. Foxnuts are also an excellent source of magnesium. Magnesium is concerned with improving insulin sensitivity and transport of glucose into the cells. Magnesium deficiency is one of the key reasons for low insulin sensitivity. Therefore, consumption of foxnuts can be helpful (Liaquat et al., 2022; Maqrashi et al., 2025).

8.5.1.1 Mechanism of action

There is a synergistic effect of bioactive compounds and nutrients in foxnuts contributing to the anti-diabetic role of foxnuts. As shown in (Fig. 8.2), compounds found in the extract like the polyphenols such as gallic acid and others noticed in foxnut seed extract have the ability to heal pancreatic cells. This increases insulin production. The pancreatic β-cells are protected from damage caused by free radicals and oxidative stress, thereby preserving insulin production. This is done by the antioxidants present in the extract. It was reported that the extract could increase the activity of superoxide dismutase (SOD), catalase (CAT), glutathione peroxidase (GPx) and reduced glutathione (GSH). This increased activity of certain endogenous antioxidants protected

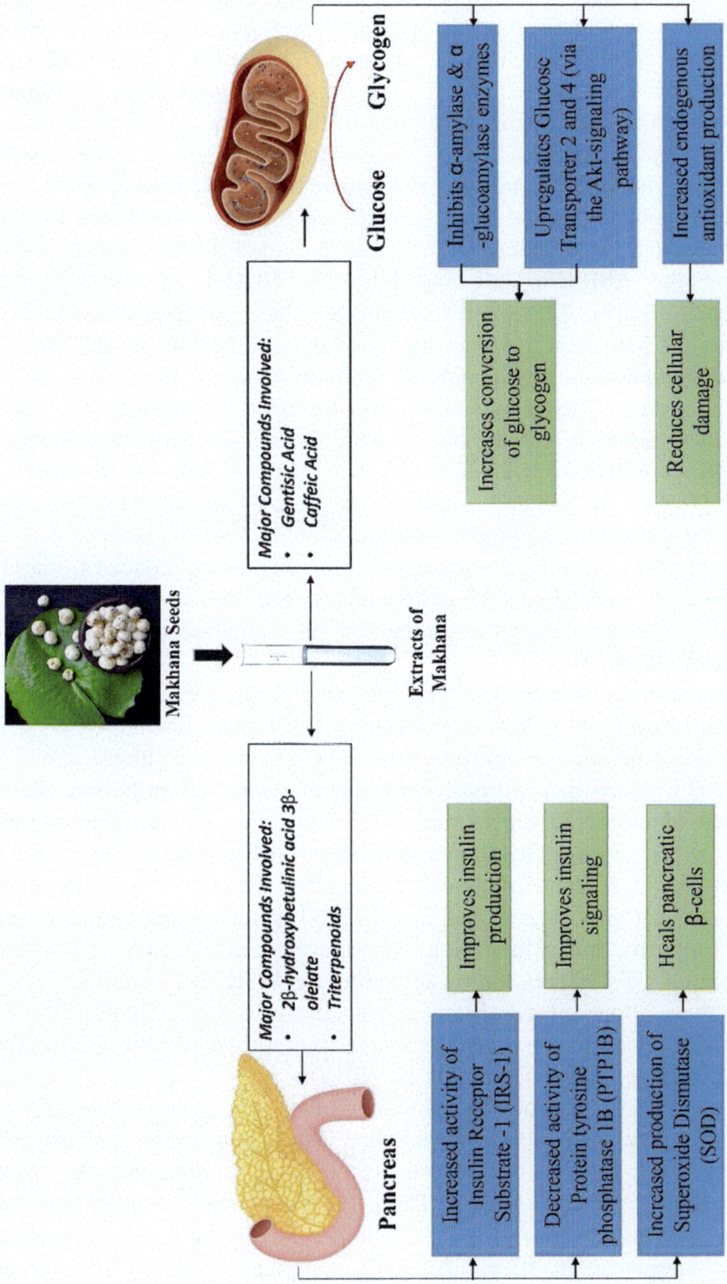

FIGURE 8.2 Mechanism of actions of makhana seeds extract in exhibiting anti-diabetic effect.

the β-cells against reactive oxygen species (ROS) mediated damage and increased the plasma insulin level in a rat model (Ahmed et al., 2015). In another study, a comparison between the effects of extracts of non-germinated foxnuts and germinated foxnuts in reducing diabetic nephropathy was assessed. It was established that both contributed significantly to reducing nephropathy. This was mainly attributed to the presence of 29 phenolic substances and one triterpenoid, 18 of which were not present in non-germinated seeds. These phenolic substances like gentisic acid, caffeic acid, and 27 other effective polyphenols found in the germinated foxnuts, contributed to higher antioxidants and renal indexes. They also provided a more stable effect in regulating the AMPK/mTOR and Keap1/Nrf2/HO-1 signaling pathways (Wang et al., 2023). These pathways are activated under physical forms of stress and can help in antioxidant-mediated defense (Balaha et al., 2022).

Some polyphenols can also block carbohydrate-digesting enzymes like alpha-glucosidase. This reduces the rate of glucose absorption from the intestine. In a research, the ethanol extract of foxnuts was administered to streptozotocin-induced diabetic rats. The role of the makhana seed extracts on the activity of α-amylase and α-glucosidase were evaluated. Bread was fortified with the extracts and a predictive simulation of digestion kinetics was used to determine the GI. In the study, the GI of fortified bread reduced from 94.61 (unfortified) to 61.66 when a 2% extract of the seed shell was introduced. This is because it could reduce the production of α-amylase and α-glucosidase (Maibam et al., 2023).

It has also been reported that the extracts from the seeds may increase the expression of essential glucose transporter (GLUT) proteins like GLUT-2 and GLUT-4. The role of these transporters is to help in moving glucose across cell membranes to allow their utilisation for metabolism. Certain polysaccharides have been identified in the extracts of foxnuts that can activate signaling pathways which upregulates the expression of GLUT proteins. In a study by Zhang et al., in 2019, a polysaccharide named EFSP-1 was isolated from foxnut seeds. It was determined that EFSP-1 could increase glucose consumption in vitro, thus exhibiting a hypoglycaemic effect. EFSP-1 could also upregulate GLUT-4 expression by activating the PI3K/Akt-signaling pathway. This pathway allows glucose uptake from the cells by facilitating the conversion of glucose to glycogen. It can also improve insulin sensitivity as observed in mice models (Zhang et al., 2019).

Triterpenoids are a group of compounds that have been isolated from the extracts. They have the ability to enhance insulin sensitivity. Increase in insulin sensitivity allows the cells to respond more efficiently to insulin, allowing them to absorb glucose from the bloodstream. Studies have also demonstrated the presence of several triterpenoids in the seeds that can have potential hypoglycaemic effects. An active compound called hydroxybetulinic acid 3β-caprylate (HBAC) was isolated using chromatographic techniques. This compound was shown to trigger the release of insulin from pancreatic β-cells and increase insulin sensitivity (Ahmed et al., 2015). Another compound, 2β-hydroxybetulinic acid 3β-oleiate (HBAO) also exhibited insulin secreting ability from pancreas. It can also reduce plasma glucose levels either by promoting uptake metabolism or by preventing hepatic gluconeogenesis. It can also increase the uptake of glucose by muscle and adipose tissues (Ahmed et al., 2018).

Insulin receptor substrate (IRS-1) is a key positive regulator of the insulin-signaling pathway whereas protein tyrosine phosphatase-1B (PTP1B) is an inhibitor of the pathway. IRS-1 helps in increasing insulin sensitivity whereas PTP1B plays a role in insulin resistance. A study evaluated the effect of triterpenoid-rich extracts from foxnut seeds on streptozocin induced diabetic rats. After 4 weeks, measurements of body weight, blood sugar, and the change in the morphology of the pancreatic tissue were checked. Additionally, the expression of proteins linked to the control of glucose metabolism in vivo, such as IRS-1 and PTP1B, was examined. Triterpenoid-rich extracts were found to be able to control glucose metabolism ($P < .01$) and restore normal body weight ($P < .05$) when compared to the model group (LD50 900 ± 2 mg/L). The expression of the insulin receptor IRS-1 protein increased, the expression of the negative regulatory protein PTP1B gene decreased, and pancreatic islet morphology improved (Yuan et al., 2014). Another study looked into using the extract as a food additive. In streptozotocin-induced diabetic mice, the blood glucose lowering and blood lipid lowering effects of triterpenoid-rich ethanolic extracts of foxnut shell were tested. For 4 weeks, gavage of 400 and 600 mg/kg E. ferox shell extract dramatically reduced insulin resistance, blood glucose, and body weight (Yuan et al., 2013; Jiang et al., 2023).

In summary, *E. ferox* can be utilized to treat diabetes primarily by protecting pancreatic β-cell against damage and apoptosis, improving insulin resistance and complications, boosting glucose uptake and utilisation, and having anti-inflammatory properties.

8.5.2 Cardio-protective activity

Foxnuts exhibit significant cardio-protective effects owing to their excellent antioxidant profile and negligible fat content. Hypertension is one of the most common cardiovascular problems found in the population. Foxnuts, being an excellent source of magnesium and potassium, prove helpful in this regard. Potassium is a vasodilator that relaxes the arteries and the heart's chambers, reducing strain on the heart (Weaver, 2013). Secondly, magnesium acts as a calcium channel blocker that prevents constriction of blood vessels, reducing the risk of arrhythmia. It also promotes the production of nitric oxide and prostacyclin, which are potent vasodilators. Lastly, the low sodium content of foxnuts makes it a good snack option for patients with cardiac problems (Houston, 2011; Jiang et al., 2023). Foxnuts contain an ample amount of phytosterols. Phytosterols reduce cholesterol by competing for absorption in the intestine with dietary and biliary cholesterol. They can also increase the expression of ATP-binding cassette (ABC) transporter A1, which is a protein that helps to move cholesterol out of the cells. They primarily reduce the concentration of low-density lipids cholesterol (LDL-C) but have minor effect on increasing high-density lipid cholesterol (HDL-C) (Marangoni and Poli, 2010). The antioxidant profile of foxnuts allows it to reduce oxidative stress that has a profound effect in damaging the heart tissues. It also helps in preventing arterial plaque formation, a major cause of cardiac arrest. Being a low-calorie snack, they aid in weight management by promoting satiety and reducing cravings for unhealthy, processed snacks that contribute to heart disease risk factors like obesity.

8.5.2.1 Mechanism of action

In a study, extracts from the seed coat of foxnuts (EFSCE) were fed to mice with non-alcoholic fatty liver disease (NAFLD) which was driven by a high-fat diet (HFD). The results demonstrated that in HFD-induced NAFLD mice, EFSCE (15 and 30 mg/kg/day) treatment efficiently decreased body weight, liver lipid deposition, and blood lipid levels such as total cholesterol (TC), triglycerides (TG), and low-density lipoprotein cholesterol (LDL) after 4 weeks. EFSCE medication may also reduce the overproduction of malondialdehyde (MDA) and increase SOD activity to combat oxidative stress compared to NAFLD mice. This was achieved by the upregulation of IRS-1, a protein that helps increase insulin sensitivity. Insulin resistance, which is associated with lipid accumulation in the body, can potentially lead to cardiac issues. Additionally, CYP2E1, a cytochrome P450 enzyme, is associated with increased production of reactive oxygen species (ROS) (Zhu et al., 2022). Polyphenols from foxnuts can help downregulate their actions (Jiang et al., 2023).

The potential of *E. ferox* seeds to mitigate myocardial ischemic reperfusion injury (MIRI) was examined in another investigation, as shown in (Fig. 8.3) MIRI is a type of damage to cardiac tissues that causes blood flow to be restored to an area after a brief period of ischemia. To assess the cardiac health protecting efficacy of *E. ferox* extract (25, 125, or 250 μg/mL), an isolated rat heart ischemia and reperfusion acute model was created. Treatment with 125 or 250 μg/mL *E. ferox* extract considerably increased aortic flow, it improved left-ventricular flow and decreased the duration of the infarct. In a chronic ischemic reperfusion model, oral dosing of *E. ferox* (250 and 500 mg/kg/day) for 21 days decreased the size of the myocardial infarct and enhanced post-ischemic ventricular function. The cardioprotective proteins were identified using the antibody array approach. Thioredoxin-1 (Trx–1) and thioredoxin-related protein-32 (TRP32) levels were higher in the hearts treated with makhana than in the control hearts.

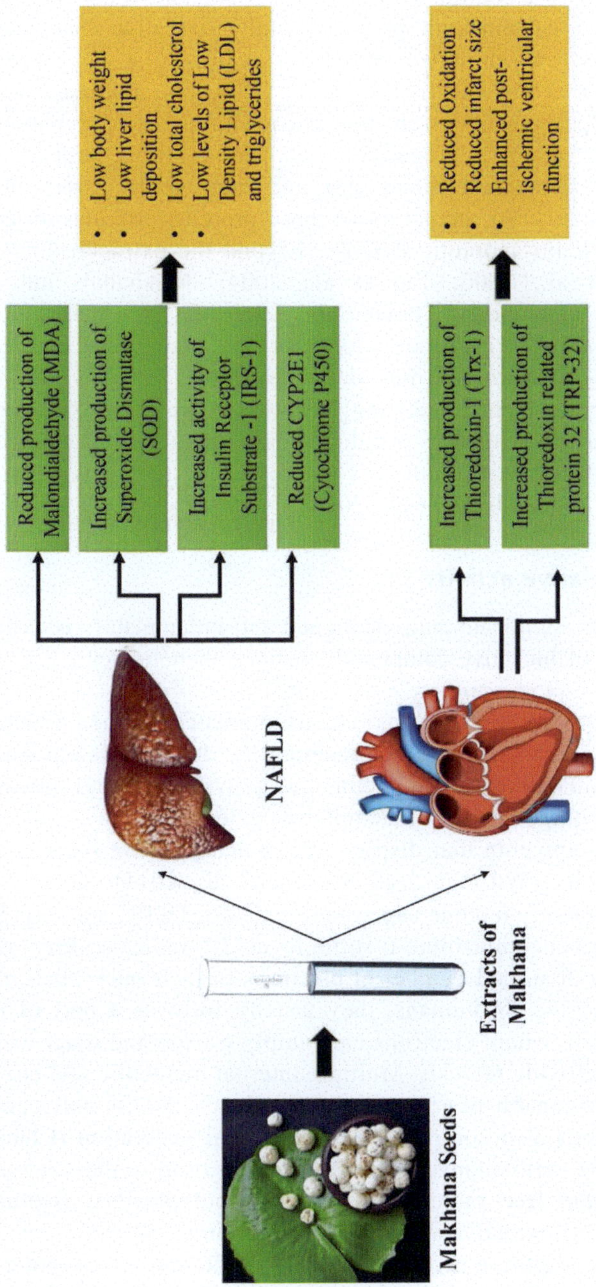

FIGURE 8.3 Effects of makhana seed extract on heart and liver that aids in cardio-protection.

Western blot analysis verified elevated thioredoxin and TRP32 protein expression.Trx-1 is a central cellular antioxidant and redox regulator. It reduces oxidized cysteines in proteins, limits oxidative inactivation of critical enzymes, and inhibits apoptosis signaling (e.g., by modulating ASK1 and other redox-sensitive kinases). TRP32 (a thioredoxin-related protein) can help keep redox-sensitive phosphatases and enzymes in their reduced, active state. Upregulation of Trx-1/TRP32 during ischemia–reperfusion decreases infarct size and improves recovery in many experimental models. So, induction of these proteins provides an endogenous antioxidant/anti-apoptotic defense, beyond the extract's direct scavenging (Das et al., 2006; Tao et al., 2004). Makhana's antioxidant phytochemicals (polyphenols, flavonoids, glucosides) are the likely scavengers. Additionally, during reperfusion after ischemia, a large burst of ROS (superoxide, H_2O_2, peroxynitrite) damages lipids, proteins, DNA, and mitochondrial components; reducing ROS limits lipid peroxidation, preserves membrane integrity, lowers mitochondrial dysfunction, and decreases the triggering of cell death pathways (necrosis and apoptosis) (Das et al., 2006; Tao et al., 2004; Jiang et al., 2023).

8.5.3 Anti-oxidative activity

Makhana exhibits significant antioxidant and anti-inflammatory activities due to a rich profile of bioactive compounds, mainly phenolic acids and specific amino acids.

Anti-oxidative properties of foxnuts are mediated by their functions to neutralize free radicals and by strengthening the defense system supported by antioxidants. Foxnuts contain numerous polyphenols and antioxidants which include compounds like kaempferol, gallic acid, chlorogenic acid, epicatechin, and apigenin that display free radical/reactive oxygen species scavenging activity (Wu et al., 2013; Ferk et al., 2011). These reactive oxygen species mainly include superoxide radicals, DPPH and ABTS [2,2′-azino-bis(3-ethylbenzothiazoline-6-sulfonic acid)] radicals. Free radicals are molecules with an odd number of electrons in their outer shell, making them highly reactive. In humans, they usually form as a part of normal cellular respiration where electrons can simply escape and react with oxygen to form superoxide radicals. Multiple internal metabolic and enzymatic reactions are also capable of generating free radicals. Additionally, pollution and lifestyle habits also contribute to free radical generation (Chandimali et al., 2025). The antioxidants contain hydroxyl group in the structure that can neutralize the free radicals. Along with polyphenols, foxnuts also contain potent anti-oxidative vitamins like vitamin E. Specific amino acids such as arginine, methionine, cysteine, and glutamine act as potent antioxidants, eliminating free radical intermediates and stopping chain reactions (Liaquat et al., 2022).

8.5.3.1 Mechanism of action

Foxnuts have the ability to inhibit lipid peroxidation. Phospholipids are a major component of the cell membranes and oxidation can compromise their integrity. However, makhana can prevent this. This is because of the number of polyphenols present in foxnuts which have anti-oxidant potential. They also upregulate the activity of SOD which converts radicals of superoxide to hydrogen peroxide. This is then degraded by catalase to water and oxygen which prevents oxidative stress (Lee et al., 2002; Jiang et al., 2023). The antioxidant activity of methanolic extract of foxnuts was tested on hamster lung fibroblasts. The tests involved DPPH scavenging activity, checking the cell viability, and the ability of the extract to reduce lipid oxidation. Total extracts of *Euryale ferox* showed relatively high DPPH radical scavenging activity, with an average IC50 value of 5.6 μg/mL.SOD activity rose by 9%, 20%, and 24%, respectively, after treatment with whole extracts at dosages of 4, 20, and 100 μg/mL. Even greater levels of SOD activity were seen in cells treated with *Euryale ferox* ethyl acetate fractions (34% increase at 100 μg/mL). SOD activity increased by 26% in cells treated with 100 μg/mL of the butanol fractions. Every examined sample showed a dose-dependent increase in SOD activity.The survival of cells increased in a dose-dependent manner after treatment with *Euryale ferox* extracts. The ethyl acetate fractions significantly improved the survival of H2O2-treated cells. The ethyl acetate fractions actually increased cell survival to a level greater than that seen in control cells at a dosage of 100 μg/mL (Lee et al., 2002; Jiang et al., 2023).

It has also been reported that extracts of foxnuts can reduce Matrix Metalloproteinase 1(MMP-1). MMP are enzymes that are required for normal physiological functions like protein breakdown. However, they are both contributors of oxidative stress and products of oxidation. Reduction in their number denotes significant anti-oxidative activity (Jiang et al., 2022). In a study, a human skin fibroblast (HSF) model of was used in order to examine the roles and processes of makhana in anti-oxidative stress and anti-aging. The effects of fermented and unfermented makhana in lowering oxidative stress was also compared. The findings demonstrated that *E.ferox* fermented by *L. curvatus*had superior ability to lower oxidative stress as compared to treatment with unfermented makhana in terms of lowering oxidative stress caused by H_2O_2 and shielding HSF from cytotoxicity caused by H_2O_2. Additionally, it was found that makhana, fermented by *L. curvatus*was capable in reducing the number of MMP-1 significantly and enhanced the Bcl-2/Bax ratio, which reduced oxidative stress-induced apoptosis and skin senescence. Following fermentation, makhana's bioactive components, polyphenols and flavonoids, were considerably elevated.This was most likely due to *Euryale ferox* polysaccharides and polypeptides being broken down into lower molecular weight polysaccharides and polypeptides by *L. curvatus* fermentation. These results could be explained by changes in the compound level structure and content of bioactive compounds during fermentation brought on by microbial metabolism (Jiang et al., 2022).

Foxnuts have a distinct antioxidant profile whichhas demonstrated to reduce the generation of ROS in the mesangial cells of kidney thus preventing diabetic nephropathy. In the study, a group of compounds were isolated form foxnuts to analyze its antioxidant potential. The analysis identified 16 known chemicals and three new sesquineolignans called euryalins A–C (1–3). The mesangial cellular assay and the DPPH radical scavenging assay were used to assess the isolates' antioxidant capacity. The DPPH assay revealed that compounds 2, rel-(2α,3β)-7-O-methylcedrusin (4), syringylglycerol-8-O-4-(sinapyl alcohol) ether (5), and (+)-syringaresinol (7) were the most active, while compounds 2, 4, 7, (1,2,5R,6S) -2-(3,4-dimethoxyphenyl) -6-(3,4-dihydroxyphenyl) Buddlenol E and −3,7-dioxabicyclo[3.3.0]octane may considerably reduce generation of ROS in the mesangial cells. According to the findings, *E. ferox* seed is a good natural antioxidant source and can help avoid diabetic nephropathy (Song et al., 2011; Jiang et al., 2023).

The chemical makeup, antioxidant potential, cytotoxic, and protein tyrosine phosphatase 1B (PTP1B) inhibitory properties of essential oil derived from *Euryale ferox Salisb.* seeds were described in one study. PTP1B is an enzyme that downregulates insulin and leptin sensitivity. The extracted oil demonstrated the ability to inhibit PTP1B. Steam distillation was used to separate the essential oil, and gas chromatography-mass spectroscopy was used to characterize it. There were 37 components found in the oil. Butylated hydroxytoluene (38.7%), palmitic acid (11.0%), linoleic acid (9.0%), α-tocopherol (5.2%), hexanoic acid (3.9%), and π-tocopherol (3.5%) were the oil's principal ingredients. The oil demonstrated exceptional toxicity in the brine shrimp bioassay (test done to check the ability of a compound to kill brine shrimp larvae), with an LC50 value of 11.48 ± 0.51 μg/mL. DPPH ($IC50 = 6.27 \pm 0.31$ μg/mL) and ABTS ($IC50 = 2.19 \pm 0.61$ μg/mL) showed notable antioxidant capacity.Additionally, PTP1B was moderately inhibited by the oil ($IC50 = 83.34 \pm 2.07$ μg/mL) (He et al., 2019).

Hence, the seeds, seed shells as well as the oil of foxnuts display significant anti-oxidant activity.

8.5.4 Hepato-protective activity

Extracts of foxnuts have also demonstrated hepato-protective effects, although the studies are limited. With respect to the composition, foxnuts contain negligible fat content making it a suitable snack for patients with liver diseases. Its rich composition of polyphenols and flavonoids add to the advantages.

8.5.4.1 Mechanism of action

The polyphenols and flavonoids can scavenge free radicals that reduce oxidative stress in the liver cells. In one study, it was found that the extract can reduce MDA (malonialdehyde) levels. MDA is a key biomarker of lipid peroxidation in the human body. Reducing its level suggests reduced damage to lipids in the liver cell membranes (Dean, 2002). Additionally, it has also been noted to downregulate the activity of CYP2E1 which is an enzyme in the cytochrome P450 family. It is concerned with the metabolism of a number of compounds and is known to generate ROS during its metabolic activity. This contributes to non-alcoholic fatty liver disease (NAFLD) (Jian et al., 2019). Downregulation of this enzyme is one of the mechanisms involved in makhana's heptao-protective effects. Another important associated mechanism is the upregulation of IRS-1 (insulin receptor substrate-1). This improves insulin sensitivity in the hepatocytes, thus reducing lipid accumulation (steatosis) in the liver cells. Insulin resistance is one of the known factors NAFLD and preventing this can significantly contribute to hepatic health.In the study, the hepatoprotective efficacy of *E. ferox* seed coat extract (EFSCE) in mice with non-alcoholic fatty liver disease (NAFLD) caused by a high-fat diet (HFD) was studied. Mice were fed an HFD for 4 weeks in order to cause NAFLD. In comparison to the control group, the NAFLD mice's body weight, adipose tissue weight, and liver-to-body weight ratio all increased. Furthermore, the NAFLD mouse model showed a rise in blood lipid markers such as total cholesterol (TC), triglycerides (TG), high-density lipoprotein cholesterol (HDL), and low-density lipoprotein cholesterol (LDL). It was demonstrated that in HFD-induced NAFLD mice, EFSCE (15 and 30 mg/kg/day) treatment significantly decreased body weight, liver lipid deposition, and blood lipid levels after 4 weeks. It was reported that EFSCE may also reduce the overproduction of malondialdehyde (MDA) and increase the activity of SOD to combat oxidative stress. Additionally, EFSCE was found to be successful in lowering the activity of aspartate aminotransferase (AST) and alanine aminotransferase (ALT) in the HFD-induced NAFLD model. These findings strongly support that *Euryale ferox* seed-coat extract has hepato-protective effects in NAFLD via metabolic and oxidative stress–modulating mechanisms (Jian et al., 2019).

8.5.5 Cytotoxic and anti-cancer activity

Understanding the anti-cancer potential of foxnuts requires first knowing its biochemical makeup. The diverse flavonoid metabolites, including flavanones, flavonols, anthocyanins and pro-anthocyanidins, make it highly antioxidant in nature. These compounds are known to have anti-proliferative and pro-apoptotic effects. In one study, 129 distinct metabolites of flavonoids were identified in the seeds (Wu et al., 2013). The seed coat has phenolic compounds like gallic acid, catechin, epicatechin, procyanidin B2, ellagic acid, epicatechin gallate (Zhang et al., 2023). These can modulate redox reactions at cellular level which is a key contributor to cancer. There are other compounds like sesquineolignans which have anti-proliferative effects as observed in certain assays (Song et al., 2011; Wu et al., 2023).

8.5.5.1 Mechanism of action

SOD, CAT and glutathione peroxidase (GSH-Px) can upregulate cellular antioxidant enzymes by influencing apoptotic pathways in cancer cells. Flavonoids and phenolics in *E. ferox* can generate mild oxidative stress (or modulate ROS signaling) in cancer cells, as demonstrated in (Fig. 8.4). Elevated ROS can damage mitochondria, depolarize the mitochondrial membrane, and promote release of cytochrome *c*, triggering intrinsic apoptosis. These compounds also stabilize or activate p53, either by direct modulation or by interfering with its regulation (e.g., via MDM2), helping drive apoptosis. Akt is a central pro-survival kinase that, when active/phosphorylated, promotes cell survival, proliferation, and inhibits apoptosis via Murine double minute 2 (MDM2)-mediated suppression of p53. By inhibiting Akt phosphorylation, extracts of foxnuts remove a survival signal, allowing p53 to accumulate and promote apoptotic pathways. The suppression of MDM2 phosphorylation is consistent with reduced MDM2-mediated p53 degradation, leading to a rise in p53 levels. The upregulation of Bax and Bak (pro-apoptotic Bcl-2 family) and downregulation of Bcl-2 (anti-apoptotic) tilts the balance toward mitochondrial outer membrane permeabilization (MOMP). This allows cytochrome *c* release, which triggers the caspase cascade (e.g., caspase-3), leading to apoptosis. The activation of executioner caspases (3/7) and PARP cleavage confirms the final stages of

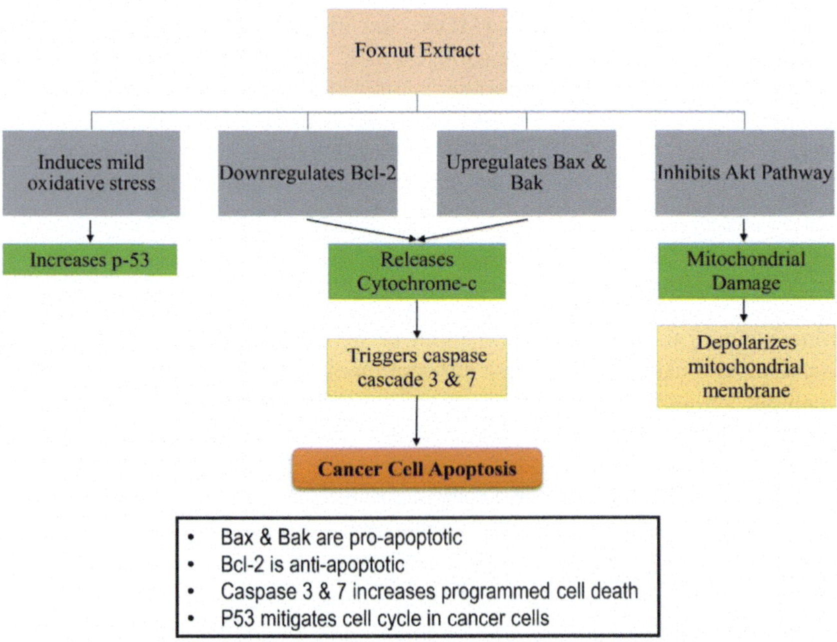

FIGURE 8.4 Mechanisms involved in inducing anti-cancer activity of makhana seeds.

apoptosis. The relative sparing of normal fibroblast cells (MRC-5) suggests that ESE's effects may be somewhat selective potentially because cancer cells often have higher basal ROS or rely more on Akt signaling, making them more vulnerable to Akt inhibition + ROS-mediated damage (Nam et al., 2019).

In a popular study, extract of foxnuts (ESE) was tested on A549 human lung carcinoma cells to observe the pro-apoptotic activity. A 50–150 μg/mL extract decreased cell viability and correspondingly increased lactate dehydrogenase (LDH). Increase in the concentration of LDH is a marker of membrane damage. It was reported that ESE was successful in disrupting the mitochondrial membrane potential and inhibiting Akt phosphorylation. It also suppressed MDM2 whose elevation is associated with tumor growth. The p53 (tumor suppressor protein) levels subsequently increased which is central to the pro-apoptotic pathway.Bax and Bak are two pro-apoptotic proteins that were upregulated while Bcl-2, which is an anti-apoptotic protein, was downregulated. Caspase three and 7 both of which are important enzymes for programmed cell death also increased. In the same study, A549 cells were implanted in immunodeficient (nude) mice, which were then treated with the extract. Tumor growth in treated mice was significantly reduced compared to controls through similar mechanisms, supporting that the extract's anti-cancer effect is not just in cultured cells but also in an animal model. The most interesting finding of the study was when the same extract was applied to normal lung fibroblast cells (MRC-5), there was minimal cytotoxicity (cell viability remained high), suggesting a degree of selectivity for cancer cells (Nam et al., 2019). In another study, the efficacy of the extract in fighting stomach cancer SGC7901 and human liver cancer HepG2 cells was studied. At a concentration of 200 μg/mL, the extract displayed inhibitory effects at a rate of 92.63% and 72.40% on SGC7901 and HepG2 cell proliferation respectively. The reason was reported to be reduction in mitochondrial membrane potential. It could arrest the cell division in the G0/G1 phase (Yin et al., 2022).

It has also been studied that makhana seed extracts are capable of degrading cellular tyrosinase which is a key enzyme involved in the production of melanin. In the study, its effect was studied on B16F10 melanoma cells and a 30 μg/mL extract could inhibit melanogenesis significantly (Baek et al., 2015; Jiang et al., 2023).

8.5.6 Neurocognitive and anti-fatigue activity

Foxnuts seeds have demonstrated notable anti-fatigue activity primarily because of the phytochemical composition. Additionally, foxnuts'high carbohydrate content allow better glycogen store build-up in the liver that can promote more sustainable energy release during exercise. The unique amino acid profile allows for improved muscle recovery, reducing muscle stress. The neurocognitive activity is also mediated by reduction in ROS generation in the brain (Jiang et al., 2023).

8.5.6.1 Mechanism of action

In a study, the anti-fatigue activity was tested on a mice model. The mice were forced to swim followed by treatment with phenolic extracts of the seed coat of foxnuts which prolonged the swimming time showing enhanced tolerance. The mice also had increased quantities of key endogenous enzymes SOD, CAT and GSH-px while malondialdehyde (MDA, a marker of lipid peroxidation) levels decreased, suggesting reduced oxidative damage. The treated mice also had higher LDH and lower blood urea nitrogen (BUN) indicating better muscle turnover. By LC-ESI-MS/MS, the researchers identified three principal phenolic compounds in the extract: *5,7-dihydroxy-2-(3,4,5-trihydroxyphenyl)-chroman-4-one*, *5,7,4-trihydroxyflavanone*, and *buddlenol E*. These molecules were responsible for the observed antioxidant and anti-fatigue effects by reducing oxidative stress during exercise (Wu et al., 2013; Jiang et al., 2023).

One researcher tested the anti-depressant activity of the seed extract in a mouse model using the chronic unpredictable mild stress (CUMS). It was reported that phosphorylation of adenosine monophosphate-activated protein kinase (AMPK – activated under low energy levels) and mammalian autophagy initiating kinase (ULK1) at Ser317, and the ratio of p-mTOR/mTOR was suppressed by ES-PE treatment were upregulated by the extracts. These help in improving mood by better cellular function. However, further studies are required to confirm the effects (Huang et al., 2018).

8.6 Conclusion

Foxnut is emerging as a highly valuable snack in India and around the world due to its rich nutritional and bioactive profile. This is supported by its unique amino acid profile and micronutrient composition. It can serve as a sustainable functional food in diets across the world.

Foxnuts are a powerhouse of nutrients, but processing plays a critical role in determining its nutritional composition and functional properties. Foxnuts are a rich source of essential amino acids with high bioavailability. As discussed in the chapter, processing conditions like roasting can improve the bioavailability of amino acids. It can also improve starch digestibility and increase the concentration of resistant starch. via increased starch gelatinization. It can also improve the bioavailability of micronutrients like iron and calcium. Roasting can also reduce the glycemic index significantly. Germination, on the other hand, has been reported to improve bioavailability of nutrients by reducing the ant-nutrient content and increasing the concentration of bioactive components especially phenols and flavonoids. The seeds have tannins, saponins and phytic acid also contributing to its antioxidant potential.

In terms of bioactive constituents, research confirms the presence of a wide array of phenolic compounds, flavonoids, tannins, and other secondary metabolites in both the seeds and the products formulated from them. The rich antioxidant content has been the sole contributing reason behind numerous health benefits exhibited by the seeds. DPPH and FRAP tests demonstrated the strong antioxidant activity of the seeds. In an animal model of diabetes (streptozotocin-induced), the ethanolic extract of *E. ferox* improved glycemic control, boosted plasma insulin, and normalized gluconeogenic enzymes. It also enhanced endogenous antioxidant enzyme activities (SOD, CAT, GSH peroxidase), thereby reducing oxidative stress in pancreas, liver, and kidney tissues. These antidiabetic properties are well complemented by its low-GI nature, especially when roasted, suggesting dual metabolic benefits. Additionally, significant cardio-protective effects have also been established. The studies confirming the hepato-protective and anti-fatigue effects are limited and require further research. Finally, the development of novel food products using foxnuts is gaining momentum. For example, a plant-based milk alternative has been developed from *E. ferox* seeds, yielding a low-calorie (~36.5 kcal/100 mL) beverage enriched with minerals (calcium, magnesium, potassium) and polyphenolic antioxidants such as gallic acid and epicatechin (Khan and Bukya, 2025). These applications align well with the increasing demand for functional, plant-based, nutrient-dense foods.

8.6.1 Future directions

Although the current corpus of study on foxnuts offers a solid basis for their use as functional foods, there are still a number of significant gaps that need to be filled in order to fully fulfill their potential to promote health. Although some studies report favorable EAAI values, comprehensive determinations of true ideal amino acid digestibility (the modern gold standard for protein quality) and PDCAAS/DIAAS evaluations for *E. ferox* are scarce. Such measurements would be valuable for accurate dietary formulation, particularly for vulnerable populations or clinical nutrition applications. Future studies should report standardized digestibility metrics alongside amino acid profiles. Additional studies may focus in these suggested areas. First, the number of in-vivo studies are limited. More animal models capturing chronic diseases (such as long-term metabolic syndrome, cardiovascular stress, and neurodegenerative illness) are required, even though preclinical antidiabetic trials offer insightful information. To establish causal relationships between foxnut bioactives and tissue-level effects, these studies should carefully evaluate molecular biomarkers, such as proinflammatory cytokines, neurotrophic factors (e.g., BDNF), and advanced glycation end products.

Second, clinical trials involving humans are crucial. Preclinical findings can be translated into useful dietary recommendations with the aid of controlled intervention studies that evaluate outcomes such as glycemic response, lipid profile, inflammatory markers, and cognitive endpoints (e.g., memory tests, mood scales). To investigate preventive and therapeutic uses, trials should involve both healthy persons and those at metabolic or neuropsychological risk. Third, more work needs to be done on the structural characterization and isolation of bioactive metabolites. Although phenols and flavonoids have been well recognized, more recent research (e.g., based on metabolomics, mass spectrometry) may reveal powerful molecules.

Makhana seeds also possesses areas of product development. The formulation of tested functional foods or supplements with well-balanced amounts of antioxidants from makhana can help in harnessing its benefits. Further research can be conducted on how the bioavailability of these nutrients are affected by processing techniques like fermentation. There are some bioactive components that have been identified for the first time in foxnut seeds, for these it is important to understand how the body metabolizes them. Further, comprehensive evaluation of toxicological profile may also be done.

To conclude, foxnuts or makhana is a functional food with numerous health benefits which have been historically used and consumed by Asians (Goren-Inbar et al., 2014). As discussed, it holds potential to mitigate the risk of cancer, diabetes and cardiac problems. Additionally, its unique nutrient profile with good EAAI and low glycemic index offers a splendid option to be used as a snack.

References

Ahmed, D., Khan, M.I., Sharma, M., Khan, M.F., 2018. Novel pentacyclic triterpene isolated from seeds of *Euryale Ferox Salisb.* ameliorates diabetes in streptozotocin induced diabetic rats. Interdiscipl. Toxicol. 11 (4), 275–288. https://doi.org/10.2478/intox-2018-0027.

Ahmed, D., Sharma, M., Kumar, V., Bajaj, H.K., Verma, A., 2015. 2β-hydroxybetulinic acid 3β-caprylate: an active principle from Euryale Ferox Salisb. seeds with antidiabetic, antioxidant, pancreas & hepatoprotective potential in streptozotocin induced diabetic rats. J. Food Sci. Technol. 52 (9), 5427–5441. https://doi.org/10.1007/s13197-014-1676-0.

Baek, S.H., Nam, I.J., Kwak, H.S., Kim, K.C., Lee, S.H., 2015. Cellular anti-melanogenic effects of a Euryale ferox seed extract ethyl acetate fraction via the lysosomal degradation machinery. Int. J. Mol. Sci. 16 (5), 9217–9235. https://doi.org/10.3390/ijms16059217.

Bagarinao, N.C., King, J., Leong, S.Y., Agyei, D., Sutton, K., Oey, I., 2024. Effect of germination on seed protein quality and secondary metabolites and potential modulation by pulsed electric field treatment. Foods 13 (11), 1598. https://doi.org/10.3390/foods13111598.

Balaha, M.F., Almalki, Z.S., Alahmari, A.K., Ahmed, N.J., Balaha, M.F., 2022. AMPK/mTOR-driven autophagy & Nrf2/HO-1 cascade modulation by amentoflavone ameliorates indomethacin-induced gastric ulcer. Biomed. Pharmacother. 151. https://doi.org/10.1016/j.biopha.2022.113200.

Bhat, M.S., Arya, S.S., 2023. Esterified unpopped foxnut (Euryale ferox) starch: molecular and rheological characterization. J. Sci. Food. Agricult. 103 (5), 2492–2501. https://doi.org/10.1002/jsfa.12440.

Birt, D.F., Boylston, T., Hendrich, S., Jane, J.L., Hollis, J., Li, L., McClelland, J., Moore, S., Phillips, G.J., Rowling, M., Schalinske, K., Paul Scott, M., Whitley, E.M., 2013. Resistant starch: promise for improving human health. Adv. Nutrit. 4 (6), 587–601. https://doi.org/10.3945/an.113.004325.

Bisht, H., Singh, K., Islam, Z., Thakur, M., Khedkar, R., Sharma, S., Shankar, N., 2025. Evaluation of the physicochemical, functional, structural, and anti-nutritional composition of germinated underutilized fox nut seed flour. In: BIO Web of Conferences, 178, p. 02020. https://doi.org/10.1051/bioconf/202517802020.

Biswas, P., Das, M., Boral, S., Mukherjee, G., Chaudhury, K., Banerjee, R., 2020. Enzyme mediated resistant starch production from Indian Fox Nut (Euryale ferox) and studies on digestibility and functional properties. Carbohydr. Polym. 237, 116158. https://doi.org/10.1016/j.carbpol.2020.116158.

Chandimali, N., Bak, S.G., Park, E.H., Lim, H.J., Won, Y.S., Kim, E.K., Park, S.I., Lee, S.J., 2025. Free radicals and their impact on health and antioxidant defenses: a review. Cell Death Discov. 11 (1). https://doi.org/10.1038/s41420-024-02278-8.

Chandrasekara, A., Shahidi, F., 2011. Antiproliferative potential and DNA scission inhibitory activity of phenolics from whole millet grains. J. Funct.Foods 3 (3), 159–170. https://doi.org/10.1016/j.jff.2011.03.008.

Choubey, S., Rachel Varughese, L., Kumar, V., Beniwal, V., 2015. Medicinal importance of gallic acid and its ester derivatives: a patent review. Pharm. Pat. Anal. 4 (4), 305–315. https://doi.org/10.4155/ppa.15.14.

Das, S., Der, P., Raychaudhuri, U., Maulik, N., Das, D.K., 2006. The effect of Euryale ferox (makhana), an herb of aquatic origin, on myocardial ischemic reperfusion injury. Mol. Cell. Biochem. 289 (1–2), 55–63. https://doi.org/10.1007/s11010-006-9147-1.

Dean, J.T., 2002. Role of malondialdehyde-acetaldehyde adducts in liver injury1,2 1 Guest editor: Arthur Cederbaum 2 This article is part of a series of reviews on "Alcohol, Oxidative Stress and Cell Injury." the full list of papers may be found on the homepage of the journal. Free Radic. Biol. Med. 32 (4), 303–308. https://doi.org/10.1016/s0891-5849(01)00742-0.

Dega, V., Barbhai, M.D., 2023. Exploring the underutilized novel foods and starches for formulation of low glycemic therapeutic foods: a review. Front. Nutr. 10. https://doi.org/10.3389/fnut.2023.1162462.

del Rosario Moguel Concha, D., Martínez, J.E.B., Velázquez, T.G.G., Martínez, C.J., Ruiz, J.C.R., 2022. Impact of germination time on protein solubility and anti-inflammatory properties of Pisum sativum L grains. Food Chem. X 13. https://doi.org/10.1016/j.fochx.2022.100219.

Devi, K.R., 2022. A comprehensive review on nutraceutical properties of Euryale ferox salisb. J. Plant Sci. Res. 9 (2), 1–6.

Devi, O.R., Moirangthem, L., Keithellakpam, O.S., Sharma, N., Singh, K.B., 2024. Phytochemical screening and evaluation of antioxidant potential in Euryale ferox salisb. And eupatorium birmanicum DC. Of Manipur, India. Pharmacogn. J. 16 (6), 1231–1237. https://doi.org/10.5530/pj.2024.16.201.

ElShamey, E., Yang, X., Yang, J., Xia, L., Zeng, Y., 2025. A systems review of grain proteins in rice and barley: biosynthesis, regulation, and impact on end-use quality. Front. Plant Sci. 16. https://doi.org/10.3389/fpls.2025.1658144.

Feng, R.-B., Wang, Y., He, C., Yang, Y., Wan, J.-B., 2018. Gallic acid, a natural polyphenol, protects against tert-butyl hydroperoxide–induced hepatotoxicity by activating ERK-Nrf2-Keap1-mediated antioxidative response. Food Chem. Toxicol. 119, 479–488. https://doi.org/10.1016/j.fct.2017.10.033.

Ferk, F., Chakraborty, A., Jäger, W., Kundi, M., Bichler, J., Mišík, M., Wagner, K.-H., Grasl-Kraupp, B., Sagmeister, S., Haidinger, G., Hoelzl, C., Nersesyan, A., Dušinská, M., Simić, T., Knasmüller, S., 2011. Potent protection of gallic acid against DNA oxidation: results of human and animal experiments. Mutat. Res. Fund Mol. Mech. Mutagen 715 (1–2), 61–71. https://doi.org/10.1016/j.mrfmmm.2011.07.010.

Goren-Inbar, N., Melamed, Y., Zohar, I., Akhilesh, K., Pappu, S., 2014. Beneath still waters – multistage aquatic exploitation of Euryale ferox (salisb.) during the Acheulian. Internet Archaeol. 37. https://doi.org/10.11141/ia.37.1.

Gupta, Dabade, A., 2021. Low glycemic index simulated flour formulation from retrograded fox nut (Euryale ferox) STARCH. Annals. Food Sci. Technol. 22 (Issue 1). www.afst.valahia.ro.

He, S., Wang, D., Zhang, Y., Yang, S., Xiang, L., Wei, D., Zhang, M., Qin, J., 2019. Chemical components and biological activities of the essential oil from traditional medicinal food, Euryale ferox Salisb., Seeds. J. Ess. Oil Bear. Plants 22 (1), 73–81. https://doi.org/10.1080/0972060x.2019.1595165.

Houston, M., 2011. The role of magnesium in hypertension and cardiovascular disease. J. Clin. Hypertens. 13 (11), 843–847. https://doi.org/10.1111/j.1751-7176.2011.00538.x.

Huang, Z., Huang, X., Wang, Q., Jiang, R., Sun, G., Xu, Y., Wu, Q., 2018. Extract of Euryale ferox Salisb exerts antidepressant effects and regulates autophagy through the adenosine monophosphate-activated protein kinase—UNC-51-like kinase 1 pathway. IUBMB Life 70 (4), 300–309. https://doi.org/10.1002/iub.1731.

Islam, Z., Bashir, K., Rastogi, M., Rikhari, A., Siddiqui, S., 2024. Evaluation of different methods of starch extraction on physico-chemical and functional properties of Makhana seed (Euryale ferox) starch. Indian J. Agricult. Biochem. 37 (2), 173–178. https://doi.org/10.5958/0974-4479.2024.00025.7.

Islam, Z., Bashir, K., Singh, S., Siddiqui, S., 2025. Effect of germination on physicochemical, functional and antinutritional characteristics of makhana seed (Euryale ferox) flour. Food Human. 5, 100717. https://doi.org/10.1016/j.foohum.2025.100717.

Jana, B.R., Bhatt, B.P., Srivastava, A., 2022. Biochemical components of makhana (Euryale ferox salisb) and its relevance to anti-obesity effect in human. Asian J. Dairy Food Res. 44 (Of). https://doi.org/10.18805/ajdfr.dr-1871.

Jayaprakash, G., Bains, A., Chawla, P., Fogarasi, M., Fogarasi, S., 2022. A narrative review on rice proteins: current scenario and food industrial application. Polymers 14 (15). https://doi.org/10.3390/polym14153003.

Jian, T., Yu, C., Ding, X., Chen, J., Li, J., Zuo, Y., Ren, B., Han, L., Li, W., 2019. Hepatoprotective effect of seed coat of *Euryale ferox* extract in non-alcoholic fatty liver disease induced by high-fat diet in mice by increasing IRs-1 and inhibiting CYP2E1. J. Oleo Sci. 68 (6), 581–589. https://doi.org/10.5650/jos.ess19018.

Jiang, Y., You, S., Zhang, Y., Zhao, J., Wang, D., Zhao, D., Li, M., Wang, C., 2022. Enhancing bioactive components of Euryale ferox with lactobacillus curvatus to reduce H2O2-induced oxidative stress in human skin fibroblasts. Antioxidants 11 (10), 1881. https://doi.org/10.3390/antiox11101881.

Jiang, J., Ou, H., Chen, R., Lu, H., Zhou, L., Yang, Z., Ethnopharmacological, T., 2023. Phytochemical, and pharmacological review of euryale ferox salisb.: a Chinese medicine food homology. Molecules 28 (11), 4399. https://doi.org/10.3390/molecules28114399.

Joseph, A., Ramesh, G., 2023. Nutrient analysis, phytochemical and antioxidant activity of a food product formulated with fox nuts (euryale ferox). Asian J. Biol. Life Sci. 12 (2), 279–285. https://doi.org/10.5530/ajbls.2023.12.38.

Kakkar, S., Bais, S., 2014. A review on protocatechuic acid and its pharmacological potential. ISRN Pharmacol. 1, 20149. https://doi.org/10.1155/2014/952943.

Kapoor, S., Kaur, A., Kaur, R., Kumar, V., Choudhary, M., 2022. Euryale ferox , a prominent superfood: nutritional, pharmaceutical, and its economical importance. J. Food Biochem. 46 (12). https://doi.org/10.1111/jfbc.14435.

Khan, A., Bukya, A., 2025. Development and evaluation of a plant-based flavored milk alternative from Euryale ferox seeds. J. Curr. Res. Food Sci. 6 (2), 35–39. https://doi.org/10.22271/foodsci.2025.v6.i2a.243.

Kumar, A., Jha, A.K., 2024. Makhana: a valuable aquatic food crop. New Era Agricult. Magaz. 3 (3).

Kumar, L., Singh, A. K., Bhatt, B. P., 2016. Nutritional status of recently developed Makhana (Gorgon Nut) variety "Swarna Vaidehi". J. AgriSearch, 3 (4). https://doi.org/10.21921/jas.v3i4.6701.

Kumar, R., Sanduja, D.M., painuly, N., 2024a. A research on phytochemical and antimicrobial investigation of euryale ferox seeds. Educ. Admin. Theor. Pract. https://doi.org/10.53555/kuey.v30i6.5884.

Kumar, V., Kumar, A., Singh, M.K., Dhyani, P., Mishra, H., Rai, D.C., 2024b. Bioactive metabolites identification of the foxnut and broken millet-based nutritional bar using HR-MS. Food Chem.: Mole. Sci. 9, 100214. https://doi.org/10.1016/j.fochms.2024.100214.

Liaquat, M., Pasha, I., Ahsin, M., Salik, A., fox nuts, R., Euryale, F.L., 2022. Contain higher concentration of phenolics, flavonoids, minerals and antioxidants, and exhibit lower Glycemic Index (GI) in human subjects. Food Prod. Proc. Nutr. 4 (1). https://doi.org/10.1186/s43014-021-00081-x.

Liu, Y., Wei, S., Liao, M., 2013. Optimization of ultrasonic extraction of phenolic compounds from Euryale ferox seed shells using response surface methodology. Ind. Crop. Prod. 49, 837–843.

Lee, S.E., Ju, E.M., Kim, J.H., 2002. Antioxidant activity of extracts from Euryale ferox seed. Exper. Mole. Med. 34 (2), 100–106. https://doi.org/10.1038/emm.2002.15.

Mahesh Kumar, M., Goudar, G., Sharma, P., Vishwakarma, R., Gogoi, P., Mahajan, A., Chandragiri, A., Kalpuri, S., Babu Geddam, J.J., Suresh, C., Chary, M., Venketrajireddy, G., Radhika, M.S., Sreenu, P., Tulja, B., 2023. Differentiating the nutrient composition, in-vitro starch digestibility, individual polyphenols and anti-oxidant properties of raw and popped makhana (Euryale ferox). J. Food Measur. Char. 17 (6), 5828–5844. https://doi.org/10.1007/s11694-023-02074-5.

Maibam, B.D., Chakraborty, S., Nickhil, C., Deka, S.C., 2023. Effect of Euryale ferox seed shell extract addition on the in vitro starch digestibility and predicted glycemic index of wheat-based bread. Int. J. Biol. Macromol. 226, 1066–1078. https://doi.org/10.1016/j.ijbiomac.2022.11.223.

Maqrashi, N.A., Busaidi, S.A., Al-Rasbi, S., Al Alawi, A.M., Al-Maqbali, J.S., 2025. Effect of magnesium supplements on improving glucose control, blood pressure and lipid profile in patients with type 2 diabetes mellitus: a systematic review and meta-analysis. Sultan Qaboos Univ. Med. J. 25 (1), 382–394. https://doi.org/10.18295/2075-0528.2848.

Marangoni, F., Poli, A., 2010. Phytosterols and cardiovascular health. Pharmacol. Res. 61 (3), 193–199. https://doi.org/10.1016/j.phrs.2010.01.001.

Mittal, R., Sharma, S., Mittal, A., 2020. A critical review on ethnobotanical and pharmacological aspects of Euryale ferox salisb. Pharmacogn. J. 12 (6), 1444–1454. https://doi.org/10.5530/pj.2020.12.199.

Murillo, S., Mallol, A., Adot, A., Juárez, F., Coll, A., Gastaldo, I., Roura, E., 2022. Culinary strategies to manage glycemic response in people with type 2 diabetes: a narrative review. Front. Nutr. 9. https://doi.org/10.3389/fnut.2022.1025993.

Nam, G.H., Jo, K.J., Park, Y.S., Kawk, H.W., Kim, S.Y., Kim, Y.M., 2019. In vitroand in vivoinduction of p53-dependent apoptosis by extract of Euryale ferox salisbin A549 human caucasian lung carcinoma cancer cells is mediated through Akt signaling pathway. Front. Oncol. 9. https://doi.org/10.3389/fonc.2019.00406.

Pruett, A., Aramouni, F.M., Bean, S.R., Haub, M.D., 2023. Effect of flour particle size on the glycemic index of Muffins made from whole Sorghum, whole corn, Brown rice, whole wheat, or refined wheat flours. Foods 12 (23). https://doi.org/10.3390/foods12234188.

Rathod, R.V., Neve, G., Jain, A., Giri, P., 2023. A comprehensive review on health benefits and nutritional aspects of foxnut (makhana). Pharm. Innov. 12 (6), 4432–4438.

Rouf Shah, T., Prasad, K., Kumar, P., Yildiz, F., 2016. Maize A potential source of human nutrition and health: a review. Cogent Food Agric. 2 (1). https://doi.org/10.1080/23311932.2016.1166995.

Sahithi, G., Thabassum, A.F., Shiva Kumara, C.S., Madhavi Reddy, M., 2012. Estimation of nutritional composition and sensory evaluation of Fero Nutri bars developed using Euryale ferox (Makhana seeds). Int. J. Food Nutr. Sci. 11 (7), 776–785.

Samtiya, M., Aluko, R.E., Dhewa, T., 2020. Plant food anti-nutritional factors and their reduction strategies: an overview. Food Prod. Proc. Nutrit. 2 (1). https://doi.org/10.1186/s43014-020-0020-5.

Shewry, P.R., Hey, S.J., 2015. The contribution of wheat to human diet and health. Food Energy Secur. 4 (3), 178–202. https://doi.org/10.1002/FES3.64.

Shukla, N., Sharma, D.J., Pankaj Kumar Sharma, D., 2024. A review on: Euryale ferox (Fox nut). J. Pharmacogn. Phytochem. 13 (5), 50–52. https://doi.org/10.22271/phyto.2024.v13.i5a.15057.

Singh, D.K., Kumar, A., Kumar, U., Singh, I.S., 2021. Production and profitability of makhana (Euryale ferox salisb.) in aquatic ecosystem of North Bihar, India. Indian J. Extens. Educ. 57 (2), 62–66. https://doi.org/10.5958/2454-552x.2021.00056.6.

Singh, B., Singh, J.P., Singh, N., Kaur, A., 2017. Saponins in pulses and their health promoting activities: a review. Food Chem 233, 540–549. https://doi.org/10.1016/j.foodchem.2017.04.161.

Sokrab, A.M., Mohamed Ahmed, I.A., Babiker, E.E., 2012. Effect of germination on antinutritional factors, total, and extractable minerals of high and low phytate corn (Zea mays L.) genotypes. J. Saudi Soc. Agricult. Sci. 11 (2), 123–128. https://doi.org/10.1016/j.jssas.2012.02.002.

Song, C.W., Wang, S.M., Zhou, L.L., Hou, F.F., Wang, K.J., Han, Q.B., Li, N., Cheng, Y.X., 2011. Isolation and identification of compounds responsible for antioxidant capacity of Euryale ferox seeds. J. Agric. Food Chem. 59 (4), 1199–1204. https://doi.org/10.1021/jf1041933.

Sun, H.l., Zhang, Y.q., Xie, X.y., Che, Y.y., 2014. Studies on chemical constituents from seeds of Euryale ferox. Zhong yao cai = Zhongyaocai = J. Chin. Med. Mater. 37 (11), 2019–2021.

Suri, K., Singh, B., Kaur, A., Singh, N., 2019. Impact of roasting and extraction methods on chemical properties, oxidative stability and Maillard reaction products of *peanut* oils. J. Food Sci. Technol. 56 (5), 2436–2445. https://doi.org/10.1007/s13197-019-03719-4.

Tao, L., Gao, E., Bryan, N.S., Qu, Y., Liu, H.-R., Hu, A., Christopher, T.A., Lopez, B.L., Yodoi, J., Koch, W.J., Feelisch, M., Ma, X.L., 2004. Cardioprotective effects of thioredoxin in myocardial ischemia and the reperfusion role of S-nitrosation. Proc. Natl. Acad. Sci. 101 (31), 11471–11476. https://doi.org/10.1073/pnas.0402941101.

Wang, Z., Yin, L., Liu, L., Lan, X., He, J., Wan, F., Shen, W., Tang, S., Tan, Z., Yang, Y., 2022. Tannic acid reduced apparent protein digestibility and induced oxidative stress and inflammatory response without altering growth performance and ruminal microbiota diversity of Xiangdong black goats. Front. Vet. Sci. 9. https://doi.org/10.3389/fvets.2022.1004841.

Wang, Y., Yuan, H., Wang, Y., 2023. Treatment of diabetes nephropathy in mice by germinating seeds of Euryale ferox through improving oxidative stress. Foods 12 (4), 767. https://doi.org/10.3390/foods12040767.

Weaver, C.M., 2013. Potassium and health. Adv. Nutr. 4 (3), 368S. https://doi.org/10.3945/an.112.003533.

Wu, C.Y., Chen, R., Wang, X., Shen, B., Yue, W., Wu, Q., 2013. Antioxidant and anti-fatigue activities of phenolic extract from the seed coat of Euryale ferox salisb. And identification of three phenolic compounds by LC-ESI-MS/MS. Molecules 18 (9), 11003–11021. https://doi.org/10.3390/molecules180911003.

Wu, M., Jiang, Y., Wang, J., Luo, T., Yi, Y., Wang, H., Wang, L., 2023. The effect and mechanism of corilagin from Euryale ferox salisb shell on LPS-induced inflammation in Raw264.7 cells. Foods 12 (5), 979. https://doi.org/10.3390/foods12050979.

Xie, X., Zeng, C., Zhang, R., Zhu, W., Li, H., Huang, Z., 2025. Profiling and discrimination of Euryale ferox seeds from different processing methods using liquid chromatography high-resolution mass spectrometry combined with molecular networking and statistical analysis. Metabolites 15 (4), 225. https://doi.org/10.3390/metabo15040225.

Yin, D., Xue, M., Yang, G., Huang, Y., Yi, Y., Wang, H., Wang, L., 2022. The alcoholic extract of Euryale ferox seed shell inhibits the proliferation of human gastric cancer SGC7901 cells and human hepatoma HepG2 cells and promotes their apoptosis. Modern Food Sci. Technol. 38 (6), 55–65. https://doi.org/10.13982/j.mfst.1673-9078.2022.6.0822.

Yu, J., Wang, S., Wang, J., Li, C., Xin, Q., Huang, W., Zhang, Y., He, Z., Wang, S., 2015. Effect of laboratory milling on properties of starches isolated from different flour millstreams of hard and soft wheat. Food Chem. 172, 504–514. https://doi.org/10.1016/j.foodchem.2014.09.070.

Yuan, H., Gong, Z., Meng, S., He, G., 2013. Hypoglycemic and hypolipidemic effects of a triterpenoid-rich extract from Euryale shell on streptozotocin-induced diabetic mice. Pharmazie 68 (3), 227–231. https://doi.org/10.1691/ph.2013.2734China.

Yuan, H., Meng, S., Wang, G., Gong, Z., Sun, W., He, G., 2014. Hypoglycemic effect of triterpenoid-rich extracts from Euryale ferox shell on normal and streptozotocin-diabetic mice. Pakistan J. Pharm. Sci. 27 (4), 859–864.

Zhang, W.N., Su, R.N., Gong, L.L., Yang, W.W., Chen, J., Yang, R., Wang, Y., Pan, W.J., Lu, Y.M., Chen, Y., 2019. Structural characterization and in vitro hypoglycemic activity of a glucan from Euryale ferox Salisb. seeds. Carbohydr. Polym. 209, 363–371. https://doi.org/10.1016/j.carbpol.2019.01.044.

Zhang, L., Zeng, J., Yuan, E., Chen, J., Zhang, Q., Wang, Z., Yin, Z., 2023. Extraction, identification, and starch-digestion inhibition of phenolics from *Euryale ferox* seed coat. J. Sci. Food Agricult. 103 (7), 3437–3446. https://doi.org/10.1002/jsfa.12460.

Zhu, L., Yang, X., Feng, J., Mao, J., Zhang, Q., He, M., Mi, Y., Mei, Y., Jin, G., Zhang, H., 2022. CYP2E1 plays a suppressive role in hepatocellular carcinoma by regulating Wnt/Dvl2/β-catenin signaling. J. Transl. Med. 20 (1), 194. https://doi.org/10.1186/s12967-022-03396-6.

Chapter 9

Phytochemicals and bioactive compounds in *Euryale ferox*: extraction, characterization, and application

Ankit Kumar[1], Chaman Kumar[1] and Rahul Kumar Rout[2]
[1]*ICAR-National Research Centre on Litchi, Rohua, Bihar, India;* [2]*ICAR-National Research Centre for Makhana, Darbhanga, Bihar, India*

9.1 Introduction

The watery rhizomatous flowering plant *Euryale ferox* (*Nymphaeacae*) is a member of the *Euryale* genus. It is the sole native plant species found in Eastern and Southern Asia, which includes portions of Japan, Korea, and India. The plant is referred to as foxnut, gorgon nut, semen euryales, Qian shi, and prickly water lily in different parts of the world. It is referred to as makhana in India and is enjoyed as a dried fruit throughout the nation. Makhana has historically been a specialty of the Bihar state's Mithila, Darbhanga, and Madhubani regions. Euryale ferox seeds have previously been used in Chinese and Ayurvedic remedies to treat hepatic dysfunction, persistent diarrhea, excessive leucorrhea, and renal disorders (Mittal et al., 2020). It was thought to somewhat boost cell-mediated immunity and humoral immunity in a daily diet. The use of halua (semisolid preparation) and laddus (balls of wheat flour) in ethnomedicine One of the primary components of the postpartum medication is Euryale ferox (Makhana). *Euryaleferox* is typically grown in lakes and ponds with stagnant water that is between 0.2 and 2.0 m deep. It is intolerant of cold and drought and likes warm, sunny weather. Fertile soil with enough organic matter is necessary, and the ideal temperature is between 20 and 30°C. The plant has eight to nine spherical leaves and vibrant purple flowers that are interwoven like octopuses in alternating rows. The ovary has 7–16 chambers with 6–8 seeds per locule, and the flower is epigynous with over 40 corollas. The seeds are spherical and have a diameter

Makhana (Euryale ferox) as a Functional Food. https://doi.org/10.1016/B978-0-443-44421-0.00011-1

of 0.5–0.8 cm. They have a concave seed cleft hilum, a brownish-red or reddish-brown inner seed coat covering their surface, and a yellow-white end that makes up around one-third of the total (Jiang et al., 2023).

According to the ICAR-NRC for Makhana, Darbhanga, *E. ferox* is grown on 15,000 ha of land in India, whereas the largest cultivation city in China is Shangrao, Jiangxi Province, with 6700 ha annually. Every year, 120,000 metric tonnes of *E. ferox* seeds are produced in India; the produce is worth Rs 250 crore to farmers. Approximately 100 t of popped *E. ferox* are sent annually from India, which exports between 1% and 2% of its total production (Kapoor et al., 2022). It is important to remember that cultivating and processing *E. ferox* requires a lot of hard effort. While seeds gathered from the pond are needed during harvesting, trimming seedlings is necessary after planting to preserve plant-to-plant spacing. Processing is then necessary, including storing, cleaning, grading, heating, and tempering—all of which are done by hand. As a result, the entire process—from planting to harvest—is difficult and painful (Jiang et al., 2023).

9.2 Key phytochemicals and their associated health advantages

Phytochemicals are naturally occurring substances in plants that give them color, flavor, and protection, although they are not necessary nutrients. They also assist human health by acting as antioxidants, lowering inflammation, and guarding against chronic illnesses like cancer and heart disease. They are also termed as secondary metabolites. The detailed classification of phytochemicals is given in Fig. 9.1.

To create complex compounds from simpler ones or break down complexes into simpler ones, living cells engage in a series of chemical events known as metabolism. As autotrophic creatures, plants have two metabolisms: the

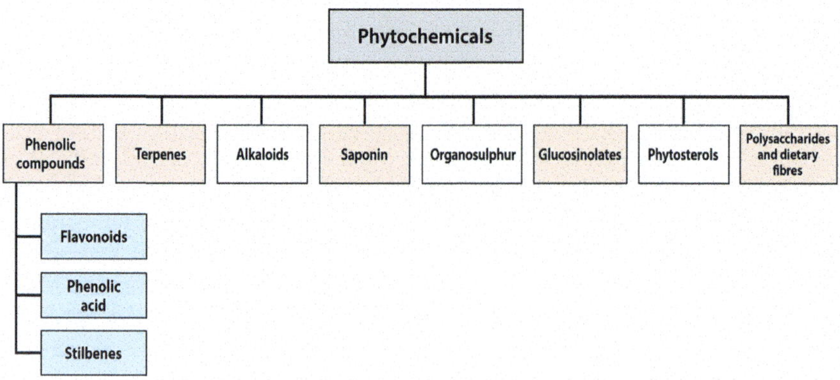

FIGURE 9.1 Classification of phytochemicals.

fundamental metabolism found in all living things and a secondary metabolism that enables them to create and store a variety of chemical substances. Most of the carbon, nitrogen, and energy end up in molecules that are shared by all cells and are essential to both their operation and the organisms to which they belong. All plants include these substances, which serve the same purposes: lipids, carbohydrates, nucleotides, and amino acids. We refer to them as main metabolites. Secondary metabolites are a wide range of organic molecules that are produced by plants using a substantial amount of assimilated carbon and energy. These molecules do not appear to be directly involved in photosynthetic processes, respiratory processes, nutrient assimilation, solute transport, protein synthesis, carbohydrates, or lipids (Mendoza and Silva, 2018). Certain plant cells generate significant secondary metabolites or phytochemicals that are relevant to the plant's reproductive process (attracting insects to facilitate pollination) or its interactions with the environment (defense against infections, predators, or environmental stress) (Kumar et al., 2023).

The details of some phytochemicals which are important in makhana are mentioned in subsequent paragraphs. Makhana has a rich reserve of phytochemicals. Terpenes, which are composed of isoprene, a hydrocarbon with five carbon atoms joined to eight hydrogen atoms (C_5H_8), can serve as an important phytochemical (Yuan and Macquarrie, 2015). Tetraterpenes are the most well-known form of carotenoids, another significant plant pigment. Important pharmaceutical terpenoids include paclitaxel (Taxol), camphor, and artemisinin. Makhana has a very small amount of another important phytochemical called carotenes which mainly gives color to fruits and vegetables. Among these, lutein, lycopene, and fucoxanthin are potent antioxidants, and α-carotene, β-carotene, and β-cryptoxanthin are the precursors of vitamin A (Cooperstone and Schwartz, 2016). Limonene, an important terpenoid, is known for its bioactive properties. Additionally, lutein is crucial for vision. Zeaxanthin is an antioxidant and fat-soluble pigment. In addition to its neuroprotective function in Alzheimer's disease, limonene is well-known for its anti-inflammatory, antioxidant, and anti-stress qualities. Myrcene possesses analgesic, anti-inflammatory, anti-aging, antioxidant, and anxiolytic qualities. Carotenoids improve gap junction communication, boost immunity, control gene transcription, and prevent lung and prostate cancer. They are also beneficial for eye health (Kumar et al., 2023).

Secondary metabolites known as phenolic compounds are created during the metabolism of phenylpropanoids in pentose phosphate and plant shikimic acid. They range from simple phenolic molecules to highly polymerized compounds and have benzene rings with one or more hydroxyl substituents (Lin et al., 2016). Simple flavonoids, phenolic acids, complex flavonoids, and colorful anthocyanins are among the many different types of chemicals found in these phenolic molecules, often known as polyphenols. These phenolic chemicals are typically associated with the plant's defense mechanisms

(Cooperstone and Schwartz, 2016). Phenolic compounds have many bioactive properties, additionally, long-term diabetes consequences such retinopathy, neuropathy, nephropathy, and cardiovascular disease can be avoided using polyphenolic substances (Vattem et al., 2005). Makhana seeds are hub of polyphenols. Gallic acid generally considered as the most abundant polyphenol in Makhana. Plant sterols, also known as phytosterols or steroid glycosides, are biochemicals that resemble cholesterol in both structure and biological roles. More than 250 distinct sterols and related chemicals have been identified in plants. Nuts, seeds, and vegetable oils are major sources of plant sterols (Kumar et al., 2023). Phytosterols include things like stigmasterol, beta-sitosterol, and campesterol (Piironen et al., 2000).

Some other phytochemicals like Alkaloids are a class of secondary metabolites that are present in various plant sections of makhana. Most of these substances are known to have antiviral, anticancer, antibacterial, antifungal, and anti-inflammatory properties (Jiang et al., 2019). They control the growth of plants and shield them from predators. Some of them are liquids that contain carbon, hydrogen, and nitrogen, while the majority, like atropine, are solids. The painkiller morphine, the muscle relaxant tubocurarine, the additive simulants, caffeine, nicotine, cocaine, etc. are a few common alkaloids (Pereira et al., 2023). Saponins are steroid or triterpene glycosides that are abundantly found in plants and are named for their "soap-like" persistent foaming when shaken with water. These substances lyse erythrocytes and are either non-volatile or surface-active. Expectorant, anti-inflammatory, antifungal, antiparasitic, and other pharmacological properties are all included in saponins (Cheok et al., 2014). Glucosinolates are sulfur and nitrogen-containing plant compounds that are primarily found in cruciferous vegetables (such as broccoli and cabbage). They function as defense chemicals, releasing biologically active isothiocyanates, nitriles, and other substances when plant tissue is damaged, potentially providing health benefits like anti-inflammatory and anti-cancer effects (Clarke, 2010). Makhana generally does not have glucosinolates. Organosulfur compounds like, Allicin (garlic), sulforaphane (broccoli), and diallyl sulfides/trisulfides, are produced when plant tissues are harmed (Goncharov et al., 2021).

9.3 Phytochemicals in *Euryle ferox*

The *Euryle ferox* plant is a great source of high-quality protein and carbohydrates. It is highly sought after in wealthy nations due to its nearly fatless nature. Secondary metabolites such as polyphenols, sequineolignans, tocopherols, cyclic dipeptides, cerebrosides, glucosylsterols, and triterpenoids are abundant in the majority of this plant. Numerous pharmacological effects, including hypoglycemic, cardioprotective, antibacterial, anticancer, antidepressant, antioxidant, and hepatoprotective qualities, are exhibited by these secondary metabolites (Kumar et al., 2025). *E. ferox's* hard seed coat, which is usually thrown away, has therapeutic qualities as well. Triterpenoid-rich extract from *E. ferox* seed shell has been shown in a recent Chinese study to have hypoglycemic effects on both normal and streptozotocin-induced diabetic rats. It has been demonstrated that the control of glucose metabolism produces the hypoglycemic effect. The experimental animals' body weight was observed to restore to normal because of the treatment. As anticipated, the morphology of the pancreatic cells' islets also recovered. The expression of the insulin receptor IRS-1 protein increased concurrently with a decrease in the expression of the negative regulation of the protein PTP1B gene (Yuan et al., 2014). The *E. ferox* seed shell's antioxidant qualities make it useful for preserving a variety of foods. It has been discovered that phenolic compounds from its seed coat significantly reduce weariness. In bone tissue engineering, the starchy kernel of *E. ferox* has become a useful tool. The starch of *E. ferox* seed has been combined with nanohydroxyapatite and chitosan to create a biomimetic, osteoinductive, tricomposite scaffold. Excellent biomineralization capacity of the nanocomposite composed of nano-hydroxyapatite-chitosan-*E. ferox* starch was demonstrated in an in vitro synthetic bone scaffold investigation, which also demonstrated relatively higher antibacterial activity against specific bacteria (Shakir et al., 2018).

Quercetin (422 μg/kg) and kaempferol (200 μg/kg), two pure flavonoids that speed up glucose metabolism in human cells, were found in Makhana seed endosperm. The endosperm of the seed contained 1.14 ± 0.02 mg/100 g of 5.4 ± 0.04 mg/100 g of vitamin C and vitamin B1. Makhana exhibited the highest levels of antioxidant activity, with DPPH and FRAP values of 288.76 ± 0.09 mg AEAC/100 g FW and 99.14 ± 0.16 mg AEAC/100 g FW, respectively. Additionally, Makhana's protein composition of dry endosperm had a notable amount of arginine (12.70 ± 0.21 g/100 g) and histidine (3.62 ± 0.11 g/100 g). Numerous macro and micronutrients were discovered to be abundant in makhana. Additionally, zinc (5.6 ± 0.24 mg/100 g) and magnesium (20.00 ± 1.09 mg/100 g), which control insulin secretion and storage, were found in significant amount. Significant levels of iron (10.9; ± 0.13 mg/100 g), selenium (0.06 ± 0.02 mg/kg), and chromium (1.05 ± 0.06 mg/kg), which serve as antioxidants, were also found in the Makhana seed kernel (Jana et al., 2024a).

Makhana regulates blood pressure since it has a high potassium content and a low sodium content. Makhana is beneficial for the human heart and low in saturated fats and cholesterol. This is an excellent source of calcium, magnesium, and sulfur. Insulin action is controlled by these divalent cations (Shankar et al., 2010). Makhana is a veritable gold mine of many micronutrients, including Zn, Cu, Fe, and Mn, and it helps people maintain good health for a long period. Makhana's rich nutritional content makes them an ideal and healthful snack for folks, especially when they are feeling peckish. Makhana's low calorie content encourages high thiamine and magnesium levels, which aid in managing weight loss. Makhana-based foods are good for bones because of their high calcium content (Kumar et al., 2025).

Makhana contains a fair amount of chromium (>1.0 ppm) and a small amount of selenium (<0.05%), which functions as an antioxidant in the human body. Chromium is necessary for the metabolism of lipids and carbohydrates as well as for the utilization of amino acids. Epidemiological studies have shown a link between a lack of Cr and cardiovascular issues. Selenium (Se) is obtained by the human body from foods that are grown in a certain area. One of the body's antioxidant defense mechanisms includes the enzyme glutathione peroxidase, which contains selenium (WHO, 1996). This enzyme serves as a gauge for the body's selenium levels. Although selenium shortage was known to cause Keshan, a condition related to cardiomyopathy that primarily affects young women and children, the significance of selenium in the human diet has only been acknowledged since the 1960s (Kumar et al., 2025) (Table 9.1).

TABLE 9.1 Plant bioactive compounds and their functional properties.

Category	Examples/Major components	Biological/Functional properties
Phenolic compounds	Gallic acid, protocatechuic acid	Antioxidant, anti-inflammatory, cardioprotective
Flavonoids	Quercetin, kaempferol derivatives	Antioxidant, anti-diabetic, anti-aging
Tannins	Ellagitannins	Antioxidant, antimicrobial
Alkaloids	Trace alkaloids	Neuroprotective, metabolic regulation
Saponins	Diosgenin-like compounds	Hypocholesterolemic, immune-modulating
Polysaccharides	Dietary fiber, mucilage	Prebiotic, gut health, antioxidant
Proteins & peptides	Albumins, globulins	Anti-fatigue, antioxidant
Amino acids	Arginine, glutamic acid, leucine	Muscle health, metabolism
Fatty acids	Linoleic acid, oleic acid	Cardioprotective, anti-inflammatory
Phytosterols	β-sitosterol	Cholesterol-lowering
Minerals (bioactive role)	Mg, K, Ca, Fe, Zn	Enzyme activation, antioxidant defense
Antioxidant compounds	Polyphenols, flavonoids	Anti-oxidative stress, anti-aging

Antioxidants are primarily characterized as polyphenolic chemicals that, by scavenging free radicals (ROS) and reducing the oxidative stress of the cell, can delay, inhibit, or prevent the oxidation of oxidizable materials. A compound's potential antioxidant activity may be indicated by its reducing capacity (Kumar et al., 2025). The reduction of ferric ions and the high chlorogenic acid content may be the cause of the DPPH free radical receiving an electron or hydrogen from antioxidant molecules to become a stable diamagnetic molecule FRAP values. With antioxidant activity 280.62 mg AEAC/100g FW and 94.27 mg AEAC/100g FW, respectively, Makhana exhibited the highest DPPH and FRAP values. Additionally, Yogesh et al. (2014) found a favorable correlation between DPPH radical scavenging capacity and increased quantities of phenolics and flavonoids. Additionally, phenolics, flavonoids, and phytic acids were discovered to be abundant in makhana bran. Bran can therefore be utilized to increase the oxidative stability of food products that are susceptible to oxidation and have high levels of polyunsaturated fatty acids (Kumar et al., 2017).

Phytochemical analysis of the extracts confirmed the occurrence of several bioactive compounds, including phenols, tannins, alkaloids, glycosides, terpenoids, steroids, and flavonoids. The total phenolic and flavonoid contents were quantified as 50.967 ± 0.107 mg GAE/g and 16.274 ± 0.73 mg QE/g, respectively (Kumar et al., 2025). Antioxidant potential of the ethanol–water extract was evaluated using the DPPH radical scavenging assay, which exhibited an IC_{50} value of 1.62 μg/mL. Antibacterial efficacy was assessed by the agar cup diffusion method against *Escherichia coli* and *Staphylococcus aureus*. Among the tested extracts, ethyl acetate and ethanol fractions demonstrated larger zones of inhibition. Furthermore, HPTLC fingerprint profiling for the separation of bioactive constituents suggested that the seed coat of *Euryale ferox* represents a promising natural source of antioxidant and antibacterial compounds (Devi et al., 2022).

Apart from bioactive secondary metabolites, makhana is also rich in starch which can be utilized for many purposes. It is necessary to physically pop the starchy kernel, the edible part of the makhana seed out of the raw nut. Makhana seeds have a high concentration of carbohydrates, which make up around 76% of the seeds, which contributes to its nutritious content and functional qualities. Starch is the primary component of these carbohydrates, and interest in exploring new sources of this polysaccharide has increased due to the growing demand for starch in a variety of industries. The production of food, medications, paper, and textiles all heavily rely on starch. Although there are potential uses for popped makhana starch (PMS) in food products, research on various starch extraction techniques and the resulting modifications under varied thermal processing conditions is necessary (Islam et al., 2024).

Roasting the seeds at 270°C–280°C, tempering them for 48 h, roasting them again at 260°C–270°C, and then breaking the hard shell to allow the seeds to break while still hot are all phases in the popping process. During the initial roasting step, the starch molecules in the seeds undergo retrogradation and gelatinization. These modifications prepare the kernels for popping, or expansion, in the subsequent roasting stage. Makhana employs a completely distinct technique for bursting rice and cocoa beans. In the case of rice, the grains are roasted until they puff up and the shell is removed after parboiling (Srisang et al., 2021). According to the study by Islam et al. (2025), the popping process dramatically alters the properties of makhana starch, improving its useful qualities. In addition to increasing its ability to absorb water and oil, the procedure decreased pasting viscosities, which suggests increased paste stability. The disruption of the starch matrix was confirmed by structural changes seen in FTIR and SEM analyses. This led to better thermal and rheological properties, making PMS a promising ingredient for a variety of food applications, including batters, chips, cakes, confections, soup, sauces, snacks, and more. Despite these positive findings, several limitations need more research. It is currently unknown what exact processes govern PMS's interaction with food matrices, particularly those pertaining to gelation, retrogradation, and textural stability. Following makhana popping, there was a notable rise in the resistant starch (RS) content and starch digestibility. Raw makhana had an estimated glycemic index (GI) of 35.63, while popped makhana had an GI of 61.52. The average granule size (diameter) was 54 μm, according to scanning electron microscopy (SEM) (Kumar et al., 2023).

9.4 Extraction of phytochemicals

The foundation of research on natural products is the extraction of chemicals from plant materials. The search for improved extraction techniques has been unrelenting. Similarly, a few promising contemporary green extraction techniques are becoming more important, including supercritical fluid, ultrasonic, accelerated solvent, microwave, and enzyme-assisted extraction techniques (Bitwell et al., 2023). In the study of natural products, extraction is crucial. There is a constant effort to develop and find more efficient and economical extraction methods. Research aimed at discovering new drugs from medicinal plants generally begins with the screening of plant extracts to detect novel chemical entities, followed by evaluation of their biological activities. Promising bioactive molecules are subsequently separated, purified, and characterized to determine their molecular structures, and are further assessed through pharmacological and toxicological studies (Altemimi et al., 2017) (Table 9.2).

This process usually follows a systematic sequence. Initially, plant materials are collected from their natural habitats. The collected samples such as leaves, flowers, or fruits are taxonomically authenticated by a qualified botanist and documented in a recognized herbarium. Selected plant parts including roots, stems, bark, leaves, or fruits are then processed using suitable solvents to obtain phytochemical extracts (Ingle et al., 2017). These extracts are

TABLE 9.2 Different extraction methods for phytochemical extraction and their advantages.

Extraction method	Key principle	Main advantages
Maceration (solvent extraction)	Passive diffusion of compounds into solvent	Simple, low cost, suitable for heat-sensitive bioactives
Soxhlet extraction	Continuous hot solvent reflux	Exhaustive extraction, good reproducibility
Ultrasound assisted extraction (UAE)	Cavitation-induced cell disruption	Faster extraction, higher yield, reduced solvent use
Microwave-assisted extraction (MAE)	Microwave heating enhances mass transfer	Rapid, efficient, short extraction time
Hot water extraction	Heat-assisted solubilization	Food-grade, safe, ideal for polysaccharides
Enzyme-assisted extraction (EAE)	Enzymatic breakdown of cell walls	Improved release of bound bioactives, eco-friendly
Supercritical fluid extraction (SFE)	Supercritical CO_2 solubilization	Solvent-free, selective, environmentally sustainable

concentrated by removing the solvent through appropriate techniques. The concentrated extracts are subjected to chromatographic methods to separate and purify the active constituents. Structural identification of the isolated compounds is carried out using spectroscopic techniques such as UV–Visible spectroscopy, infrared spectroscopy, proton and carbon NMR, and mass spectrometry. Subsequently, the compounds undergo pharmacological and toxicological evaluations using chemical or biological assays. Ultimately, the identified bioactive molecules may be synthesized or semi-synthesized for further development and application (Bitwell et al., 2023).

Phytochemicals differ in their solubility characteristics and thermal stability, which significantly influence their extraction behavior. The choice of extraction solvent plays a critical role in determining both the yield and quality of the recovered compounds, as well as their suitability for use in food and nutraceutical applications (Ingle et al., 2017). Extraction solvents are broadly categorized into green solvents such as water, ethanol, glycerol, fatty acids or oils, acetic acid, ionic liquids, carbon dioxide (CO_2), deep eutectic solvents, and natural deep eutectic solvents (NADES) and conventional organic solvents including acetone, chloroform, butanol, methanol, ethyl acetate, methyl acetate, benzene, hexane, and cyclohexane. The use of inappropriate solvents or exposure to unsuitable temperature conditions may result in degradation or loss of functional properties of phytochemicals (Kumar et al., 2023).

Extraction efficiency is also strongly influenced by the nature of the plant matrix containing the target compounds. Matrix-related factors such as botanical origin, structural characteristics, pre-treatment methods, particle size, and solid-to-liquid ratio significantly affect phytochemical recovery and the performance of extraction techniques. To obtain high-quality products, phytochemicals must be extracted in a manner that preserves their native structure and bioactivity (Bitwell et al., 2023). Therefore, careful selection of an appropriate extraction strategy is essential. Commonly employed conventional extraction techniques include maceration, percolation, decoction, reflux extraction, and Soxhlet extraction. In contrast, advanced or novel approaches comprise pressurized liquid extraction (PLE), high hydrostatic pressure extraction (HHP), microwave-assisted extraction (MAE), ultrasound assisted extraction (UAE), pulsed electric field extraction (PEF), vibro-cavitation extraction, vacuum-oscillating boiling extraction, milling-based extraction, rotary pulsation extraction (RPE), liquid–gas extraction (LGS), enzyme-assisted extraction (EAE), supercritical fluid extraction (SFE), and natural deep eutectic solvent-based extraction (NADES) (Kumar et al., 2023).

The amino acid profile of makhana (fox nut) endosperm indicates a well-balanced composition with a predominance of non-essential amino acids such as glutamic acid and aspartic acid, which contribute to metabolic functions and flavor enhancement. Among essential amino acids, leucine, valine, isoleucine, and phenylalanine are present in appreciable amounts, highlighting its nutritional quality. Notably, arginine is found in high concentration, which is beneficial for cardiovascular health and metabolic regulation. However, sulfur-containing amino acids like cysteine and methionine are comparatively lower, which may be considered a limiting factor. Overall, the profile suggests that makhana can serve as a valuable plant-based protein source with functional and health-promoting potential (Jana et al., 2024).

Makhana is known for its bioactive compounds and antioxidant properties (Table 9.3). In a study by Mahesh Kumar et al., (2023), makhana samples were analyzed for their bioactive compounds both in raw form as well as popped form. Both free and bound polyphenols were extracted using the conventional extraction technique of agitation. The samples (1.0 g) were extracted for free polyphenol using 10 mL of 80% methanol three times while being stirred for 3 hours at room temperature in the dark. After centrifuging the sample extracts at 2236×g for 15 min at 10°C, the resulting supernatants were combined. A rotary evaporator operating under vacuum (R-215, Buchi, Flawil, Switzerland) was used to evaporate the combined supernatant. For estimating free polyphenols, the dried residue of the sample extract was reconstituted using 5 milliliters of methanol. Following the extraction of free polyphenols, the sample residue was dried at 40°C in a hot air oven until it was fully dry. For bound polyphenol, the residue from free polyphenol extract, was added with 2 M NaOH in an agitator for 2 h followed by reducing the pH of extract to 2 using 1 N HCL. The acidic extract was then decanted with ethyl acetate to

TABLE 9.3 **Bioactive Constituents and Antioxidant Capacity of Raw Makhana Kernel.**

Quality Attribute	Unit	Value
Total phenolics	mg GAE g^{-1}	2.22–2.28
Total flavonoids	mg RE 100 g^{-1}	3.23–3.51
Antioxidant capacity (DPPH assay)	% inhibition	25.57–27.70

GAE = Gallic Acid Equivalents; RE = Rutin Equivalents. Kumar, R., Singh, I. S., Jha, V., 2025. Bioactive compounds in *Euryale ferox* Salisb. and their nutritional and therapeutic actions-a review. Vegetos 1–9; Significant differences ($p < 0.05$) are indicated by consecutive means ± ±SD with distinct superscripts.

separate the organic portion. The decanted fraction was dried to obtain the bound phenol which was redissolved in Methanol for further analysis (Qiu et al., 2010).

The analysis results showed that the popped makhana has 31.69% less polyphenols compared to raw makhana. This shows that popping decreases bioactive compounds in makhana. The major polyphenols detected in makhana samples were gallic acid, protocatechuic acid, catechin, ferulic acid, 4-coumaric acid and 2-coumaric acid, luteolin-7-O-glucosides with highest amount of catechin in free polyphenol and gallic acid in bound polyphenols (Mahesh Kumar et al., 2023). Quercetin (422 μg/kg) and kaempferol (200 μg/kg), two pure flavonoids that speed up glucose metabolism in human cells, were found in Makhana seed endosperm in a study by Jana et al. (2024). Vitamin B_1 (1.14 ± 0.02 mg/100 g) and vitamin C (5.4 ± 0.04 mg/100 g) were found in seed endosperm. Makhana exhibited the highest levels of antioxidant activity, with DPPH and FRAP values of 288.76 ± 0.09 mg AEAC/100 g FW and 99.14 ± 0.16 mg AEAC/100 g FW, respectively. Additionally, Makhana's protein composition of dry endosperm was found to include a significant amount of arginine (12.70 ± 0.21 g/100 g) and histidine (3.62 ± 0.11 g/100 g). Numerous macro and micronutrients were discovered to be abundant in makhana. Significant amounts of zinc (5.6 ± 0.24 mg/100 g) and magnesium (20.00 ± 1.09 mg/100 g), which control insulin secretion and storage, were also found. Significant levels of the antioxidants iron (10.9 ± 0.13 mg/100 g), selenium (0.06 ± 0.02 mg/kg), and chromium (1.05 ± 0.06 mg/kg) were also found in the Makhana seed kernel. The amino acid profile of makhana seed is shown in Table 9.4, which represents that makhana seed is full of essential amino acids.

Fatty acids play a crucial role in human nutrition, as they contribute to energy metabolism, cellular function, and the prevention of various chronic diseases. In particular, unsaturated fatty acids are associated with several health benefits, including improved cardiovascular health and reduced

TABLE 9.4 Amino acid profile of Makhana.

Amino acid	% of total protein
Aspartic acid	8–8.5
Glutamic acid	19.7–19.8
Serine	5.45–5.70
Histidine	3.62–3.70
Glycine	3.90–3.95
Arginine	12.70–12.90
Alanine	7.12–7.30
Tyrosine	2.40–2.50
Cysteine	0.12–0.15
Valine	6.24–6.50
Methionine	3.43–3.53
Phenylalanine	5.60–5.73
Isoleucine	5.03–5.20
Leucine	9.51–9.58

inflammation. A total of 11 fatty acids were identified in raw makhana, whereas nine fatty acids were detected in popped makhana. Among these, linoleic acid, palmitic acid, and oleic acid were the major fatty acids present in the samples, accounting for approximately 42.83%–53.15%, 23.81%–27.71%, and 12.39%–14.97% of total fatty acid methyl esters (FAME), respectively. The proportion of polyunsaturated fatty acids (PUFA) increased from 44.20% in raw makhana to 54.41% in popped makhana. In contrast, the level of monounsaturated fatty acids (MUFA) decreased after popping, declining from 17.36% in raw samples to 13.29% in popped makhana. Additionally, eicosenoic acid and erucic acid were detected in raw makhana at levels of 0.73% and 0.88%, respectively; however, these fatty acids were not observed in the popped samples, which may be attributed to their degradation during the popping process (Kumar et al., 2023). The bioactive potential and health benefits properties of makhana are represented in Fig. 9.2, which is mainly due to micro and macro nutrients present in Makhana.

Apart from seeds of makhana, other parts of makhana plants are also rich in bioactive compounds like leaves. The bioactive compounds from makhana leaves can also be extracted suing different extraction techniques like ultrasound assisted extraction, microwave assisted extraction etc. Lower surfaces of

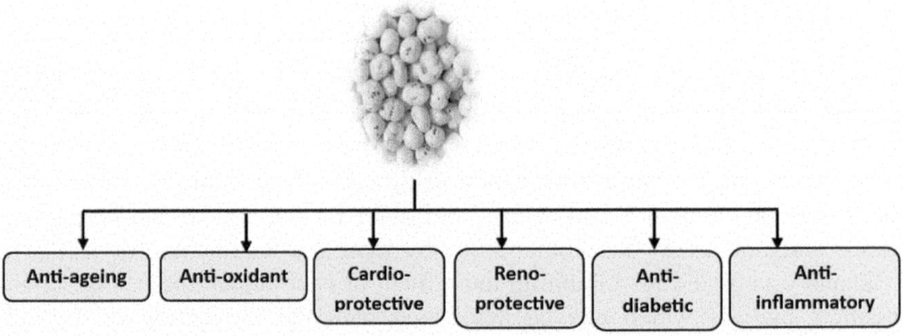

FIGURE 9.2 Bioactive potential and health benefits of Makhana.

E. ferox leaves contains dark purple color. Up till now, some earlier research has shown that a plant's color is related to its anthocyanin level and composition. Its dark purple hue is most likely caused by the abundance of anthocyanins it contains. One type of flavonoid compound that gives many leaves, flowers, and fruits their red, blue, and purple hues is anthocyanins. Due to their nontoxicity, anthocyanins, which are naturally occurring water-soluble colorants, have the potential to replace synthetic colorants in food coloring (Delgado-Vargas et al., 2000). Additionally, because of their antioxidant activity, anthocyanins have been demonstrated to offer potential health advantages, including the prevention of diabetes, hyperlipidemia, cancer, and cardiovascular disorders. In addition to being used as natural colorants, anthocyanins are now frequently used as components in medications, functional foods, and dietary supplements. The discovery of natural anthocyanins from novel sources has drawn more interest due to the rise in anthocyanin demand (Santos-Buelga et al., 2014). In a study by Wu et al. (2020), response surface methodology-based optimization technique was used to find the optimum extraction condition for ultrasound assisted extraction (UAE) of anthocyanins from *E. ferox* leaves. The ideal extraction conditions were identified as an extraction temperature of 40°C, an extraction time of 38 min, an ethanol concentration of 62% (v/v), and a solvent-to-solid ratio of 27 mL/g following optimization using the single factor test and RSM. In validation trials, the experimental yield of anthocyanins at optimal conditions was 281.56 ± 3.02 mg/100 g, which was in good agreement with the expected value of 289.45 mg/100 g. Furthermore, comparative tests showed that the yield obtained with UAE was much higher than that obtained with traditional solvent extraction under ideal conditions (181.81 ± 1.94 mg/100 g), suggesting that the UAE increased the yield of anthocyanins extracted from *E. ferox* leaves.

The seeds of the makhana, are prized for their nutritional and therapeutic properties. Alkaloids, flavonoids, tannins, saponins, and phenolic compounds are among the bioactive phytochemicals that provide the plant its anti-inflammatory, anti-cancer, and antioxidant properties. Proteins, essential amino acids, fiber, polyunsaturated fatty acids (particularly omega-3), vitamins (B-complex), and minerals (Ca, Mg, P) are all abundant in *E. ferox* seeds. Modern analytical techniques, in particular Gas Chromatography-Mass Spectrometry (GC-MS), can precisely detect and quantify the intricate phytochemical composition of *E. ferox*. The plant's volatile and non-volatile components may be better understood using GC-MS technology, which also sheds light on the plant's anti-cancer, anti-diabetic, and antioxidant properties (Sinha and Jha, 2025). In another study by Ekal and Kumbamoorthy (2025), makhana was prized for inhibiting the growth of islet autoantibodies using an in-silico analysis which can be associated with the prevention of Type-2 diabetes. Makhana's polysaccharides stimulate the release of insulin and aid in the active absorption of glucose by insulin-resistant cells. Euryale ferox which contained about 49 phytochemicals that are being studied and evaluated through docking studies. These chemicals' canonical SMILES were prepared using Marvin sketch and Pubchem. To anticipate the target proteins that are found to be interacting with Makhana chemicals, the SMILES were then uploaded to the SwissADME database and put into the Stitch database. The three-dimensional structures of the ligands were constructed for investigation using the Chem3D Ultra 11.0 tool. The Docking results indicated that Type 1 diabetes mellitus was most closely linked to PPARG. The binding energy of Ferotocotrimer E was found to be −11 when all 49 of Makhana's compounds were screened against PPARG (Uniprot ID: A0A0S2Z4K5) using autodock vina. The active ingredient Ferotocotrimer E of Makhana has a greater ability to prevent the development of islet auto-antibodies linked to Type I diabetes and maintain the health of pancreatic islet beta cells, according to the binding affinity.

Foxnut may be essential for enhancing male reproductive health because it has been shown to increase the body's sperm count in specific formulations. Additionally, makhana, a traditional Chinese powdered remedy for spermatorrhea, is used. Fox nuts help to keep our bodies hydrated and in a balanced water balance (Tehseen et al., 2020). This maintains the body's secretions hydrated. By increasing body moisture and making secretions stickier, foxnut supports the male reproductive system. Makhana is also believed to assist the reproductive system since it increases the number and quality of sperm. According to Ayurvedic medicine, 5–10 g of the powdered seed should be taken twice daily. The dosage can be raised to 30 g daily for adults in order to improve their reproductive system. It has a high amino acid index but a low glycemic value. Because of its spermatogenic and aphrodisiac qualities, makhana has long been used to cure male infertility, making it a crop of cultural significance (Shaw et al., 2025).

Makhana is prized for its nutritional richness as well as its many health advantages, such as its ability to control blood sugar, protect the heart, prevent cancer, protect the liver, reduce inflammation, and scavenge free radicals. Cerebrosides, polysaccharides, tocopherols, polyphenols, cyclic dipeptides, glucosylsterols, triterpenoids, and sesquineolignans are among the several nutrients found in Makhana seeds (Jiang et al., 2023). Due to its high dietary fiber content, Makhana is also thought to be beneficial for controlling blood pressure, blood sugar, anemia, heart disease, and constipation. Additionally, a number of investigations have documented the traditional use of *E. ferox* to treat liver failure, enuresis, chronic diarrhea, and diabetic nephropathy. *E. ferox* kernel flour is utilized in several Ayurvedic remedies to support the kidney and spleen. Numerous amino acids (such as arginine, glutamine, and histidine), minerals (such as trace levels of chromium and selenium), and flavonoids (such as quercetin and kaempferol) are found in *E. ferox* and help to stimulate hormonal immunity. Additionally, *E. ferox* is high in folate, which aids in fetal growth during the first trimester of pregnancy and has been shown to benefit postpartum women (Mittal et al., 2020).

9.5 Utilization of makhana for bioactive rich products

One of India's most significant aquatic nut crops is makhana (*Euryale ferox* Salisb.). It is well-known for having more carbohydrates and protein. It is an excellent medicinal plant that was utilized in ancient Chinese and Indian medicine 3000 years ago, and it typically yields tasty nuts. Ayurvedic remedies make use of fox nut seeds. It has 75.04% carbs and 11.16% protein. It is highly helpful for anemia and strengthens the heart. Makhana is a crucial component that helps to fortify the kidneys and spleen. Its high potassium and low salt levels lower blood pressure, while its extremely low monosaturated fat content keeps blood sugar levels from rising (Jana et al., 2019). The high nutritional content of makhana makes it suitable for the preparation of many value-added products (Fig. 9.3). Research indicates that after makhana flour is created, it may be successfully mixed with other edible flours in varying proportions to create a variety of snacks, including makhana wheat chapatti (1:1) and makhana cabbage pakora, as well as desserts, including makhana barfi and makhana kalakand. Products made from makhana have a modest calorific value and little sugar. They also have excellent expansion and a color that may appeal to health-conscious consumers.

Makhana being known for its high value bioactive compounds finds a lot of food applications. It is used to treat chronic conditions like diarrhea, excessive leucorrhea, spleen hypofunction, and renal issues. It is also said to be a detoxifier. It is possible to create a nutrition bar by using broken foxnut seeds in the manufacturing of any product with added value (Vikram and Mishra, 2021). In a study by Kumar et al. (2024), Makhana flour was utilized for the preparation of nutritional bars. The bar with the highest rheological and sensory scores was made with 15% makhana flour and barnyard flour each. The metabolome that was retrieved showed a total of 29 bioactive compounds. Palmitoyl serinol, glycitein, persin, bufagargarizin, apigenin, carvone, and other major metabolites that covered a large mass spectrum were identified.

FIGURE 9.3 Value added products from makhana.

These substances have anti-inflammatory, anti-carcinogenic, anti-diabetic, anti-microbial, and heart health boosting properties.

In another study by Kathuria et al. (2025), cookies were prepared using spent makhana flour. Commercial cookies have a poor protein level in addition to their nutritional profile. Consequently, adding popped makhana made from the fox nut or gorgon nut (*Euryale ferox*) offers yet another way to enhance the nutritional value of baked foods. Popped makhana is a good supply of important minerals like magnesium, phosphate, and potassium as well as readily digestible protein. It is also low in cholesterol and fat. Spent makhana flour includes flat or irregularly shaped makhana with smaller size, generally not suitable for marketing. The incorporation of spent makhana improved the functional as well as nutritional properties of cookies. In order to nutritionally improve the disrupted lipid parameters, a study focused on the development, sensory evaluation, standardization, and nutritional calculation of baked soy and makhana cookies enhanced with a hypolipidemic mixture was performed by Grewal and Kaur (2023). Using chia seeds, almonds, wheat bran, and powdered mango seeds, a hypolipidemic nutritional blend was created. The produced hypolipidemic nutritional mixture was added to baked soy and makhana cookies at varying levels of integration. RDA 2020 and IFCT 2017 were used to determine the nutritional content of these goods, and a 9-point hedonic scale was used to measure their appropriateness. At 12% and 15% inclusion levels, the baked soy and whole wheat cookies enhanced with the hypolipidemic mixture had the highest acceptable mean sensory scores, 7.75 ± 0.75 and 7.59 ± 0.91, respectively. As per statistical analysis, baked cookies with hypolipidemic qualities contained significantly more dietary fiber, linoleic acid, and monounsaturated fatty acids. High levels of nutrients included in baked soy and makhana cookies with hypolipidemic qualities are scientifically responsible for nutritionally controlling and enhancing the disturbed lipid parameters along with respectable organoleptic scores.

Though the makhana flour have high number of functional compounds, still they have some antinutrients like phytic acid. Germination is a well pronounced technique that reduces the antinutrients in grains. In a study by Islam et al. (2025), the physicochemical, anti-nutritional, and structural characteristics of makhana seed flour as affected by 72-hour germination was evaluated. The amount of moisture, protein, and ash increased as a result of germination. Additionally, it greatly increased the capacity to absorb water (2.48–5.65 g/g) and oil (1.33–2.63 g/g). Germination significantly reduced antinutrients, which led to a drop in tannins (3.69–2.65 mg/100 g) and phytic acid (168.64–82.94 mg/100 g). While structural analyses using XRD and FTIR revealed changes in the crystalline and molecular structures, scanning electron microscopy revealed that the starch granules in the germinated flour had been disturbed. These results show that germination improves the nutritional content and usefulness of makhana seed flour, increasing its suitability for a range of culinary applications.

Makhana flour has found another application in preparation of functional pasta (Mridula et al., 2024). Pasta is a food item that appeals to people of all ages due to its tasty flavor, ease of preparation, adaptability, shelf stability, and affordability. Due to its high gluten protein level, which is necessary for effective dough formation, durum wheat flour or semolina is the most widely used raw ingredient for pasta (Phongthai et al., 2017). Gluten, which is necessary for the formation of viscoelasticity and chewability, affects the pasta's cooking quality. By adding enough protein, hydrocolloid, and water, the lack of gluten in the pasta dough matrix can lessen technological quality problems (high solid loss, stickiness, and less firm texture) (Mridula et al., 2017). Due to its capacity to interact and reassociate with other food ingredients, starch plays a structural role in pasta compositions, minimizing cooking loss. The nutritional value of semi-popped makhana (SPM), a by-product of popping makhana seeds, may be comparable to that of popped makhana. Furthermore, popping at a high temperature enables its starch to gelatinize, which makes it useful for creating gluten-free items. By substituting semi-popped makhana (SPM), water chestnut (WCN), and potato flour (PF) for wheat flour, cereal gluten-free pasta was prepared. The research assessed the impact of these ingredients on the nutritional, technological, phytochemical, and sensory qualities of CGF pasta. SPM flour (30–50 g), WCN flour (20–30 g), and PF (30–50 g) were used as variables in the Box–Behnken design of response surface methods for optimization and statistical analysis, while the amount of carrot juice remained constant (Mridula et al., 2024).

Although makhana starch is a healthy food source, it also has processing and neutraceutical value when it is transformed into resistant starch, poorly digesting starch, and changes in amylose and amylopectin. Foods high in starch are practically essential to people's everyday diets all throughout the world. Starch's quick digestion results in high blood glucose levels, which are clearly linked to a number of illnesses. Amylopullulanase treatment improved the resistant starch level of the nutrient-dense but untested popping makhana (*Euryale ferox*) in order to reduce its quick digestion and heightened glycemic response. The physicochemical and functional properties of amylopullulanase-treated makhana flour (MM) and native makhana flour (NM) were compared in the study by Biswas et al. (2020). The increases in amylose content, resistant starch, and crystallinity were found to be 12.33%, 14.88%, and 11.32%, respectively, while the decreases in readily digestible starch and oil holding capacity were 13.01% and 3.12 g/g, respectively. During the in vitro digestibility analysis, this guaranteed the reduced and sustained release of glucose. These results highlight the increased potential of makhana flour treated with amylopullulanase for the creation of therapeutic foods Fig. 9.4.

Makhana, due to its richness in protein and carbs, has several health advantages, such as anti-inflammatory, antifungal, anti-diabetic, and stress-relieving qualities, as well as boosting liver and heart health. The sensory analysis showed that adding Foxnut protein powder greatly improved the flavor, consistency, and allure of several recipes. 93.33% of participants thought the experimental version of the banana strawberry milk shake was "Excellent," but just 40% thought the normal version was. Similarly, 80% of participants gave the shrikhand with Foxnut protein powder a "Excellent" rating, but just 33.33% gave the conventional version the same rating. Additionally, the oats phirni did remarkably well, receiving a "Excellent" rating from 93.33% of participants, compared to only 20% who gave the normal version the same rating (Mittal and Dhakar, 2024).

9.6 Conclusion and future suggestions

Because of its wide range of phytochemicals and bioactive substances, *Euryale ferox*, has become a pharmacologically promising and nutritionally rich aquatic crop. The presence of important bioactive compounds, including phenolic compounds, flavonoids, polysaccharides, bioactive proteins, and vital micronutrients, has been thoroughly discussed in this chapter. These bioactive compounds work together to provide the plant's anti-inflammatory, anti-

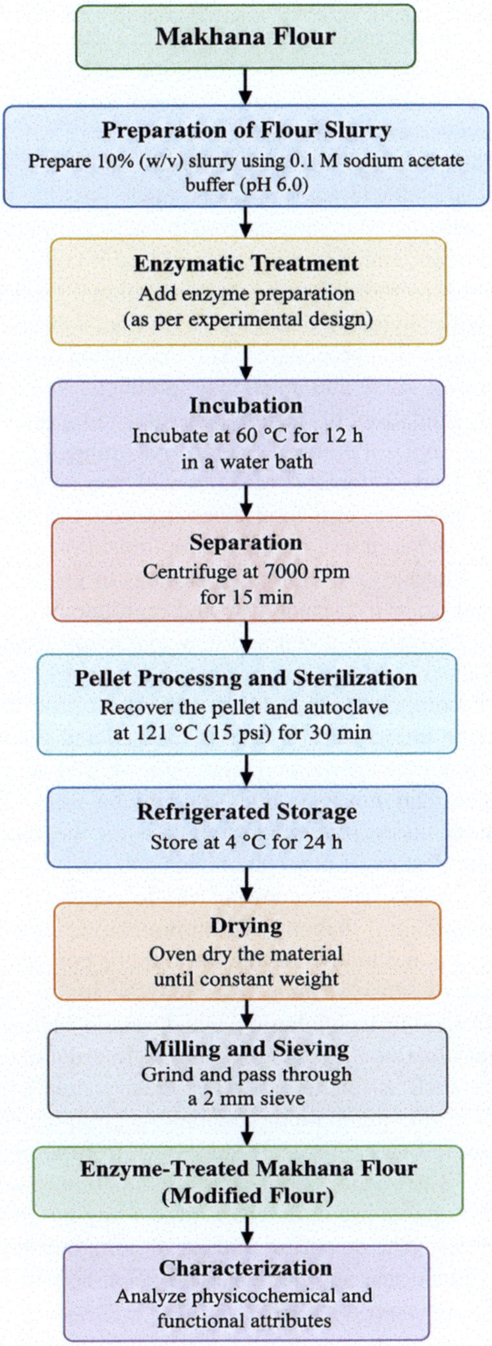

FIGURE 9.4 Enzyme modification process of makhana.

diabetic, antioxidant, cardioprotective, and gut-health-promoting qualities. The efficiency, selectivity, and sustainability of bioactive recovery have been greatly enhanced by developments in extraction technologies, which range from traditional solvent-based procedures to green and aided approaches like ultrasonic, microwave, enzyme-assisted, and supercritical fluid extraction. Furthermore, a deeper comprehension of these compounds structural, functional, and biological characteristics has been made possible by contemporary characterization methods. Makhana-derived bioactive compound's potential for use in nutraceuticals, functional meals, and health-conscious formulations highlights their worth as more than just a conventional food item. Still unresolved are issues with phytochemical bioavailability, stability, scalability, and standardization of extraction procedures. The integration of makhana bioactive compounds into novel food and medicinal products, synergistic effects of bioactive, clinical validation of health benefits, and process improvement should be the main areas of future research. All things considered, Euryale ferox is an underutilized, sustainable resource with a great deal of potential for value addition that promotes agri-food diversification and human health.

Standardization and extraction process optimization should be the top priorities of future Euryale ferox research in order to guarantee the bioactive recovery's industrial usability, scalability, and reproducibility. For large-scale use, a comparative analysis of green and new extraction technologies will be essential, with a focus on solvent safety, energy efficiency, and environmental sustainability. To accomplish thorough phytochemical profiling and to clarify the structure–function interactions of both individual and synergistic bioactive substances, advanced characterization and omics-based techniques (metabolomics, proteomics, and nutrigenomics) should be used. To support their functional usefulness, more research is required on the bioavailability, stability, and metabolic destiny of phytochemicals obtained from makhana under processing and gastrointestinal circumstances. To support health claims pertaining to antioxidant, anti-diabetic, cardioprotective, and neuroprotective qualities, more clinical and in vivo research should be conducted. The stability and tailored release of sensitive bioactives may be improved by integrating delivery and encapsulating technologies, which would increase their efficacy in nutraceutical and functional food compositions. In order to create innovative value-added goods, such as functional foods, dietary supplements, and therapeutic substances, while maintaining regulatory compliance and customer acceptance, interdisciplinary efforts are necessary from an application standpoint. Furthermore, waste valorization and circular bioeconomy models can benefit from the sustainable use of makhana processing by-products as sources of bioactive chemicals. The scientific, industrial, and societal significance of Euryale ferox as a functional and nutraceutical resource will be reinforced by these future directions taken together.

References

Altemimi, A., Lakhssassi, N., Baharlouei, A., Watson, D.G., Lightfoot, D.A., 2017. Phytochemicals: extraction, isolation, and identification of bioactive compounds from plant extracts. Plants 6 (4). https://doi.org/10.3390/plants6040042.

Biswas, P., Das, M., Boral, S., Mukherjee, G., Chaudhury, K., Banerjee, R., 2020. Enzyme mediated resistant starch production from Indian Fox Nut (*Euryale ferox*) and studies on digestibility and functional properties. Carbohydr. Polym. 237, 116158. https://doi.org/10.1016/j.carbpol.2020.116158.

Bitwell, C., Indra, S.S., Luke, C., Kakoma, M.K., 2023. A Review of Modern and Conventional Extraction Techniques and Their Applications for Extracting Phytochemicals from Plants, 19. Elsevier B.V., Zambia Scientific African https://doi.org/10.1016/j.sciaf.2023.e01585.

Cheok, C.Y., Salman, H.A.K., Sulaiman, R., 2014. Extraction and quantification of saponins: a review. Food Res. Int. 59, 16–40. https://doi.org/10.1016/j.foodres.2014.01.057.

Clarke, D.B., 2010. Glucosinolates, structures and analysis in food. Anal. Methods 2 (4), 310–325. https://doi.org/10.1039/b9ay00280d.

Cooperstone, J.L., Schwartz, S.J., 2016. Recent Insights into Health Benefits of Carotenoids. Elsevier BV, pp. 473–497. https://doi.org/10.1016/b978-0-08-100371-8.00020-8.

Delgado-Vargas, F., Jiménez, A.R., Paredes-López, O., 2000. Natural pigments: carotenoids, anthocyanins, and betalains—characteristics, biosynthesis, processing, and stability. Crit. Rev. Food Sci. Nutr. 40 (3), 173–289. https://doi.org/10.1080/10408690091189257.

Devi, N., Rani, K., Kharb, P., Kaushik, P., 2022. Bio-fabrication of euryale ferox (makhana) leaf silver nanoparticles and their antibacterial, antioxidant and cytotoxic potential. Plants 11 (20), 2766. https://doi.org/10.3390/plants11202766.

Ekal, V., Kumbamoorthy, S., 2025. Bioactivity of makhana in inhibiting the development of islet autoantibodies: an insilico study. Research Journal of Pharmacy and Technology, India 18 (5), 2017–2022. https://doi.org/10.52711/0974-360X.2025.00288.

Goncharov, N.V., Belinskaia, D.A., Ukolov, A.I., Jenkins, R.O., Avdonin, P.V., 2021. Organosulfur compounds as nutraceuticals nutraceuticals: efficacy, safety and toxicity. Elsevier, Russian Federation, pp. 911–924. https://doi.org/10.1016/B978-0-12-821038-3.00054-9.

Grewal, R., Kaur, T.J., 2023. Product development of baked soya and makhana cookies with hypolipidemic properties. Ind. J. Nutri. Diet. 60, 3. https://doi.org/10.21048/IJND.2023.60.3.33110.

Ingle, K.P., Deshmukh, A.G., Padole, D.A., Dudhare, M.S., Moharil, M.P., Khelurkar, V.C., 2017. Phytochemicals: extraction methods, identification and detection of bioactive compounds from plant extracts. J. Pharmacogn. Phytochem. 6 (1), 32–36.

Islam, Z., Bashir, K., Rastogi, M., Rikhari, A., Siddiqui, S., 2024. Evaluation of different methods of starch extraction on physico-chemical and functional properties of Makhana seed (*Euryale ferox*) starch. Indian J Agricult Biochem 37 (2), 173–178. https://doi.org/10.5958/0974-4479.2024.00025.7.

Islam, Z., Bashir, K., Singh, S., Rastogi, M., Siddiqui, S., Punia Bangar, S., 2025a. Structural and functional properties of starch from popped seeds of makhana (*Euryale ferox*). Starch - Stärke 77 (7). https://doi.org/10.1002/star.70030.

Islam, Z., Bashir, K., Singh, S., Siddiqui, S., 2025b. Effect of germination on physicochemical, functional and antinutritional characteristics of makhana seed (*Euryale ferox*) flour. Food and Humanity 5, 100717. https://doi.org/10.1016/j.foohum.2025.100717.

Jana, B.R., Srivastava, A., Idris, M., 2019. New makhana (*Euryale ferox* Salisb.) processed products for health benefit. J. Food Process. Technol. 10 (4), 789.

Jana, B.R., Kumar, M., Raut, S.M., 2024. Antidiabetic biomolecules and nutrient elements in Makhana (*Euryale ferox* Salisb.). Ann Phytomed Int J 13 (2). https://doi.org/10.54085/ap.2024.13.2.52.

Jiang, Z.M., Wang, L.J., Gao, Z., Zhuang, B., Yin, Q., Liu, E.H., 2019. Green and efficient extraction of different types of bioactive alkaloids using deep eutectic solvents. Microchem J 145, 345–353. https://doi.org/10.1016/j.microc.2018.10.057.

Jiang, J., Ou, H., Chen, R., Lu, H., Zhou, L., Yang, Z., 2023. T. ethnopharmacological, phytochemical, and pharmacological review of *Euryale ferox* Salisb.: a Chinese medicine food homology. Molecules 28 (11), 4399. https://doi.org/10.3390/molecules28114399.

Kapoor, S., Kaur, A., Kaur, R., Kumar, V., Choudhary, M., 2022. Euryale ferox, a prominent superfood: nutritional, pharmaceutical, and its economical importance. J Food Biochem 46 (12). https://doi.org/10.1111/jfbc.14435.

Kathuria, C., Ranjan, S., Singh Tomar, M., Kumar, V., Ahmad, W., Dobhal, A., Joshi, S., Chand, K., Kumar, S., 2025. Development and quality assessment of *Chenopodium album*, Spinacia oleracea, and spent makhana (*Euryale ferox*) enriched vegetable cookies. Food Chemistry Advances, 101040. https://doi.org/10.1016/j.focha.2025.101040.

Kumar, R., Ahmad, J., Abdin, M.Z., Jha, V., Kumar, A., Singh, I.S., 2017. Optimization of seed germination in Makhana (*Euryale ferox* Salisb) under controlled conditions. HortFlora Res Spectr 6 (1), 55–58.

Kumar, A., Nirmal, P., Kumar, M., Jose, A., Tomer, V., Oz, E., Proestos, C., Zeng, M., Elobeid, T., Sneha, K., Oz, F., 2023. Major phytochemicals: recent advances in health benefits and extraction method. Molecules 28 (2), 887. https://doi.org/10.3390/molecules28020887.

Kumar, V., Kumar, A., Kumar Singh, M., Dhyani, P., Mishra, H., Chandra Rai, D., 2024. Bioactive metabolites identification of the foxnut and broken millet-based nutritional bar using HR-MS. Food Chem Mole Sci 9. https://doi.org/10.1016/j.fochms.2024.100214.

Kumar, R., Singh, I.S., Jha, V., 2025. Bioactive compounds in *Euryale ferox* Salisb. and their nutritional and therapeutic actions-a review. Vegetos 1–9.

Kumar Sinha, C.B., Jha, V., 2025. Identification of nutraceutically important metabolites in raw and popped makhana (*Euryale ferox* Salisb.) seeds using GCMS technique. Int J Plant Environ 11 (03), 514–524. https://doi.org/10.18811/ijpen.v11i03.09.

Lin, D., Xiao, M., Zhao, J., Li, Z., Xing, B., Li, X., Kong, M., Li, L., Zhang, Q., Liu, Y., Chen, H., Wen, Q., Wu, H., Chen, S., 2016. An overview of plant phenolic compounds and their importance in human nutrition and management of type 2 diabetes. Molecules 21 (10), 1374. https://doi.org/10.3390/molecules21101374.

Mahesh Kumar, M., Goudar, G., Sharma, P., Vishwakarma, R., Gogoi, P., Mahajan, A., Chandragiri, A., Kalpuri, S., Babu Geddam, J.J., Suresh, C., Chary, M., Venketrajireddy, G., Radhika, M.S., Sreenu, P., Tulja, B., 2023. Differentiating the nutrient composition, in-vitro starch digestibility, individual polyphenols and anti-oxidant properties of raw and popped makhana (*Euryale ferox*). J. Food Meas. Char. 17 (6), 5828–5844. https://doi.org/10.1007/s11694-023-02074-5.

Mendoza, N., Silva, E.M.E., 2018. Introduction to phytochemicals: secondary metabolites from plants with active principles for pharmacological. In: Phytochemicals: Source of antioxidants and role in disease prevention, 25.

Mittal, M., Dhakar, A., 2024. Organoleptic evaluation of foxnut protein powder. Int J Nutr Lifestyle 5 (3), 147–158. https://doi.org/10.69859/ijnl.2025.v5i3003.

Mittal, R., Sharma, S., Mittal, A., 2020. A critical review on ethnobotanical and pharmacological aspects of *Euryale ferox* Salisb. Pharmacogn J 12 (6), 1444–1454. https://doi.org/10.5530/pj.2020.12.199.

Mridula, D., Gupta, R.K., Khaira, H., Bhadwal, S., 2017. Groundnut meal and carrot fortified pasta: optimization of ingredients level using RSM. Proc. Natl. Acad. Sci. India B Biol. Sci. 87 (2), 277–288. https://doi.org/10.1007/s40011-015-0586-x.

Mridula, D., Vishwakarma, R.K., Arora, S., Bala, M., 2024. Formulation and characterization of novel cereal gluten-free pasta from semi-popped makhana, water chestnut, and potato. Food Bioprocess Technol. 17 (11), 4262–4277. https://doi.org/10.1007/s11947-024-03349-5.

Pereira, A.G., Cassani, L., Garcia-Oliveira, P., Otero, P., Mansoor, S., Echave, J., Xiao, J., Simal-Gándara, J., Prieto, M.A., 2023. Plant alkaloids: production, extraction, and potential therapeutic properties natural secondary metabolites: from nature, through science, to industry. Springer International Publishing, Spain, pp. 157–200. https://doi.org/10.1007/978-3-031-18587-8_6.

Phongthai, S., D'Amico, S., Schoenlechner, R., Homthawornchoo, W., Rawdkuen, S., 2017. Effects of protein enrichment on the properties of rice flour based gluten-free pasta. LWT 80, 378–385. https://doi.org/10.1016/j.lwt.2017.02.044.

Piironen, V., Lindsay, D.G., Miettinen, T.A., Toivo, J., Lampi, A.M., 2000. Plant sterols: biosynthesis, biological function and their importance to human nutrition. J. Sci. Food Agric. 80 (7), 939–966. https://doi.org/10.1002/(SICI)1097-0010(20000515)80:7<939::AID-JSFA644>3.0.CO;2-C.

Qiu, Y., Liu, Q., Beta, T., 2010. Antioxidant properties of commercial wild rice and analysis of soluble and insoluble phenolic acids. Food Chem. 121 (1), 140–147. https://doi.org/10.1016/j.foodchem.2009.12.021.

Santos-Buelga, C., Mateus, N., Freitas, V.De, Anthocyanins, 2014. Plant pigments and beyond. J. Agric. Food Chem. 62 (29), 6879–6884. https://doi.org/10.1021/jf501950s.

Shakir, M., Mirza, S., Jolly, R., Rauf, A., Owais, M., 2018. Synthesis, characterization and in vitro screening of a nano-hydroxyapatite/chitosan/*Euryale ferox* nanoensemble-an inimitable approach for bone tissue engineering. J Chem 42 (1), 363–371. https://doi.org/10.1039/c7nj02953e.

Shankar, M., Chaudhary, N., Singh, D., 2010. A review on gorgon nut. Int J Pharmaceut Biol Archiv 1 (2), 101–107.

Shaw, R., Das, S., Chaube, R., 2025. Reproductive health benefits of *Euryale ferox* Salisb.: a short comprehensive review. J. Endocrinol. Reprod. 183–190. https://doi.org/10.18311/jer/2024/44756.

Srisang, N., Prachayawarakorn, S., Soponronnarit, S., Chungcharoen, T., 2021. An innovative hybrid drying technique for parboiled rice production without steaming: an appraisement of the drying kinetics, attributes, energy consumption, and microstructure. Food Bioprocess Technol. 14 (12), 2347–2364. https://doi.org/10.1007/s11947-021-02729-5.

Tehseen, S., Sarfraz, F., Muntaham, S., Ateeq, N., Ashfaq, F., Yasmin, I., Mehmood, T., 2020. Foxnut (*Euryale ferox* Salisb.): a health promising fruit. Acta Sci Agricult 4 (12), 68–72.

Vattem, D.A., Randhir, R., Shetty, K., 2005. Cranberry phenolics-mediated antioxidant enzyme response in oxidatively stressed porcine muscle. Process Biochem. 40 (6), 2225–2238. https://doi.org/10.1016/j.procbio.2004.09.001.

Vikram, V., Mishra, S., 2021. A review on *Euryale ferox*: daily dietary regime for Covid -19 patients. Int J Appl Res 7 (6S), 29–34. https://doi.org/10.22271/allresearch.2021.v7.i6sa.8607.

WHO, 1996. Trace Elements in Human Nutrition and Health. World Health Organization, Geneva.

Wu, C.Y., Wang, H., Fan, X.H., Yue, W., Wu, Q.N., 2020. Waste *Euryale ferox* Salisb. Leaves as a potential source of anthocyanins: extraction optimization, identification and antioxidant activities evaluation. Waste Biomass Valor 11 (8), 4327–4340. https://doi.org/10.1007/s12649-019-00762-2.

Yogesh, K., Jha, S.N., Ahmad, T., 2014. Antioxidant potential of aqueous extract of some food grain powder in meat model system. J. Food Sci. Technol. 51 (11), 3446–3451. https://doi.org/10.1007/s13197-012-0804-y.

Yuan, Y., Macquarrie, D., 2015. Microwave assisted extraction of sulfated polysaccharides (Fucoidan) from ascophyllum nodosum and its antioxidant activity. Carbohydr Polym 129, 101–107. https://doi.org/10.1016/j.carbpol.2015.04.057.

Yuan, H., Meng, S., Wang, G., Gong, Z., Sun, W., He, G., 2014. Hypoglycemic effect of triterpenoid-rich extracts from *Euryale ferox* shell on normal and streptozotocin-diabetic mice. Pak J Pharm Sci 27 (4), 859–864.

Chapter 10

Starch chemistry and functional properties of makhana flour; protein quality and functional attributes of makhana, fatty acid profile and lipid stability in makhana products

Selva Kumar T.[1], Sriranjani Kumar[2, 3], Manukonda Maheswari[4], Saranya Devi Subburaj[5] and Anand Kumar[6]

[1]*Vel Tech Rangarajan Dr. Sagunthala R&D Institute of Science and Technology, Chennai, Tamil Nadu, India;* [2]*Institute of Plant and Microbial Biology, Academia Sinica, Taipei City, Taiwan;* [3]*Department of Life Sciences, National Central University, Taoyuan City, Taiwan;* [4]*Dr. NTR College of Food Science & Technology, Acharya N.G. Ranga Agricultural University, Bapatla, Andra Pradesh, India;* [5]*Computational Biology and Therapeutics Lab, Saveetha College of Physiotherapy, Saveetha Institute of Medical and Technical Sciences (SIMATS), Saveetha University, Chennai, Tamil Nadu, India;* [6]*College of Food Science and Technology, Guangdong Ocean University, Zhanjiang, China*

10.1 Introduction

Makhana (*Euryle ferox* Salisb.), sometimes referred to as foxnut or gorgon nut, is a seed derived from an aquatic crop that is a member of the Nymphaeaceae family. It typically grows in stagnant water bodies, such as lakes, ponds, and low-lying depressions. Since it is typically used as "prasad" during religious offerings, it is also regarded as the "food of god" in India. It is primarily grown in lowland ponds in West Bengal, Bihar, Odisha, and Assam. The districts of Darbhanga, Madhubani, Purnia, and Katihar in Bihar account for nearly 80% of all makhana in India (Kumar et al., 2011). However, only 13,000 to 15,000 ha are used for makhana agriculture (Singh et al., 2017). Due to the

Makhana (Euryale ferox) as a Functional Food. https://doi.org/10.1016/B978-0-443-44421-0.00019-6

presence of the "Mallah" community, which cultivates Makhana in ponds, Madhubani and Dharbanga districts are the main contributors. Makhana cultivation and harvesting are labor-intensive, tedious, and time-consuming processes (Khadatkar et al., 2015).

According to Kumar et al. (2011), the main obstacles to cultivating makhana are a lack of ownership of the pond; high operational drudgery; lack of financing facilities; lack of scientific knowledge of cultivation; absence of better variety; short lease time; and labor-intensive method. The Mallah community's sole source of revenue is makhana cultivation, but its improvement has received little attention. Since most of them do not own ponds, they do not receive fair returns. Since experienced laborers from the Mallah community typically carry out the operation from generation to generation, attention must be paid to improving makhana harvesting practices to lessen drudgery and enhance their standard of living. Mud gets into the workers mouths, noses, eyes, and ears as they are harvesting. In addition, workers frequently have skin and occupational health issues, including rashes, itching, etc., as well as injuries from the pond's sharp thorns and prickles, leech bites, etc. (Khadatkar et al., 2015).

Makhana is a healthy source of minerals, protein, and carbohydrates. In contrast to popped makhana, which has carbohydrate, moisture, protein, and fat. In addition, it helps the cottage industry, has nutritional and therapeutic value, and is advised for the treatment of respiratory, circulatory, digestive, excretory, and reproductive systems (Rathod et al., 2023). Fresh, mature seeds are used in an arboretum to conduct ex-situ culture investigations of makhana. Fresh fruit production was found to be 3.05 t/ha. The makhana seeds have higher protein and fat contents, according to the biochemical study (Alfasane et al., 2008). It was verified as better than dry fruits due to their higher protein and lower fat content (Rathod et al., 2023). The kernel comprises 16 diverse varieties of amino acids, including makhana's biological value (BV), essential amino acid index (EAAI), and chemical score (CS), which are comparable to fish (Garg et al., 2021). Makhana's protein and amino acid composition has greater alanine (Ala), arginine (Arg), and Tyrosine (Tyr) than eggs, as well as the FAO/WHO pattern. On the other side, its net protein utilization, true and apparent digestibility rate, was found to be less than that of soybean, egg, human, and cow milk. Makhana is rich in both macro and micronutrients (Rathod et al., 2023). Shankar et al. (2010) noted a decreasing trend in the purification of starch and revealed the presence of trace metals such as Cu, Na, Ca, Fe, and Mg. In addition, it has good levels of N, P, and K in its vegetative sections. The entire makhana plant is a good source of minerals. Because to the muddy and damp field conditions, the makhana crop also collects significant amounts of P, Fe, Mn, and Zn elements in biomass. According to Singh et al. (2003), makhana can grow other cereals (rice and wheat) to refill the depleted nutrients like N, P, and Zn. Because of the largest tissue concentrations of the less soluble Fe and Mn, it also significantly aids in the sustainable management of natural resources.

Moreover, the range of vitamin A and C concentration was 62.23–63.84 IU/gm and 0.18%–0.20%, respectively. Because makhana contributes organic matter to the soil, water, and plants, it is a preferable alternative crop to include in cropping systems. In both Chinese and Indian medicine, makhana is a crop that has a significant medicinal value. Numerous illnesses affecting the respiratory, circulatory, digestive, excretory, and reproductive systems are treated with it (Choudhury et al., 2026).

It can be employed in the pharmacological and therapeutic industries because it is an easily accessible source of natural antioxidants (Ugoeze and Odeku, 2025). Moreover, it is utilized to treat renal issues, chronic diarrhea, extreme leucorrhea, polyuria, gonorrhea, articular aches, micturition, seminal loss, and beriberi. It is used to treat chronic diarrhea, vaginal discharge, renal failure, ineffectiveness, and nocturnal discharges. A study by Jana and Idris, (2018a,b), makhana comprises greater concentrations of antiaging amino acids like methionine, glutamine, arginine, and cystine. Song et al. (2011) showed that several antioxidant-active components of the seed, as well as their combined antioxidant activities, may be significant factors in the seeds' capability to manage proteinuria. Consequently, it might be exploited to avert diabetic nephropathy since it is a good source of natural antioxidants (Al-Waili et al., 2017). This chapter inspects the protein quality and techno-functional features of makhana flour, its starch chemistry, and functional features. It also enlarges the lipid stability and fatty acid profile, highlighting its significance for food processing and storage as well as their nutritious value. Furthermore, it examines new advances in processing technology, the development of value-added products, and possible applications of makhana in functional foods and nutraceutical formulations, prominence its involvement to economic growth, sustainability, and well-being.

10.2 Starch chemistry of makhana flour

10.2.1 Composition and structure of starch granules

Makhana (*Euryale ferox*), also referred to as fox nut or gorgon nut, is an aquatic cash crop with considerable economic and nutritional value, especially in South Asia. It is a useful component of both conventional and functional diets because of its high carbohydrate, moderate protein, and extremely low-fat content. Despite its higher market value, a significant portion of its production is lost after harvest, roughly 1.72% of the total production is unused, while around 80% of the supply is sold as surplus, signifying room for further processing and value accumulation (Khadatkar et al., 2020a,b). Makhana's discrete nutritional profile, health benefits, and adaptability in food uses are crucial factors contributing to its increasing claim.

10.2.2 Proximate structure and nutritious profile

The chemical composition of makhana differs depending on the way it is treated. The usual configuration of raw makhana contains carbs as 76.9%, moisture as 12.8%, protein as 9.7%, fat as 0.1%, and trace amounts of minerals like iron (0.0014%), calcium (0.02%), and phosphorus (0.9%) (Khadatkar et al., 2020a,b). Popping is a common treatment method that transforms the composition to 84.9% carbs, 4% moisture, 9.5% protein, and 0.5% fat, which is evidence of dehydration and structural expansion of starch granules. Also, the calorific value is quite different, with raw makhana comprising about 362 kcal/100 g and puffed makhana comprising 328 kcal/100 g, indicating that it is a low-fat and high-energy food (Khadatkar et al., 2020a,b).

Makhana seeds (~78%) are made up of carbohydrates, and these are mostly in the form of starch granules. This is because of its high starch, and it helps to support its functional properties, such as gelatinization, viscosity development, and digestibility behavior. In addition to carbs, makhana comprises 10%–12% protein and possesses an excellent amino acid composition, which makes it have a high nutritional value even though it is lower than that of some legumes (Tehseen et al., 2020a,b). The 16 amino acids that are in makhana kernels include significant levels of arginine, alanine, and tyrosine required by the body to complete physiological functions (Khadatkar et al., 2020a,b). The protein quality indexes of makhana are significant, and the essential amino acid index (EAAI) of 89%–93% compares favorably to some of the staple foods such as wheat and rice (Tehseen et al., 2020a,b). Similarly, other measures, such as net protein utilization (NPU), chemical score (CS), and biological value (BV), are good indicators of protein efficiency, although to a lesser degree than animal sources, such as milk and eggs. Its biological usefulness, however, can be compromised to some extent by the relatively large leucine-to-isoleucine ratio (Kumari et al., 2019a,b,c).

10.2.3 Composition of mineral and vitamin

Makhana also has essential minerals and micronutrients, and its nutritional quality depends on such trace elements as copper (Cu), sodium (Na), calcium (Ca), iron (Fe), and magnesium (Mg), which are found in measurable quantities (Tehseen et al., 2020a,b). Higher levels of phosphorus (P), iron (Fe), and zinc (Zn) are often caused by farming in nutrient-rich, muddy aquatic environments, which enhance mineral uptake. In addition, vitamins in makhana increase immune response and antioxidant ability, including vitamin C (0.18 IU/g) and vitamin A (62.23–63.84 IU/g) (Khadatkar et al., 2020a,b).

10.2.4 Starch granule's structure

Amylose (linear chains) and amylopectin (branched chains) make up the semicrystalline granules that make up makhana's starch. The hierarchical structure with alternating crystalline and amorphous regions of these polymers determines the physicochemical properties of starch. Although amorphous regions allow water to permeate and swell during heating, crystalline regions that are mainly made of amylopectin double helices give structural rigidity (Khadatkar et al., 2020a,b). The functional activity of makhana starch depends mainly on the granular structure. Gelatinization is a process of heating the granules in the presence of water that involves swelling, loss of crystallinity, and leaching of amylose. This enhances the viscosity and paste formation, which is very important in many food uses. When the starch cools down, amylopectin and amylose chains will rearrange to form a more structured form, which helps in the formation of resistant starch and gel firmness (Khadatkar et al., 2020a,b).

10.2.5 Assessment with other nuts and their functional implications

Due to its higher nutritional value, makhana is occasionally equated to other dry nuts like almonds and walnuts, even though they are very different in composition. Makhana is also a better low-fat choice since it is very low in fat, unlike these high-fat nuts (Tehseen et al., 2020a,b; Lokendrakumar et al., 2016a). It is appropriate in functional foods, nutraceuticals, and beneficial diets because of higher carbohydrate and minimal protein content, in combination with crucial micronutrients. Also, due to its high nutritional and economic value, makhana is often called black diamond (Kumar et al., 2016a,b,c). It plays a significant role in balanced diets due to its great digestibility and intermediate efficiency in protein use (Khadatkar et al., 2020a,b). In conclusion, the functional, nutritional, and industrial value of makhana starch grains highly relies on their composition and structure. High-energy, low-fat, and health-promoting food products utilize starch because of its abundance and a balanced protein profile and a plenteous micronutrient profile. It possesses a semicrystalline starch structure that regulates key activities such as retrogradation and gelatinization, which influence the shelf stability, texture, and digestibility (Kumari et al., 2019a,b,c). Conclusively, these features make makhana a promising ingredient for value-added and food processing, especially in the growing market.

10.2.6 Physicochemical features

The semicrystalline structure of granules, amylose–amylopectin ratio, and molecular structure play a significant role in determining the physicochemical properties of makhana starch, which, in turn, influence its rheological, digestibility, and thermal properties. Total starch constitutes about 65%–70% of the makhana starch, and the crystalline and amorphous regions of amylose (23%–25%) and amylopectin (75%–77%) alternate. This structural arrangement regulates its functional reaction to heat and hydration. Although amylopectin contributes to the swelling of the granules and viscosity building, the amylose content is relatively mild, which helps the formation of the gel and retrogradation process (Zhu, 2015; Biswas et al., 2020). Differential scanning calorimetry (DSC) studies indicate makhana starch has high enthalpy and gelatinization temperatures compared with several commercial starches, which means it has greater structural stability. This behavior is caused by strong intra- and intermolecular hydrogen bonds in crystalline regions; the destruction of these bonds requires more heat energy (Zhao et al., 2016a,b). Water initially penetrates the amorphous spaces in the process of gelatinization, and the granules become swollen and lose their crystallinity. This is then followed by leaching amylose, which significantly increases the viscosity of the paste. It is due to the swelling behavior that is controlled by makhana starch that moderate and stable viscosity profiles suitable for food application are achieved through avoiding over-rupture of the granules (Qu et al., 2025a,b).

The pasting properties of makhana starch, including the peak, breakdown, and setback viscosity, also underscore its usefulness. Setback viscosity designates its capability to re-associate during cooling, while minor peak viscosity and minor interruption reflect stability to heat and shear stress (Qu et al., 2025a,b). Makhana-based starch is suitable for food systems, including gels and thickened formulations, due to its characteristics. Starch is retrograded when it cools down to the gelatinized state. In the makhana-based starch, the reaction is bi-phasic, where the amylose substances reassociate rapidly, and amylopectin chains recrystallize at a lesser rate. It causes reduced solubility, higher crystallinity, and hardness of the gel during storage (Huang et al., 2021a,b; Bhat and Arya, 2023a,b). Most importantly, retrogradation yields resistant starch, which reduces the glycemic index by inhibiting the digestion of enzymes. This characteristic enhances the nutritional value of foods made up of makhana by promoting a controlled release of glucose and enhanced gut health (Huang et al., 2021a,b). Processing conditions have a great impact on these physicochemical characteristics. Traditional methods, such as popping, lead to major structural changes, shifting the viscosity and digestibility characteristics, and raising the extent of gelatinization (Mal and Pandey, 2025a,b). Even though this kind of processing makes the water more absorbent and digestible, it might also ruin the ordered crystalline space, which might affect the level of resistant starch. The results indicate that makhana starch is sensitive to mechanical and thermal treatment and can be used in personalized food production. Makhana-based starch can be used as a functional ingredient in nutraceutical preparations and as an ingredient of functional foods due to its physicochemical properties, which are high thermal stability, controlled swelling, lower viscosity, and high retrogradation capabilities. Its increasing significance in food science research is underscored by its capability to create a compromise between processing stability and nutritional benefits.

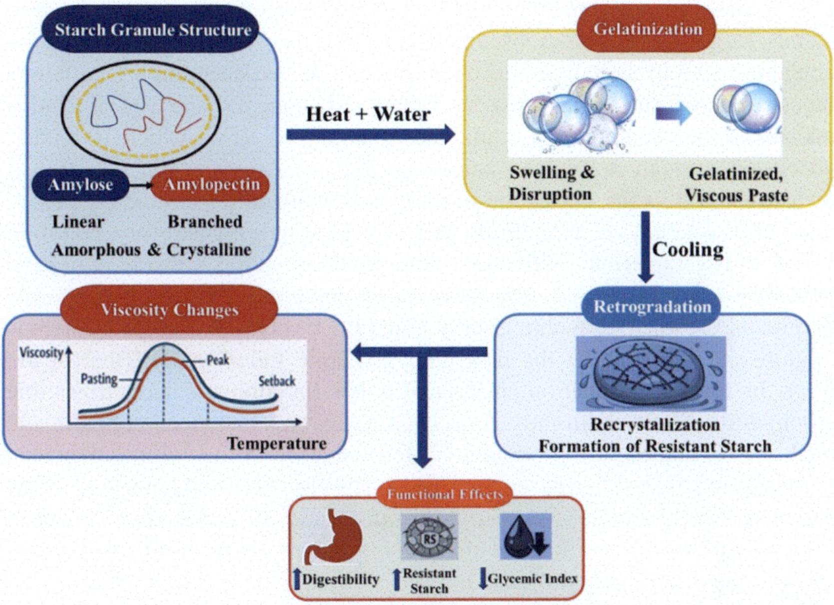

FIGURE 10.1 Structural and functional relationship of makhana starch.

Starch in makhana flour consists of amylose and amylopectin organized in semicrystalline granules. On heating with water, granules swell and gelatinize, increasing viscosity, while cooling leads to retrogradation and formation of resistant starch (Huang et al., 2021a,b). These structural changes affect digestibility, enhance resistant starch content, and help lower the glycemic index (Fig. 10.1). Retrogradation in makhana starch follows the classical biphasic mechanism, rapid amylose reassociation followed by slower amylopectin recrystallization, leading to increased crystallinity, gel firmness, and decreased solubility during storage (Bhat and Arya, 2023a,b). Notably, processing methods like popping change the viscosity and digestibility profiles and dramatically increase the degree of gelatinization (up to ~10-fold), further highlighting how sensitive makhana starch functionality is to structural and thermal changes (Mal and Pandey, 2025a,b). Altogether, these physicochemical features highlight makhana starch's applicability for usages in food and nutraceutical compositions that call for regulated viscosity, gel strength, and storage firmness.

10.2.7 Functional effects

Makhana starch's digestibility profile, resistant starch (RS), as well as glycemic response, are directly linked to the functional effects, which are controlled by their molecular architecture and structural variations stimulated by processing. Target-based alterations significantly decrease RDS while growing RS content (~14.88%), reducing enzymatic hydrolysis and encouraging sustained glucose release during in vitro assimilation (Zhang et al., 2021). Native makhana starch comprises a significant quantity of RDS, which is readily hydrolyzed through α-amylase, releasing glucose. An improved quantity of amylose as well as the expansion of more systematized crystalline structures, which obstruct enzyme availability and decline starch digestibility, resulting in RDS to RS transition. With steric hindrance, as well as enzyme inhibition mechanisms, sophisticated structural transformations, for example, starch, polyphenol complexation, can greatly boost RS levels (~50 to 55%). There is a significant reduction in the projected glycemic index (pGI) (Bhat et al., 2022). In terms of nutrition, makhanas's RS functions as an indigestible carbohydrate. Short-chain fatty acids (SCFAs), which enhance gut health and metabolic regulation, are produced via fermentation in the colon rather than the small intestine (Sharma et al., 2016). RS-enriched makhana has a low estimated pGI (less than 50) and significantly lower digestion rates (less than 20%), which makes it a low-GI dietary ingredient appropriate for diabetic and functional food preparations, and it is studied in both in vitro and in vivo (Mal and Pandey, 2025a,b). Moreover, the human studies of roasted makhana prove that the product has a low GI compared with most snacks made from cereals. This is likely due to its starch-protein matrix, phenolics, and low lipid content, which affect the kinetics of starch hydrolysis (Kapelko-Żeberska et al., 2024). Also, the changes of granule integrity and lessening in antinutritional factors, other processing options like roasting, germination, can improve digestibility, without compromising a controlled glycemic response (Ngo et al., 2022). Makhana is a low-to-moderate glycemic functional food with a higher potential of forming slow-release carbohydrate systems, diabetic-friendly formulations, and gut-health-promoting products because of the interaction of digestible starch fractions, RS formation, and structural changes.

Makhana-based starch has important physiological and techno-functional advantages that expand its use in functional food systems, in addition to its ability to regulate blood sugar levels and enrich RS. Makhana can be used in low-glycemic bread and extruded products because the regulated processing produces RS3, which improves textural qualities, decreases swelling strength, and increases thermal stability (Zhang et al., 2021). Also, lipids react with amylose during heat treatment to form amylose lipid complexes, which further inhibit enzymatic hydrolysis and block the release of glucose (Bhat et al., 2022). Makhana starch is also added to enhance the functional performance of composite food systems. The interactions with dietary proteins and fiber reduce the starch digestibility and increase the stability by forming a compact matrix impeding the access of enzymes (Singh et al., 2010). These structural modifications enhance the stability of products and extend their shelf life, besides enhancing metabolic reactions. The matrices are able to regulate viscosity and water absorption, which are significant in food processing and sensory quality. In terms of nutrition, makhana's RS offers prebiotic nutrient support that particularly encourages the development of advantageous gut flora. During its colonic fermentation, SCFAs are produced, in particular, butyrate, which is required to maintain colonic health, reduce inflammation, and increase insulin sensitivity (Sharma et al., 2016; Fuentes-Zaragoza et al., 2011). This goes to show that it can be used to treat metabolic disorders like type 2 diabetes and obesity. Other alternatives to alter the structural and digestibility of makhana starch are offered by the emerging processing technologies, including enzymatic and high-pressure processing. These methods allow for accurate RS generation management without sacrificing sensation quality (Ngo et al., 2022). All these properties ensure that makhana starch is a potential ingredient in the development of the next-generation functional meals that focus on gut health, glycemic regulation, and the prevention of chronic diseases.

Table 10.1 shows that differences in starch composition and functional behavior of different flours are largely determined by the amylose and amylopectin content of their composition. The viscosity of makhana flour is good, and gluten-free option as the starch content is moderate with the amylase and amylopectin in balanced proportions. Rice flour contains starch and amylopectin at a high level but has low shear stability, which makes it develop strong swelling as well as a soft texture. Wheat flour is unique due to the presence of gluten that reacts with starch to form elastic dough structures required in baking (Zhu, 2015). Unlike potato starch, which is a source of amylopectin and therefore has impressive swelling and high viscosity, maize flour, which is relatively higher in amylose content, gives harder gels. Flours such as peas and lentils, however, contain greater amounts of amylose and less starch, which elevates RS, retards digestion, and enhances nutritional value. The proportion of amylose to amylopectin has a significant impact on the texture, digestibility, and application of these types of flours in food systems, which are in equal amounts. Makhana flour is an intermediate amylase with a moderate starch content that provides it with even gel-forming properties and viscosity compared with other flours. Legume flours (pea and lentil) contain less starch and more amylose, which enhances RS and slows digestion, compared with cereal flours (rice and wheat), which have more starch (Table 10.1). Overall, makhana stands out as a gluten-free substitute with useful qualities that fall between high-starch cereals and high-amylose legumes (Biswas et al., 2020).

TABLE 10.1 Comparative starch composition of makhana and other flours.

Flour source	Total starch (%)	Amylose (%)	Amylopectin (%)	Notable characteristics	References
Makhana (*Euryale ferox*)	65–70	23–25	75–77	Moderate amylose; good viscosity; gluten-free	Zhu (2015), Biswas et al. (2020)
Rice flour	80–85	15–25	75–85	High swelling; low shear stability (varies with type)	Singh et al. (2003)
Wheat flour	70–75	20–30	70–80	Gluten present; elastic dough structure	Tester et al. (2004)
Maize (Corn) flour	70–75	25–28	72–75	Higher amylose; firmer gels	Singh et al. (2003)
Potato starch	75–80	20–23	77–80	High swelling power; high viscosity	Singh et al. (2003)
Pea flour	45–50	30–40	60–70	High amylose; high resistant starch	Tester et al. (2004)
Lentil flour	45–50	30–35	65–70	Protein-rich; slower digestibility	Tester et al. (2004)

10.3 Functional properties of makhana flour

The functional properties of Makhana play a pivotal role in determining its suitability for various industrial food applications, specifically in baking, extrusion-based snack production, and nutraceutical formulations. Most of its properties are primarily influenced by starch-rich composition, unique microstructure, structural and physiochemical transformations that occur during thermal processing (popping), a high-temperature short-time treatment (HTST) method. The biochemical attribute underlying this is starch gelatinization, the disruption of crystalline regions resulting in the formation of a porous matrix, which enhances hydration properties and thus alters rheological behavior. Studies with (Singh et al., 2020a,b) demonstrated that popped makhana exhibit alteration in starch granule morphology, including swelling, rupture, and increased amorphous regions that directly influence water interaction and pasting characteristics.

Water absorption capacity is one of the most critical functional attributes that determine the applicability of makhana in food systems. Quantitative studies of Makhana flour reported high water absorption capacity (WAC). Research findings indicate the WAC of pure makhana flour is approximately 6.39 g gel/g, which is comparatively higher than that of wheat (1.39 g/g) or besan (1.34 g/g). Whereas the moisture content and the bulk density were found to be 9.15% and 696.74 kg/cm^3 (Jana et al., 2019). This elevated WAC to that of the conventional sources is attributed to the presence of exposed hydroxyl groups in gelatinized starch and increased surface area due to its porosity, which facilitates hydrogen bonding with water molecules (Jha et al., 2011). Also, the capillary structure formed during popping helps as a water retention reservoir, which is very useful in doughs requiring moisture stability.

The water absorption index (WAI) further reflects the ability of makhana flour to imbibe water and swell under heating conditions. It is also crucial in its role as a thickener and binder in food systems. This property arises from the porous structure of the endosperm in nature and the melting of starch crystallites during the popping process, which creates capillaries that readily associate with water (Wardhan et al., 2025). The WAI also increases proportionally to the popped makhana flour in composite blends, and this property is underlying the starch content in makhana, predominantly composed of amylopectin and moderate amylose content, where amylose leaching and depolymerization during popping (Wardhan et al., 2025). Swelling capacity was also found to increase at high temperatures, indicating that heat-moisture treatments improve flour's hydration efficiency (Kumar and Saini, 2016). Experimental evaluations using standard gravimetric methods have demonstrated that WAI increase is proportional to temperature, indicating progressive gelatinization and molecular relaxation of starch chains. This swelling behavior is primarily controlled by amylopectin branching and disruption of crystalline lamellae, which allows water molecules to penetrate the granule matrix (Kaur et al., 2015). Furthermore, differential scanning calorimetry studies have shown that the gelatinization onset temperature of makhana starch lies between 65–70 °C, where it peaks at 80–85 °C, confirming thermal sensitivity and hydration efficiency. Upon popping, due to partial gelatinization and weakening of intermolecular hydrogen bonds, the swelling capacity is enhanced. Experimental data show improvement by 40% comparing native starch that highlights thermal processing on functional performance (Singh et al., 2016). The water solubility index (WSI) is another important parameter that reflects the extent of molecular degradation and solubilization of starch components. Studies find that WSI substantially increased in popped makhana due to amylose leaching and formation of low molecular weight dextrins (Verma et al., 2020).

The pasting profile of makhana flour is typically analyzed using Rapid Visco Analyzer (RVA), providing insights into its gelatinization behavior and suitability for thermal processing. In this testing, it possesses a high pasting temperature (around 67.8–95°C, depending on variety and blend) and low viscosity (Maibam and Deka, 2022). This high temperature suggests EFKS (Euryale ferox kernel starch) could be prioritized for food products requiring high-temperature cooking without losing structural integrity. Further research on wheat makhana blends has shown that there is a drop in the peak of viscosity, trough viscosity, and breakdown viscosity with the increased makhana flour ratio. Accounting for its rheological behavior, this property could be modified by substitution levels. In composite blends, the final range lies between 3042–3651 cP, picturing the stability of starch granules after gelatinization (Kumar and Saini, 2016). The peak viscosity in popped makhana is lower compared with native starch since the starch is partially gelatinized during popping (Jana et al., 2019).

Rheological characterization of makhana flour dispersions reveals non-Newtonian shear thinning behavior. The apparent viscosity decreased with increasing shear rate, exhibiting pseudo-plastic behavior, which facilitates mixing, extrusion, and pumping operations. Oscillatory rheological studies have found that makhana gels exhibit a higher storage modulus (G') than a lower modulus (G''). This proves a weak gel network dominated by elastic behavior (Kumar et al., 2020).

Makhana flour's functional properties have been studied most thoroughly in bakery systems, as highlighted in Fig. 10.2. Partial substitution of wheat flour improves mineral and antioxidant content without compromising sensory quality. This flour serves as a versatile ingredient across many different food sectors. In the baking industry, makhana flour is used to fortify cookies, cakes, and bread (Patel et al., 2024), wherein found that replacing up to 20% of makhana flour in bread maintains natural sensory attributes to that of the normal composition while enhancing nutritional value. In cookies, it increases protein and mineral content, but the drawback is that it also increases hardness

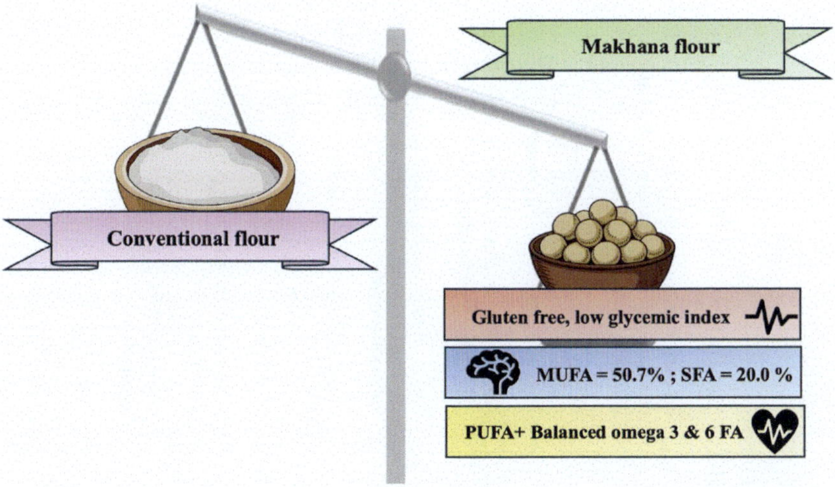

FIGURE 10.2 Makhana flour as a stable staple alternative for conventional flour sources.

and chewiness due to the thickness of the resulting dough (Kapoor et al., 2023; Kaur et al., 2023; Kumar, 2015). Makhana flour is profoundly found in making flatbreads (chapatis), fried snacks (pakoras), and puffed snacks due to its high expansion capacity and low oil absorption (a capacity of 2.09 g gel/g) (Jana et al., 2019). Extrusion trials show makhana-based snacks reach higher expansion ratios and lower bulk densities, the pairing responsible for light, crispy textures. Low fat content also limits oil uptake during frying, reducing the final lipid load compared with conventional snack bases (Devi et al., 2020). Makhana wheat chapati with a 1:1 ratio is an excellent high-calorie (calorific value of 317.24 cal/100g), low sugar option prevailing for diabetic and elderly patients. Due to its low glycemic index (GI) and high antioxidant properties, it is used in formulating health supplements and functional sweets such as Barfi (19.33% sugar, 5.40% protein) and Kalakand (low free sugar 16.66%, high protein—11.53%), often valued for their high protein-low free sugar and bioactive compound content (Jana et al., 2019).

10.4 Protein quality and functional attributes of makhana

The protein content in makhana typically ranges from 9.15%–11.16%, this variations depend on genotype, environmental conditions, and processing methods. Also, unlike cereal proteins, makhana proteins are globular in nature with a low prolamin fraction that mainly influence its gluten free status and enhanced digestibility (Jha and Sharma, 2010). Table 10.2 describes the amino acid profiles of makhana. Although the protein availability range is slightly lower to that of other cereals, the quality of the protein is superior due to an

TABLE 10.2 Amino acids profiling of makhana (Jha et al., 1991).

Amino acid	Quantity		Classification	FAO/ WHO	Significance
	Raw	Fried			
Arginine	15.19	16.07	Essential	5.2	Precursor to nitric acid and creatine—increases the elasticity of arteries and veins and blood flow
Glutamic acid	16.64	17.06	Nonessential	14.7	Antiaging
Leucine	8.34	8.85	Essential	7	Body growth and development
Methionine	3.06	2.95	Essential	3.5	Precursor to creatinine with arginine, aids hair, skin, nails health
Lysine	3.79	4.69	Essential	5.4	Fortify cereals lacking lysine
Isoleucine	4.18	4.8	Essential	4	Supports growth
Valine	5.18	5.49	Essential	5	Branched-chain amino acid
Cysteine	0.75	1.21	Semi essential		A precursor to taurine, which reduces diabetes
Threonine	3.34	3.51	Essential	4	Component for a unique protein assemblage
Phenylalanine	5.78	6.12	Essential	3.05	Component for a unique protein assemblage

Tyrosine	6.38	2.91	Nonessential	3.05	
Histidine	3.15	3.12	Essential	2.5	Support immune function
Alanine	5.50	5.84	Nonessential	6.1	
Asparatic acid	5.76	5.05	Essential/Nonessential	7.7	Chemical messenger in the brain; antioxidant
Glutamic acid	16.64	17.06	Nonessential	14.7	Antiaging
Glycine	3.01	3.28	Nonessential	2.2	Component for unique protein assemblage
Proline	4.00	3.24	Nonessential	10.7	Body growth and development
Serine	5.05	5.64	Nonessential	7.7	Component for unique protein assemblage
Essential amino acid index (EAAI)	93.63%	89.97%	Protein quality index		Indicates high protein utilization potential
Biological value (BV)		55	Protein availability and usage measurement		Better to soybean (50.7 g/16 g N) (Sodi and Kumar, 2019)

unique assemblage of amino acids with high levels of arginine (significant for nitric oxide synthesis, enhancing blood flow elasticity and vascular function), glutamic acid (neurotransmission and protein metabolism), leucine (also isoleucine in the ratio 1.9% (Nath and Chakraborty, 1985), supports muscle protein synthesis through activation of the mTOR signaling pathway and metabolic regulation) and phenylalanine, contributing to its nutritional superiority among plant based sources.

Makhana protein digestibility is primarily a function of its structural conformation and what it lacks. Studies find that almost all the In vitro assays on makhana had consistently return values above 80%–85%, indicating efficient enzymatic hydrolysis and high bioavailability. The main reasons behind this are the absence of gluten and relatively low concentrations of trypsin inhibitors and phytates, antinutritional factors that reduce digestibility in many other plant protein sources (Chandra et al., 2015). Popping pushes digestibility higher still: heat denatures protein structures, exposing peptide bonds and improving proteolytic enzyme access.

And also, the high amounts of arginine and methionine, which are precursors to creatine, aid in skin health (Nath and Chakraborty, 1985). Makhana-derived protein is attributed with high digestibility (values exceeding 80%) and gluten-free, making it a staple food source for celiac patients (Joye, 2019). Whereas the amino acid profiling is more structured compared with cashew or almond, it also provides essential antioxidants that neutralize free radicals to fight aging, attributed by the glutamine/glutamic acid ratio = raw makhana (16.64%) and in popped seeds (17.06%) (Jha et al., 1991), which has a remarkable antiaging effect. As a very good immune stimulant, *Euryale* raw powder or popped stimulates hormonal immunity and is effective for women at the postnatal stage. It is effective against arthritis, impotence, and premature aging (Puri et al., 2000).

Protein solubility is a key determinant of functional performance and is highly pH-dependent, where the maximum solubility is observed in alkaline conditions, pH around 9–11, and minimum solubility is observed near the isoelectric point (pH 4–5). This behavior is typical in plant-based protein sources, governed by electrostatic interactions and protein charge distribution that influence protein-water interactions and aggregation behavior. This is also crucial for maximizing protein recovery during extraction (Singh et al., 2016)

Makhana proteins exhibit significantly higher emulsifying activity and stability, where it absorbs at oil-water interfaces, reduce interfacial tension, forms viscoelastic films, thus making them suitable for plant-based meat and dairy alternatives. But the foaming capacity is lower here compared with wheat gluten, thus struggling to resist rapid unfolding at the <>air-water interface. This is due to the globular structure that resists the rapid unfolding, which is required for the formation of stable foam films. Thus, the capacity decreases as the substitutional level increases in composite flours (Kumar and Saini, 2016). This is a real limitation for applications where aeration matters. On standard nutritional indices, makhana protein holds up reasonably well. PER is comparable to legumes, indicating capacity for growth and tissue maintenance. Biological Value sits in the 50–60 range, reflecting moderate efficiency of protein retention after absorption (FAO, 2013). Specific PDCAAS data for makhana are sparse in the published literature, but digestibility figures and the amino acid profile together suggest a moderately high score. Lysine is the limiting amino acid, though its concentration exceeds typical cereal levels, which improves the overall balance. While explicit PDCAAS numbers are not detailed in all traditional reviews, the high EAAI and chemical score (CS) of 70 for raw makhana indicate excellent protein utilization potential for growth and repair (Jana et al., 2021).

This is caused by the globular protein in makhana that may resist surface denaturation more than wheat gluten (Griffith et al., 1998). Makhana consistently outperforms standard cereals in protein quality indices, where the protein quality in makhana is reflected in its high essential amino acid index (EAAI), typically around 93.63%, surpassing many predominantly available traditional cereals such as rice (83%), wheat (65%), and Bengal gram (81.5%). The biological value of fried makhana is approximately 55%, which is lower than milk (85.5%) (Joye, 2019). In food processing, makhana protein has found application in protein-enriched bakery products, dairy analogues, and nutraceutical formulations. High solubility and emulsifying capacity suit it to beverage systems and protein isolates. The foaming limitation is real but addressable; pH adjustment and enzymatic treatment have both been shown to partially compensate for the globular structure's resistance to interfacial unfolding.

10.5 Fatty acid profile of makhana

Although the lipid fraction of makhana is quantitatively minimal, it exhibits a nutritionally significant fatty acid composition that contributes to its main attribute as a low-fat functional food. The total lipid estimate in makhana seeds ranges about 0.1%–0.5%, as reported by Nath et al. (2015) in their compositional analysis of aquatic seed crops, making it stand out from the conventional oil nuts and seeds. Despite the low lipid profile, the important factor is the finding of the predominant presence of unsaturated fatty acids, particularly polyunsaturated fatty acids (PUFAs), in the qualitative analysis, which plays a major role in human health and metabolic regulation. Table 10.3 shows the fatty acid and lipid profile of makhana.

Furthermore, Gas chromatography-mass spectrometry (GC-MS) profiling of makhana seed extracts reveals a spectrum of fatty acids that include palmitic acid (C16:0), stearic acid (C18:0), oleic acid (C18:1), and linolenic acid (C18:2), with linolenic acid as the major proportion of the unsaturated lipid fraction (Jha et al., 2017). The availability of these fatty acids indicates even traces of lipid content in makhana, contributing to its membrane fluidity, lipid

TABLE 10.3 Fatty acid composition and comparative lipid profile.

Fatty acid type	Quantity	Classification	Significance	References
Total fat Raw makhana	0.46–0.51 g	Proximate lipid composition	Weight management, cardio health maintenance	Wardhan et al. (2025)
Total fat Popped	0.33 g	Proximate lipid composition	Ideal for heat patients, weight management	Wardhan et al. (2025)
Linolenic acid	Confirmed with GC-MS	PUFA	Blood pressure & cardiovascular function regulation,	Sinha and Jha (2025)
Linolenic acid (omega-3)	Identified (GC-MS)	PUFA	Antiinflammatory and cardioprotective benefits	Sinha and Jha (2025)
SFAs	Minor components (palmitic, myristic)	SFA	Prevent arterial clogging	Sinha and Jha (2025)
Palmitic acid (C16:0)	Identified (GC-MS)	SFA	Energy source and structural component in the cell membrane	Sinha and Jha (2025)
MUFA	Very low	USFA	Prevent a spike in blood sugar level	Jana et al. (2019)
Linolenic acid (omega-6)	Identified (GC-MS)	PUFA	Regulate blood pressure, cardiovascular function	Sinha and Jha (2025)

signaling, and metabolic homeostasis. Here, the ratio of unsaturated fatty acids is higher than that of saturated fatty acids (SFAs), making it more advantageous for cardiovascular health.

This minimal concentration of SFAs in makhana is primarily represented by palmitic acid and stearic acid, which together constitute a minor fraction in total lipid availability. Although palmitic acid is controversial since it is associated with hyperlipidemic effects with high intake amounts, its presence in makhana is in near-negligible quantities, and therefore, the physicochemical impacts from this compound are minimal. Stearic acid, on the contrary, is considered metabolically neutral due to its property of rapid conversion to oleic acid in vivo, thus reducing cardiovascular risk further (Mensink, 2016). This low SFA content perfectly aligns with the dietary recommendations of nutritionists, since reduced intake prevents atherosclerosis and coronary heart disease.

The next important class of fatty acids is monounsaturated fatty acids (MUFAs), available as oleic acid in makhana in moderate amounts. Oleic acid is known for its role in improving lipid profile by maintaining high-density lipoprotein (HDL) but significantly lowers low-density lipoprotein (LDL) cholesterol levels, making it advantageous with cardioprotective effects. In the analytical studies foreseen by Kumari et al. (2019a,b,c), demonstrated that the MUFA fraction of makhana is lower than that of the oil-rich nuts, but significantly contributes to the overall functional lipid profile of makhana. And also, the presence of oleic acid enhances oxidative stability, as MUFAs are less prone to lipid peroxidation compared with PUFAs.

The most nutritionally significant component of lipid class in makhana is the PUFAs fraction, which is dominated by linolenic acid (omega-6) and trace levels of alpha-linolenic acid (omega-3). This particular fatty acid is in high demand since it cannot be synthesized endogenously but has various functions, including eicosanoid synthesis, inflammatory regulation, and maintenance of epidermal barrier function. Studies on makhana's lipid composition have consistently reported the predominant availability of linolenic acid that accounts for a significant proportion of total lipids despite the low absolute fat content (Singh et al., 2018).

The presence of omega-3 fatty acids, although minimal, still contributes to the antiinflammatory and cardioprotective potential of makhana. Alpha linolenic acid serves as a precursor molecule for the synthesis of long-chain omega-3 fatty acids such as EPA and DHA (eicosapentaenoic acid and docosahexaenoic acid), two main fatty acids involved in neuroprotection, anti-inflammatory signaling, and cardiovascular regulation. This balance between omega-3 and omega-6 fatty acids is crucial in determining health status, where a healthy human should always require an optimal ratio, thus preventing chronic inflammatory conditions. Comparing the other processed snack foods, makhana exhibits a favorable omega6-omega3 ratio, thus aiding in providing functional nutrition while also quenching the craving for a better snack (Simopoulos, 2016).

The breakthrough in makhana lipidomics comes from understanding its metabolomic profiling studies, which helped in identifying over 30 bioactive metabolites that include fatty acid derivatives and lipid-associated compounds. These metabolites are implicated in antioxidant activity and cellular protection against oxidative stress, further enhancing the nutraceutical value of makhana (Dutta et al., 2020). The antioxidant potential of these lipid-associated compounds is crucial in mitigating lipid peroxidation and maintaining cellular integrity. Figure 10.3 illustrates the nutritional comparison of makhana in terms of conventional resources.

From another perspective, from the physiological view, the low-fat content combined with a favorable fatty acid profile together make makhana stand as an ideal staple for weight management and metabolic health. Where the conventional resources, such as almonds and cashews, uphold a significant proportion of calorific value from the available fats, makhana provides energy from the carbohydrates with minimal contribution from lipids. This unique macronutrient distribution reduces calorie density while preserving the benefits from essential fatty acids, a perfect choice for hypocaloric diets and diabetic nutrition (Gopalan et al., 2017).

The comparative analysis of makhana with that of other nuts reveals that cashew and almonds, even though containing higher PUFAs and MUFAs, the substantial amounts of total fat lead to high calorie intake. But in the case of makhana, the same is achieved with low-fat, high-micronutrient and bioactive compound content, reducing the potential risks of higher lipid consumption. Studies confirm that makhana contains less than 1% fat,

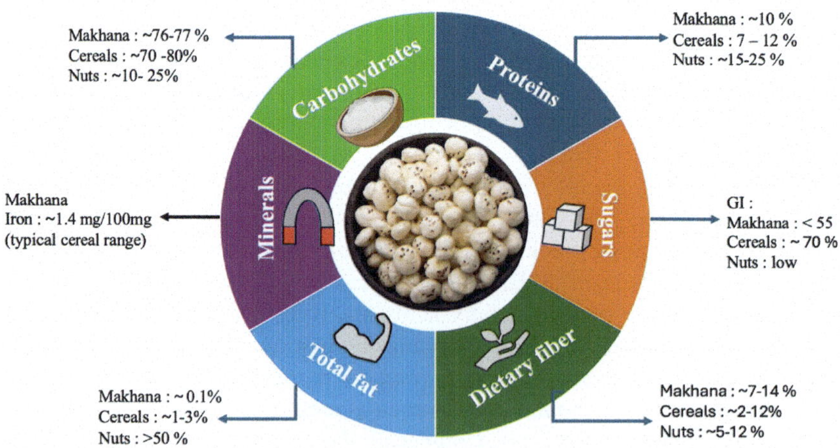

FIGURE 10.3 Nutritional comparison wheel of makhana versus conventional sources.

compared to 50%–60% in almonds and cashews, thus recognized as a heart-healthy diet (USDA food Data Central, 2020).

In terms of digestibility, enhancement comes from two factors: one from the low concentration and the other from the association with easily digestible starch matrices. These simple lipid protein interactions slow down enzymatic hydrolysis and absorption, making it rapid and facilitating easy digestion with less metabolic burden. This is one of the notable properties in the elderly diet chart, and individuals with compromised digestive function can reduce gastrointestinal discomfort. This has drawn the attention of many industries toward makhana in developing low-fat snack formulations. Also, inclusion of makhana into extruded products has been shown to reduce oil uptake during frying, reducing lipid content and improving nutritional quality. This property is attributed to the hydrophilic nature of makhana starch, limiting oil absorption and enhances product crispiness, making it a base for healthier snack alternatives (Maibam et al., 2021).

Makhana's fatty acid composition also tends to have antiinflammatory and antioxidant relevance, as well as cardio-metabolite. Linoleic acid, being the major constituent, modulates inflammatory pathways; associated phenolic compounds raise antioxidant capacity further. Studies by Verma et al. (2020) found that makhana-based diets reduced oxidative stress markers and improved lipid profiles in experimental subjects. The fatty acid profile of Euryale ferox is, in short, better than the raw numbers suggest. Low total lipid content, unsaturated fatty acid components predominance, and a much more favorable omega-6 to omega-3 ratio combined with low caloric density and high digestibility altogether make it a functionally useful ingredient not just in the food industry, but also a promising element in clinical and preventive nutrition research.

10.6 Lipid stability in makhana products

Makhana are the edible seeds of Euryale Ferox Self- Pollinated Aquatic cash crop that belongs to the family of Nymphaeaceae, grown in stagnant Perennial water bodies like ponds, lakes, and also in the shallow wetlands of India, China, and Southeast Asia. In India, approximately 80% of the whole production of processed Makhana is produced in the districts of Darbhanga, Madhubani, Purnia, and Katihar in Bihar. The other makhana cultivated states are West Bengal, Manipur, Assam, Rajasthan, Madhya Pradesh, and Uttar Pradesh (Kumar et al., 2011). There are several names of makhana in various regions of the world. It is commonly called the Fox Nut or Gorgon nut. In Punjabi, it is known as kamala da bija, Manipuri as Thangjing, Nikori in Assamese, Padma Bija in Bengali. In Korea named them as Ga-Si-Yeon, Hasu no mi in Japanese, Chinese call them Qian Shi. However, in spite of several names, makhana is a well-known, popular term used frequently worldwide.

The ICAR Research Complex for the Eastern Region in Patna, Bihar, developed the first improved variety, named Swarna Vaidehi (Kumar et al., 2016a,b,c). Another high-yielding variety, Sabour Makhana-1 (Kumari et al., 2016), was also developed by ICAR. The Cultivation of Makhana begins with the transplanting method or by another method of self-growth in the ponds through left out seed during harvesting starts in the months of January–February and it starts growing in the months of March to August. Flowering starts in the month of April. After 35–40 days of flowering, the fruits become fully developed and mature large berries with a diameter of 5–8 cm. It is spongy, spiny, and has persistent sepals on top. About 20 to 200 seeds are present in each fruit. A bright red aril is streaked around the lumpy seeds. The aril of fresh seeds decomposes and turns black after 3–4 days. The diameter of the seed ranges from 0.5 to 1.5 cm, and they are sufficiently bold with a hard outer covering. The white, starchy perisperm is the edible portion of the seed. The genus *Euryale* is monotypic, consisting of only one species. Makhana is a diploid plant with a chromosome constitution of $2n = 2x = 58$. The harvesting of raw makhana seeds is traditionally carried out by the "MALLAH" Community and involves a highly labor-intensive manual process, typically conducted during the months of August to November. The diver dwell into the pond or lakes by collecting the ripe seeds, which are located at the bottom. The traditional basket tool used to separate the seeds from mud is called GAJA, with a capacity of 2–5 kg of raw makhana seeds.

10.6.1 Processing of makhana products

The raw material used for makhana products is raw makhana seeds of round shape with an outer layer that ranges in color from black to brown (Rathod et al., 2023). These seeds are sundried, and later it undergoes several unit operations like roasting, frying, and extrusion.

Due to its high protein and low-fat content, the makhana is similar to almonds, walnuts, cashew nuts, and coconut. Makhana reported 16 types of amino acids in the kernel. The essential amino acid index (EAAI) ranges of 89%–93%, chemical score (CS), and biological value (BV) of <<>>Makhana are similar to fish (Khadatkar et al., 2020a,b). Compared with eggs, the higher content of arginine, alanine, and tyrosine is present profoundly in makhana, which has antiaging properties. Net protein utilization (NPU), true digestibility (TU), and apparent digestibility (CE) of makhana are given as 49.3, 89.3, and 69.1. These NPU, TU, and CE of makhana were lower than those of soybean, egg, and cow milk (Khadatkar et al., 2020a,b). Makhana is rich in vitamins such as vitamin A and vitamin C, ranging from 62.23 to 63.84 IU/g and 0.18%–0.20% respectively (Khadatkar et al., 2020a, b). The biological value of makhana is very low; it contributes a high leucine to isoleucine ratio in makhana seed (Kumari et al., 2019a,b,c).

Roasting is an important unit operation carried out in makhana processing at a temperature of 300–335 °C for 3–9 min, either in a single-stage or in a two-stage roasting method with a tempering stage for 23 h in between two roasting stages, and is consumed globally in roasted or popped form of makhana. Moisture content plays a vital role in processing; low moisture content yields a highly porous and crunchy texture product with an appealing flavor, which can be used for direct consumption. The raw makhana seeds are cleaned, dried, and heated in iron or earthen pans at high temperatures by continuous stirring. This high temperature softens the outer shell and partial gelatinization of starch, reducing the moisture content of kernels contribute to the puffed structure that develops during popping. The mild nutty flavor that develops during roasting makes makhana a popular ingredient in snack products (Jha et al., 1999). Table 10.4 shows the nutritional composition of both the raw and popped makhana.

Roasting significantly increases phenolic compounds, flavonoids, and antioxidant activity. However, Roasting conditions are monitored for every batch. Excessive roasting leads to undesirable changes such as burning and discoloration, and also promotes oxidation reactions that affect product quality. Hence, maintaining appropriate roasting temperature and time is essential to produce high-quality makhana. Frying is one of the commonly used processing methods for preparing value-added makhana products. Fried makhana products are widely marketed as convenient, ready-to-eat snacks with different seasonings such as spices, herbs, and flavored coatings. During frying, Makhana kernels are exposed to hot oil at high temperatures for a short period of time. This high temperature causes rapid evaporation of moisture from the kernels, resulting in a light and crispy texture, and it enhances the sensory quality of makhana products. However, exposure to high temperatures leads to the degradation of heat-labile compounds and accelerates lipid oxidation, which results in off-flavors, thereby reducing the shelf life (Liaquat et al., 2022).

TABLE 10.4 Nutritional composition of raw makhana and popped makhana

Nutritional composition	Raw makhana	Popped makhana	Reference
Carbohydrate	76.9%	84.9%	Singh et al. (2020a,b)
Protein	9.7%	9.5%	
Fat	0.1%	0.5%	
Moisture	12.8%	4%	
Calorific value	362 kcal.	328 Kcal	

Extrusion is an advanced food processing technology widely used for the production of expanded snack foods. Makhana is suitable for extrusion technology because of its high content of starch and moderate protein, which makes it suitable for the desired shape, size, and texture. Raw makhana seeds are first cleaned, followed by drying. The dried seeds are milled into flour. The flour can be used or mixed with other flours to improve the nutritional composition and functional properties of the final product. The mixture is then fed into an extruder where it is subjected to high temperature, pressure, and mechanical shear, thereby starch undergoes gelatinization and partial degradation (Guy, 2001). Protein denatures and interacts with other components of the food matrix. The cooked material exists through the die of the extruder, thereby a sudden drop in pressure causes rapid moisture evaporation and expansion results in the formation of a porous structure which yields a light and crispy texture.

10.6.2 Oxidative stability and shelf life of makhana products

The oxidative stability of makhana-based products is influenced by their lipid composition, storage conditions, and processing methods such as roasting, frying, and extrusion (Frankel, 2014). Although makhana seeds derived from Euryale ferox contain relatively low-fat content, lipid oxidation can still occur during storage, leading to changes in flavor, aroma, and overall product quality (Singh et al., 2020a,b). The presence of natural antioxidants such as phenolic compounds and flavonoids in makhana may contribute to improved oxidative stability and delayed rancidity in processed products (Das et al., 2006).

The shelf life of makhana products depends on factors such as moisture content, packaging conditions, and exposure to oxygen, which can accelerate oxidative reactions and quality deterioration during storage (Kumar et al., 2016a,b,c). Understanding lipid oxidation kinetics is therefore important for predicting shelf life and maintaining the quality and stability of makhana-based food products (Frankel, 2014). The low levels of lipid still undergo oxidation by reacting with oxygen or high temperatures (as in the case of roasting and frying), leading to the formation of primary and secondary oxidation products such as peroxides, aldehydes, and ketones. During storage, these compounds lead to deterioration in flavor, aroma, and nutritional quality. The shelf life of makhana depends on internal and external factors. Temperature, light exposure, and oxygen availability, and the latter factors are moisture content, processing conditions, and packaging conditions. The oxidative stability of makhana products is commonly evaluated using analytical methods such as peroxide value, Thio barbituric acid reactive substances (TBARS), free fatty acid content (FFA), and P-anisidine value. These parameters provide useful information about the extent of lipid oxidation to determine the storage stability of the product over time.

10.6.3 Antioxidant activity and protective compounds

Makhana is well recognized for its antioxidant activity due to the presence of phenolic compounds, flavonoids, and other phytochemicals. These compounds reduce oxidative reactions and have the ability to neutralize free radicals. Apart from its macro and micronutrient composition, it's the phenolic constituents and antioxidant properties that impart specific medicinal attributes to fox nuts (Lee et al., 2002). Nuts rank third after spices and fruits for containing the most phytochemicals (Chang et al., 2016). Phenolic and flavonoid compounds are important determinants of the antioxidant activity and overall nutritional quality of foodstuffs. Several studies have supported their role in managing human diseases (Lin et al., 2016). Antioxidant properties exhibited by the phenolic compounds are very potent as compared with those of other nutrients (Shahidi and Naczk, 2003). The presence of natural antioxidants such as phenolic compounds and flavonoids in makhana may contribute to improved oxidative stability and delayed rancidity in processed products (Das et al., 2006). Phenolic compounds act as hydrogen donors, which stabilize free radicals and prevent chain reactions in the lipid oxidation mechanism. Flavonoids present in makhana exhibit strong antioxidant activity and protect cellular components from oxidative stress. As a result, makhana products often show improved oxidative stability compared with many other snack foods. In addition to these compounds, vitamins and minerals present in the seeds contribute to the overall antioxidant capacity of makhana products.

10.6.4 Lipid oxidation kinetics in makhana products

Lipid oxidation kinetics refers to the rate and pattern by which lipids undergo oxidative deterioration during processing and storage. As makhana is low in fat, but unsaturated fatty acids are still susceptible to oxidative reactions when exposed to heat, oxygen, and light. Lipid oxidation in makhana products generally follows a free radical chain reaction mechanism consisting of three stages: Initiation, propagation, and termination. Lipid oxidation is a major factor affecting the quality and shelf life of food products, particularly during storage and thermal processing (Frankel, 2014). Although seeds of Euryale ferox contain relatively low lipid content, oxidative reactions can still occur during processing and storage, resulting in changes in flavor, aroma, and nutritional quality of makhana products (Kapoor et al., 2022).

In the initiation stage, reactive oxygen species or high processing temperatures may cause the formation of lipid radicals from unsaturated fatty acids. During propagation, these radicals react with molecular oxygen to produce lipid peroxides and hydroperoxides. These primary oxidation products are unstable and can further decompose into secondary compounds such as aldehydes, ketones, and alcohols, which are responsible for rancid flavors and deterioration of sensory quality. The termination stage occurs when radicals combine to form stable nonreactive compounds, thereby slowing down the oxidation process. In many food systems, lipid oxidation follows zero-order or first-order reaction kinetics, depending on factors such as lipid concentration, oxygen availability, and storage conditions. By applying kinetic models to experimental data, it is possible to estimate reaction rate constants and predict the shelf life of makhana products under different storage environments. Understanding lipid oxidation kinetics is therefore important for predicting the stability and shelf life of makhana-based food products, particularly when subjected to processes such as roasting, frying, and extrusion (Frankel, 2014).

Kinetic models such as zero-order and first-order reactions are commonly used to describe the rate of lipid oxidation and to estimate the storage stability of lipid-containing food systems (Nguyen et al., 2024). In zero-order reaction kinetics, the rate of oxidation remains constant and does not depend on the concentration of the reacting substances. This frequently occurs during the early propagation phase and in the presence of high amounts of oxygen. Parameters such as peroxide value and free fatty acid content increase linearly with storage time. Zero-order kinetics is commonly applied in studies involving high-temperature processing. Lipid oxidation in makhana typically follows first-order kinetics, where the rate of oxidation depends on the concentration of available oxygen and fatty acids. The reaction rate is proportional to the concentration of the reacting molecules. The formation of secondary oxidation products is described as a first-order reaction, where the concentration of reactants decreases exponentially over time. Overall, understanding lipid oxidation kinetics is essential for improving the quality and stability of makhana products. The presence of natural antioxidants and bioactive compounds in makhana seeds may also contribute to delaying lipid oxidation and improving the oxidative stability of makhana products during storage (Kapoor et al., 2022). Strategies such as optimizing processing conditions, incorporating natural antioxidants, and using oxygen-barrier packaging materials can effectively slow oxidative reactions and enhance the shelf life of makhana-based snack foods.

10.7 Integration of functional attributes

As discussed above in the section 10.6.1 makhana contains a high amount of carbohydrates, moderate amounts of protein, very small amounts of lipids, dietary fiber, minerals, and it contains several bioactive compounds (Singh et al., 2020a,b). These components interact with each other and contribute to the overall functional performance of makhana-based products during processing and consumption. Lipids contribute to the retention of flavor compounds and enhance the sensory perception of makhana-based value-added products by influencing mouthfeel and aroma stability during processing and storage (Frankel, 2014; Singh et al., 2020a,b). These complexes alter the

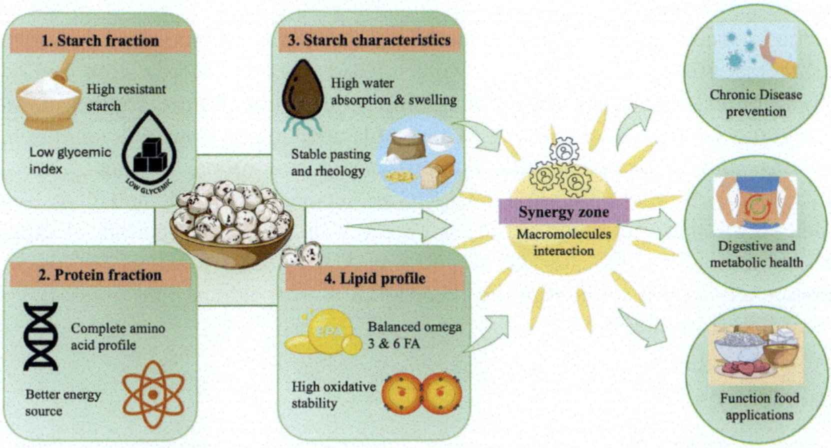

FIGURE 10.4 Makhana's functional attributes integration model.

gelatinization behavior of starch and slow down its digestion, contributing to a more gradual release of glucose during digestion. Lipids also help retain flavor compounds and improve the sensory perception of makhana value-added products. Makhana's functional attributes integration model is depicted in Fig. 10.4.

Processing methods affect the interaction of functional attributes; germination improved the functional properties of makhana flour. Roasting modifies the internal structure of the seeds, thereby enhancing texture and digestibility. The interaction of these components results in the development of functional foods, which are termed novel foods that provide health benefits beyond basic nutrition. Makhana seeds are widely used and have been found to have therapeutic effects used in Chinese and Ayurvedic remedies for the treatment of a number of illnesses, including renal disease, ongoing diarrhea, severe leucorrhea, and splenic hypofunction (Kumari et al., 2019a,b,c). Makhana can be used in various processing methods, either as a seed or flour, thereby incorporating it into cereals and millets to enhance its nutritional content. Makhana has great potential for the development of functional foods.

10.8 Conclusion

The biochemical composition of makhana strengthens as a functional ingredient in modern food systems. Makhana represents an underutilized aquatic crop with immense potential for innovation in value-added products. Lipid oxidation kinetics play a crucial role in determining the storage stability of makhana products. Understanding zero-order and first-order reaction kinetics is important for predicting the shelf life and affecting the quality during storage. Makhana shows great promise for its adaptability to various processing techniques, allowing the creation of a diverse range of products such as roasted snacks, extruded foods, bakery items, nutraceutical powders, and therapeutic foods. Despite growing interest in makhana research, several opportunities remain for further scientific exploration. Future studies could focus on a deeper understanding of the molecular interactions among starch, proteins, and lipids under different processing conditions. Advanced techniques such as microstructural analysis, molecular modeling, and digestion studies. Another promising area of research is the investigation of lipid oxidation kinetics and storage stability in value-added makhana products. Researchers may also explore the incorporation of makhana into novel functional food formulations.

References

Al-Waili, N., Al-Waili, H., Al-Waili, T., Salom, K., 2017. Natural antioxidants in the treatment and prevention of diabetic nephropathy; a potential approach that warrants clinical trials. Redox Rep. 22 (3), 99–118.

Alfasane, M.A., Khondker, M., Begum, Z.T., Banu, L.A., Rahman, M.M., Shahjadee, U.F., 2008. Fruit production and biochemical aspects of seeds of Euryale ferox Salisb. under ex-situ conditions. Bangladesh J. Bot. 37 (2), 179–181.

Bhat, F.M., Arya, S.S., 2023a. Starch retrogradation and its impact on food quality: a review. J. Food Sci. Technol. 60 (2), 345–357.

Bhat, M.S., Arya, S.S., 2023b. Esterified unpopped foxnut (Euryale ferox) starch: molecular and rheological characterization. J. Sci. Food Agric. 103 (5), 2492–2501.

Bhat, F.M., Riar, C.S., Srivastava, B., 2022. Effect of starch modification and polyphenol complexation on digestibility and glycemic response of starch-based foods. Food Chem. 374, 131712.

Bhat, M.S., Bashir, O., Jan, K., 2022. Resistant starch formation and physicochemical properties of modified fox nut (*Euryale ferox*) starch. Food Chem. 374, 131709.

Biswas, P., Das, M., Boral, S., Mukherjee, G., Chaudhury, K., Banerjee, R., October 2020. Enzyme mediated resistant starch production from Indian Fox Nut (Euryale ferox) and studies on digestibility and functional properties. Carbohydr. Polym. 237, 116–158. https://doi.org/10.1016/j.carbpol.2020.116158.

Chandra, S., et al. (2015). Food Res. Int. https://doi.org/10.1016/j.foodres.2015.03.012.

Chang, S.K.C., Alasalvar, C., Bolling, B.W., 2016. Nutrient composition and health benefits of nuts. Food Chem. 221, 1–10.

Choudhury, S.G., Basak, N., Banerjee, T., Mukhopadhyay, S., Ghosh, B.N., Das, K., Patil, N.G., 2026. Foxnut (makhana): a productive and economic crop providing good niche for soil carbon storage and stability in lower Gangetic wetlands. Sci. Rep.

Das, S., Der, P., Raychaudhuri, U., Das, D.K., 2006. The effect of Euryale ferox on myocardial ischemic reperfusion injury. Mol. Cell. Biochem. 289, 55–63.

Devi, M., Sharma, K., Jha, S.N., Arora, S., Patel, S.K.Y., Vishwakarm, R.K., 2020. Effect of popping on physicochemical, technological, antioxidant, and microstructural properties of makhana seed. J. Food Process. Preserv., e14787. https://doiorg/101111/fpp14787.

Dutta, S., et al., 2020. Metabolomic profiling of makhana. Sci. Rep. https://doi.org/10.1038/s41598-020-64888-3.

FAO, 2013. Dietary Protein Quality Evaluation in Human Nutrition: Report of an FAO Expert Consultation. https://doi.org/10.4060/i3124e.

Frankel, E.N., 2014. Lipid Oxidation, second ed. Woodhead Publishing.

Fuentes-Zaragoza, E., Riquelme-Navarrete, M.J., Sánchez-Zapata, E., Pérez-Álvarez, J.A., 2011. Resistant starch as functional ingredient: a review. Food Res. Int. 44 (6), 1433–1440.

Garg, N., Chauhan, K., Baria, K., & Mehan, M. B. (2021). Foxnut (Makhana): The Powerpack Seed. Everyman's Science, 165–168.4.

Gopalan, C., et al. (2017). Nutritive value of Indian foods. National Institute of Nutrition.

Griffith, L.D., Castell-Perez, M.E., Griffith, M.E., 1998. Effects of blend and processing method on the nutritional quality of weaning foods made from select cereals and legumes. Cereal Chem. 75, 105–112.

Guy, R., 2001. Extrusion Cooking: Technologies and Applications. Woodhead Publishing.

Huang, J., Li, X., Wang, Y., 2021a. Structural and digestibility characteristics of starch during retrogradation. Food Hydrocoll. 112, 106–115.

Huang, S., Chao, C., Yu, J., Copeland, L., Wang, S., 2021b. New insight into starch retrogradation: the effect of short-range molecular order in gelatinized starch. Food Hydrocoll. 120, 106921.

Jana, B.R., Idris, M., 2018a. Anti-aging amino acids in *Euryale ferox* (Salisb.): a review. Adv. Plants Agricult. Res. 8 (1), 39–43. https://doi.org/10.15406/apar.2018.08.00289.

Jana, B.R., Idris, M., 2018b. Anti-aging amino acids in *Euryale ferox* (Salisb.). Adv. Plants Agric. Res 8, 43–48.

Jana, B.R., Kumar, U., Singh, I.S., Srivastava, A., 2021. Makhana: a nutritious and fabulous functional food. Biot. Res. Today 3 (7), 605–607.

Jana, B.R., Srivastava, A., Idris, M., 2019. New makhana (*Euryale ferox* Salisb.) processed products for health benefit. J. Food Process. Technol. 10, 789. https://doi.org/10.4172/2157-7110.1000789.

Jha, V., Barat, G.K., Jha, U.N., 1991. A nutritional evaluation of *Euryale ferox* Salisb (Makhana). J. Food Sci. Technol. 8 (5), 326–328.

Jha, S. N., et al. (2011). J. Food Process. Preserv. https://doi.org/10.1007/s13197-011-0268-3.

Jha, N., Jha, N., Jha, N., et al., 2017. Effect of green tea powder on the quality attributes and antioxidant activity of whole-wheat flour pan bread. LWT – Food Sci. Technol. 79, 342–348.

Jha, S.N., Sharma, R., 2010. Physical, gravimetric and functional characterization of various milling fractions of popped gorgon nut (Euryale ferox). J. Food Sci. Technol. 47 (5), 564–570.

Jha, S.N., Srivastav, P.P., Mishra, H.N., 1999. Processing and quality evaluation of popped gorgon nut (Euryale ferox). J. Food Sci. Technol. 36 (5), 419–422.

Joye, I., 2019. Protein digestibility of cereal products. Foods 8 (6), 199.

Kapelko-Żeberska, M., Zięba, T., Meisel, M., Buksa, K., Gryszkin, A., 2024. Production of resistant starch by roasting retrograded starch with glucose. Molecules 29 (12), 2883.

Kapoor, S., Kaur, A., Choudhary, M., Kumar, V., 2023. Effect of incorporating foxnut (*Euryale ferox*) powder on physico-chemical attributes and amino acid composition of cookies. Acta Sci. Nutr. Health 7 (8), 19–27. https://doi.org/10.31080/ASNH.2023.07.1276.

Kapoor, R., Sharma, A., Singh, P., 2022. Oxidative stability and quality changes in makhana (Euryale ferox) during processing and storage. J. Food Sci. Technol. 59 (4), 1452–1460.

Kaur, M., et al., 2015. Carbohydr. Polym. https://doi.org/10.1016/j.carbpol.2015.02.012.

Kaur, M., Sandhu, K.S., Arora, A., 2023. Effect of lotus seed (makhana) flour incorporation on the physicochemical and sensory properties of cookies. J. Food Measur. Character. 17 (2), 1450–1460.

Khadatkar, A., Gite, L.P., Gupta, V.K., 2015. Interventions to reduce drudgery of workers in the traditional method of harvesting Makhana (Euryale ferox salisb.) seeds from ponds. Curr. Sci. 1332–1337.

Khadatkar, A., Mehta, C.R., Gite, L.P., 2020a. Makhana (Euryale ferox Salisb.): a high-valued aquatic food crop with emphasis on its agronomic management: a review. Sci. Hortic. 261, 108995. https://doi.org/10.1016/j.scienta.2019.108995.

Khadatkar, A., Singh, R., Pandey, R.K., 2020b. Nutritional composition and health benefits of makhana (*Euryale ferox*). J. Food Sci. Technol. 57 (6), 2021–2030.

Kumar, S., 2015. Application and effect of addition of popped makhana flour on the properties and qualities of cookies. Int. J. Process. Post Harvest Technol. 6 (1), 80–86.

Kumar & Saini (2016). Study of various characteristics of composite flour prepared from the blend of wheat flour and gorgon nut flour. Int. J. Agric. Environ. Biotechnol., 9(4), 679–689. https://doi.org/10.5958/2230-732X.2016.00089.9.

Kumar, L., Gupta, V.K., Jha, B.K., Singh, I.S., Bhatt, B.P., Singh, A.K., 2011. Status of Makhana (Euryale Ferox Salisb.) Cultivation in India, vol 32. Technical Bulletin No.

Kumar, A., Sharma, S., Singh, B., 2020. Effect of processing on physicochemical and functional properties of makhana (Euryale ferox) flour. J. Food Measur. Character. 14 (6), 3384–3392.

Kumar, L., Singh, A.K., Bhatt, B.P., 2016a. Nutritional status of recently developed makhana (gorgon nut) variety "Swarna Vaidehi. J. AgriSearch 3 (4). https://doi.org/10.21921/jas.v3i4.6701.

Kumar, A., Yadav, P.K., Nath, P., Nath, S., Kumari, U., 2016b. Euryale ferox Salisb.: an important aquatic crop—a review. Int. J. Adv. Agricult. Sci. Technol. 3 (6), 1–10.

Kumar, S., Singh, A., Sharma, P., 2016c. Makhana (Euryale ferox): a potential aquatic crop with high nutritional and economic value. Indian J. Agric. Sci. 86 (7), 891–896.

Kumari, R., Jakhar, D.S., Kumar, P., 2019a. Nutritional and medicinal importance of makhana (euryale ferox salisb.). Marumegh 4 (2), 53–55.

Kumari, M., Verma, R., Patel, S., 2019b. Protein quality evaluation of makhana and its nutritional implications. Food Chem. 278, 234–240.

Kumari, R., Jakhar, D.S., Kumar, P., 2019c. Nutritional, and medicinal importance of makhana (Euryale ferox Salisb.). Murumegh, Kisaan E-Patrika 4 (2), 53–55.

Lee, S.E., Ju, E.M., Kim, J.H., 2002. Antioxidant activity of extracts from Euryale ferox seed. Experimen. Mol. Med. 34 (2), 100–106.

Liaquat, M., Pasha, I., Ahsin, M., Salik, A., 2022. Roasted fox nuts (Euryale Ferox L.) contain higher concentration of phenolics, flavonoids, minerals and antioxidants, and exhibit lower glycemic index (GI) in human subjects. Food Prod. Proc. Nutr. 4 (1), 1–12. https://doi.org/10.1186/s43014-021-00081-x.

Lin, D., Xiao, M., Zhao, J., Li, Z., Xing, B., Li, X., Kong, M., Li, L., Zhang, Q., Liu, Y., Chen, H., Qin, W., 2016. An overview of plant phenolic compounds and their importance in human nutrition and management of diseases. Molecules 21 (10), 1374.

Lokendra Kumar, L.K., Singh, A.K., Bhatt, B.P., 2016. Nutritional Status of Recently Developed Makhana (Gorgon Nut) Variety-Swarna Vaidehi.

Maibam, D., Choudhury, A.J., Das, A., 2021. Effect of extrusion processing on physicochemical, functional and sensory properties of makhana (Euryale ferox) based extruded snacks. J. Food Process. Preserv. 45 (10), e15835.

Maibam, D.B., Deka, S.C., 2022. Physicochemical properties and structure of starches of foxnut (*Euryale ferox* Salisb.) from India and its application. J. Food Process. Preserv. 46 (2), e16262.

Mal, P., Pandey, R.K., 2025a. Effect of thermal processing on physicochemical properties of makhana starch. Food Chem. Adv. 4, 100–112.

Mal, T., Pandey, A., 2025b. Isolation and effect of physical modification on the release characteristics Euryale ferox Salisb polysaccharide. Int. J. Biol. Macromol. 305, 141243.

Mensink, R.P., 2016. Effects of saturated fatty acids on serum lipids and lipoproteins. Am. J. Clin. Nutr. https://doi.org/10.3945/ajcn.115.126035.

Nath, B.K., Chakraborty, A.K., 1985. Studies on the amino acid composition of *Euryale ferox* Salisb. J. Food Sci. Technol. 22, 293.

Nath, A., et al., 2015. Nutritional composition of makhana. J. Food Sci. Technol.

Ngo, T.V., Kusumawardani, S., Kunyanee, K., Luangsakul, N., 2022. Polyphenol-modified starches and their applications in the food industry: recent updates and future directions. Foods 11 (21), 3384.

Nguyen, K.A., Hennebelle, M., van Duynhoven, J.P., Dubbelboer, A., Boerkamp, V.J., Wierenga, P.A., 2024. Mechanistic kinetic modelling of lipid oxidation in vegetable oils to estimate shelf-life. Food Chem. 433, 137266.

Patel, B.M., Vala, K.V., Ravrani, K.A., 2024. Makhana as a Traditional Medicine and Modern Food: A Review. International Journal of Scientific Development and Research. https://ijsdr.org/papers/IJSDR2502011.pdf.

Puri, A., Sahai, R., Singh, K.L., 2000. Immunostimulant activity of dry fruits and plant materials used in Indian traditional medical system for mothers after childbirth and invalids. J. Ethnopharmacol. 71 (1–2), 89–92.

Qu, C., Yu, D., Jing, Z., Gu, S., Wang, Y., Xie, W., Wu, Q., 2025a. Comparative study on structural characterization, physicochemical properties, and in vitro probiotic activities of resistant starch from different varieties of Euryale ferox. Food Chem. X 28, 102572.

Qu, J., Zhang, L., Chen, H., 2025b. Rheological and pasting properties of lotus seed (Euryale ferox) starch. Int. J. Biol. Macromol. 240, 124–132.

Rathod, R.V., Neve, G., Jain, A., Giri, P., 2023. A comprehensive review on health benefits and nutritional aspects of foxnut (makhana). Pharm. Innov. 12 (6), 4432–4438.

Shahidi, F., Naczk, M., 2003. Phenolics in food and nutraceuticals. CRC Press.

Shankar, M., Chaudhary, N., Singh, D., 2010. A review on gorgon nut. Int. J. Pharm. Biol. Arch. 1 (2), 101–107.

Sharma, P., Gujral, H.S., Singh, B., 2016. Effect of roasting on physicochemical, nutritional and antioxidant properties of fox nut (*Euryale ferox*). J. Food Sci. Technol. 53 (5), 2314–2323.

Simopoulos, A.P., 2016. An increase in the omega-6/omega-3 fatty acid ratio increases the risk for obesity. Biomed. Pharmacother. https://doi.org/10.1016/j.biopha.2016.03.020.

Singh, N., Singh, J., Kaur, L., Sodhi, N.S., Gill, B.S., 2003. Morphological, thermal and rheological properties of starches from different botanical sources. Food Chem. 81 (2), 219–231.

Singh, R., et al. (2016). J. Sci. Food Agric. https://doi.org/10.1002/jsfa.7421.

Singh, J., Dartois, A., Kaur, L., 2010. Starch digestibility in food matrix: a review. Trends Food Sci. Technol. 21 (4), 168–180.

Singh, I.S., Kumar, L., Bhatt, B.P., Thakur, A.K., Choudhary, A.K., Kumar, A., 2017. Integrated aquaculture with Fox nut-A case study from north Bihar, India. Int. J. Curr. Microbiol. Appl. Sci. 6 (10), 4906–4912.

Singh, G., et al., 2018. Nutritional and functional properties of makhana. J. Food Proc. https://doi.org/10.1007/s11694-018-9865-6.

Singh, D.K., Kumar, U., Kumar, A., Singh, I.S., 2020a. Strategies for Export Promotion of Popped Makhana in India, pp. 30–33.

Singh, R.K., Chaudhary, A., Singh, D., 2020b. Makhana (*Euryale ferox*): a high-valued aquatic crop with emphasis on its agronomic management and utilization. Sci. Hortic. 261, 108995.

Sinha, C.B.K., Jha, V., 2025. Identification of nutraceutically important metabolites in raw and popped makhana (*Euryale ferox* Salisb.) seeds using GCMS technique. Int. J. Plant Environ. 11 (03), 514–522. https://doi.org/10.18811/ijpen.v11i03.09.

Sodi, M., Kumar, S., 2019. Makhana in nutritional, medicinal and socio-cultural uses in the Koshi–Mithila region of Bihar. Scholars Acad. J. Biosci. 7 (6), 276–287. https://doi.org/10.36347/sajb.2019.v07i06.002.

Song, C.W., Wang, S.M., Zhou, L.L., Hou, F.F., Wang, K.J., Han, Q.B., et al., 2011. Isolation and identification of compounds responsible for antioxidant capacity of euryale ferox seeds. J. Agric. Food Chem. 59 (4), 1199–1204. https://doi.org/10.1021/jf1041933.

Tehseen, S., Sarfraz, F., Ateeq, N., Ashfaq, F., Yasmin, I., Mehmood, T., 2020a. Acta scientific agriculture (ISSN: 2581-365X) foxnut (euryale ferox salisb.): a health promising fruit. Acta Sci. Agric. 4 (12), 68–72.

Tehseen, S., Ali, M., Ahmad, N., 2020b. Nutritional evaluation and mineral composition of fox nut (Euryale ferox). Food Res. Int. 137, 109–115.

Tester, R.F., Karkalas, J., Qi, X., 2004. Starch—composition, fine structure, and architecture. J. Cereal. Sci. 39 (2), 151–165.

USDA FoodData Central. (2020). Nutrient database. https://fdc.nal.usda.gov.

Ugoeze, K.C., Odeku, O.A., 2025. Antioxidants in Infectious Disease Management. *Antioxidants*. Nature's Defense Against Disease, pp. 169–218.

Verma, S., et al. (2020). Antioxidant and lipid-lowering effects of makhana. Food Res. Int.

Wardhan, H., Suntwal, T., Bansal, A., Laxmikant, Gulati, A., 2025. Strategies to Boost India's Makhana Exports. Academic Foundation/ICRIER.

Zhang, H., Xu, F., Wu, Y., Hu, H., Dai, X., 2021. Structural, physicochemical and digestive properties of *Euryale ferox* resistant starch prepared by autoclaving–retrogradation. Food Chem. 359, 129924.

Zhang, Y., Li, M., Wang, L., 2021. Effect of autoclaving–retrogradation treatment on resistant starch formation and digestibility of lotus seed (makhana) starch. Int. J. Biol. Macromol. 183, 1842–1850.

Zhao, L., Huang, J., Man, J., Huai, H., Chen, Y., Wei, C., 2016a. Physicochemical properties of Euryale ferox kernel starches from two different regions. Int. J. Food Prop. 19 (2), 289–299.

Zhao, Y., Liu, C., Li, Q., 2016b. Thermal and structural properties of Euryale ferox starch by DSC and XRD analysis. Carbohydr. Polym. 150, 230–238.

Zhu, F., 2015. Composition, structure, physicochemical properties, and modifications of lotus and makhana (*Euryale ferox*) seed starches. Carbohydr. Polym. 123, 452–463.

Chapter 11

Makhana plant protection measures and nutrient requirements for successful crop production

Vinod Kumar Padala[1], Akhileshwar Vishwakarma[1], N. Ramya[2] and Manoj Kumar[1]
[1]*ICAR-National Research Centre for Makhana, Darbhanga, Bihar, India;* [2]*ICAR-Indian Institute of Agricultural Biotechnology, Ranchi, Jharkhand, India*

11.1 Introduction

Makhana, also known as gorgon nut (*Euryale ferox* Salisb.), is a perennial aquatic crop (Haines, 1925; Biswas and Calder, 1936; Subramaniam, 1962) of the family Nymphaeaceae that is cultivated primarily in shallow, stagnant waters and ponds. It is also referred to by a number of regional names around the world, such as Kenjitsu, Komsil, Qian Shi, Ji Tou Shi, prickly lily seeds, etc. Archeobotanical research shows that the species was a significant wild food source in the Yangtze River basin during the Neolithic era, with numerous remains found at archeological sites like Kuahuquiao, Hemudu, and Tianluoshan (Goren-Inbar et al., 2014). It is a high-value minor crop with nutritive and medicinal significance, rich in carbohydrates, proteins, and micronutrients, and has long been an important source of livelihood for wetland communities, especially in the Mithila region of north Bihar, India (Jha et al., 1991). The crop's traditional cultivation system, based on natural and man-made ponds combined with its expanding commercialization, has increased both its economic importance and the attention given to its crop health and pest issues.

Although makhana was traditionally considered relatively hardy with few major pest outbreaks, recent surveys and reviews indicate a diverse assemblage of insect species associated with different plant parts (leaves, petioles, ribs, flowers, and developing fruits) and with harvested/processed seed material. Yield losses due to the insect pests range from approximately 17%–23% (Nath

Makhana (Euryale ferox) as a Functional Food. https://doi.org/10.1016/B978-0-443-44421-0.00021-4

et al., 2018a) in seed yield under typical growing conditions without effective pest control measures. Key field pests reported repeatedly in the literature include aphids (*Rhopalosiphum nymphaeae*), lepidopteran case-worms (*Elophila depunctalis* and *E. crisonalis*), leaf midge/larval midges (Chironomidae), and stem/rib borers (Nath et al., 2018b; Mishra et al., 2003a); in addition, gastropods ad sap-feeing Hemiptera have been recorded as occasional or emerging pests. These insect groups attack at different crop stages and can cause reductions in photosynthetic area, direct feeding losses, or fruit/seed mortality that translate to measurable yield reductions (Table 11.1). In this chapter, we discussed the major and minor insect pests infesting the makhana crop, their effects on crop yield, and management practices to protect the crop.

11.2 Water lily aphid, *Rhopalosiphum nymphaeae* (L.) (Order: Hemiptera; family: Aphididae)

Water lily aphid, *Rhopalosiphum nymphaeae* (Linnaeus) (Hemiptera: Aphididae), also known as reddish-brown plum aphid, has a wide geographic distribution. In India, it is a major pest of Makhan in North Bihar, particularly in Darbhanga, Madhubani, Saharsa, Supaul, Purnea, and Katihar, along with the adjoining districts where pond-based cultivation predominates (Mishra et al., 1992; Nath et al., 2018b). The water lily aphid is a highly polyphagous insect, with host plants belonging to at least 49 plant groups across both aquatic and terrestrial habitats (Blackman and Eastop, 2006). American lotus (*Nelumbo lutea* Willd.), water hyacinth (*Eichhornia crassipes* Mart.), water lettuce (*Pistia stratiotes* L.), water spinach (*Ipomoea aquatica* Forssk.), crested floating-heart (*Nymphoides cristata* Roxb.), common duckweed (*Lemna minor* L.), rice (*Oryza sativa* L.), and a number of cultivated crop and fruit plants are among its host range (Center et al., 1999; Mishra et al., 1992). Recent studies reported two new aquatic weed hosts for the *Rhopalosiphum nymphaeae* in India: Blue water lily (*Nymphaea nouchali* Burm. f.) and Four-leaf clover (*Marsilea quadrifolia* L.) (Fig. 11.1). Genetic diversity analyses indicated low variation within Indian and USA populations, whereas populations from China, Canada, and Lithuania exhibited moderate diversity. These findings suggest that the Indian population of *R. nymphaeae* is genetically uniform, while global populations show higher variability (Padala et al., 2023). The species is a known vector of various plant viruses, including abaca mosaic virus, cabbage black ring spot virus, cauliflower mosaic virus, cucumber mosaic virus, and onion yellow dwarf virus, and it also causes dieback of water lettuce (Center et al., 1999). In addition to causing significant harm in water gardens and nurseries, this pest is considered a potential biological control agent to manage some aquatic weeds. It is considered a major pest of makhana, especially during the early vegetative stage (Nath et al., 2018b). Heavy infestation results in poor plant growth, reduced leaf size, and lower photosynthetic efficiency, ultimately leading to significant yield

TABLE 11.1 Major and minor insect pests of makhana (*Euryale ferox*) and their damage characteristics.

Sl. No.	Insect pest	Order/Family	Plant part affected	Damaging stage	Nature of damage	Key symptoms	Economic significance	Key references
1	Water lily aphid (*Rhopalosiphum nymphaeae*)	Hemiptera/ Aphididae	Leaves, petioles	Nymphs and adults	Sap sucking, honeydew secretion, and reduced photosynthesis	Leaf yellowing, curling, sooty mold	Major pest causing growth suppression and yield loss	Mishra et al. (1992); Nath et al. (2018b)
2	Case worms (*Parapoynx crisonalis* and *Parapoynx stagnalis*)	Lepidoptera/ Crambidae	Leaves	Larvae	Skeletonization and defoliation by larval feeding	Skeletonized leaves, larval cases	Major defoliators affecting leaf area and yield	Mishra et al. (1992); Nath et al. (2018b)
3	Root borer (*Donacia delesserti*)	Coleoptera/ Chrysomelidae	Roots, stems, leaves, flowers	Larvae and adults	Root tunneling by larvae and foliar feeding by adults	Stunted growth, root damage, leaf notching	A major pest reduces plant vigor and yield	Jha et al. (1991); Mishra et al. (1992); Jolivet and Hawkeswood (1995)
4	Singhara beetle (*Galerucella birmanica*)	Coleoptera/ Chrysomelidae	Leaves, tender shoots	Larvae and adults	Scraping and chewing of leaf tissue causes defoliation	Scraped leaf surfaces, defoliation	Potentially serious defoliator under high populations	Ding et al. (2006a, b); Pemberton (1999); Halder et al. (2019)
5	Rib borer (*Chironomus* spp.)	Diptera/ Chironomidae	Leaf ribs, petioles	Larvae	Internal boring disrupting nutrient flow	Wilting, sudden leaf decay	Serious pest causing leaf collapse and reduced vigor	Jha et al. (1991); Mishra et al. (1992)

FIGURE 11.1 Water lily aphid, *R. nymphaeae, collected from different host plants.* (A) Makhana *Euryale ferox*, (B) Water chestnut, *Trapa natans*, (C) *Blue* water lily, *Nymphaea nouchali*, (D) Four leaf clover, *Marsilea quardrifolia*, (E) *Red* water lily, *Nymphaea rubra.*

reduction if unmanaged (Nath et al., 2018c). This species has both apterous and alate viviparous females.

Apterous viviparous females: The body is reddish-brown to dark olive. With low lateral frontal tubercles, the head is dark brown. Six segmented antennae. Each abdominal tergite has a center clump of spinules surrounded by spinules organized in polygonal patterns; Tergite VIII has two setae. On segments I and VII, marginal tubercles are clearly visible and always present. Siphunculi are dark brown and slightly enlarged. There is a basal constriction in the cauda (Palmer, 1953).

Alate viviparous females: The ventral side has a creamy bloom, and the body is lustrous black to dark brown. dark brown head. There are 14–18 sensoria in antennal segment III and three auxiliary sensoria in segment IV. Pale abdominal dorsum with border sclerites that are pale brown. Siphunculi are pale brown, about 2.25× the length of the dusky cauda, and enlarged on the apical half (Palmer, 1953).

The aphid thrives in humid pond ecosystems and reproduces mainly through parthenogenesis, leading to rapid population buildup. The polyphagous species occurs in multiple morphs, with both wingless and winged viviparous females commonly observed in the plains of India (Biswas and Ghosh, 1981). Winged forms facilitate dispersal between ponds. Population peaks are generally observed during cool and moderately warm months (Blackman and Eastop, 2006).

Females give birth to live nymphs. The life cycle includes four nymphal instars, and a generation may be completed within 7–10 days under favorable conditions (Fig. 11.2). Both winged and wingless morphs occur depending on population density and environmental conditions. Multiple overlapping generations occur during the crop season (Nath et al., 2018b). This aphid has also been recorded to survive under submerged conditions (Holman, 2009). The species is polyphagous and polymorphic, producing different morphs under temperate conditions, while parthenogenetic apterous and alate viviparous females predominate in the plains of India (Biswas and Ghosh, 1981). On water lilies, large aphid colonies frequently form, congregating along leaf veins and infesting blossoms as well. There have been reports of this species' sexual forms on *Prunus* spp. Both nymphs and adults feed by sucking sap from tender leaves and petioles, which weakens plants and reduces their vigor. Infested plants show yellowing and curling of leaves, restricted development, and reduced leaf expansion. The secretion of sticky honeydew promotes the development of sooty mold, further impairing photosynthesis and overall plant health (Mishra et al., 2003b).

Previous studies have reported that the aphid, *Rhopalosiphum nymphaeae* L., exhibits a distinctly aggregated spatial distribution on makhana (*Euryale ferox*). Both nymphs and adults were found to be clumped in their distribution across seasons and years, as indicated by multiple dispersion indices and confirmed using Taylor's power law and Iwao's mean crowding regression

a b

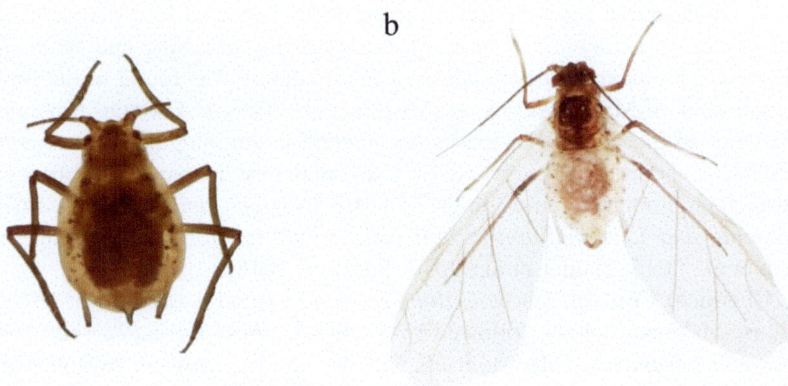

FIGURE 11.2 (a and b) Nymph and adult of water lily aphid, *R. nymphaeae* under stereozoom microscope.

analyses. Aggregation parameters (b) and density contagiousness coefficients (β) greater than unity consistently supported the aggregated distribution pattern. It has also been reported that optimal sample size requirements decrease with increasing aphid density and with lower levels of precision, reflecting the patchy nature of aphid infestations (Padala et al., 2023).

11.2.1 Management

Aphid management requires an integrated strategy that minimizes financial losses while preserving aquatic biodiversity. This strategy includes monitoring, cultural practices, biological control, and selective intervention.

Monitoring: Regular scouting of leaves is essential to detect early infestations of aphid colonies, especially during cool, humid periods that favor aphid build-up. Thus, any visible colony should prompt closer observation and possible action (NIPHM, 2025)

Cultural Practices: Maintaining good water quality and reducing weed hosts that can harbor aphid populations before they spread to makhana plants (Wenninger, 2011). Removing heavily infested leaves or plants early in the season reduces colony density and slows population growth. Encouraging pond hygiene to reduce alternate host plants and debris where aphids may overwinter or persist (Kuhar et al., 2005).

Biological Control: Natural biological control agents, including Spiders, Coccinellid beetles (ladybirds), lacewings, anthocorids, nabids, predatory midges, syrphid flies, carabids, staphylinids, and ants, play a critical role in suppressing aphid populations (Sunderland, 1988; Rabasse and van Steenis, 1999; Dedryver et al., 2010; Ali et al., 2016; Nath et al., 2018c). Predator like *Harmonia axyridis* exhibits strong predation of aphids, indicating their potential for conservation biological control in pond systems (Li et al., 2024). The aphid, *R. nymphaeae,* was predated by several insects, of which Scymnus sp. was considered the most effective predator, followed by *Coccinella septempunctata*, *Menochilus sexmaculatus*, *Micraspis discolor*, and *Brumoides suturalis*. The maximum population of the predators was found in the first or second week of March every year (Mishra et al., 2003b). A serious problem of resistance to insecticides increases the interest in entomopathogenic fungi as biocontrol agents in the context of integrated pest management strategies. Various entomopathogenic fungi like *Metarhizium anisopliae, Beauveria bassiana, and Lecanicillium lecanii* can be cultured for aphid management (Lee et al., 2015; Halder et al., 2020; Saruhan, 2018).

Chemical Control: Garlic (*Allium sativum*) extract was the most effective against Makhana aphids, followed by neem (*Azadirachta indica*) and kanail (*Thevetia peruviana*) (Mishra et al., 2003b). Imidacloprid 70 WS @ 5 g/kg seed treatment or root dipping in Imidacloprid 70 WS @ 5 g/Lt of water for 30 min at the time of transplanting should be used as a preventive measure. If an aphid population is seen following transplanting, foliar spraying with

NSKE @ 0.5% or Acetamaprid 20 SP @ 0.5 gm/Lt of water (Nath et al., 2018b).

11.3 Case worms *Parapoynx crisonalis* and *Parapoynx stagnalis* (Order: Lepidoptera; family: Crambidae)

The aquatic lepidopteran pest *Parapoynx crisonalis* (Walker, 1859) (Lepidoptera: Crambidae: Nymphulinae) is found across Asia, including China, Japan, Indonesia, Myanmar, Thailand, Sri Lanka, India and Australia (Regier et al., 2012; Chen et al., 2006; Huang and Li, 2013). The species has recently been reported from the British Isles, indicating an expansion of its geographical range (Davis, 2012). *P. crisonalis* infests a wide range of aquatic vegetables, cultivated ornamental species, and wild aquatic plants, such as *Trapa natans* L. and *Nymphoides peltatum* (Gmel.) *Nymphaea tetragona* Georgi and *Euryale ferox* Salisb (Chen et al., 2017; Yen, 2004). *Parapoynx stagnalis* is distributed across many regions of the world, with notable occurrences in South and Southeast Africa, China, Japan, Australia, and parts of South America, including Argentina, Brazil, Uruguay, and Venezuela, as well as Central Africa and several West African countries (Pathak, 1975; Brenière, 1983; Reissig et al., 1985; Hill, 1987; Dale, 1994). *Parapoynx stagnalis* (Lepidoptera: Crambidae), a severe paddy pest, mainly targets young rice plants in wet areas (Gogoi and Bora, 2012; Sison, 1938; Alum, 1967; Grist and Lever, 1969; Chi et al., 1995; Vromant et al., 1998). In addition to South America (Argentina, Brazil, Uruguay, and Venezuela), Central Africa (Madagascar, Malawi, Mozambique, Rwanda, and Zaire), East Africa, and several West African nations, where it is thought to exist throughout much of the region, the species is widely distributed throughout South and Southeast Asia, China, Japan, and Australia (Pathak, 1975; Brenière, 1983; Reissig et al., 1985; Hill, 1987; Dale, 1994), these pests are commonly reported from makhana-growing regions of Bihar and other parts of eastern India where aquatic vegetation is abundant (Mishra et al., 1992; Nath et al., 2018b). Caseworms are considered major pests of makhana (*Euryale ferox* Salisb.), causing severe foliar damage and significant yield losses (Nath et al., 2018b).

The snowy white adult moth of *P. stagnalis* has 6–8 mm length and its wings expand about 15 mm. There are two or three submarginal fulvous bands and a few light brown to black flecks on the white wings (Kattali et al., 2013). Caseworms *Parapoynx crisonalis* and *Parapoynx stagnalis* first appear on Makhana crops in February, with peak infestation occurring from March to June. Subsequently, the population declines during July and August. Early instar larvae feed by scraping the leaf surface, either along the margins or in the central portions of the leaf. The larvae construct an oval-shaped case from excised leaf tissue and remain within it, moving across the leaf surface while enclosed in the case (Kumar et al., 2021; Nath et al. 2018a,b,c; Mishra et al., 2003a). Kalita et al. (2020) found that *Parapoynx stagnalis* infestation was significantly positively correlated with rainfall, maximum and minimum temperatures, and morning relative humidity in the rice crop. Similarly, Gogoi and Bora (2013) observed a positive association between *Parapoynx stagnalis* incidence and total rainfall (Fei et al., 1995), further demonstrating that climatic factors, including relative humidity, temperature, and number of rainy days, play a crucial role in the population buildup of *Parapoynx stagnalis*.

Adult moths are active at night and begin depositing eggs 1–3 days after emergence. Each female deposits roughly 300–400 eggs on the second day, 200–300 on the third day, and 50–70 on the fourth day following emergence. Egg laying mostly takes place at night. After an average of 4 days of incubation, the eggs hatch. While unmated females may live up to a week following oviposition, mated females often die within 24 h. Parapoynx stagnalis goes through five larval instars; fully formed larvae are around 14 mm long and 1.6 mm in diameter, whereas newly hatched larvae are pale in color and measure about 1.2 mm in length. Using fragments of leaves, larvae build portable cases and stay underwater while consuming leaf tissues. To feed, early instar larvae scrape the leaf's surface, either in the center or along its margins. The larvae construct an oval-shaped case from excised leaf tissue and remain within it, moving across the leaf surface while enclosed in the case. The larval stage is most destructive and lasts about 2–3 weeks. Pupation occurs within the case attached to plant material (Mishra et al., 1992). Larvae scrape and consume the leaf tissue, leaving behind only the veins and skeletonized portions of the leaves. This damage appears as irregular holes and extensive skeletonization of the foliage, often reducing the leaf surface significantly. In many cases, larval cases can be seen attached to the leaves or floating nearby, indicating active infestation (Nath et al., 2018a; Padala et al., 2022).

The study found that *Parapoynx stagnalis* larvae predominantly exhibited a highly aggregated spatial distribution across rice fields and cultivars, regardless of the analytical method used. Uniform distribution was rare and occurred only at specific crop growth stages. In farmers' fields, uniformity was observed on just two occasions during the booting stage, indicating stronger aggregation under natural field conditions. In contrast, aggregation was comparatively weaker in the controlled experimental field, where uniform distribution appeared more frequently—once during the stem elongation stage and five times during the booting stage, suggesting that managed conditions reduced larval clustering (Gogoi and Bora, 2012).

11.3.1 Management

Proper field sanitation should be maintained by keeping the field and bunds free from weeds. Fertilizers should be applied in split doses rather than all at once. For efficient biological management, natural enemies, including spiders, dragonflies, and birds that prey on the adults, as well as hydrophilid and dytiscid water beetles that feed on the larvae, should be preserved and promoted. Flubendiamide 39.35% SC at 250 g/ha or Cartap hydrochloride 50% SP at 1000 g/ha or Spinosad 45 SC at 100 g/ha, or Emamectin benzoate 5% SC at 250 g/ha are the suggested chemicals for caseworm management (Patel et al., 2024).

11.4 Root borer, *Donacia delesserti* (Order: Coleoptera; family: Chrysomelidae)

It is abundantly distributed in Thailand, Sri Lanka, Bangladesh, and India. It is the pest of the wetland pond system of makhana cultivation, which generally occurs in North Bihar of India, where Makhana is commercially grown. It is grouped among the major insects attacking makhana in surveys from the Koshi and Mithila regions (Mishra et al., 1992; Jha et al., 1991). The genus *Donacia* includes both monophagous and polyphagous species. Their host plants are primarily members of the Nymphaeaceae and various monocotyledonous plants inhabiting marshlands, riverbanks, and the shallow, muddy margins of lakes. Larvae develop within the submerged, muddy roots of these plants, whereas adults are typically found on the emergent and floating leaves of their hosts (Fig. 11.3) (Houlihan 1969; Jach, 1998). Adults and larvae of *Donacia*

FIGURE 11.3 (A) Larva; (B) Pupae and (C) Adults of root borer, *Donacia delesserti*.

spp. feed on parts of aquatic macrophytes, including pondweeds and emergent water plants. Host plants include *Nelumbo nucifera* and *Euryale ferox*. In Makhana, *D. delesserti* attacks the fibrous roots, leaves, and flowers of *E. ferox*. Within makhana ponds, *D. delesserti* is recognized as a major pest, contributing to reduced plant vigor and yield by damaging both rooting structures and above-ground foliage or floral parts (Nath et al., 2018b).

Members of this group have a head marked by one central and two deep lateral grooves. The eyes are large and convex, and the antennae are slender and thread-like. The thorax is nearly square in shape, with straight sides and usually pointed angles. The elytra are broader at the base than the thorax and show distinct punctured striations, with the spaces between them finely transversely wrinkled. The third tarsal segment is deeply bilobed, and the claws are simple (Jacoby, 1908). Adults are typically found on the surfaces or emerging parts of host aquatic plants, often feeding on leaves and flowers. Larvae are adapted to the aquatic environment and feed on roots and lower stem tissues submerged in water. These are capable of flight, which facilitates dispersal among adjacent ponds (Jolivet and Hawkeswood, 1995) (Fig. 11.3).

Females lay eggs on or near host plants in aquatic environments. Larvae typically feed on submerged stems and roots, obtaining oxygen via air tubes or plant aerenchyma. Development proceeds through multiple larval instars underwater before pupation near or within plant tissues or detritus. Adults emerge and feed on aerial parts of plants (Jolivet and Hawkeswood, 1995). In Makhana, larvae feed on the root, disrupting nutrient and water uptake. Adult feeding on leaves reduces photosynthetic area. The dual feeding weakens plant health, leading to lower vigor and potential loss of reproductive output (Mishra et al., 1992). Affected plants commonly exhibit yellowing of leaves accompanied by a loss of turgor, giving them a wilted or weakened appearance. Growth is often stunted, and plants may wilt in severe cases due to damage to the root system, which compromises water and nutrient uptake. Adult beetles cause visible feeding notches and surface damage on leaves and flower parts, further reducing plant vigor. Under heavy infestations, the combined impact of root injury and foliar feeding can cause plants to collapse prematurely as the roots are no longer able to adequately support the plant's physiological needs (Mishra et al., 1992). Usually, root borer infestation in makhana fields generally occurs from around April to July, coinciding with the period of active vegetative growth of the crop (Mishra et al., 1992).

11.4.1 Management

Effective management of root borer can be achieved through an integrated approach. Cultural practices such as maintaining field sanitation by removing weeds and crop residues, practicing crop rotation with water chestnut or fodder crops during winter, and deep summer plowing help break the pest life cycle.

Chemical control should be used only as a last resort, with preference given to neem-based products such as oil @ 0.5%.

11.5 Singhara beetles, *Galerucella birmanica* (Order: Coleoptera; family: Chrysomelidae)

South Asian countries, including India, Pakistan, Sri Lanka, and Myanmar, have reported cases of *G. birmanica*, where it infests aquatic crops such as water chestnut (*Trapa natans*) and is also reported in the Makhana crop (Halder et al., 2019). Although mostly studied on water chestnut but is also reported in the pest list associated with Makhana, indicating that it feeds on foliage of related aquatic plants. Its feeding is primarily on leaves and tender shoots of aquatic macrophytes (Ding et al., 2006a; Pemberton, 1999; Khatib, 1934; Lu et al., 1984). Water shield, *Brasenia schreberi*, which is native to North America (Anonymous, 2007), has also been reported as a secondary host in China. It is considered a significant defoliator of aquatic crops. While most quantitative yield loss data are from water chestnut research, *G. birmanica* is described as one of the most damaging insects in water chestnut cultivation (Ding et al., 2006a). *G. birmanica* is thought to be the most promising biological control agent for T. natans in North America because of its damage potential (Ding et al., 2006b).

Adults are typical leaf beetles, small (around 3–5 mm), with a body that is elongated and convex, featuring elytra covering the abdomen. Larvae are cruciform, elongate, initially yellowish-brown but darkening to greyish-brown with dark transverse bands and setae on warts, head and legs black. Eggs are globose and reddish-yellow (Jacoby, 1908). The beetle completes its entire life cycle on water chestnut plants, with all developmental stages occurring on the foliage. Both larvae and adults feed mainly on leaves and petioles by scraping the upper epidermal surface, which leads to the formation of numerous tiny holes. Under heavy infestation, leaves become extensively perforated, greatly reducing the photosynthetic area. Although the fruit integument is generally not attacked under natural conditions, laboratory observations show that the insects can feed on and survive on it. Larvae inflict far greater damage than adults, with third-instar grubs being the most destructive stage. An adult consumes an average of 17.4 mg of green plant material within 24 h. Continuous feeding weakens plants, retards growth, and significantly reduces flowering and nut formation. Under heavy infestation, the leaf blades are completely eaten, leaving only the network of main veins and the spongy petiole intact (Ding et al., 2006b). In severe cases, damaged plants may submerge in water, resulting in substantial yield loss or complete crop failure, sometimes forcing farmers to abandon or replant the crop (Srivastava et al., 1969; Halder et al., 2019).

Singhara beetle, *G. birmanica*, completes its life cycle within 33–46 days under laboratory conditions. On the upper surface of leaves, gravid females laid their eggs in tiny clusters of five to 13, with an average fecundity of 90.4 ± 14.67 eggs. Although oviposition was occasionally seen during the day, egg laying mostly took place at night. The range of egg viability was 53%–64%. Before hatching, the tiny, spherical, bright yellow eggs progressively turned reddish brown. The duration of the incubation phase was between 3.5 and 5.5 days (Halder et al., 2019). The newly hatched larvae were light brown and progressively darkened with age. They were cruciform in shape, robust, elongated, and possessed a well-developed, sclerotized head. Shortly after emergence, larvae remained inactive for a brief period before dispersing over the leaf surface and feeding on the upper epidermis. The larval stage consisted of three instars before pupation, with a total larval period of 11.75–14 days (Halder et al., 2019; Jolivet and Verma, 2002). Before pupation, fully grown third-instar larvae ceased feeding and became stationary by attaching themselves to the leaf surface with a gummy secretion from the anus. This prepupal stage lasted 2.5–3.5 days. The pupa was exarate, bright orange-yellow in color. The pupal period varied from 3.5 to 4.75 days (Halder et al., 2019; Jolivet and Verma, 2002). Adults emerged bright yellow and gradually turned greyish brown. They were generally sluggish and flew only occasionally. Females were slightly larger and lived longer than males, with average lifespans of 21.10 and 16.38 days, respectively (Halder et al., 2019). There are five to nine overlapping generations per year observed in the crop (Srivastava et al., 1969; Halder et al., 2019).

Under conditions in central and southern China, *G. birmanica* reportedly produces up to six generations annually (Lu et al., 1984), whereas in northeastern China, *G. birmanica* normally completes only one to two generations, with adults disappearing from *T. natans* by mid to late August. On the contrary, populations in southeast China may experience estivation during the hottest part of the summer, yet they are still able to generate at least six generations (Ding et al., 2006a). Adults are mobile and may fly short distances between aquatic patches, while larvae remain associated with plant foliage. In *Brasenia schreberi* cultivation in China, *G. birmanica* larvae and adults also consume leaf tissue, indicating potential economic impact on related aquatic crops (Wang et al., 2025).

11.5.1 Management

Management of the Singhara beetle involves regular monitoring to detect early infestations and maintaining pond hygiene, along with effective weed control to minimize beetle habitats. Studies have shown that dipping seedlings' roots in Chlorpyrifos at 0.05% can effectively manage Singhara beetle populations (Padala et al., 2022). In addition, combining *Metarhizium anisopliae* IIVR strain or *Beauveria bassiana* with 1% neem oil in a 1:1 ratio has been reported to provide rapid and efficient control of *G. birmanica* grubs (Halder et al., 2019).

11.6 Rib borer, *Chironomus* spp. (Order: Diptera; family: Chironomidae)

The Chironomidae, a family of dipteran insects, are found in every region of the world, even Antarctica. Only a few areas lack representatives of this family. While the immature stages of most species develop in freshwater, numerous species are also known to inhabit terrestrial, marine, and brackish environments (Ashe et al., 1987). It is extensively distributed in Makhana ponds of North Bihar and other aquatic crop systems. Recognized as a serious internal feeder, causing rapid decline of leaves and affecting overall plant health productivity (Mishra et al., 1992; Jha et al., 1991).

Although many chironomid species initially resemble mosquitoes, they can be identified by their extended mouthparts and lack of wing scales, which are traits of the Culicidae family. The larvae have a slender, cylindrical body with opposing mandibles attached to a rigid, nonretractable head capsule. They have one pair of unjointed anterior parapods, or "prolegs," on the first segment of their bodies, and one pair of unjointed posterior parapods on the last segment, which also has one or two pairs of anal tubules and a pair of procerci that contain setae (Sharma and Gupta, 2014).

Larvae of chironomid midges are widespread and common members of epiphytic invertebrate communities in wetlands and lake ecosystems (Woodcock et al., 2005; Langdon et al., 2010). Many species function as primary consumers, feeding on algae and detritus, and they, in turn, provide an important food source for other organisms, such as invertebrates, fish, and waterfowl. Through this role, they represent a key connection between energy inputs and higher trophic levels in food webs (Taylor and Batzer, 2010; Anderson et al., 2012; Kovalenko et al., 2013). Living macrophytes mainly act as a substrate, while plant detritus serves both as habitat and as a food resource (Williams et al., 2007; Bhat et al., 2012). Their strong ability to colonize new areas, broad feeding flexibility, and capacity to use a variety of substrates enable chironomids to inhabit a wide range of environments (De Lima Behrend et al., 2013; Rosa et al., 2013). Numerous studies have shown that the composition and structure of plant-associated chironomid communities are influenced by multiple environmental factors, including food availability (periphytic algae), habitat complexity (the spatial arrangement and structure of macrophytes), nutrient levels, salinity, oxygen concentration, and pH (Canedo-Argüelles and Rieradevall, 2009; Maasri et al., 2010; Tarkowska-Kukuryk, 2011; Galizzi et al., 2012; Van Der Valk, 2012; Wazbinski and Quinlan, 2013).

Chironomid midges mainly infest aquatic plants, with severe damage commonly observed on the leaf ribs and petioles of *Euryale ferox*. The larvae tunnel within the leaf ribs and aerenchyma tissues, which makes their presence difficult to detect. The larval stage is responsible for causing damage, while adults are short-lived flies, and population levels usually peak during the middle stage of crop growth. Injury occurs when larval boring interferes with nutrient transport, resulting in early leaf deterioration. Typical symptoms include yellowing and wilting of leaves, hollowed or damaged ribs, and the collapse of affected foliage (Mishra et al., 1992). The rib borer (*Chironomus* spp.) first appeared in February, after which its population increased steadily until July, and infestation persisted up to crop harvest in August. Larval incidence was lowest in February, but it began to rise from March onward and remained high through July. Thereafter, the population declined from August onward (Nath et al., 2018b). Often occurs in association with aphids and case worms, increasing combined damage (Mishra et al., 1992; Jha et al., 1991).

In general, *Chironomus* spp. can be effectively managed through preventive measures. These include reducing the field slope to ensure uniform, shallow water levels, which helps suppress bloodworm larval infestations. Appropriate land tillage before planting should be undertaken to expose larvae in the soil to natural enemies. Sound soil management practices, such as crushing and removing plant residues with tractors equipped with paddle wheels, further contribute to lowering larval populations. In addition, flooding and sowing smaller field sections are recommended to reduce bloodworm infestation and minimize crop damage. For efficient control of chironomid midges in rice, fipronil is advised. Fipronil seed treatment at 12.5 g a.i. per ha offers superior control for 9–14 days following sowing. Remaining control is extended to 14–19 days at higher rates (25–50 g a.i. per hectare). Depending on crop damage and monitoring, a postsowing chlorpyrifos spray at 75 g a.i. per ha (10–15 DAS) may be employed in cases of high infestation (Stevens et al., 1998).

11.7 Leaf blight disease

Leaf blight is a widely occurring disease of makhana, caused by *Alternaria alternata*, characterized by dark brown to black, irregular or nearly circular necrotic spots on the leaves. The spots typically show concentric rings, producing a distinct target-board pattern, and often coalesce to form larger blighted areas. Under severe infection, the leaves are completely damaged, and their long stalks turn yellow and dry (Mishra et al., 2003a). In infected cells, *Alternaria alternata* hyphae occurred both intercellularly and intracellularly, causing severe cytoplasmic degradation, disruption of the plasma membrane, and disorganization of cellular organelles. Chloroplasts showed reduced size, altered shape, disrupted envelopes, deformed grana, and complete absence of starch grains, though plastoglobules and lipid bodies remained unaffected. Mitochondria increased in number but exhibited membrane damage, while vacuoles became numerous with disrupted tonoplasts (Daly, 1976; Wheeler, 1975; Gray et al., 1983). In advanced stages, cellular structures collapsed, making organelles indistinguishable. As a weak necrotrophic pathogen, *A. alternata* initially damages host cell structures and later causes complete cellular disintegration, producing ultrastructural changes similar to stress-induced chlorosis and senescence, likely mediated by fungal toxins such as T-toxin. Its incidence can be reduced by up to 90.5% with three sprays of copper oxychloride applied at 15-day intervals (Haidar and Nath, 1987).

11.8 Nutrient management

On account of producing huge foliage and total plant biomass rich in nutrient content, Makhana is considered a heavy feeder crop. Conventionally, no external fertilization was given to Makhana growing in ponds, as the pond systems are considered self-sustaining in nutrient supply due to decomposition of the leftover plant biomass post Makhana harvesting. However, under intensive cultivation in field conditions, wherein more than a single crop (Makhana) is to be taken, external application of nutrients is required to sustain nutrient supply to the growing plants and the crop(s) that follow. For a high-yielding Makhana variety, such as *Swarna Vaidehi*, 100:60:40 kg of N: P_2O_5: K_2O is recommended for 1.0 ha of Makhana farming in field conditions (Singh et al., 2020). Organic matter is extremely important in Makhana cultivation. Farmyard manures or other organic sources of nutrients should therefore be added sufficiently (10-20 t/ha) to reduce the reliance on mineral fertilizers.

As the soils of eastern and north-eastern India are mostly deficient in micronutrients like zinc and boron, which are of critical importance to crop production in the region, additional yield benefit can be harvested by micronutrient fertilization in Makhana farming. Recent experimental observations have rather confirmed the positive effects of secondary nutrient (Mg) application and micronutrient fertilization (Zn, B, Cu) on yield and quality of Makhana in north Bihar, India (Kumar, 2022). Recommended NPK application can increase Makhana seed yield by 20%–25%. Additional 25% yield advantage can be obtained by micronutrient application. Foliar spray of $CuSO_4$ (0.1%) has been found particularly useful in improving the yield and quality of Makhana. Copper Sulfate spray leads to the formation of bigger and bolder seeds, and also reduces pest and disease incidences in the Makhana crop. Incidents of major insect pests of Makhana are also observed to be less in the crops sufficiently fertilized with secondary and micronutrients, including Mg, Zn, B, and Cu. Thus, proper nutrient management along with mindful pest control measures is the key to successful Makhana production.

11.9 Conclusion

Makhana cultivation, once considered relatively resilient to insect attack, is increasingly challenged by a complex assemblage of field and storage pests that threaten crop productivity and farmer livelihoods. The diversity of insect species exploiting different plant parts reflects both the ecological richness of aquatic agro-ecosystems and the vulnerabilities associated with pond-based cultivation. Key pests such as aphids, case worms, rib and root borers, and leaf-feeding beetles can cause substantial damage individually, while their combined occurrence often amplifies yield losses. The emergence of post-harvest pests further underscores the need to extend pest management beyond the field. Given the ecological sensitivity of aquatic systems, pest management in makhana must rely on integrated approaches that emphasize monitoring, cultural practices, biological regulation, and judicious use of chemicals. Strengthening research on pest ecology, economic thresholds, and environmentally safe control measures will be essential for sustaining makhana production under expanding commercial demand. This chapter highlights the importance of informed pest management as a critical component of sustainable makhana agro-ecosystems.

References

Ali, A., Desneux, N., Lu, Y., Liu, B., Wu, K., 2016. Characterization of the natural enemy community attacking cotton aphid in the Bt cotton ecosystem in northern China. Sci. Rep. 6. https://doi.org/10.1038/srep24273.

Alum, A.Z., 1967. Insect pests of rice in East Pakistan. In: Major Insect Pests of Rice Plant. Johns Hopkins Press, pp. 633–655.

Anderson, T.J., Stelzer, R.S., Drecktrah, H.G., Eggert, S.L., 2012. Secondary production of Chironomidae in a large eutrophic lake: implications for lake sturgeon production. Freshw. Sci. 31, 365–377.

Anonymous. (2007). Floating-leaved Rooted Plants: Brasenia schreberi Gmel., Water-Shield. Washington State Department of Ecology. http://www.ecy.wa.gov/PROGRAMS/wq/plants/plantid2/descriptions/brasch.html.

Ashe, P., Murray, D.A., Reiss, F., 1987. The zoogeographical distribution of Chironomidae (Insecta: Diptera). Annales de Limnologie – Int. J. Limnol. 23 (1), 27–60. https://doi.org/10.1051/limn/1987003.

Bhat, S.U., Dar, G.A., Sofi, A.H., Dar, N.A., Pandit, A.K., 2012. Macroinvertebrate community associations on three different macrophyte species in Manasbal Lake. Res. J. Environ. Sci. 6, 62–71.

Biswas, K., Calder, C., 1936. Handbook of Common Water and Marsh Plants of India. Government of India Press.

Biswas, S., Ghosh, A.K., 1981. A preliminary study on life system of *Rhopalosiphum nymphaeae* (Linn.) (Homoptera: Insecta). J. Univ. Gauhati 26–27, 129–134.

Blackman, R. L., & Eastop, V. F. (2006). Aphids on the World's Herbaceous Plants and Shrubs. John Wiley & Sons. (in association with the Natural History Museum).

Brenière, J. (1983). The Principal Insect Pests of Rice in West Africa and Their Control. West Africa Development Association.

Canedo-Argüelles, M., Rieradevall, M., 2009. Quantification of environment-driven changes in epiphytic macroinvertebrate communities associated with *Phragmites australis*. J. Limnol. 68, 229–241.

Center, T. D., Dray, F. A., Jr., Jubinsky, G. P., & Grodowitz, M. J. (1999). Insects and Other Arthropods that Feed on Aquatic and Wetland Plants (Technical Bulletin No. 1870). U.S. Department of Agriculture, Agricultural Research Service.

Chen, F., Song, S., & Wu, C. (2006). A review of the genus Parapoynx Hübner in China (Lepidoptera: Pyralidae: Acentropinae). Aquat. Insects, 28(4), 291–303.

Chen, Q., Chen, Z.S., Gu, X.S., Ma, L., Wang, X., Huang, G.H., 2017. The complete mitogenome of *Parapoynx crisonalis* (Walker, 1859) (Lepidoptera: Crambidae), with phylogenetic relationships amongst three acentropine larval forms. Aquat. Insects 38 (1–2), 79–91.

Chi, T.T.N., Tam, B.T.T., Dau, H.X., Khoa, N.T., Lan, N.T.P., Paris, T.R., 1995. Current status of rice pest management by farmers in direct-seeded rice and transplanted rice areas. Omonrice 4, 42–50.

Dale, D., 1994. Insect pests of the rice plant: their biology and ecology. In: Heinrichs, E.A. (Ed.), Biology and Management of Rice Insects. Wiley Eastern, pp. 363–485.

Daly, J.M., 1976. The carbon balance of diseased plants: changes in respiration, photosynthesis and translocation. In: Heitefuss, R., Williams, P.H. (Eds.), Physiological Plant Pathology (Encyclopedia of Plant Physiology, New Series, vol. 4. Springer-Verlag, pp. 450–479.

Davis, A.M. (2012). reportA Review of the Status of Microlepidoptera in Britain (Butterfly Conservation Report No. S12-02). Butterfly Conservation.

De Lima Behrend, R.D., Teixeira, M.C., Prata-Fernandes, S.E., Carvalho-Camargo, J., Rosin, G. C., Michiyo-Takeda, A., 2013. Effects of a native and a nonnative macrophyte species of Hydrocharitaceae on Chironomidae and Oligochaeta assemblage structure. Acta Sci. Biol Sci. 35, 351–359.

Dedryver, C.A., Le Ralec, A., Fabre, F., 2010. The conflicting relationships between aphids and men: a review of aphid damage and control strategies. C. R. Biol. 333 (6–7), 539–553. https://doi.org/10.1016/j.crvi.2010.03.009.

Ding, J., Blossey, B., Du, Y., Zheng, F., 2006a. *Galerucella birmanica* (Coleoptera: Chrysomelidae), a promising potential biological control agent of *Trapa natans*. Biol. Control 36 (1), 80–90. https://doi.org/10.1016/j.biocontrol.2005.08.001.

Ding, J., Blossey, B., Du, Y., Zheng, F., 2006b. Impact of *Galerucella birmanica* (Coleoptera: Chrysomelidae) on growth and seed production of *Trapa natans*. Biol. Control 37 (3), 338–345. https://doi.org/10.1016/j.biocontrol.2005.12.003.

Fei, H., Su, Q., Zhang, X., Fei, H., Su, Q., Zhang, X., 1995. Effects of immigrated population and climatic factors on field population dynamics of rice caseworm. Acta Phytophylacica Sin. 22 (3), 193–197.

Galizzi, M.C., Zilli, F., Marchese, M., 2012. Diet and functional feeding groups of Chironomidae (Diptera) in the middle Paraná river floodplain (Argentina). Iheringia, Série Zoologia 102, 117–125.

Gogoi, H., Bora, D., 2012. Spatial distribution of *Nymphula depunctalis* Guenée larvae (Lepidoptera: Pyralidae), an early vegetative pest of *Oryza sativa* L. Acad. J. Entomol. 5 (1), 41–46.

Gogoi, H., Bora, D., 2013. High yielding rice cultivars, high rainfall and high humidity favour *Nymphula depunctalis* (Lepidoptera: Pyralidae) to reach major pest status: a study in Dhemaji District, Assam. Natl. Acad. Sci. Lett. 36 (5), 469–475.

Goren-Inbar, N., Melamed, Y., Zohar, I., Akhilesh, K., Pappu, S., 2014. Beneath still waters—Multistage aquatic exploitation of *Euryale ferox* (Salisb.) during the Acheulian. Internet Archaeol. 37, 1–42.

Gray, D.J., Amerson, H.V., Van Dyke, C.G., 1983. Ultrastructure of the infection and early colonization of *Pinus taeda* by *Cronartium quercuum* formae speciales *fusiforme*. Mycologia 75, 117–130.

Grist, D.H., Lever, R.J.A.W., 1969. Pests of Rice. Longmans.

Haidar, M.G., Nath, R.P., 1987. Chemical control of *Alternaria* leaf blight of makhana (*Euryale ferox*). Nat. Acad. Sci. Lett. 10 (9), 301–302.

Haines, H.H., 1925. The Botany of Bihar and Orissa (Part II). Government Press.

Halder, J., Rai, A.B., Kushwaha, D., 2019. Bionomics and biorational management of Singhara beetle, *Galerucella birmanica* Jacoby, a potential threat to water chestnut production in India. Int. J. Entomol. Res. 4 (4), 77–80.

Halder, J., Rai, A.B., Chakrabarti, S., Dey, D., 2020. Distribution, host range and bionomics of *Rhopalosiphum nymphaeae* (Linnaeus, 1761), a polyphagous aphid in aquatic vegetables. Defence Life Sci. J. 5 (1), 49–53.

Hill, D.S., 1987. Agricultural Insect Pests of the Tropics and Their Control, second ed. Cambridge University Press.

Holman, J., 2009. Host Plant Catalog of Aphids: Palearctic Region. Springer. https://doi.org/10.1007/978-1-4020-8286-3.

Houlihan, D.F., 1969. Respiratory physiology of the larva of *Donacia simplex*, a root-piercing beetle. J. Insect Physiol. 15, 1517–1536.

Huang, G.H., Li, J.H., 2013. Color Handbook of Insect Pests of Aquatic Vegetables in China.

International Plant Health Management, 2025. Principles and Concept of Integrated Pest Management. National Institute of Plant Health Management (NIPHM).

Jach, M.A., 1998. Annotated checklist of aquatic and riparian/littoral beetle families of the world (Coleoptera). In: Jäch, M.A., Ji, L. (Eds.), Water Beetles of China, vol II. Zoologisch-Botanische Gesellschaft & Wiener Coleopterologenverein, pp. 25–42.

Jacoby, M., 1908. Coleoptera: Chrysomelidae, vol 1. Taylor & Francis.

Jha, V., Kargupta, A.N., Dutta, R.N., Jha, U.N., Mishra, R.K., Saraswati, K.C., 1991. Utilization and conservation of *Euryale ferox* Salisbury in Mithila (North Bihar), India. Aquat. Bot. 39 (3–4), 295–314.

Jolivet, P., Hawkeswood, T.J., 1995. Host-Plants of Chrysomelidae of the World. Backhuys Publishers.

Jolivet, P., Verma, K.K., 2002. Biology of Leaf Beetles. Intercept Ltd.

Kalita, S., Hazarika, L.K., Deka, R.L., Gayon, J., 2020. Effect of weather parameters and varieties on occurrence of insect pests and natural enemies of rice. J. Pharmacogn. Phytochem. 9, 2429–2438.

Kattali, M., Kottickal, L.V., Sudheer, S., Ramesh, N., 2013. Genetic analysis of rice caseworm, *Parapoynx stagnalis*, to determine its phylogenetic status. Int. J. Cell Mol. Biol. 2, 1–4.

Khatib, M.H., 1934. The life-history and biology of *Galerucella birmanica* Jacoby (Coleoptera: Chrysomelidae) and the external morphology of larva and pupa. In: Proceedings of the Indian Academy of Sciences, 3, pp. 715–732.

Kumar, M., Padala, V.K., Ramya, N., Chakraborty, A., Dey, J.K., 2021. Plant health issues in fox nut/makhana (*Euryale ferox*): an agronomic perspective. J. Plant Health Iss. 2 (1), 30–35.

Kumar, M., 2022. Effect of secondary and micronutrients on yield and quality of fox nut (*Euryale ferox* Salisb.) in subtropical climate. J. Plant Nutr. 46 (2), 212–218.

Kovalenko, K.E., Ciborowski, J.J.H., Daly, C., Dixon, D.G., Farwell, A.J., Foote, A.L., Frederick, K.R., Gardner-Costa, J.M., Kennedy, K., Liber, K., Roy, M.C., Slama, C.A., Smits, J.E.G., 2013. Food web structure in oil sands reclaimed wetlands. Ecol. Appl. 23, 1048–1060.

Kuhar, T.P., Reiter, S., & Doughty, H. 2005. Green Peach Aphid on Vegetables. Virginia Cooperative Extension.

Langdon, P.G., Ruiz, Z., Wynne, S., Sayer, C.D., Davidson, T.A., 2010. Ecological influences on larval chironomid communities in shallow lakes: implications for paleolimnological interpretations. Freshw. Biol. 55, 531–545.

Lee, W.W., Shin, T.Y., Bae, S.M., Woo, S.D., 2015. Screening and evaluation of entomopathogenic fungi against the green peach aphid, *Myzus persicae*, using multiple tools. J. Asia Pac. Entomol. 18 (4), 607–615. https://doi.org/10.1016/j.aspen.2015.08.004.

Li, C., Yu, J., Mao, R., Kang, K., Xu, L., Wu, M., 2024. Functional and numerical responses of *Harmonia axyridis* (Coleoptera: Coccinellidae) to *Rhopalosiphum nymphaeae* and their potential for biological control. Insects 15 (9). https://doi.org/10.3390/insects15090633.

Lu, Z.Q., Zhu, J., Zhu, S., Chen, Z.D., 1984. Preliminary studies on the beetle *Galerucella birmanica* Jacoby—an insect pest of water chestnut and water shield. Acta Entomol. Sin. 27, 73–76.

Maasri, A., Fayolle, S., Franquet, E., 2010. Algal foraging by a rheophilic chironomid (*Eukiefferiella claripennis* Lundbeck) extensively encountered in high nutrient-enriched streams. Fundament. Appl. Limnol. 177, 151–159.

Mishra, R.K., Jha, V., Dehadrai, P.V. (Eds.), 2003a. Makhana. Directorate of Information and Publications on Agriculture. Indian Council of Agricultural Research.

Mishra, R.K., Jha, V., Jha, B.P., 2003b. Associations of insects and gastropods on makhana with reference to pests and their control. In: Mishra, R.K., Jha, V., Dehadrai, P.V. (Eds.), Makhana. ICAR, pp. 124–144.

Mishra, R.K., Singh, S.P., Jha, D.N., 1992. Insect associations of *Euryale ferox* Salisb. In the ponds of Darbhanga, north Bihar. J. Freshw. Biol. 4 (3), 199–208.

Nath, P., Kumar, A., Yadav, P.K., Kumar, R., 2018a. Estimation of losses caused by aphid (*Rhopalosiphum nymphaeae*), caseworms (*Elophila depunctalis* and *E. crisonalis*), and rib borer (*Chironomus* sp.) in makhana crop. J. Pharmacogn. Phytochem. SP4, 199–201.

Nath, P., Kumar, A., Yadav, P.K., Kumar, R., 2018b. Emerging pests of makhana (*Euryale ferox* Salisb.) crop in Koshi region of Bihar. Int. J. Curr. Microbiol. Appl. Sci. 7, 4605–4609.

Nath, P., Kumar, A., Yadav, P.K., Kumar, R., 2018c. Estimation of losses caused by aphid (*Rhopalosiphum nymphaeae*), caseworms (*Elophila* spp.) and rib borer (*Chironomus* sp.) in makhana (*Euryale ferox*) crop. Pharm. Innov. 7 (4), 630–634.

Padala, V.K., Ramya, N., Kumar, M., Rajna, S., Rajashekhar, M., Venkatesh, Y.N., Raut, S.M., Chander, S., Singh, I.S., Monobrullah, M., 2023. Spatial distribution and optimum sample size for monitoring of water lily aphid *Rhopalosiphum nymphaeae* (L.) in makhana *Euryale ferox* Salisb. Int. J. Trop. Insect Sci. 43 (6), 2167–2177.

Padala, V.K., Ramya, N., Rajashekhar, M., Keerthi, M.C., Kumar, M., Raut, S.M., Singh, I.S., Monobrullah, M., 2022. Chlorpyrifos root dip treatment effect on population of Singhara beetle, *Galerucella birmanica* Jacoby, natural enemies and yield of water chestnut. Natl. Acad. Sci. Lett. 45 (5), 379–382.

Palmer, M.A.M., 1953. Aphids of the rocky mountain region. Ann. Entomol. Soc. Am. 46 (3), 365. https://doi.org/10.1093/aesa/46.3.365.

Patel, G., Chauhan, S., Kumar, R., Mourya, P.K., 2024. Management of rice caseworm (Nymphula depunctalis) in rice in eastern region of Uttar Pradesh, India. J. Exper. Agricult. Int. 46 (8), 866–874.

Pathak, M.D., 1975. Insect Pests of Rice. International Rice Research Institute.

Pemberton, R.W., 1999. Natural enemies of *Trapa* spp. in northeast Asia and Europe. Biol. Control 14 (3), 168–180. https://doi.org/10.1006/bcon.1998.0687.

Rabasse, J.M., van Steenis, M.J., 1999. Biological control of aphids. In: Integrated Pest and Disease Management in Greenhouse Crops. Springer, pp. 235–243.

Regier, J.C., Mitter, C., Solis, M.A., Hayden, J.E., Landry, B., Nuss, M., Simonsen, T.J., Yen, S.H., Zwick, A., Cummings, M.P., 2012. A molecular phylogeny for the pyraloid moths

(Lepidoptera: Pyraloidea) and its implications for higher-level classification. Syst. Entomol. 37 (4), 635–656.

Reissig, W.H., Heinrichs, E.A., Litsinger, J.A., Moody, K., Fiedler, L., Mew, T.W., Barrion, A.T., 1985. Illustrated Guide to Integrated Pest Management in Rice in Tropical Asia. International Rice Research Institute.

Rosa, B.F.J.V., Dias-Silva, M.V.D., Alves, R.G., 2013. Composition and structure of the Chironomidae (Insecta: Diptera) community associated with bryophytes in a first-order stream in the Atlantic Forest, Brazil. Neotrop. Entomol. 42, 15–22.

Saruhan, İ., 2018. Efficacy of some entomopathogenic fungi against *Aphis fabae* Scopoli (Hemiptera: Aphididae). Egypt. J. Biol. Pest Contr. 28. https://doi.org/10.1186/s41938-018-0090-3.

Singh, I.S., Kumar, M., Raut, S.M., Thakur, A.K., Singh, S.P., 2020. Integrated nutrient management package for field cultivation of Makhana in North Bihar. J. AgriSearch 7 (3), 138–141.

Sison, P. (1938). Some observations on the life-history, habits, and control of the rice caseworm, Nymphula depunctalis Guen. 273-301.

Sharma, M.R., Gupta, V., 2014. Morphological identification of Chironomus larvae in Jaipur district (Rajasthan) India. Microbiology 3 (9), 411–413.

Srivastava, A.S., Gupta, B.P., Awasthi, G.P., Pande, B.P., Saxena, H.P., 1969. Bionomics, nature of damage and control of *Galerucella birmanica* Jacoby—pest of waternut (Coleoptera: Chrysomelidae). Beiträge Entomol. 19 (3–6), 607–616.

Subramaniam, K., 1962. Aquatic Angiosperms, pp. 6–8.

Stevens, M.M., Helliwell, S., Warren, G.N., 1998. Fipronil seed treatments for the control of chironomid larvae (Diptera: Chironomidae) in aerially-sown rice crops. Field Crops Res. 57 (2), 195–207.

Sunderland, K.D., 1988. Carabidae and other invertebrates. In: Minks, A.K., Harrewijn, P. (Eds.), World Crop Pests: Aphids—Their Biology, Natural Enemies and Control, B. Elsevier, pp. 293–310.

Tarkowska-Kukuryk, M., 2011. Composition and distribution of epiphytic midges (Diptera: Chironomidae) in relation to emergent macrophyte cover in shallow lakes. Pol. J. Ecol. 59, 141–151.

Taylor, A.N., Batzer, D.P., 2010. Spatial and temporal variation in invertebrate consumer diets in forested and herbaceous wetlands. Hydrobiologia 651, 145–159.

Van der Valk, A.G., 2012. The Biology of Freshwater Wetlands, second ed. Oxford University Press.

Vromant, N., Rothuis, A.J., Cuc, N.T.T., Ollevier, F., 1998. The effect of fish on the abundance of rice caseworm *Nymphula depunctalis* (Guenee) (Lepidoptera: Pyralidae) in direct-seeded concurrent rice–fish fields. Biocontrol Sci. Technol. 8, 539–546.

Wang, Y., Wang, Y., Zhou, C., 2025. Population dynamics of *Galerucella birmanica* and its aggregation behavior in *Brasenia schreberi* aquaculture systems. Insects 16 (4), 371. https://doi.org/10.3390/insects16040371.

Wazbinski, K.E., Quinlan, R., 2013. Midge (Chironomidae, Chaoboridae, Ceratopogonidae) assemblages and their relationship with biological and physicochemical variables in shallow, polymictic lakes. Freshw. Biol. 58, 2464–2477.

Wenninger, E.J., 2011. *Sugar beet Root Aphids: Identification, Biology, and Management* (CIS No. 1176). University of Idaho Extension.

Wheeler, H., 1975. Plant Pathogenesis. Springer-Verlag.

Williams, D.D., Heeg, N., Magnusson, A.K., 2007. Habitat background selection by colonizing intermittent pond invertebrates. Hydrobiologia 592, 487–498.

Woodcock, T., Longcore, J., McAuley, D., Mingo, T., Bennatti, C.R., Stromborg, K., 2005. The role of pH in structuring communities of marine wetlands macrophytes and chironomid larvae (Diptera). Wetlands 25, 306–315.

Yen, S. H. (2004). Insecta: Lepidoptera, Crambidae, Acentropinae. In Freshwater Invertebrates of the Malaysian Region. (pp. 545–554).

Chapter 12

Makhana processing waste valorization

Dharmender, Rahul Kumar Rout, Anupam Sahoo, Akhileshwar Vishwakarma, Shivani Jha and Manoj Kumar
ICAR-National Research Centre for Makhana, Darbhanga, Bihar, India

12.1 Introduction

With the increasing awareness of the nutritional and medicinal values of Makhana, its demand is growing globally. Considering its commercial, medicinal, and nutritional importance, the seeds of Makhana are also known as 'black diamond,' and the puffed product as 'white gold.' Makhana can be considered a superfood due to its numerous health benefits. Numerous studies have been documented, citing the health benefits of Makhana consumption, thereby giving the crop a global platform. However, once the edible portions are utilized, remaining non-edible portions like fruit sheath, leaves, stems, roots, shells, etc. remain under-utilized. Even after popping, the semi popped Makhana (locally called *Thurri*) fails to achieve market value. These *thurri* are generally converted into powder form to be incorporated into other value-added products like Makhana kheer mix, pasta, cookies, sweets, bakery items, etc. The non-edible parts are mostly used as farmyard manure, or simply discarded. But few studies documented the beneficial compounds and functional properties in these wastes and by-products, which, if processed properly, can fetch good market values. Surprisingly, every part of the plant has some important compounds, which can be used for various non-feed applications. The following sections explain the wastes and by-products generated from Makhana, their useful compounds, functional properties shown by various components, and industrial opportunities. The various constraints observed in these sectors are also briefed, and possible solutions to sustain the market have also been discussed in depth.

Makhana (Euryale ferox) as a Functional Food. https://doi.org/10.1016/B978-0-443-44421-0.00002-0

12.2 Makhana wastes and by-products

12.2.1 The waste paradigm

The Makhana industry generates a huge amount of waste that has been perceived as a weakness in the past instead of a strength. Nevertheless, as there has been a rise in the focus on sustainability and circular bioeconomy strategies, these residues are also being taken into consideration as nutritional, biochemical, and industrial resources. Starch fractions, lignocellulosic materials and bioactive compounds are makhana by-products which are generally utilized to achieve product valorization. Makhana processing gives rise to several categories of by-products and residues, and each one has its own physico-chemical and functional nature that may be used in the value-added applications:

(a) **Seed Shells (Pericarp):** One of the major by-products of popping and polishing processes is the seed coat (pericarp) of Euryale ferox (makhana), which happens to be the hard, black, stony outer covering of the seed and makes up about 35–40% of the fresh seed weight. This exterior layer breaks during the process of popping and is the most abundant processing residue. It has long been a low-value material, either utilized as a low-grade fuel or discarded. The scientific evidence, however, is increasing that this fraction is an active and chemically diverse fraction that is difficult to classify as waste. According to phytochemical studies, the makhana seed coat has a broad range of bioactive molecules with different classes such as phenols, flavonoids, tannins, alkaloids, glycosides, terpenoids, and steroids. The most diverse range of these compounds is solubilized in polar solvents, but mixtures of ethanol and water are particularly more effective, with a recovery of approximately 8–9% under the conditions of maceration, and even greater under the Soxhlet solvent extraction condition, suggesting that polar phytoconstituents are the most abundant. According to the quantitative analyses, the total phenolic content (approximately 51 mg GAE/g dry extract) and significant flavonoid concentration (approximately 16 mg QE/g) are reported and relate to the presence of a high level of antioxidant activity. The presence of very low values of IC0 in DPPH assays serves as evidence of an efficient free-radical scavenging activity. Moreover, seed coat extracts are also antibacterial against Escherichia coli and Staphylococcus aureus, but tend to have a stronger activity against Gram-positive bacteria. HPTLC fingerprinting also allows confirming the existence of various bioactive components and provides an effective instrument of quality control and compound profiling. All these results indicate that makhana seed coat can be used as a promising source of natural antioxidants and antimicrobial agents in nutraceutical and functional applications, which will help in valorization instead of disposal.

(b) **Unpopped (Grade-B) Seeds:** A large proportion of the seeds do not pop when roasted because of an uneven distribution of moisture or structural flaws or variation of size. Even though they are classified as low grade in the trade, they are still composed like intact kernels. They are mainly starch, moderate protein and very low-fat contents. Physio-chemically, they have good hydration and swelling ability as in starchy seeds. Carbohydrates and small bioactive compounds are stored in them biochemically. They are millable functionally into flour, starch recoveries, extrusion products, or fermentation feedstocks. The fact that they are now diverted to animal feed constitutes a waste of a nutritionally good resource.

(c) **Vegetative Biomass:** The plant of the makhana grows large, thorny, and orbicular leaves (usually more than 1 meter across), as well as a thick rhizomatous root system, and thus produces a large vegetative biomass. This biomass can be deposited in the water body during harvest to decompose, and lead to nutrient loading and eutrophication, or can be removed and discarded manually (Rai et al., 2002). Physico-chemically, this biomass is full of fibrous structural carbohydrates and organic substance. It is mostly composed of plant polysaccharides and small phytochemicals biochemically. It can be used as a source of composting, vermicomposting, or bioenergy generation in terms of functionality.

(d) **Broken Kernels:** Broken kernels occur during the processing of roasting, popping, grading, and handling processes because of mechanical stress and thermal expansion. Despite being classified as a low-grade fraction in commerce, their composition is not much different compared to intact kernels since they are formed of the edible endosperm. They mostly contain a high amount of starch, average amount of protein, and low amount of fat. Through structural disruption, broken kernels have increased surface area and increased water absorption, and this can enhance the digestibility and techno-functional behavior of processed foods. Valorization wise, this fraction is particularly favorable to flour production, starch separation, extrusion products, and fermentation feeds. They contain resistant starch which also increases their functionality as well as prebiotic foods.

(e) **Sludge Processing:** Sludge processing occurs when the sludge is suspended in the process water resulting during washing, soaking, and polishing processes whereby grains of suspended solids, fine starch granules and soluble organic material are introduced into the process water. This fraction is usually dispersed starch particles, residual minor protein, and water-soluble phytochemicals. No special compositional research on makhana sludge has been conducted, but the experience of similar cereal and seed-processing plants indicates that these residues are biodegradable and are rich in organic load. When released into the environment, they could increase the biological oxygen demand (BOD) of wastewater. On the other hand, this stream presents the possibilities of valorization in the form of starch recovery, anaerobic digestion to produce biogas and compost.

(f) **Shell Residues:** Shell residues are several rigid outer fragments which are detached in the process of popping. These are hard and fibrous residues where the main type of materials is lignocellulosic material or cellulose, hemicellulose, and lignin. They have low digestible nutrients and exhibit good mechanical strength and porosity. These properties qualify them to be used in non-food applications such as biochar, making of activated carbon, preparation of biodegradable composite fillers and bio-mass fuel. They also biodegrade at a relatively slow rate promoting their application in compost blends in need of structural stability.

12.2.2 The circular economy

The "Waste to Wealth" concept posits that these by-products are not refuse but rather misplaced resources. Recent advancements in food engineering, metabolomics, and material science have illuminated the potential of these wastes. The shells are rich in lignocellulose and polyphenols; the unpopped seeds are reservoirs of unique starch; and the leaves are abundant in flavonoids. This chapter explores the technological interventions that allow these materials to be upcycled into adsorbents, functional food ingredients, active packaging materials, and pharmaceutical bioactives.

12.2.3 Phytochemistry and metabolomics of waste streams

To effectively valorize waste, a deep understanding of its chemical constitution is essential. The non-edible parts of *E. ferox* are reservoirs of secondary metabolites, which the plant produces for defense and adaptation.

12.2.3.1 Metabolomic diversity in tissues

A recent comparative metabolomic profiling using UHPLC-MS/MS mapped the distribution of 589 secondary active substances across five different tissues: roots, rhizomes, leaves, seed shells, and kernels. The study revealed that polyphenols account for 87% of these compounds, but their distribution is highly tissue-specific (Liu et al., 2023):

(a) **Leaves:** The leaves were found to accumulate significant amounts of flavonoids. Specifically, they are rich in flavonoid O-glycosides, such as quercetin-3-O-galactoside (hyperoside) and quercetin-3-O-rutinoside (rutin). The accumulation of these compounds is regulated by the expression of genes like *PAL* (phenylalanine ammonia-lyase), 4CL (4-coumarate: CoA ligase), and *CHS* (chalcone synthase) (Liu et al., 2023).

(b) **Seed Shells:** The shells are distinct in their accumulation of phenolic acids rather than flavonoids. Metabolomic analysis reveals that shells contain high levels of gallic acid and ellagic acid (Liu et al., 2023). These phenolic compounds are known for their strong antioxidant, anti-inflammatory, and antimicrobial activities.

(c) **Roots and Rhizomes:** These tissues are rich in lipids and specific phenolic acids, contributing to the plant's defense mechanisms and offering potential for the extraction of lipid-soluble bioactives (Liu et al., 2023).

This differential accumulation suggests that a biorefinery approach could be adopted: leaves could be processed for flavonoid extraction (cardioprotective agents), while shells could be targeted for phenolic acid extraction (antioxidants/preservatives) Table 12.1.

12.2.4 Valorization of makhana shells

The hard shell of the Makhana seed is the most abundant and chemically resilient by-product. Its high carbon content and lignocellulosic nature make it an excellent precursor for carbon-based materials, while its polyphenol content makes it suitable for bioactive extraction.

12.2.4.1 Production of activated carbon

Activated carbon (AC) is a versatile adsorbent used widely in water treatment, air purification, and catalysis. Makhana shells, with their natural hardness and density, are ideal precursors for producing high-quality AC.

(a) Phosphoric acid (H_3PO_4) activation

Chemical activation using H_3PO_4 promotes the depolymerization of lignin and hemicellulose at lower temperatures, preserving the carbon structure. Optimal conditions reported involve an impregnation ratio of 1.5 (H_3PO_4: Shells) and an activation temperature of 700°C. Under these conditions, the carbon exhibited a BET surface area of 2636 m^2/g and a total pore volume of 1.53 cm^3/g (Kumar and Jena, 2016). Fourier Transform Infrared (FTIR) spectroscopy confirmed the development of abundant surface functional groups, including hydroxyls (-OH) and carbonyls (C=O), which enhance the adsorption of polar molecules (Kumar and Jena, 2016).

(b) Zinc chloride ($ZnCl_2$) activation

$ZnCl_2$ acts as a dehydrating agent, restricting the formation of tar and favoring pore development. Research indicates that an impregnation ratio of 2.0 and a carbonization temperature of 600°C yields microporous carbon with a maximum BET surface area of 3073 m^2/g and a micropore volume of 1.25 cm^3/g (Kumar and Jena, 2015). While both agents produce high-quality carbon, $ZnCl_2$ activation tends to yield slightly higher surface areas suitable for gas storage (like hydrogen or methane) or capturing small molecules, whereas H_3PO_4 activation is often preferred for environmental safety and developing mesopores for larger organic molecules Table 12.2.

TABLE 12.1 Major metabolites from the metabolomics of raw makhana seeds and popped makhana (Sinha and Jha, 2025).

S. No.	Metabolite name	Group	MW	RT (min)	Formula	Bioactive function
1	Phosphoric acid	Oxo acid	98	6.48	H_3PO_4	Storage and consumption of energy and help kidney function
2	Glycerol	Sugar alcohol	92.09	7.539	$C_3H_8O_3$	Skin hydration, wound healing, and treating high intracranial pressure
3	Glyceric acid	Sugar Acid	106.08	8.196	$C_3H_6O_4$	Role in metabolic activity, energy production
4	Butanoic acid	Fatty acid	88.11	9.207	$C_4H_8O_2$	Helps in digestion, anti-cancerous, anti-inflammation
5	Decanoic acid	Fatty Acid	172.26	9.612	$C_{12}H_{24}O_2$	Anti-bacterial, AIF, manufacture of ester, fruit flavor, volatile component
6	Dodecanoic acid	Fatty acid	200.31	11.918	$C_{12}H_{24}O_2$	Anti-microbial property
7	Methyl tetradecanoate	Fatty acid	242.40	12.654	$C_{15}H_{30}O_2$	Transport anti-cancer drug, flavoring agent and fragrance
8	Xylitol	Sugar alcohol	152.15	12.973	$C_5H_{12}O_5$	Reduce bacteria in tooth decay, reduce ear infection, help saliva production
9	Myristic acid	Fatty acid	228.38	14.035	$C_{14}H_{28}O_2$	Post-translation protein changes, control metabolic process
10	Methyl L-sorboside	Sugar	212.20	14.063	$C_7H_{14}O_6$	Production of Vit. C, Sweeteners
11	Hexadecanoic acid	Fatty acid	256.42	14.834	$C_{16}H_{32}O_2$	Antioxidants, hypocholesterolemic

12	Methyl galactoside	Sugar	194.18	14.834	$C_7H_{14}O_6$	Hydrolyze the enzyme, regulate metabolic process
13	D-Sorbitol	Sugar alcohol	182.17	15.204	$C_6H_{14}O_6$	Laxative, humectants used as indomethacin, relieve constipation
14	Gallic acid	Phenolic acid	170.12	15.322	$C_7H_6O_5$	Anti-inflammatory, antioxidants, anti-tumor, anti-bacterial, anti-obesity
15	D-Glucose	Sugar	180.15	15.617	$C_6H_{12}O_6$	The primary source of energy
16	Heptadecanoic acid	Fatty acid	270.45	15.719	$C_{17}H_{34}O_2$	Promoting mammalian metabolite
17	Palmitic acid	Fatty acid	256.43	15.904	$C_{16}H_{32}O_2$	Cell function and structural integrity
18	D-allofuranose	Sugar	180.16	13.811	$C_6H_{12}O_6$	Energy provider and help in biological process
19	Methyl stearate	Lipids	298.50	16.498	$C_{19}H_{38}O_2$	Help in neuroprotection, CVD, and mitochondrial function
20	Myo-Inositol	Sugar	180.16	16.502	$C_6H_{12}O_6$	Co-factor of enzyme, messenger molecule in signal transduction
21	Linoleic acid/9,12-Octadecadienoate	Polyunsaturated omega-6 fatty acid, Lipids	280.44	17.108	$C_{18}H_{30}O_2$	Prevention of nerve blood flow, decrease risk of CVD, prevent arrhythmias, decrease thrombosis
22	Stearic acid	Fatty acid	284.48	17.254	$C_{18}H_{36}O_2$	Food additive, negatively regulates type-2 diabetes
23	Farnesol	Sesquiterpenoid alcohol	222.37	17.36	$C_{15}H_{26}O_2$	Anti-cancer, anti-inflammatory
24	Docosanoic acid	Fatty acid	340.58	18.721	$C_{22}H_{44}O$	Cause heart lipidosis, neurotransmitter
25	2-Palmitoylglycerol	Fatty acid ester	330.51	18.891	$C_{19}H_{38}O_4$	Modulate pain sensitivity

Continued

TABLE 12.1 Major metabolites from the metabolomics of raw makhana seeds and popped makhana (Sinha and Jha, 2025).—cont'd

S. No.	Metabolite name	Group	MW	RT (min)	Formula	Bioactive function
26	1-Monopalmitin	Lipids	330.50	19.044	$C_{19}H_{38}O_4$	Drug delivery, lipid-based product inhibits cell proliferation synthesis
27	13-Docosenoic acid	Fatty acid	338.56	19.162	$C_{22}H_{42}O_2$	Impair heart function
28	Tricosanoic acid	Fatty acid	354.61	19.198	C23H46O2	Human metabolism, help in hair growth and cardiovascular health
29	Tetracosanoic acid	Fatty acid	368.63	19.575	$C_{24}H_{48}O_2$	Human metabolite, reduce the risk of neurocarcinogenesis
30	Lactulose	Sugar	342.29	19.552	$C_{12}H_{22}O_{11}$	Treatment of constipation and hepatic encephalopathy
31	Lactose	Sugar	342.3	19.603	$C_{12}H_{22}O_{11}$	Energy source, inhibit pathogenic bacteria
32	2-Monostearin	Lipid	358.56	19.682	$C_{21}H_{42}O_4$	Drug delivery vehicle

TABLE 12.2 Surface area and pore characteristics of activated carbon prepared from makhana shells.

Activation agent	Optimum Temp (∘C)	Impregnation ratio	BET Surface Area (m2/g)	Total Pore Volume (cm3/g)	References
H_3PO_4	700	1.5	2636	1.53	Kumar and Jena (2016)
$ZnCl_2$	600	2.0	3073	1.40	Kumar and Jena (2015)
$ZnCl_2$	700	2.0	2404	1.13	Kumar and Jena (2015)

Data derived from Kumar, A., Jena, H.M. 2015. High surface area microporous activated carbons prepared from Fox nut (Euryale ferox) shell by zinc chloride activation. Appl. Surf. Sci. 356, 753–761. https://doi.org/10.1016/j.apsusc.2015.08.074; Kumar, A., Jena, H.M. 2016. Preparation and characterization of high surface area activated carbon from Fox nut (Euryale ferox) shell by chemical activation with H3PO4. Results Phys. 6 2016. 651–658. https://doi.org/10.1016/j.rinp.2016.09.012.

12.2.5 Biotechnological and biochemical valorization pathways

Fermentation: Fermentation has become a useful bioprocessing technique to enrich the functional and nutraceutical value of plant-based resources. Fermentation with *Lactobacillus curvatus* has been found to extend oxidative stress-removing capacity in Euryale ferox so much. The process enhances the release of bioactive compounds and the degradation of macromolecules into small polysaccharides and polypeptides, which is usually linked with improved bioavailability and biological actions. Fermented makhana products functionally have been shown to have an improved intracellular and extra-cellular antioxidant capacity in human skin fibroblast (HSF) models in the presence of H_2O_2. This is indicated in the lower levels of reactive oxygen species (ROS) and higher activity of antioxidant enzymes like superoxide dismutase (SOD), glutathione peroxidase (GSH-Px), and catalase (CAT) which are important elements in cellular defense against oxidative stress. These effects are molecularly associated with the activation of the Nrf_2 and MAPK signaling pathways, which are some of the most important regulators of cellular redox homeostasis. Changes in Bax/Bcl-2 ratio, decreasing matrix metalloproteinase-1 (MMP-1) activity, and conserving type I collagen (COL-1) contribute to the cytoprotective and anti-aging effects of fermented makhana (Jiang et al., 2022).

Enzymatic Hydrolysis: Lack of slow digestion of starch-containing foods frequently leads to acute postprandial glycemic reactions that are directly linked to metabolic disorders. Increasing the resistant starch is an accepted method of controlling glucose release. Another targeted approach of the debranching enzyme amylopullulanase has been explored to alter the structure of starch in popped makhana. Amylopullulanase stimulates the debranching of amylopectin and subsequent molecular rearrangement, which stimulates amylose reassociation and resistant starch. The presence of amylose (12.33%), resistant starch (14.88%) and crystallinity (11.32%) has also significantly risen in the presence of enzyme treated makhana flour implying the presence of a higher structural organization. The same increases in the easily digestible starch and the oil carrying capacity are indicative of the changes in the structure and functionality of starch granules. In vitro digestibility analyses corroborate reduced starch hydrolysis and more prolonged glucose release profile, which is a favorable characteristic of low-glycemic and therapeutic foods. Therefore, the enzymatic restructuring of makhana starch can serve as an important valorisation strategy toward the deployment of functional ingredients that can help prevent glycemic regulation and metabolic health (Biswas et al., 2020).

Microbial Bioconversion and Nanobiotechnology: Nanobiotechnology provides new avenues for transformation of biological resources into high-value functional materials. Antimicrobial and antioxidant effects of metallic nanoparticles, especially silver nanoparticles are well known. Biological synthesis methods continue to receive more preference over traditional chemical and physical methods as they are cost effective and environment friendly. Biological materials in green synthesis systems serve as natural reducing and stabilizing forces. Although microbial synthesis may be effective, it needs controlled culture conditions and biosafety. Plant-mediated synthesis is a more sustainable alternative that is much easier. Euryale ferox leaves in this case are barely used compared to the use of seeds; among the bioactive compounds found in these leaves include flavonoids and anthocyanins, which render the formation of nanoparticles. The process of converting biomass of makhana into functional nanomaterials illustrates how agricultural residues may be converted into useful antioxidant and antimicrobial products (Devi et al., 2022).

Biopolymers and Prebiotic Potential: The Makhana (Euryale ferox) has great prospects of valorisation in the shape of biopolymers and prebiotic components. It is possible to extract polysaccharides associated with seed coat and starch fractions and physically alter them to produce functional biopolymers. These materials have a potentially promising future as the controlled-release excipients, especially in the colon-targeted drug delivery systems because of their favorable swelling and erosion characteristics. Even though massive production of microbial biopolymers using makhana substrates is currently minimal, it has starch-rich structure with which developments can be made in future. Nutritionally, the makhana resistant starch has significant prebiotic potential. Resistant starch cannot be digested in the upper gastrointestinal tract but can become a fermentable food source to the microbiota that are beneficial in the gut. It has been shown that some makhana types promote the growth of probiotic genera, including Lactobacillus and Bifidobacterium, that leads to the production of short-chain fatty acids and the inhibition of pathogenic microbes. Fermentability can also be increased by further processing methods to decrease molecular complexity. The following characteristics make makhana a promising source of gut health-based functional ingredients (Qu et al., 2025; Mal and Pandey, 2025).

Scaling Up Bioprocesses Constraints: Although laboratory-scale makhana bioprocessing has been promising, there are several technical and socio-economic challenges in the process of scaling up makhana bioprocessing. The larger systems may be challenging to maintain uniform pH, temperature, and oxygen transfer, which may influence process consistency and yield. Seasonal and geographical differences in the quality of raw materials also make standardization more difficult. Economic factors involve large capital investment in special equipment and process validation. Even cultivation of makhana is sometimes labour-intensive and subject to poor levels of mechanization and skilled manpower. Food and nutraceutical applications require regulatory requirements that are very strict in terms of quality and safety and add up to the compliance costs. Also, storage, transportation, and aggregation of raw materials may impose supply chain constraints in terms of industry-wide implementation. These issues will be solved through concerted efforts in the processing technology, supply chain management, and policy-support.

12.2.6 Environmental remediation applications

The activated carbon and even the raw shell powder have demonstrated efficacy in removing pollutants from aqueous solutions, offering a low-cost alternative to commercial adsorbents.

Dye Removal: The raw hard shell of *E. ferox* was utilized for the removal of Basic Fuchsin, a carcinogenic cationic dye. The adsorption data fitted best with the Langmuir isotherm model, yielding a maximum adsorption capacity of 14.28 mg/g at 303 K.

Kinetics: The process followed a pseudo-second-order kinetic model, indicating that chemisorption (involving electron sharing or exchange) was the rate-limiting step (Kalita et al., 2017).

Thermodynamics: The negative value of Gibbs free energy and positive enthalpy indicated that the adsorption was spontaneous and endothermic, meaning efficiency increases at higher temperatures (Kalita et al., 2017).

12.2.7 Food applications: functional ingredients

Beyond industrial carbon, the shells can be processed to extract "Euryale ferox seed shell extract" (EFSSE), which has potent bioactive properties relevant to human health.

12.2.7.1 Enzyme inhibition and glycemic control

Diabetes management often involves controlling post-prandial hyperglycaemia. EFSSE has shown dose-dependent inhibition of alpha-amylase and alpha-glucosidase, the key enzymes responsible for hydrolyzing starch into glucose (Maibam et al., 2023a,b).

Functional Bread: When EFSSE was incorporated into wheat bread at concentrations of 0.25%–2%, it significantly reduced the in vitro starch digestibility (IVSD) and the predicted glycemic index (pGI). The pGI dropped from 88.4 (Control) to 69.8 (2% fortification). The polyphenols in the extract bind to the starch granules or the enzymes, retarding the hydrolysis rate. This effectively converts rapid-digesting starch into slow-digesting fractions. While fortification increased total phenolic content (TPC) and antioxidant activity (DPPH scavenging), concentrations above 1% negatively impacted sensory attributes like color (darkening) and taste (astringency) (Maibam et al., 2023a, b) Table 12.3.

TABLE 12.3 Effect of Makhana shell extract (EFSSE) fortification on functional properties of wheat bread.

Fortification Level (%)	Total phenolic Content (mg GAE/g)	Antioxidant Activity (DPPH %)	Predicted Glycemic Index (pGI)
0 (control)	0.23 pm 0.01	12.5 pm 0.5	88.4 pm 1.2
0.5	0.85 pm 0.02	45.3 pm 1.1	82.1 pm 0.9
1.0	1.42 pm 0.03	68.7 pm 1.5	76.5 pm 1.1
2.0	2.65 pm 0.05	89.2 pm 1.8	69.8 pm 0.8

Data derived from Maibam, B.D., Nickhil, C., Chandra Deka, S. 2023a. Preparation, physicochemical characterization, and in vitro starch digestibility on complex of Euryale ferox kernel starch with ferulic acid and quercetin. Int. J. Biol. Macromol. 250, 126178. https://doi.org/10.1016/j.ijbiomac.2023.126178; Maibam, B.D., Chakraborty, S., Nickhil, C., Deka, S.C. 2023b. Effect of Euryale ferox seed shell extract addition on the in vitro starch digestibility and predicted glycemic index of wheat-based bread. Int. J. Biol. Macromol. 226, 1066–1078. https://doi.org/10.1016/j.ijbiomac.2022.11.223.

12.2.7.2 Active packaging systems

In the domain of food safety, shell extracts are being integrated into biodegradable packaging to extend shelf life.

Bilayer Films: A novel bilayer film was developed using hydrophobic ethyl cellulose (outer layer) and hydrophilic gelatin-carboxymethyl chitosan (inner layer) doped with Makhana shell polyphenols. When applied to ready-to-eat cooked beef, these films released polyphenols that inhibited the growth of *Listeria monocytogenes* and retarded lipid oxidation (Cai et al., 2024).

Nanoliposomes: To improve stability, complex tannins from Makhana shells were encapsulated in nanoliposomes. These nanoparticles effectively inhibited tyrosinase activity. They prevented enzymatic browning in fresh-cut produce (specifically asparagus lettuce), maintaining visual quality and nutritional value during storage (Dai et al., 2024).

12.2.8 Valorization of unpopped seeds

Unpopped seeds, often referred to as "rejects," represent a direct economic loss. However, they are rich in starch, which possesses unique physicochemical properties compared to conventional starches like corn or potato.

12.2.8.1 Isolation and modification of starch

Starch isolated from unpopped Foxnut has an amylose content of approximately 22.9%. To expand its industrial applicability, it can be chemically modified, for instance, using Octenyl Succinic Anhydride (OSA). OSA modification (at 3% concentration) introduces hydrophobic octenyl groups. This modification decreased the amylose content to 13.7% but significantly increased the oil absorption capacity (0.61 g/g) and swelling power (13.60 g/g). Scanning Electron Microscopy (SEM) showed that while native granules were smooth and polyhedral, modified granules exhibited a rougher surface with small cavities due to the esterification reaction (Shweta and Saxena, 2021).

12.2.8.2 Stabilization of pickering emulsions

The amphiphilic nature of OSA-modified starch makes it an excellent stabilizer for Pickering emulsions (emulsions stabilized by solid particles). Emulsions stabilized with 3% OSA-modified starch showed a high emulsion index (85%) even after 15 days of storage, compared to only 5% for native starch. These emulsions exhibited shear-thinning (pseudoplastic) behavior, which is highly desirable for food products like mayonnaise, salad dressings, and sauces, allowing them to flow when poured but retain shape when resting (Shweta and Saxena, 2021).

12.2.8.3 Enzyme-resistant starch (prebiotics)

Unpopped seeds can also be processed to enhance their Resistant Starch (RS) content. RS escapes digestion in the small intestine and ferments in the colon, acting as a prebiotic. Treatment with the enzyme amylopullulanase debranches the starch structure. Subsequent retrogradation allows the formation of highly crystalline "Type III" resistant starch. Studies show this treatment can significantly enhance crystallinity and RS content, making the flour a functional ingredient for low-calorie or diabetic-friendly foods (Biswas et al., 2020).

12.2.8.4 Vegetative biomass and safety considerations

While the leaves and rhizomes are rich in flavonoids, their utilization requires careful safety assessment due to the plant's interaction with its aquatic environment.

(a) Heavy metal bioaccumulation

Euryale ferox is known to be a hyperaccumulator of heavy metals. A study in the biotic environment of Darbhanga, Bihar, revealed that the seeds and vegetative parts can bioconcentrate toxic metals such as Lead (Pb), Chromium (Cr), Cadmium (Cd), and Copper (Cu) from polluted water bodies (Rai et al., 2002). Before any vegetative biomass (or even shell-based products) can be used in food or pharmaceutical chains, raw materials must undergo rigorous heavy metal testing. This bioaccumulation trait, however, suggests the plant could be used for phytoremediation of polluted water bodies, provided the resulting biomass is disposed of safely (e.g., via incineration in controlled energy recovery units) rather than consumed.

12.3 Advanced analytical techniques for waste evaluation

The analysis of the processing residues of makhana (Euryale ferox) necessitates the combination of both the sophisticated analytical analysis supplementing the traditional proximate analysis. Although proximal composition (moisture, protein, fat and ash, crude fiber) gives a baseline nutritional data, it is not adequate to describe structural, physicochemical, and functional characteristics of bioresources that are intended to be used in nutraceutical or industrial applications. An example is that proximate analysis would not be able to determine the molecular structure of polysaccharides, or the nature of interactions between phenolic and fibrils, or the stability of bioactive compounds in processing. Thus, sophisticated analysis tools are becoming useful in characterization that is more detailed and reliable. Fourier Transform Infrared Spectroscopy (FTIR) is extensively employed to determine functional groups and assure the existence of bioactive substances including phenolics, flavonoids, and polysaccharides. FTIR spectra can be used to give useful data on hydroxyl, carbonyl and aromatic functional groups which are frequently linked to antioxidant and antimicrobial activities. Moreover, to track chemical changes produced by processing treatment (roasting, drying or enzymatic hydrolysis) FTIR may be used. These observations can be applicable in the forecasting of ingredient behavior and interaction in complex food systems. Scanning Electron Microscopy (SEM) offers the descriptive details of the morphology, porosity, and fiber structure at both micro- and nano-levels. SEM imaging is used to view the structural integrity, particle size distribution, and pore architecture, which have a direct effect on hydration behavior, oil-holding capacity, and adsorption efficiency. These structural characteristics have been especially significant in instances where the makhana residues are investigated as dietary fiber supplements, encapsulation agents, or environmental adsorbents. Functional performance can also be linked with changes in microstructure following processing. X-ray Diffraction (XRD) helps in ascertaining crystalline and amorphous in starch and fiber fractions. The method is used to measure the patterns of starch crystallinity and reorganization of the structure during thermal or mechanical treatment. The proportion between crystalline and amorphous areas will determine the digestibility, formation of resistant starch and the textural behavior. Thus, XRD analysis can be used to give vital data about designing low-glycemic or gut-health-advantageous functional foods based on by-products of makhana (Sharma et al., 2024). Differential Scanning Calorimetry (DSC) is an assessment of the thermal transition, including gelatinization, melting, and degradation temperature. These thermal characteristics are required to determine processing tolerance, storage stability, and compatibility in food formulations. DSC information can influence the choice of appropriate processing conditions to retain bioactive contents and retain functional quality. Further, the thermal analysis assists in estimating the shelf stability and any possible structural changes throughout the product development. In addition to proximate analysis, sophisticated compositional profiling, such as total phenolic content, flavonoid content, antioxidant capacity (DPPH, ABTS, FRAP analysis), and resistant starch measurement, gives a better and practical estimation of bioresource potential. This type of profiling enables the correlation of chemical properties and biological action, therefore giving force to functional assertions.

12.4 Methods of utilization

E. ferox seeds are nutritionally rich and versatile, therefore utilized in multiple sectors such as food, nutraceuticals, pharmaceuticals, and traditional health systems. Currently, Euryale ferox seed is used to make a variety of sweet meals, such as pudding, halva, kheer, and vermicelli. Additionally, it enhances the flavor and ease of digestion of chicken bites (Biswas et al., 2016). When making a soup with E. ferox, red dates, and walnuts, the seed flour of E. ferox is utilized in place of arrowroot (Jun 2007). In Chinese cuisine, rice dumplings are made by combining seeds with sticky rice in a 1:3 ratio (Liu, 2016). Additionally, it has reportedly been utilized to make low-alcohol wine (Zhang, 2011). In addition to the seeds, the leaves are used to make tea (Changan and Wei, 2016; Xue, 2017) and in salads and chutneys in many parts of the world (Kumar et al., 2020). A promising step in investigating the functional side of E. ferox for other food-based and non-food-based applications is the extraction of starch from raw seed. In numerous baked goods, such as bread, starches have been employed in place of fat (Balic et al., 2017). E. ferox can be used to separate the modified starch that keeps Pickering emulsion stable. Octenyl succinic anhydride (OSA) was used in this investigation to esterify E. ferox starch, which may have improved the emulsifying capacity. Physicochemical, functional, pasting, rheological, thermal, and structural properties were significantly impacted by the esterified E. ferox starch, which had a degree of substitution between 0.0055 (1% OSA) and 0.0185 (3% OSA). These starches can separate the water and oil with storage, regardless of the size of the droplets (Shweta and Saxena, 2021) Euryale ferox is said to have therapeutic value in Chinese and Indian medicine. In addition to beriberi, polyuria, persistent diarrhea, spermatorrhea, leucorrhea, gonorrhea, seminal loss, micturition, and articular pains, E. ferox is used to treat conditions pertaining to the digestive, respiratory, excretory, reproductive, and circulatory systems (Jha et al., 1991; Puri et al., 2000). Because E. ferox is a readily available natural source of antioxidants, it can be utilized in the pharmaceutical industry (Ho et al., 2012; Lin et al., 2012). Additionally, medicinal liquor with anti-rheumatoid qualities is made using E. ferox. According to the liquor contains E. ferox seed (25%–38%), Chinese medicinal liquor (20%–40%), and yeast (5%–13%). Because E. ferox is an immunostimulant that boosts humoral immunity, it is also advised that pregnant women utilize it (Puri et al., 2000). E. ferox was shown to contain higher quantities of antiaging amino acids such as arginine, cystine, glutamine, and methionine (Jana and Idris, 2018). E. ferox has several non-foods uses in addition to its use in food items. It is combined with Rosa multiflora and Euphorbia lathyrism L. to make an oil that relieves skin irritation (Nho et al., 2015). E. ferox is also employed in skin-whitening compositions alongside Poriacocus, Dioscorea oppositifolia, and Nelumbo nucifera because of its antimelanogenic properties. This composition's primary function is to prevent the synthesis of melanin and the tyrosinase enzyme. The seed shell of E. ferox has been utilized as an adsorbent to remove dangerous substances from industrial wastes to prepare activated carbon (Kumar and Jena, 2015). Table 12.4.

12.5 Potential bioactive compounds

Bioactive compounds are the naturally occurring substances found in Euryale ferox that have positive impacts on human health include anti-inflammatory, anti-diabetic, and antioxidant qualities. The primary nutritional components found in Euryale ferox seed is Carbohydrates (80% starch), high-quality protein, vitamins (C and E), and minerals magnesium, potassium, phosphorus, iron, and zinc (Bhatt et al., 2017; Devi et al., 2020; Kumar and Jena, 2016; Vikram and Mishra, 2021; Mishra et al., 2003; Jana and Idris, 2018). Apart this primary nutritional component Euryale ferox is a rich source of diverse bioactive compounds that contribute to their multifunctional health benefits. Phytochemical analyses have shown the presence of several classes of bioactive compounds, including phenolic compounds, flavonoids, polyphenols, essential oils, alkaloids, saponins, tannins, glycosides, triterpenoids, and polysaccharides. Various biological activities, including antioxidant, antidiabetic, anticancer, anti-fatigue, antimicrobial, antidepressant, cardioprotective, and hepatoprotective effects, are displayed by these compounds. Additionally, their synergistic interaction enhances the physiological effects, making E. ferox a promising ingredient in nutraceuticals (Kashyap and Shukla, 2024). Hepatoprotective action has been reported for substances found in E. ferox, including 1,2,3-trigalloyl glucose, gallic acid, corilagin, 1,6-digalloyl glucose, 1,2,3,6-tetragalloyl glucose, 1,2,6-trigalloyl glucose, and ethyl gallate. These substances aid in lowering lipid deposition and removing oxidative stress, which otherwise damages DNA, proteins, lipids, and cell membranes and exacerbates the onset of non-alcoholic fatty liver disease (Sumida et al., 2013). Hypoglycemic agents include pentacyclic triterpenes like 2β-hydroxybetulinic acid 3β-caprylate and 2β-hydroxybetulinic acid 3β-oleiate that are isolated from E. ferox. The extract contributes to the plant's antidiabetic impact by stimulating the release of insulin from the pancreatic β-cells (Ahmed et al., 2015, 2019). Glycosylsterols with cardioprotective

TABLE 12.4 Methods of utilization of *Euryale ferox*.

Material	Application/ Utilization	Effect/Outcomes	References
Seed kernel	Preparation of enzyme-modified resistant starch and preparation of rice dumplings	Exhibits low glycemic response, excellent dietary fiber attributes and considered a "functional diet food" in traditional therapy	(Biswas et al., 2020; Liu, 2016)
Seeds	Used for gastro-retentive non-effervescent tablet formulation and used in fermented medicinal liquor	Replaced 50% HPMC K4M with foxnut powder; extended floating time by 4 h over the standard and reported to have anti-rheumatoid effects	(Negi et al., 2011; Lin et al., 2012)
Leaves	Used to prepare vinegar and Preparation of herbal tea rich in *Eurotium cristatum and i*ngredient for oil preparation along with *Euphorbia lathyris* and *Rosa multiflora*	Acts as a heat-clearing, diuretic, blood-purifying agent, supports immunity, clears heat, and improves digestion	(Changan and Wei, 2016; Xue, 2017)
Seed kernel	Preparation of kheer	Quick preparation suitable for hot or cold consumption	Verma et al. (2008)
Seed kernel	Used in powdered health mix (oats, sesame, lily, rice, walnuts, *Euryale ferox*, lotus seeds) and also used in soup along with red dates and walnuts	Helps calm nerves and promotes better sleep and used for reducing nocturnal enuresis	(Yaopeng, 2018; Jun (2007)
Seed kernel	Used in making low-alcohol Foxnut wine and added to chicken meat bites	Contains high polyphenols responsible for strong antioxidant activity and increases total phenolics, flavonoids, β-carotene, tocopherol, and vitamin C; improves nutritional profile of meat	(Zhang, 2011; Biswas et al., 2016)
Seed kernel	Preparation of resistant starch and starch production for stabilizing Pickering emulsion	Slow-digesting, low-GI starch with improved oil uptake and solubility, reduced water absorption — perfect for stable emulsions	(Qu et al., 2025; Shweta and Saxena, 2021)

TABLE 12.4 Methods of utilization of *Euryale ferox*.—cont'd

Material	Application/ Utilization	Effect/Outcomes	References
Euryale ferox extract	Component in a cosmetic formulation with *Nelumbo nucifera*, *Dioscorea oppositifolia*, and *Poria cocos*	Helps in reducing skin sensitivity	Nho et al. (2015)
Euryale ferox extract	Activated carbon production using chemical activation with H_3PO_4	Reduces melanin formation by inhibiting tyrosinase	
Seed shell	Activated carbon production using chemical activation with H_3PO_4	Works as a coloring agent for textile fibers	
Seed shell	High surface microporous activated carbon using zinc chloride activation	Carbon obtained shows a high number of functional groups on the surface	Kumar and Jena (2016)
Seed shell	Used in textile finishing as a functional finishing agent	Activated carbon suitable for adsorbing and removing industrial contaminants	Kumar and Jena (2015)
Popped makhana	Roasted and flavored snack (onion, chocolate, tomato, chilly, masala, etc.), Instant Makhana Kheer Mix (powdered mixture of dry fruits, sugar, and milk powder), Makhana flour or powder is combined with wheat flour for chapati and used as a thickening and alternative to arrowroot, Breakfast cereal that is ready to eat (flaked and combined with flavors and dried fruits), Baby food or porridge (powdered and combined with banana powder or Sattu), Preparing Makhana Barfi, Preparation for Makhana Kalakand, Formulation of breads and buns (nutritious and gluten-free breads), Formulation of cookies and biscuits		National Institute of Food Technology Entrepreneurship and Management (NIFTEM) (2022)

properties include 24-ethylcholesta-5,22E-dienyl-3β-O-pyranoglucoside, 24-ethylcholest-5-enyl-3β-O-pyranoglucoside, 24-methylcholest-5-enyl-3β-O-pyranoglucoside. By upregulating thioredoxin and associated proteins, these substances are thought to help scavenge reactive oxygen species produced during ischemia myocardial reperfusion. E. ferox extract scavenged O_2 radicals by 93% and 100% at 0.005% and 0.01% concentrations, respectively, in an in-vitro reactive oxygen species scavenging assay. Additionally, these substances can prevent ulcers (Azhar et al., 2015; Das et al., 2006). Buddlenol E, 5,7-dihydroxy-2-(3,4,5-trihydroxyphenyl) chroman-4-one, and 5, 7, four trihydroxy flavanone that were extracted from E. ferox are examples of polyphenols that have antifatigue activity (Wu et al., 2013). The phytochemicals found in E. ferox, such as gallic acid, resveratrol, and alliin, cause human Caucasian lung cancer cells to undergo dose-dependent apoptosis, suggesting anticancer potential (Nam et al., 2019). By preventing the proliferation and angiogenesis of cancer cells while maintaining normal cell viability, the flavonol kaempferol has also been shown to lower the risk of cancer (Tehseen et al., 2020). Arginine, glutamic acid, leucine, isoleucine, methionine, and cysteine are the main amino acids found in E. ferox. In raw E. ferox, the ratios of arginine and lysine to proline and leucine to isoleucine were 1.7% and 9.6%, respectively. The building blocks of creatine, which is good for healthy skin, hair, and nails, are arginine and methionine. Amino acids have been shown to have antioxidant activity, which aids in the removal of free radicals and stops chain reactions, delaying the aging process (Jana and Idris, 2018). Flavonoids are associated with a broad spectrum of health-promoting effects and are an indispensable component in a variety of nutraceutical, pharmaceutical, medicinal, and cosmetic applications, because of their antioxidative, anti-inflammatory, anti-mutagenic and anti-carcinogenic properties coupled with their capacity to modulate key cellular enzyme functions. They are also known to be potent inhibitors for several enzymes, such as xanthine oxidase (XO), cyclo-oxygenase (COX), lipoxygenase, and phosphoinositide 3-kinase (Metodiewa et al., 1997; Hayashi et al., 1988). They have beneficial biochemical and antioxidant properties linked to several illnesses, including atherosclerosis, cancer, and Alzheimer's disease (Burak and Imen, 1999; Castañeda-Ovando et al., 2009; Lee et al., 2009). The high content of phytosterol (0.16 mg) in perisperm and phenols (0.28 mg) in the endosperm of E. ferox may be accountable for the medical properties of the crop (Praween and Singh, 2015). Selenium in low quantity (1–2 mg/kg) acts as antioxidant which helps in glucose metabolism (Wang et al., 2017). Makhana regulates blood pressure because it has a low sodium content and a high potassium content. Makhana has fewer saturated fats and cholesterol, and it is also better for heart of a human. Ca, Mg, and S are commonly found in this (Bhatt et al., 2017; Mishra et al., 2003) Figs. 12.1 and 12.2.

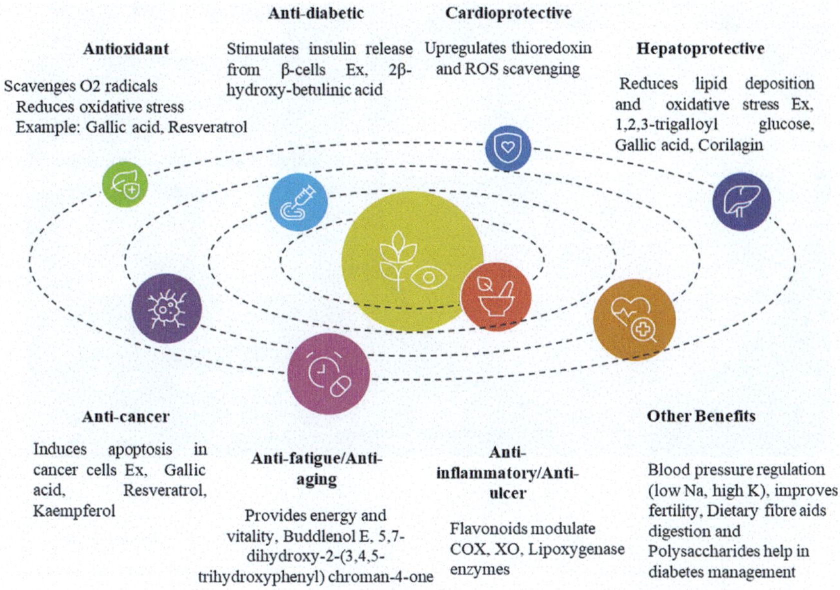

FIGURE 12.1 Comprehensive health benefits of makhana.

In addition, the crude E. ferox polysaccharides were observed to have the highest free radical scavenging activity (97.39%) as well as notable reduction power.

The E. ferox polysaccharide fraction also extracted by DEAE sepharose fast flow and Superdex 75 gel chromatography showed hypoglycemic properties, therefore could be used to make healthy food for diabetic people (Zhang et al., 2019). E. ferox is thought to be a helpful food for regulating blood pressure, blood sugar, heart disease, and anemia because it has a fair quantity of dietary fiber, which helps relieve constipation. E. ferox has also been used traditionally to treat liver dysfunction, enuresis, diabetic nephropathy, and chronic diarrhea (Mittal et al., 2020). E. ferox provides several therapeutic benefits for the respiratory, circulatory, digestive, renal, and reproductive systems.

It also aids in improving fertility in both sexes (Jana and Idris, 2018; Singh et al., 2017). To reduce oxidative stress, they also aid in boosting superoxide dismutase activity and reducing the overproduction of malondialdehyde (Jian et al., 2019; Yang and Lee, 2015) Table 12.5.

12.6 Other valuable components

Although the popped seed kernel of *Euryale Ferox* (Makhana) is the principal commercial product, several other parts of the plant that are usually discarded

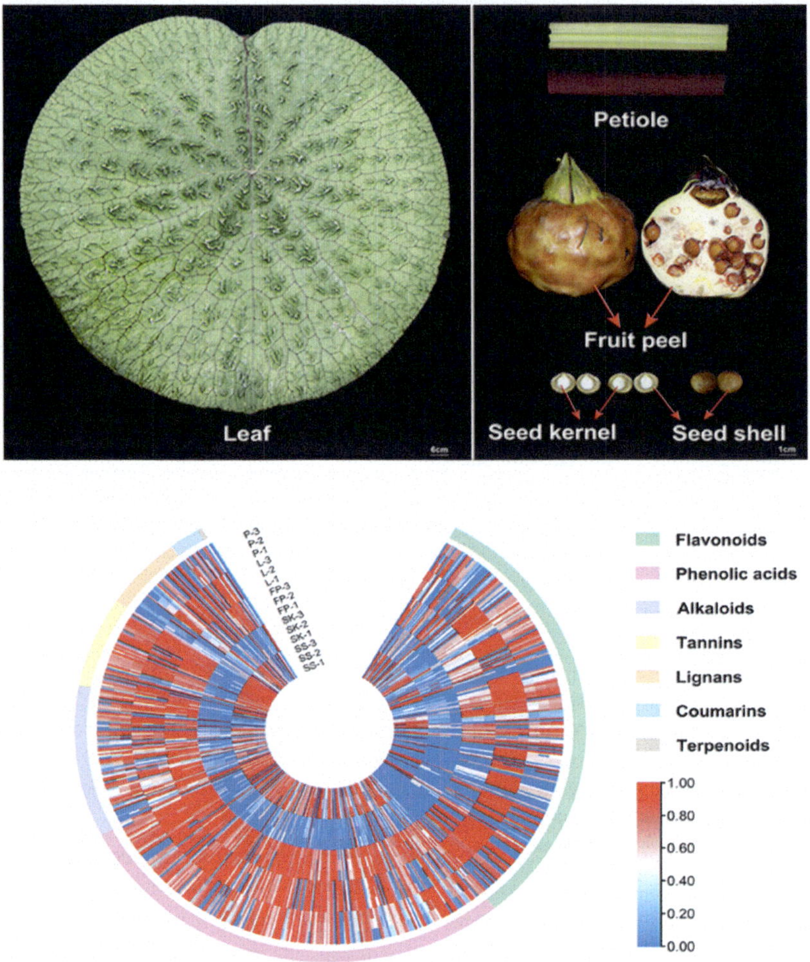

FIGURE 12.2 Tissue specific sampling layout and metabolite distribution patterns of Euryale ferox. *Data and visualization adapted from Liu, A.L., Wang, Y.H., Wang, T.Y., Zhu, Y., Wu, P. and Li, L.J., 2023. Comparative metabolomic profiling of secondary metabolites in different tissues of Euryale ferox and functional characterization of phenylalanine ammonia-lyase. Ind. Crops Prod. 195, p.116450; figure not reproduced.*

before harvesting of makhana seeds also contain important chemical and bioactive compounds. These "other valuable components" including (a) the phenolic rich outer seed coat (outer husk) and shell, (b) unpopped kernel (*Thurri*) that are modified through processing such as roasted or germinated

TABLE 12.5 Bioactive compounds and health benefits of makhana (Euryale ferox) residues.

Component/ Property	Presence in makhana by-products	Health-benefits	Nutraceutical/Functional applications	Comparison with other Agro-waste
Phenolic compounds	High (seed coat & husk)	Antioxidant, anti-inflammatory, reduces oxidative stress	Natural antioxidant supplements, functional beverages	Comparable to rice bran; higher stability than fruit peels
Flavonoids	Moderate to high	Anti-diabetic, cardioprotective, anti-aging	Anti-diabetic nutraceuticals, wellness foods	Similar to citrus peel; lower bitterness
Dietary fiber	Very high	Gut health, glycemic control, cholesterol reduction	Fiber-rich flour blends, weight-management foods	Higher fiber than rice bran and wheat bran
Antioxidant activity	Strong DPPH & ABTS scavenging	Protects against chronic diseases and aging	Natural food preservatives, antioxidant formulations	Comparable to grape pomace
Prebiotic potential	Present in carbohydrate residues	Supports gut microbiota	Prebiotic powders, synbiotic products	Similar to cereal brans
Low anti-nutritional factors	Advantageous	Better nutrient bioavailability	Safe functional ingredient	Lower anti-nutrients than many seed coats

(unable to pop) flours, (c) vegetative materials including leaves and petioles, and (d) carbon enrich popping waste can be converted into activated carbon or adsorbents. Each of these fractions has shown potential for nutritional, bioactive, or industrial applications based on recent research findings.

Recent metabolomic studies, which further supports the value-adding potential of these by-products in a thorough profiling investigation using UPLC-MS/MS, highlighted the biochemical diversity found in different parts of the makhana tissue (Liu et al., 2023). Figure 12.1a shows the appearance and structure of the five main parts that is, leaf, petiole, fruit peel, seed shell, and seed kernel, which were systematically analyzed. Results shows that the accumulations of the important phytochemicals varied by tissue, including phenolic acids, flavonoids, lignans, tannins, alkaloids, coumarins, and terpenoids. Figure 12.1b presents a circular heatmap demonstrated how these metabolites vary across plant parts. This metabolic specialization provides a scientific basis for the future use of makhana by products in food, health, cosmetic, and industrial applications.

12.6.1 Seed husk and shell: Sources of phenolics, antioxidants, and enzyme-inhibitory extracts

The seed coat and shell, a major by-product generated during popping and dehulling process, and it contains a high concentration of phenolic compounds and flavonoids. Phenolic extracts derived from the seed coat exhibit strong radical-scavenging activity in vitro (Wu et al., 2013) confirming the high antioxidant potential of this waste. Complementing this finding (Martins et al., 2016) reported that phenolic extracts forms makhana shells showed significant antioxidant and anti-fatigue effects in vivo using animal models, indicating pharmaceutical ayurveda relevance. In addition to their antioxidant properties, seed coat phenols have been found to inhibit key carbohydrate-digesting enzymes such as α-amylase and α-glucosidase. When incorporated into bakery industry, these extracts reduced in-vitro starch breakdown and lowered the predicted glycemic index of the product (Maibam et al., 2023a,b). These results suggest seed coat extracts can be valorized as functional food additives for glycemic control.

12.6.2 Unpopped or germinated kernel products: Roasted and germinated flours

Roasted unpopped makhana kernel (*thurri)* and the surplus winter-season germinated seeds that fail to pop serve as secondary by products. Both kernels can significantly influence both nutrient availability and functional behavior during therapeutic food preparation (Dega and Barbhai, 2023). Roasted kernels have been shown to increase measurable levels of phenolics and flavonoids, thereby enhancing the overall antioxidant capacity of the product. In a controlled human study reported that roasted makhana produced a lower post-prandial glycemic response compared with unroasted seeds (Liaquat et al., 2022). This makes roasted makhana particularly suitable for developing health-oriented snack foods with improved functional properties. Biological processing, such as seed sprouting prior to milling, also leads to notable improvements in nutritional quality. Germination increases protein content and modifies techno-functional properties, including water- and oil-holding capacities, which are important for product formulation (Kathuria et al., 2025). further observed that germinated makhana flour contains reduced levels of antinutritional factors such as phytic acid and tannins. These enhancements make sprouted makhana flour a promising ingredient for composite flour blends, bakery applications, and formulations aimed at boosting protein and nutrient density.

12.6.3 Leaves, fruit peels, and other vegetative biomass: Polyphenols, anthocyanins, and cosmeceutical potential

Large volumes of vegetative materials including leaves and fruit peels are discarded during makhana cultivation and harvesting. However, multiple studies reveal that this biomass contains valuable secondary metabolites. Leaf aqueous extracts have been used successfully as environmentally friendly reducing and stabilizing agents for the synthesis of silver nanoparticles (Ag NPs) Fig. 12.3.

The resulting EL-Ag NPs exhibit strong antimicrobial and antioxidant properties, demonstrating their potential applications in biomedical fields and antimicrobial packaging materials derived from agricultural waste (Devi and Deka, 2022). Such uses highlight the possibility of transforming discarded biomass into valuable Phyto material resources. Recent research highlights that Euryale ferox contains important secondary metabolites within its vegetative tissues, especially anthocyanins and other polyphenols (Wu et al., 2020). reported that the leaves of *E. ferox* hold measurable amounts of anthocyanins, confirmed through HPLC-QTOF-MS/MS analysis, and exhibit notable antioxidant potential. The distinctive purple pigmentation of *E. ferox* flowers strongly suggests the existence of anthocyanins and related flavonoids, indicating a viable subject for future research, even though no focused biochemical studies on the flowers have been published up to date. Furthermore, recent research by (Wu et al., 2025) demonstrated that tannin compounds derived from *E. ferox* fruit peels efficiently suppress tyrosinase activity and

FIGURE 12.3 Parts of makhana plant and its processing.

melanogenesis, providing unambiguous proof of their suitability for use in skin-lightening and cosmetic formulations. Overall, the growing body of research indicates that *E. ferox* leaves, flowers (as possible sources), and fruit peels are rich bioactive resources with substantial promise for the creation of natural food colors and cosmetics products.

12.6.4 Shell-derived carbonaceous adsorbents: Conversion of waste into remediation materials

The hard seed shell generated after popping represents one of the largest lignocellulosic waste streams in makhana processing. Because of its high fixed-carbon content and porous structure, this shell material has shown considerable promise as a low-cost adsorbent for the removal of contaminants from water. Studies report that untreated shell powder is capable of adsorbing synthetic dyes such as basic fuchsin and other organic pollutants from aqueous solutions, demonstrating its natural affinity for pollutant binding (Kalita et al., 2017). Chemical or physical activation greatly expands surface area and pore availability, allowing for further improvement. It has been demonstrated that activated carbons made from *E. ferox* shells have a high absorption efficiency for hazardous heavy metals, such as Pb (II), suggesting a promising application in both home and commercial water purification systems (Devi et al., 2023). When taken as a whole, these results show a clear waste-to-wealth opportunity: a plentiful agricultural by-product may be converted into a useful, environmentally acceptable substance for environmental rehabilitation.

12.7 Industrial applications

Euryale ferox, popularly known as gorgon nut or makhana, is an economically significant crop cultivated in India (mainly in Bihar) and Southeast Asia. The processes including cultivation and post-harvest processing are very difficult and are labor intensive. And furthermore, the farmer focuses on main products for their profit and income due to lack of knowledge about the importance of other by-products. Here we will discuss the industrial importance of those by-products Fig. 12.4.

12.7.1 Food and nutraceutical industries

The seed coat and husk contain 40%–60% of dietary fiber. They produce functional food ingredients, fiber-rich flour supplements, and anti-obesity & diabetic food formulations upon milling (Kadu et al., 2020). The seed coat powder is blended with wheat flour (5%–15%) to create low-calorie, high-fiber bakery products. The polyphenol- and flavonoid-rich extracts form husks serve as Natural food preservatives, additives for functional beverages and also use as replacement for synthetic antioxidants (BHA/BHT) (Zhang et al., 2023). The carbohydrate-rich waste supports: prebiotic powder development, symbiotic blends with probiotics and Gut-microbiome enhancing nutraceuticals.

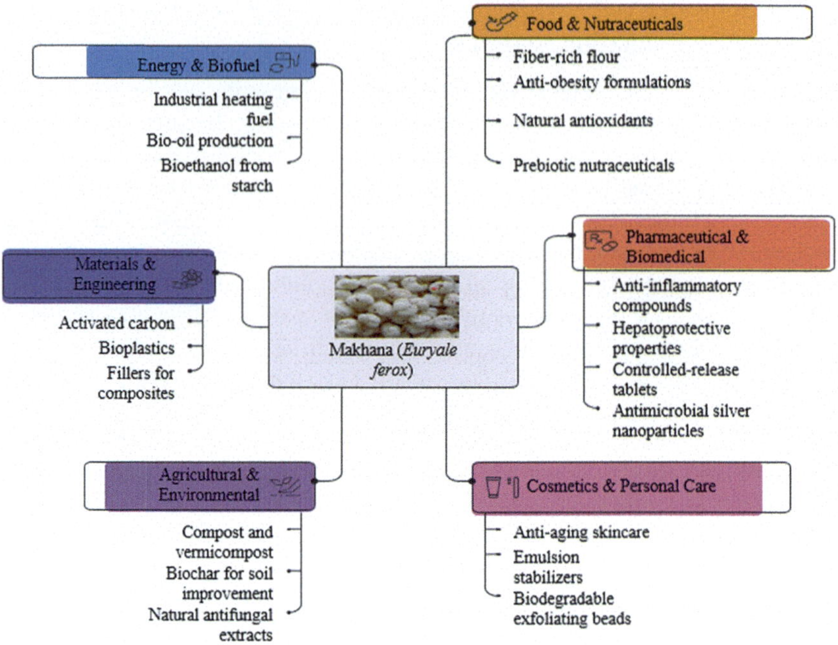

FIGURE 12.4 Industrial applications of makhana (Euryale ferox) by-products.

12.7.2 Pharmaceutical and biomedical applications

Makhana waste contains- Polyphenols, Tannins, Flavonoids which exhibits anti-inflammatory, antioxidants, hepatoprotective and anti-diabetic activities (Kadu et al., 2020). Shell derived biopolymers are studied for controlled-released matrix tablets, low-cost excipient formulations and hydrogel composites. Seed coat extracts are rich in polyphenols and show antioxidants activity and protective effects in liver models, which supports development of hepatoprotective nutraceuticals or adjunct therapies against oxidative-stress related disorders (Jian et al., 2019). Husks are also be used in ayurveda and herbal products as blood purifying, anti-aging and digesting agents. Leaves have been used to bio fabricate silver nanoparticles (AgNPs) that show antimicrobial and cytotoxic activities which can be applied to wound care, coatings or antimicrobial device coatings (Devi et al., 2022).

12.7.3 Cosmetics and personal care industry

Phenolic extracts stabilize emulsion and protect formulations from oxidation. Used as anti-aging creams, anti-pigmentation serums and moisturizers with natural antioxidants for skin care. Finely milled shells serves as biodegradable exfoliating beads. Also used as eco-friendly alternatives to microplastics.

12.7.4 Agricultural and environmental applications

The husk and shells can be used as compost or vermicompost to get nutrient-rich manure, organic carbon enhancers and also to increase aeration and fertility of soil (National Institute of Food Technology Entrepreneurship and Management (NIFTEM), 2022). Pyrolysis converts waste into biochar with applications in soil improvement and carbon sequestration, water purification, and soil absorption (Kumar et al., 2016). Seed coat extracts exhibit anti-fungal, anti-microbial and insect repellent activity.

12.7.5 Materials and engineering applications

Makhana shells are an excellent precursor for high-surface-area activated carbon, industrial adsorbents and water purification and dye-removal systems (Kumar et al., 2016). Husk fibers can be blended with polymers to form bioplastics, packaging materials and other biodegradable cutlery and containers.Finely powdered husk serves as a filler in rubber, plastics and polyurethane foams which improves strength and reduces manufacturing costs.

12.7.6 Energy and biofuel sectors

Makhana shells have a high calorific value, as they are used for Industrial heating and co-firing with coal.Thermal conversion of Makhana shell yields bio-oil, biochar and syngas (Kumar et al., 2016). Processing residues are rich in starch which supports fermentation to bioethanol and production of hydrogen to rich syngas Table 12.6.

12.8 Opportunities and limitations

Scale-up, commercial viability, regulatory safety (toxicology of extracts and nanomaterials), and life-cycle assessments are still lacking for many suggested uses, despite the encouraging results of the valorization studies. In particular:

1. Thorough toxicological profiling is needed before food- or pharma-grade adoption of shell or leaf extracts.
2. Pilot-scale activated-carbon production methods and cost analyses are required to validate environmental remediation pathways.
3. Standardization of extraction/processing protocols is necessary to deliver reproducible functional ingredients.

Addressing these gaps will be essential to move from laboratory demonstrations to commercial value chains.

TABLE 12.6 Industrial applications of makhana wastes and by-products.

Waste/By-product	Key extract/ Ingredient	Summary of industrial applications	References
Seed coat (testa)	Polyphenols, flavonoids, tannins	Antioxidant, antimicrobial, and antidiabetic potential; suitable for nutraceutical and pharmaceutical formulations.	(Kadu et al., 2020; Zhang et al., 2023)
Seeds/starch (unpopped or popped)	Starch, resistant starch	Can be used in low-GI foods, prebiotic formulations, functional foods, pharmaceutical excipients, and biodegradable materials.	(Yang et al., 2021; Islam et al., 2021)
Shell/husk residues	Lignocellulosic fiber	Useful for activated carbon production for wastewater treatment and environmental remediation.	(Kalita et al., 2017)
Leaves/ peduncles	Phenolics, antioxidants	Green synthesis of antibacterial/antioxidant nanoparticles for biomedical and cosmetic uses.	(Devi, 2022)
Whole-plant residues (biomass in ponds)	Organic matter, nutrients	Enhances aquaculture productivity when used in integrated farming systems; improves water quality and nutrient cycling.	Verma et al. (2008)

12.9 Conclusions and future directions

The transition of the Makhana industry toward a circular economy is not just an environmental imperative but an economic opportunity. The hard shells, currently a disposal nuisance, possess the structural integrity to become premium activated carbon and the chemical richness to serve as anti-diabetic food additives. The unpopped seeds can be transformed into advanced emulsifiers and prebiotics. By adopting these valorization pathways, stakeholders can turn "waste" into "wealth," ensuring the sustainability of this unique wetland crop while providing additional income streams for farmers and processors.

Self-declaration

Artificial Intelligence (AI) tools were used only for English language editing in the preparation of this manuscript. No AI tools were used to generate any kind of content, including text, tables, or figures.

References

Ahmed, D., Kumar, V., Verma, A., Shukla, G.S., Sharma, M., 2015. Retracted article: antidiabetic, antioxidant, antihyperlipidemic effect of extract of Euryale ferox salisb. with enhanced histopathology of pancreas, liver and kidney in streptozotocin induced diabetic rats. SpringerPlus 4 (1). https://doi.org/10.1186/s40064-015-1059-7.

Ahmed, D., Khan, M.I., Sharma, M., Khan, M.F., 2019. Novel pentacyclic triterpene isolated from seeds of Euryale Ferox Salisb. ameliorates diabetes in streptozotocin induced diabetic rats. Interdiscip. Toxicol. 11 (4), 275–288. https://doi.org/10.2478/intox-2018-0027.

Azhar, M., Anjum, N., Quddusi, N., 2015. Pharmacologically active cardioprotective plants at a glance. Hamdard Med. 58 (1), 51–83.

Balic, R., Miljkovic, T., Ozsisli, B., Simsek, S., 2017. Utilization of modified wheat and tapioca starches as fat replacements in bread formulation. J. Food Process. Preserv. 41 (1), e12888. https://doi.org/10.1111/jfpp.12888.

Bhatt, B.P., Gupta, V.K., Kumar, L., Singh, I.S., Sarkar, B., 2017. Euryale ferox (salisb): promising aquatic food crop of eastern indo gangetic plains. Int. J. Curr. Microbiol. Appl. Sci. 6 (6), 1914–1921. https://doi.org/10.20546/ijcmas.2017.606.223.

Biswas, A.K., Beura, C.K., Sagar, M.P., 2016. Effect of different cooking methods on retention of antioxidant activity of bioactive compounds of fox nut (Euryale ferox) seeds incorporated in chicken meat bites. Indian J. Poultry Sci. 51 (1), 88. https://doi.org/10.5958/0974-8180.2016.00002.7.

Biswas, P., Das, M., Boral, S., Mukherjee, G., Chaudhury, K., Banerjee, R., 2020. Enzyme mediated resistant starch production from Indian Fox Nut (Euryale ferox) and studies on digestibility and functional properties. Carbohydr. Polym. 237, 116158. https://doi.org/10.1016/j.carbpol.2020.116158.

Burak, M., Imen, Y., 1999. Flavonoids and their antioxidant properties. Turkiye Klin Tip Bil Derg 19 (1), 296–304.

Cai, M., Zhang, X., Zhong, H., Li, C., Shi, C., Cui, H., Lin, L., 2024. Ethyl cellulose/gelatin-carboxymethyl chitosan bilayer films doped with Euryale ferox seed shell polyphenol for cooked meat preservation. Int. J. Biol. Macromol. 256. https://doi.org/10.1016/j.ijbiomac.2023.128286.

Castañeda-Ovando, A., de Lourdes Pacheco-Hernández, M., Páez-Hernández, M.E., Rodríguez, J.A., Galán-Vidal, C.A., 2009. Chemical studies of anthocyanins: a review. Food Chem. 113 (4), 859–871.

Changan, P., Wei, C., 2016. Process for Making Solid Food Particle Healthy Vinegar from Sellaginella Unicata and Euryale Ferox Leaves.

Dai, Y., Shi, C., Qin, Z., Song, W., Ding, B., Wei, S., Chen, H., 2024. Potential application of nanoliposomes loaded with complex tannins from the seed shell of Euryale ferox in the anti-browning of fresh-cut asparagus lettuce. Int. J. Biol. Macromol. 275 (Pt 2), 133669. https://doi.org/10.1016/j.ijbiomac.2024.133669.

Das, S., Der, P., Raychaudhuri, U., Maulik, N., Das, D.K., 2006. The effect of Euryale ferox (makhana), an herb of aquatic origin, on myocardial ischemic reperfusion injury. Mol. Cell. Biochem. 289 (1–2), 55–63. https://doi.org/10.1007/s11010-006-9147-1.

Dega, V., Barbhai, M.D., 2023. Exploring the underutilized novel foods and starches for formulation of low glycemic therapeutic foods: a review. Front. Nutr. 10. https://doi.org/10.3389/fnut.2023.1162462.

Devi, K.R., 2022. A comprehensive review on nutraceutical properties of Euryale ferox Salisb. J. Plant Sci. Res. 9 (2), 01–06.

Devi, M.B., Deka, S.C., 2022. Physicochemical properties and structure of starches of foxnut (Euryale ferox Salisb.) from India and its application. J. Food Proc. Preserv. 46 (2). https://doi.org/10.1111/jfpp.16262.

Devi, M., Sharma, K., Narayan Jha, S., Arora, S., Patel, S., Kumar, Y., Kumar Vishwakarma, R., 2020. Effect of popping on physicochemical, technological, antioxidant, and microstructural properties of makhana seed. J. Food Proc. Preserv. 44 (10). https://doi.org/10.1111/jfpp.14787.

Devi, N., Rani, K., Kharb, P., Kaushik, P., 2022. Bio-fabrication of euryale ferox (makhana) leaf silver nanoparticles and their antibacterial, antioxidant and cytotoxic potential. Plants 11 (20), 2766. https://doi.org/10.3390/plants11202766.

Devi, B., Goswami, M., Rabha, S., Kalita, S., Sarma, H.P., Devi, A., 2023. Efficacious sorption capacities for Pb(II) from contaminated water: a comparative study using biowaste and its activated carbon as potential adsorbents. ACS Omega 8 (17), 15141–15151. https://doi.org/10.1021/acsomega.3c00142.

Hayashi, T., Sawa, K., Kawasaki, M., Arisawa, M., Shimizu, M., Morita, N., 1988. Inhibition of cow's milk xanthine oxidase by flavonoids. J. Natural Prod. 51 (2), 345–348. https://doi.org/10.1021/np50056a030.

Ho, Y.L., Huang, S.S., Deng, J.S., Lin, Y.H., Chang, Y.S., Huang, G.J., 2012. In vitro antioxidant properties and total phenolic contents of wetland medicinal plants in Taiwan. Botan. Stud. 53 (1), 55–66.

Islam, Z., Bashir, K., Singh, S., Rastogi, M., Siddiqui, S., Bangar, S.P., 2021. Structural and functional properties of starch from popped seeds of Makhana (Euryale ferox). Starch – Stärke. https://doi.org/10.1002/star.202000300.

Jana, B.R., Idris, M., 2018. Anti-aging amino acids in Euryale ferox (Salisb.). Adv. Plants Agric. Res 8, 43–48.

Jha, V., Kargupta, A.N., Dutta, R.N., Jha, U.N., Mishra, R.K., Saraswati, K.C., 1991. Utilization and conservation of euryale ferox salisbury in Mathila (North Bihar), India. Aquat. Bot. 39 (3–4), 295–314. https://doi.org/10.1016/0304-3770(91)90005-.

Jian, T., Yu, C., Ding, X., Chen, J., Li, J., Zuo, Y., Ren, B., Han, L., Li, W., 2019. Hepatoprotective effect of seed coat of*Euryale ferox* extract in non-alcoholic fatty liver disease induced by high-fat diet in mice by increasing IRs-1 and inhibiting CYP2E1. J. Oleo Sci. 68 (6), 581–589. https://doi.org/10.5650/jos.ess19018.

Jiang, Y., You, S., Zhang, Y., Zhao, J., Wang, D., Zhao, D., Li, M., Wang, C., 2022. Enhancing bioactive components of euryale ferox with Lactobacillus curvatus to reduce H_2O_2-induced oxidative stress in human skin fibroblasts. Antioxidants 11 (10), 1881. https://doi.org/10.3390/antiox11101881.

Jun, W., 2007. Soup Containing Euryale Ferox and Walnut Powder.

Kadu, M., Maknojia, R., Maharana, S., 2020. Preliminary study of euryale ferox salisb. Seed coat as a potential antioxidant and antibacterial source. Asian J. Biol. Life Sci. 9 (3), 313–320. https://doi.org/10.5530/ajbls.2020.9.47.

Kalita, S., Pathak, M., Devi, G., et al., 2017. Utilization of Euryale ferox shell for removal of basic fuchsin dye from water: equilibrium and kinetics investigation. RSC Adv. 7, 27248–27259. https://doi.org/10.1039/C7RA03972G.

Kashyap, A., Shukla, O.J., 2024. Sustainable food supply chain: exploration, identification, and analysis of the critical drivers for the foxnut (Makhana) industry. J. Glob. Oper. Strateg. Sourc. https://doi.org/10.1108/JGOSS-05-2023-0042.

Kathuria, C., Ranjan, S., Singh Tomar, M., Kumar, V., Ahmad, W., Dobhal, A., Joshi, S., Chand, K., Kumar, S., 2025. Development and quality assessment of *Chenopodium album*, Spinacia oleracea, and spent makhana (Euryale ferox) enriched vegetable cookies. Food Chem. Adv. 8, 101040. https://doi.org/10.1016/j.focha.2025.101040.

Kumar, A., Jena, H.M., 2015. High surface area microporous activated carbons prepared from Fox nut (Euryale ferox) shell by zinc chloride activation. Appl. Surf. Sci. 356, 753–761. https://doi.org/10.1016/j.apsusc.2015.08.074.

Kumar, A., Jena, H.M., 2016. Preparation and characterization of high surface area activated carbon from Fox nut (Euryale ferox) shell by chemical activation with H3PO4. Results Phys. 6, 651–658. https://doi.org/10.1016/j.rinp.2016.09.012.

Kumar, A., Yadav, P.K., Nat, P., Nath, S., Kumari, U., 2016. Euryale ferox Salisb. Is an immense potential aquatic cash fruit crop: a review. Int. J. Adv. Agricult. Sci. Technol. 3 (5), 35–49.

Kumar, M., Shekhar, D., Kumari, A., Prasad, R.P., Bhatt, B.P., 2020. Makhana farming for maximizing farm income: a success story. Biot. Res. Today 2 (9), 962–965.

Lee, M.R., Kim, J.H., Son, E.S., Park, H.R., 2009. Protective effect of extracts from Euryale ferox against glutamate-induced cytotoxicity in neuronal cells. Nat. Prod. Sci. 15 (3), 162–166.

Liaquat, M., Pasha, I., Ahsin, M., Salik, A., fox nuts, R., Euryale, F.L., 2022. Contain higher concentration of phenolics, flavonoids, minerals and antioxidants, and exhibit lower Glycemic Index (GI) in human subjects. Food Prod. Proc. Nutr. 4 (1). https://doi.org/10.1186/s43014-021-00081-x.

Lin, H., Wang, Y., Zhang, W., 2012. Health Care Medicinal Liquor for Ameliorating Rheumatic Arthritis and Methods for Preparing the Same.

Liu, T. 2016. Euryale ferox rice dumplings and preparation method thereof. Chinese Patent CN106261716. 22.

Liu, A.L., Wang, Y.H., Wang, T.Y., Zhu, Y., Wu, P., Li, L.J., 2023. Comparative metabolomic profiling of secondary metabolites in different tissues of Euryale ferox and functional characterization of phenylalanine ammonia-lyase. Ind. Crops Prod. 195. https://doi.org/10.1016/j.indcrop.2023.116450.

Maibam, B.D., Nickhil, C., Chandra Deka, S., 2023a. Preparation, physicochemical characterization, and in vitro starch digestibility on complex of Euryale ferox kernel starch with ferulic acid and quercetin. Int. J. Biol. Macromol. 250, 126178. https://doi.org/10.1016/j.ijbiomac.2023.126178.

Maibam, B.D., Chakraborty, S., Nickhil, C., Deka, S.C., 2023b. Effect of Euryale ferox seed shell extract addition on the in vitro starch digestibility and predicted glycemic index of wheat-based bread. Int. J. Biol. Macromol. 226, 1066–1078. https://doi.org/10.1016/j.ijbiomac.2022.11.223.

Mal, T., Pandey, A., 2025. Isolation and effect of physical modification on the release characteristics Euryale ferox Salisb polysaccharide. Int. J. Biol. Macromol. 305 (Pt 2), 141243. https://doi.org/10.1016/j.ijbiomac.2025.141243.

Martins, N., Barros, L., Ferreira, I.C.F.R., 2016. In vivo antioxidant activity of phenolic compounds: facts and gaps. Trends Food Sci. Technol. 48, 1–12. https://doi.org/10.1016/j.tifs.2015.11.008.

Metodiewa, D., Kochman, A., Karolczak, S., 1997. Evidence for antiradical and antioxidant properties of four biologically active N,N-Diethylaminoethyl ethers of flavaone oximes: a comparison with natural polyphenolic flavonoid rutin action. IUBMB Life 41 (5), 1067–1075. https://doi.org/10.1080/15216549700202141.

Mishra, R.K., Jha, V., Dehadrai, P.V. 2003. MAKHANA. ICAR, New Delhi.

Mittal, R., Sharma, S., Mittal, A., 2020. A critical review on ethnobotanical and pharmacological aspects of euryale ferox salisb. Pharmacogn. J. 12 (6), 1444–1454. https://doi.org/10.5530/pj.2020.12.199.

Nam, G.-H., Jo, K.-J., Park, Y.-S., Kawk, H.W., Kim, S.-Y., Kim, Y.-M., 2019. In vitro and in vivo induction of p53-dependent apoptosis by extract of euryale ferox salisb in A549 human caucasian lung carcinoma cancer cells is mediated through Akt signaling pathway. Front. Oncol. 9. https://doi.org/10.3389/fonc.2019.00406.

National Institute of Food Technology Entrepreneurship and Management (NIFTEM), 2022. Handbook of Makhana (Euryale ferox) — Processing and value addition [Extension/handbook PDF]. NIFTEM.

Negi, J.S., Jugran, V., Kasliwal, N., 2011. Development of non-effervescent floating matrix tablets based on Euryale ferox seeds. Asian J. Pharm. 5 (2), 93–100. https://doi.org/10.4103/0973-8398.84549.

Nho, Y. H., Youm J.K., Kim, H. K. 2015. Cosmetic composition for preventingor alleviating sensitive skin, containing mixture extraction oil of Euryale ferox Salisb., Euphorbia Lathyris L. And Rosa Multiflora Thumb. Korean patent WO/2015/152564.

Praween, N.N.P., Singh, A.K., 2015. Effect of Kinetin on Protein Content of Euryale Ferox Salisb (Makhana) during Fruit Development.

Puri, A., Sahai, R., Singh, K.L., Saxena, R.P., Tandon, J.S., Saxena, K.C., 2000. Immunostimulant activity of dry fruits and plant materials used in Indian traditional medical system for mothers after child birth and invalids. J. Ethnopharmacol. 71 (1–2), 89–92. https://doi.org/10.1016/S0378-8741(99)00181-6.

Qu, C., Yu, D., Jing, Z., Gu, S., Wang, Y., Xie, W., Wu, Q., 2025. Comparative study on structural characterization, physicochemical properties, and in vitro probiotic activities of resistant starch from different varieties of Euryale ferox. Food Chem. X 28, 102572. https://doi.org/10.1016/j.fochx.2025.102572.

Rai, U.N., D Tripathi, R., Vajpayee, P., Jha, V., Ali, M.B., 2002. Bioaccumulation of toxic metals (Cr, Cd, Pb and Cu) by seeds of Euryale ferox Salisb. (Makhana). Chemosphere 46 (2), 267–272. https://doi.org/10.1016/s0045-6535(01)00087-x.

Sharma, K., Patel, S., Jha, S.N., Devi, M., Nandan, S., Vishwakarma, R.K., 2024. Change in popping, structural, functional, and crystalline properties of roasted makhana seeds in prolong storage: an FTIR, GC-MS, SEM, XRD study. Res. Square. https://doi.org/10.21203/rs.3.rs-3910663/v1.

Shweta, Y.K., Saxena, D.C., 2021. Valorization of unpopped Foxnut starch in stabilizing Pickering emulsion using OSA modification. Int. J. Biol. Macromol. 191, 657–667. https://doi.org/10.1016/j.ijbiomac.2021.09.148.

Singh, I.S., Kumar, L., Bhatt, B.P., Thakur, A.K., Choudhary, A.K., Kumar, A., 2017. Integrated aquaculture with fox nut- A case study from North Bihar, India. Int. J. Curr. Microbiol. Appl. Sci. 6 (10), 4906–4912. https://doi.org/10.20546/ijcmas.2017.610.461.

Sinha, C.B.K., Jha, V., 2025. Identification of nutraceutically important metabolites in raw and popped makhana (euryale ferox salisb.) seeds using GCMS technique. Int. J. Plant Environ. 11 (03), 514–524. https://doi.org/10.18811/ijpen.v11i03.09.

Sumida, Y., Niki, E., Naito, Y., Yoshikawa, T., 2013. Involvement of free radicals and oxidative stress in NAFLD/NASH. Free Radic. Res. 47 (11), 869–880. https://doi.org/10.3109/10715762.2013.837577.

Tehseen, S., Sarfraz, F., Muntaham, S., Ateeq, N., Ashfaq, F., Yasmin, I., Mehmood, T., 2020. Foxnut (Euryale ferox Salisb.): a health promising fruit. Acta Scientific Agriculture 4 (12), 68–72.

Verma, A.M., Jha, V., Ahmad, S.H., 2008. Fish-Makhana (Euryale ferox Salisb.) integration — a case study of sustainable aquafarming system in North Bihar. J. Indian Fish. Assoc. 35.

Vikram, V., Mishra, S., 2021. A review on Euryale ferox: daily dietary regime for Covid -19 patients. Int. J. Appl. Res. 7 (6S), 29–34. https://doi.org/10.22271/allresearch.2021.v7.i6sa.8607.

Wang, N., Tan, H.Y., Li, S., Xu, Y., Guo, W., Feng, Y., 2017. Supplementation of micronutrient selenium in metabolic diseases: its role as an antioxidant. Oxid. Med. Cell. Long. 2017. https://doi.org/10.1155/2017/7478523.

Wu, C.Y., Chen, R., Wang, X.S., Shen, B., Yue, W., Wu, Q., 2013. Antioxidant and anti-fatigue activities of phenolic extract from the seed coat of Euryale ferox Salisb. and identification of three phenolic compounds by LC-ESI-MS/MS. Molecules 18 (9), 11003–11021. https://doi.org/10.3390/molecules180911003China.

Wu, C.Y., Wang, H., Fan, X.H., Yue, W., Wu, Q.N., 2020. Waste Euryale ferox salisb. leaves as a potential source of anthocyanins: extraction optimization, identification and antioxidant activities evaluation. Waste Biomass Valoriz. 11 (8), 4327–4340.

Wu, L., Song, W., Jiang, Y., Dai, Y., Qin, Z., Liu, L., Chen, H., 2025. Structural characterization of complex tannins from Euryale ferox fruit peels and their inhibitory mechanisms against tyrosinase activity and melanogenesis. Int. J. Biol. Macromol. 298, 139909.

Xue, W. 2017. Euryale Ferox Tea. Chinese Patent CN109645177. 12.

Yang, H.Y., Lee, T.H., 2015. Antioxidant enzymes as redox-based biomarkers: a brief review. BMB Rep. 48 (4), 200–208. https://doi.org/10.5483/BMBRep.2015.48.4.274.

Yang, Y., Chen, Q., Yu, A., Tong, S., Gu, Z., 2021. Study on structural characterization, physicochemical properties and digestive properties of euryale ferox resistant starch. Food Chem. 359, 129924. https://doi.org/10.1016/j.foodchem.2021.129924.

Yaopeng, W. 2018. Inventor; Yaopeng W, assignee. Nerve soothing and sleep assisting powder. Chinese Patent CN109527352. 16.

Zhang, B. (2011). Preparation method of low-alcohol foxnut wine. https://patents.google.com/patent/CN102337191B/en.

Zhang, W.N., Su, R.N., Gong, L.L., Yang, W.W., Chen, J., Yang, R., Wang, Y., Pan, W.J., Lu, Y.M., Chen, Y., 2019. Structural characterization and in vitro hypoglycemic activity of a glucan from Euryale ferox Salisb. seeds. Carbohydr. Polym. 209, 363–371. https://doi.org/10.1016/j.carbpol.2019.01.044.

Zhang, L., Zeng, J., Yuan, E., Chen, J., Zhang, Q., Wang, Z., Yin, Z., 2023. Extraction, identification, and starch-digestion inhibition of phenolics from Euryale ferox seed coat. J. Sci. Food Agric. 103 (7), 3437–3446. https://doi.org/10.1002/jsfa.12460.

Further reading

Zhao, H.R., Zhao, S.X., Sun, C.Q., Guillaume, D., 1989. Glucosylsterols in extracts of Euryale ferox identified by high resolution NMR and mass spectrometry. J. Lipid Res. 30 (10), 1633–1637.

Chapter 13

Decoding makhana bioactives: Proteomic discovery, physiological relevance, and storage stability

Sidharth Dash Sharma[1, a], Sreemoyee Sensharma[1, a], Ishita Debnath[2], Palak[2], Soumit Thakur[2], Chinmayee Dash[2], Mahesh Chandak[3], Gopal Pramanik[2] and Subrata Pramanik[4]

[1]*Jyoti and Bhupat Mehta School of Health Sciences and Technology, Indian Institute of Technology Guwahati, Guwahati, Assam, India;* [2]*Pharmacology, Department of Pharmaceutical Sciences and Technology, Birla Institute of Technology, Ranchi, Jharkhand, India;* [3]*Wockhardt Research Center, Wockhardt Limited, Chhatrapati Sambhajinagar, Maharashtra, India;* [4]*Department of Biological Sciences, Indian Institute of Science Education and Research Kolkata, Mohanpur, West Bengal, India*

13.1 Introduction

Euryale ferox Salisbury, also known as makhana, belongs to the Nymphaeaceae family and is found in the subtropical and tropical regions of southeast and east Asia (Shaw et al., 2025). The plant is grown as an aquatic cash crop in shallow water bodies in Bihar and the northeastern regions of India (Fig. 13.1A–C). The economic and cultural value led to the granting of a geographical indication (GI) tag, recognizing its unique quality and origin. It has been used as food and medicine in traditional systems, such as Ayurveda and Traditional Chinese Medicine (TCM) (Shaw et al., 2025).

Historically, makhana has been used in Ayurvedic medicine as a Rasayana, a rejuvenator that enhances vitality and slows the aging process, and in TCM as Qian Shi, primarily for tonifying the kidneys and strengthening the spleen to treat chronic diarrhea and renal dysfunction (Shukla et al., 2024). Modern pharmacological validation has confirmed that the seeds and their secondary metabolites possess a wide range of bioactivities, including hypoglycemic,

---

[a]Sidharth Dash Sharma and Sreemoyee Sensharma are equal contributors in this chapter.

Makhana (Euryale ferox) as a Functional Food. https://doi.org/10.1016/B978-0-443-44421-0.00013-5

FIGURE 13.1 Overview of makhana (*E. ferox*), cultivation system, and processing. (A–C) Makhana plant, pond system, and cultivation are shown (images are retrieved from https://www.makhana.org/). (D) The sequential stages involved in makhana production begin with the flowering stage, followed by cultivation in aquatic environments and subsequent fruit (seed) development. Mature seeds are then harvested from water bodies, after which they undergo roasting to produce the characteristic popped makhana. The processed seeds are subsequently graded, stored, and packaged for commercial distribution. This workflow highlights the traditional processing steps that contribute to the quality and market value of makhana (Semwal et al., 2021; Wardhan et al., 2025).

cardioprotective, hepatoprotective, and neuroprotective effects (Devi et al., 2022). Over the past decade, makhana has been recognized as a potent source of bioactive compounds, having antioxidant, antidiabetic (Jana et al., 2024), antiinflammatory, and neuroprotective properties (Kapoor et al., 2022). The recent surge in its popularity is also linked to its status as an allergen-free, gluten-free snack with a low glycemic index, making it highly suitable for patients with metabolic disorders (Liaquat et al., 2022). Furthermore, advances in proteomics and metabolomics have enabled bioactive profiling and analytical characterization, with implications for the molecular mechanisms underlying these therapeutic effects. Consequently, makhana has gained attention as a potential nutraceutical resource for disease prevention and health promotion (Liaquat et al., 2022).

The preparation of makhana is a complex process carried out using ancient methods. The steps involved in the process are drying, size reduction, heating to elevated temperatures, roasting, and popping (Fig. 13.1D) (Singh et al., 2020). These processing steps significantly influence its physicochemical, nutritional, and bioactive composition. Thermal processing improves digestibility and modifies the structures of starch and protein while also altering the fiber, mineral, and phytochemical content. Although some heat-sensitive phytochemicals may decline during processing, roasting and popping can improve palatability, digestibility, and the bioavailability of selected nutrients (Zhao et al., 1989).

Makhana is a nutritious alternative to crops, such as wheat and rice, but its cultivation is difficult due to high production costs and low yields. Consumer awareness of the importance of healthy diets and functional foods has boosted global demand for makhana, making it an emerging commodity in both national and international markets. Thus, it shows that makhana has the potential to be used in the development of functional foods, nutraceuticals, and health foods, as it has both traditional and modern importance (Ranjit Kumar et al., 2024).

This chapter provides an updated overview of the bioactive components, proteomic insights, physiological relevance, medicinal potential, nutraceutical applications, and storage stability of makhana, highlighting its growing importance in food science, nutrition, and therapeutic research.

13.2 Source of makhana

13.2.1 *E. ferox* plants

E. ferox, known by various names, including makhana, fox nut, or gorgon nut, is one of the most important aquatic noncereal crops in the Nymphaeaceae family (Singh et al., 2020). The plant has large leaves with spines, a circular shape, quilted surfaces, a diameter of 1–2 m, a short, thick rhizome, and nutritious seeds (Wu et al., 2013). *E. ferox* plant thrives in still or slowly moving water at depths of 0.2–5 m, such as ponds, lakes, marshes, and ditches, in warm, humid tropical to subtropical environments with temperatures ranging from 20 to 35°C. The edible part of the *E. ferox* plant is the starchy kernel inside its seeds, which, when popped at high temperatures, becomes makhana. The popped form is consumed as a nutritious food supplement and works as a healthy alternative to other fried snacks, owing to low fat and cholesterol content (Ahmad, 2020).

13.2.1.1 Traditional and cultural significance

Makhana cultivation started approximately 3000 years ago in India and China (Jiang et al., 2023). Recent policy documents also recognize Mithila Makhana as a GI, highlighting its strong link to local identity and traditional production practices. Beyond Mithila, makhana occurs in wild or semiwild stands in other parts of northern and northeastern India, including Uttar Pradesh, West Bengal, Assam, and Manipur, where it supplements local diets and, in some areas, contributes modestly to household income (Fig. 13.2) (Semwal et al., 2021). Archaeobotanical work has even documented remains of *E. ferox* from Pleistocene archaeological sites in the Levant, suggesting that humans have exploited this plant as a starchy, energy-dense food resource for many

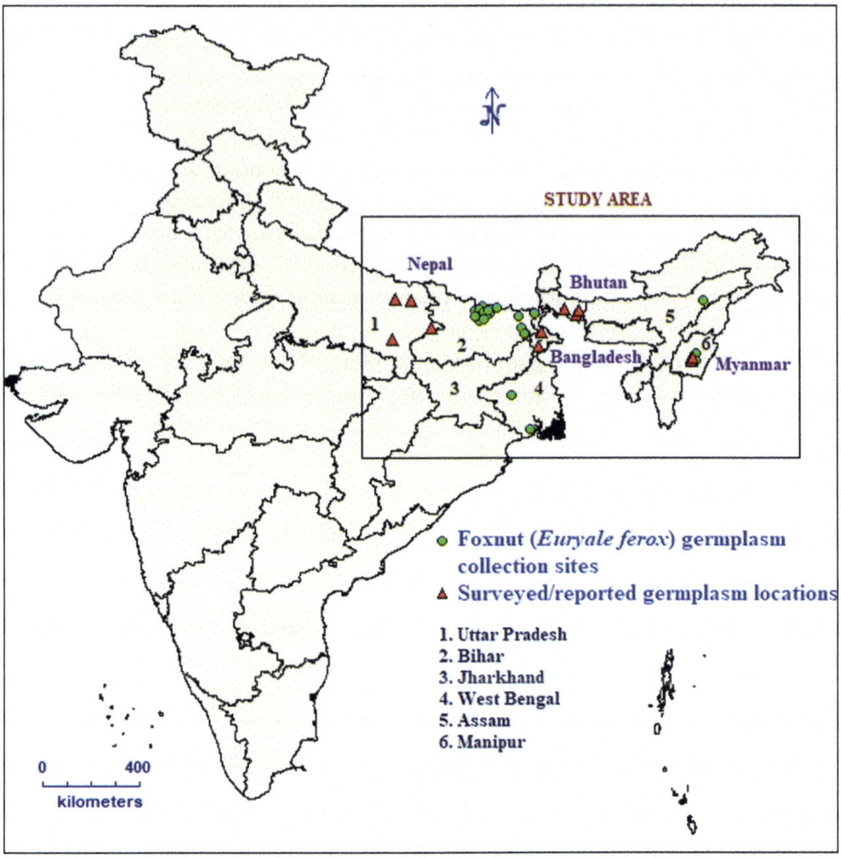

FIGURE 13.2 Geographic distribution of *E. ferox* (makhana) germplasm collection sites in India and neighboring regions. The map illustrates the study area across eastern and northeastern India, along with adjacent countries, including Nepal, Bhutan, Bangladesh, and Myanmar. Green circles represent foxnut (*E. ferox*) germplasm collection sites, while red triangles indicate surveyed or previously reported germplasm locations. The numbered regions correspond to major cultivation zones: (1) Uttar Pradesh, (2) Bihar, (3) Jharkhand, (4) West Bengal, (5) Assam, and (6) Manipur. The distribution highlights the concentration of makhana genetic resources in flood-prone and wetland ecosystems of the Indo-Gangetic plains and northeastern regions. *The figure is extracted from Semwal, D.P., Pandey, A., Gore, P.G., Ahlawat, S.P., Yadav, S.K., Kumar, A., 2021. Habitat prediction mapping using BioClim model for prioritizing germplasm collection and conservation of an aquatic cash crop "makhana" (*Euryale ferox *Salisb.) in India. Genetic Res. Crop Evol. 68 (8), 3445–3456. https://doi.org/10.1007/s10722-021-01265-7.* ***Map lines delineate study areas and do not necessarily depict accepted national boundaries.***

millennia, although continuous cultural traditions are best documented in South Asia (Singh et al., 2018).

Makhana is associated with the ceremonial, religious, social, cultural, medical, and nutritional aspects of the people of the Koshi-Mithila region (Singh et al., 2018). In this region, makhana pops find various uses, such as using makhana pops as a roasted makhana snack, especially in the evenings; to prepare various delicious and rich food items; as a noncereal food for Hindu devotees to observe fasting; and to break roza (fasting) for Muslim communities during Ramzan Sharif (Singh et al., 2018). Devotional narratives from Mithila specifically promote makhana as an excellent vrata food, that is, light, nonirritant, and spiritually clean, and emphasize its relationship with purity and self-discipline (Ekal and Kumbamoorthy, 2024).

Makhana pops are also used as offerings to gods and goddesses in temples due to their perceived divine nature. They are further used in decorations and garlands, particularly for ceremonial and political purposes. Makhana is rich in essential macro- and micronutrients, as well as essential amino acids. It is also reported to contain higher levels of sugars, proteins, ascorbic acid, and phenolic compounds than several other dry fruits, such as almonds, walnuts, cashews, and coconuts (Sonu and Jha, 2026). Raw makhana outperforms both protein-rich sources, such as soybean and fish, and staple foods such as rice and wheat, highlighting its high nutritional value (Idris and Jana, 2018). Overall, these findings support the view that makhana is an energy-dense food with low-fat and high-quality protein, enriched with key minerals, which explains its centuries-long use as a strengthening and health-promoting ingredient by traditional societies, as well as its transformation into a superfood. It is used as a remedy to treat a wide range of conditions, including digestive, respiratory, excretory, reproductive, and circulatory problems (Kapoor et al., 2022). It has also been used in the treatment of beriberi, polyuria, diarrhea, spermatorrhea, and joint discomfort. In TCM, makhana is used to prepare medicinal liquors with antirheumatoid properties, formulated from a combination of ingredients, including seed extract, Chinese liquor, and yeast (Biswas et al., 2020). It is used in Ayurvedic and TCM to treat renal, gastrointestinal, hepatic, and reproductive system disorders and as a nutrient-rich food to boost energy levels and support postpregnancy recovery (Jiang et al., 2023).

Makhana is classified as a *madhura rasa* (sweet taste) substance with *guru* (heavy) and *snigdha* (unctuous) *guna*, *shita virya* (cool potency), and *madhura vipaka* (sweet postdigestive effect), and is considered beneficial in pacifying *vata* and *pitta* doshas, as well as nourishing *rasa* and *shukra* dhatus (Jha et al., 2018). Makhana is described in classical and contemporary ayurvedic commentary as having qualities, such as *balya* (strength-increasing), *brihmana* (body-building), and *Vrishya* (aphrodisiac), and it is used in formulation for chronic complaints, burning sensation, and reproductive deficiency. Powdered makhana, typically in combination with milk or ghee, is recommended as part of dietary regimens for individuals with weariness, recurring diarrhea, or urinary complaints, owing to its perceived cooling, stabilizing, and tissue-building benefits (Jha et al., 2018).

Parallel traditions in Chinese classics describe makhana seeds as tonics employed in compound formulations for joint discomfort, weakness, or growth stimulation, demonstrating the clear transregional acceptance of this aquatic nut as a food-to-medicine bridge (Kumari and Jha, 2017). Further pharmacognostic reviews indicate that many of these indications, such as antidiarrheal, cardioprotective, and neurotonic, also have partial experimental support documenting antioxidant, antiinflammatory, and metabolic effects of seed extracts, though rigorous clinical trials remain limited (Kumar, 2025; Mittal et al., 2020). From a cultural aspect, the main issue is that makhana fills a critical gap between diet and medication, appearing in both folk cures and commercial formulations. Makhana's traditional and cultural value is inextricably linked to the knowledge systems and labor that support it (Nisar et al., 2024). Studies on specific communities in North Bihar, particularly the Mallah and other fishing-related groups, show that they have developed sophisticated, multistaged techniques for harvesting, drying, roasting, and popping makhana that significantly improve yield and quality (Kumari and Jha, 2017). These techniques involve harvesting prickly fruits underwater, sequential drying and storage of seeds, and popping kernels using specialized iron pans and wooden mallets, all of which are physically demanding and technically complex.

13.2.1.2 Agricultural importance

Makhana (*E. ferox*) is increasingly recognized as a nutritionally rich alternative to conventional cereal crops, such as wheat and rice; however, its cultivation is constrained by high production costs and relatively low yields (Khadatkar et al., 2020; Bhatt et al., 2017). Despite these limitations, crop diversification studies highlight the potential of shifting from traditional cereals to high-value aquatic crops such as makhana, which possesses significant economic potential. One of the key advantages of makhana cultivation is its suitability for flood-prone and waterlogged ecosystems, as it thrives in wetlands, ponds, and shallow water bodies, thereby enabling the productive utilization of otherwise unproductive areas (Gupta et al., 2022). Its ecological adaptation to wetland habitats further enhances its agricultural importance and sustainability.

In addition to its agronomic benefits, makhana cultivation plays a vital role in supporting rural livelihoods, particularly in Bihar, where it serves as a major source of income for farmers (Kumar, 2025). The production and consumption of makhana are widespread across Asian countries, including India, China, Japan, and Korea, reflecting its cultural and nutritional significance (Jain et al., 2010). Furthermore, its cultivation has expanded beyond Asia to regions, such as Taiwan, North America, and Russia. Within India, makhana is predominantly cultivated in Bihar, followed by West Bengal, Uttar Pradesh, and Tripura, underscoring its regional agricultural importance (Jain et al., 2010; Kumar, 2025).

To minimize these challenges, scientists developed a new hybrid-line method, selective breeding, which results in higher yields (Guo and Min,

2020). The biotechnological techniques are micropropagation and tissue culture. The studies show that makhana has agricultural importance due to its roles in agriculture and nutrition. It contains compounds, such as polysaccharides and essential oils, which meet market demand (Jiang et al., 2023). The properties make it an important food crop with valuable applications in treating conditions, such as digestive disorders. In the agricultural aspect, the high medicinal and nutritional value increases its cultivation. The study also suggests that more analysis is needed to fully understand its biological activities, enhance its utilization, and develop processing techniques. Makhana supports sustainable agriculture, improves agro-based industries, and enhances the farmer economy due to its medicinal value.

13.2.1.3 Traditional medicine applications

Makhana has been used for centuries in traditional medicine, particularly Ayurveda and TCM, to treat a variety of ailments. It is a highly nutritious and versatile food that has long been used in Ayurvedic practice and traditional Asian cuisines. Makhana has also been extensively used in traditional Eastern medicine as a natural remedy for various conditions, including kidney disorders, chronic diarrhea, excessive leucorrhea, and splenic hypofunction (Sinha and Jha, 2025). It is recognized for its nourishing, astringent, and balancing effects on digestive, reproductive, and overall health. Due to its low glycemic index, it helps regulate blood sugar levels and is a suitable choice for individuals with diabetes. Its rich content of essential minerals, including calcium (Ca^{2+}), iron ($Fe^{2+/3+}$), phosphorus (P, PO_4^{3-}), potassium (K^+), and magnesium (Mg^{2+}), supports heart health and overall well-being (Farooqi et al., 2025). Makhana's antioxidant capabilities make it an important source and possible medicinal ingredient (Idris and Jana, 2018). The plant offers numerous health benefits, including antioxidant, anticancer, and antiinflammatory effects, due to its bioactive phytochemicals, such as alkaloids, flavonoids, tannins, saponins, and phenolic compounds. The remarkable therapeutic potential and strong antioxidant activity of fox nut are attributed to the presence of these antioxidant compounds (Zhao et al., 1989).

13.2.1.4 Edible and nonedible plant parts

Makhana is cultivated mainly for its edible and economically important seeds. The seeds, kernels, and endosperm are widely consumed raw or roasted and are rich in starch, protein, essential amino acids, minerals, vitamins C and E, and carotenoid pigments, such as β-carotene. They also contain bioactive compounds, including polyphenols, flavonoids, lignans, tocopherols, glycolipids, and cyclic peptides (Shukla et al., 2024; Xie et al., 2025). The popped or roasted seeds are also popular as snacks. In addition to the edible and economically important seeds, the stems are also edible and can be consumed as a nutrient-rich vegetable (Wang et al., 2025).

Traditionally, it is claimed that all parts of the plant except the leaves are edible, as the leaves are generally not used for human consumption. However, the leaves are the significant sources of glycosylated flavonoids and other polyphenols with potential medicinal and industrial value (Mahesh Kumar et al., 2023). Likewise, the seed coat, which is usually removed and discarded during processing, is not consumed on its own but is exceptionally rich in polyphenols, such as gallic acid, catechin, ellagic acid, and procyanidins (Joseph and Ramesh, 2023). Recent studies have highlighted the development of value-added food products from fox nuts (*E. ferox*), demonstrating their rich nutritional profile, presence of bioactive phytochemicals, and notable antioxidant potential (Zhang et al., 2023). Although the seed kernel remains the primary edible component, other plant parts—including the stems, leaves, and seed shells—also hold considerable importance due to their potential applications in nutrition, medicine, and various industrial uses (Table 13.1).

TABLE 13.1 Nutrient-rich part of *E. ferox*.

Plant part	Nutrient	Details	Reference(s)
Rhizome	Dietary fiber, vitamin C, vitamin B, K^+, Mg^{2+}, Fe^{2+}/Fe^{3+}, and Ca^{2+}	High fiber supports digestion; vitamin C and flavonoids give antioxidant and antiinflammatory effects	Ranjit Kumar et al. (2024)
Seed (whole kernel)	Complex carbohydrate, complete protein, vitamin B, Mg^{2+}, K^+, fiber, and low fat	16%–18% protein, ~44% carbohydrate; good amino acid balance; useful minerals that are required for the human diet	Sinha and Jha (2025)
Seed embryo (plumule)	Polyphenols, flavonoid, and alkaloid	Exhibit strongest antioxidant effect among all plant parts	Liu et al. (2023)
Seed coat	Phenolic compound, tannin, and flavonoids	Mainly valued for functional properties and antioxidant, rather than bulk nutrients	Liu et al. (2023); Song et al. (2011)
Leaf	Phenolic acid, dietary fibers, and alkaloids	Used as an herbal ingredient with anti-obesity and cardiometabolic benefits	Kapoor et al. (2022)
Stems	Flavonoid, phenolic acids, and aroma compounds	Used as a flavoring agent; high in essential oils	Ranjit Kumar et al., (2024)

This table highlights the different parts of ***E. ferox***, which serve as the primary nutritional source and provides a detailed explanation of their related therapeutic potential.

13.2.2 Related plants

Within the family Nymphaeaceae, *E. ferox* is most closely related to other genera, including *Victoria* and *Nymphaea*, with genomic and chloroplast data indicating a close evolutionary relationship between *E. ferox* and *Victoria* (Yang et al., 2020) (Fig. 13.3). They are closely related to the night-blooming species of *Nymphaea*. Another close relative of *E. ferox* is *Brasenia schreberi*, which belongs to a sister lineage of Nymphaeales, and is often used as a model for comparative genomics (Wen et al., 2025).

Nutritionally, closely related species, such as *Nymphaea* and *B. schreberi*, have a carbohydrate-rich, low-fat composition with moderate protein content and considerable levels of minerals, including K^+, Mg^{2+}, and Ca^{2+} (Qian et al., 2022). *Nymphaea* seeds have a similar energy content to cereal grains, with 7%–8% protein and 74%–78% carbohydrate content, while the leaves and stems have a higher protein content, up to 25%–27%, and can be used as a source of minerals and as a vegetable. These plants contain considerable amounts of bioactive compounds, including phenolics, flavonoids, and tannins, as well as vitamin E, which contribute to their antioxidant activity. *B. schreberi* is notable for its mucilage, which is rich in polysaccharides, such as galactose and arabinose, and exhibits antioxidant and hypoglycemic activities (Sharma et al., 2024).

Trapa natans, scientifically known as the water chestnut, and *Nelumbo nucifera*, the lotus, are some of the nutritionally important aquatic food plants (Das et al., 2021). They have a high carbohydrate content, moderate protein content, and low lipid content, along with a good number of micronutrients and phytochemicals. In all these underused aquatic food plants, the macronutrient content varies from 4.4%–32.1% protein, 38.7%–77.5% carbohydrates, and 0.93%–6.74% lipid, giving these aquatic plants an approximate

FIGURE 13.3 Closest relatives of Nymphaeales (prickly waterlily). Phylogenetic relationship of *E. ferox* within the Nymphaeales family, where it is closely related to *Nymphaea*, *Victoria*, and *Brasenia*. The diagram illustrates the close evolutionary relationship of *E. ferox* with early diverging angiosperms (Fern, 2026). *The figure image is adapted from Fern, K., 2026. Useful tropical plants. Tropical plants database. http://tropical.theferns.info/viewtropical.php.*

energy content of 318–377 kcal/100g (Rehman et al., 2024). For instance, the seeds of the water chestnut, *Trapa natans*, have a high content of starch and carbohydrates and a very low percentage of lipid content, approximately 0.36%. These seeds can be used as a flour and as a source of energy, and the flour is gluten-free (Mondal et al., 2024). Similarly, *Nelumbo nucifera* seeds are rich in complex carbohydrates and proteins, with a balanced amino acid content, while the rhizomes contain dietary fiber, vitamin C, and minerals, such as Fe^{2+}/Fe^{3+} and copper (Cu^{2+}) (Suraiya et al., 2024). Apart from their nutrient content, these two aquatic plants have gained importance due to the presence of bioactive compounds, including phenolics, flavonoid, and vitamin B complex and C, as well as free amino acids with 4.0–10.66 mg/g, which contribute to the health-promoting effects. These aquatic plants, therefore, have the potential to serve not only as functional foods—providing the required nutrients—but also serve as nutraceuticals (Shukla et al., 2024).

13.3 Bioactive and medicinal components

In recent times, makhana seeds have been widely recognized for their rich bioactive compounds and medicinal properties. Makhana has been classified as a functional and nutraceutical food product due to its high carbohydrate content (e.g., starch, proteins, dietary fiber), essential amino acids, vitamins, and minerals, such as Ca^{2+}, Mg^{2+}, K^{+}, and P/PO_4^{3-} (Kumar, Singh and Jha, 2026).

Makhana seeds contain 65%–75% carbohydrates, 9%–15% protein, and a very low amount of fat (0.1%–1.5%), along with 0.5%–1.5% crude fiber, making it a low-fat, energy-rich functional food. These seeds are also rich in minerals, including Ca^{2+}(0.02%–0.05%), P/PO_4^{3-} (0.25%–0.3%), K^{+} (0.3%–0.5%), and Mg^{2+} (0.04%–0.07%), which are important for metabolic and neuromuscular functions (Idris and Jana, 2018; Jha et al., 2018). Makhana also contains antiaging amino acids, such as arginine, cystine, glutamine, and methionine, which enhance its therapeutic advantages. Additionally, these seeds are rich in bioactive compounds, including phenolics, alkaloids, flavonoids, polysaccharides, proteins, and essential minerals, which confer antioxidant, antiinflammatory, and therapeutic properties associated with various health benefits (Ramesh et al., 2026), such as cardioprotective, hepatoprotective, antifatigue, antiaging, and antidiabetic effects, thereby enhancing their status as a "superfood" in recent times (Shukla et al., 2024; Sharma et al., 2024).

Makhana has been found to be gluten-free and to have a low glycemic index, making it beneficial for those with gluten intolerance or blood sugar issues (Jana et al., 2019). Processing techniques, especially popping, play an important role in modifying the product's physicochemical and nutritional characteristics, including improving starch digestibility, reducing antinutritional factors, and enhancing acceptability (Mahesh Kumar et al., 2023).

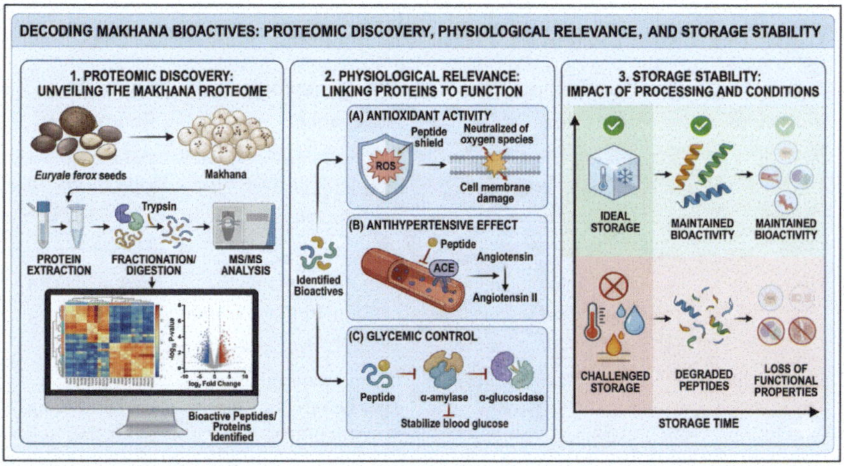

FIGURE 13.4 Decoding the bioactives of makhana: From proteomic identification to functional relevance and stability. Proteins derived from *E. ferox* (makhana) seeds are enzymatically digested and analyzed using LC–MS to identify bioactive peptides. These peptides exhibit key physiological functions, including antioxidant activity, antihypertensive effects, and glycemic regulation through inhibition of α-amylase and α-glucosidase. The stability of these bioactives is strongly influenced by storage conditions, where optimal environments preserve peptide integrity and functionality, while adverse conditions promote degradation and loss of bioactivity (BioRender, 2026). *The figure is generated using Biorender.com.*

The therapeutic efficacy of makhana has been investigated using gas chromatography–mass spectrometry (GC–MS) and liquid chromatography–high-resolution mass spectrometry (LC–HRMS), which have identified over 400 distinct compounds in the seeds (Fig. 13.4). Metabolomic profiling of popped makhana has revealed 36 primary metabolites, including glycerol, decanoic acid, myristic acid, and farnesol, as summarized in Table 13.2 (Sinha and Jha, 2025; Xie et al., 2025).

13.3.1 Proteomic insights into makhana bioactives

13.3.1.1 Macronutrient profile

Makhana seeds are characterized by an exceptionally high starch content (approximately 78%–85%), moderate protein levels (9%–11%), and negligible fat content (typically below 0.3%–0.4%), rendering them a quintessentially starch-based food material distinct from most other edible nuts and seeds (Devi et al., 2020). This macronutrient profile positions makhana uniquely as a low-fat, gluten-free, high-carbohydrate food with potential applications in weight management, dietary formulations, and sports nutrition. The protein fraction, though quantitatively moderate, is qualitatively distinguished by the presence of all essential amino acids, making it a complementary protein source for vegetarians (Jha et al., 2018; Jiang et al., 2023).

TABLE 13.2 List of bioactive molecules of makhana (Jiang et al., 2023).

Type of molecules	Specific bioactive molecules	Pharmacological relevance	Reference(s)
Phenolic acids	Gallic acid, Chlorogenic acid, Caffeic acid, and Gentisic acid	Antioxidant, antifatigue, and antidiabetic	Devi et al. (2024); Xie et al. (2025)
Flavonoids	Kaempferol, Quercetin, Rutin, and Epicatechin	Neuroprotection, antiinflammatory, and anticancer	Li et al. (2026)
Polysaccharides	EPJ, EFSP-1, and RS3 (Resistant Starch)	Immunomodulation, gut health, and blood sugar control	Qu et al. (2025); Zhang et al. (2025)
Sterols	β-sitosterol, Glucosyl sterols, and Cerebrosides	Cholesterol-lowering, cardiovascular health	Liaquat et al. (2022)
Alkaloids	2-Hydroxyphenylalanine, Hydromorphinol (derivatives)	Antimicrobial, metabolic regulation	Liu et al. (2023); Wu et al. (2021)
Tocopherols	Vitamin E, Vitamin E acetate	Antiaging, neuroprotection, and antioxidant	Lee et al. (2009)

The presence of Mg^{2+} and zinc (Zn^{2+}) minerals—both critical regulators of insulin secretion and storage—along with chromium, a cofactor in glucose tolerance, signifies the potential relevance of makhana in the dietary management of metabolic disorders (Mahesh Kumar et al., 2023).

13.3.1.2 Protein characterization: From classical to proteomic approaches

The analytical technologies applied to makhana bioactive research have evolved significantly over the past 2 decades, progressing from basic colorimetric assays and High-Performance Liquid Chromatography (HPLC) Ultraviolet (UV) detection to sophisticated mass spectrometry-based platforms (Fig. 13.4).

The protein content of makhana varies with seed developmental stage—predevelopment seeds, green seeds, and old (mature) seeds represent distinct developmental states with progressively changing protein profiles as starch accumulation increases and the seed approaches dormancy. Early characterization of makhana proteins relied on classical biochemical methods, including Kjeldahl nitrogen determination and amino acid analysis by HPLC (Jana et al., 2019). These studies established the total protein content and amino acid composition of makhana seed extracts but provided limited information on individual protein identities, functions, and regulatory roles.

The transformative advance in makhana protein characterization came with the application of isobaric tags for relative and absolute quantitation (iTRAQ) proteomic technology to *E. ferox* seeds at four developmental stages (10, 20, 30, and 40 days after anthesis) (Wu et al., 2022). This technology enables simultaneous relative quantification of proteins across multiple experimental conditions within a single mass spectrometric run, making it ideal for temporal profiling of the seed developmental proteome. A landmark study using liquid chromatography–mass spectrometry (LC-MS) identified 2433 credible proteins from the seed proteome (Wu et al., 2022). Differential expression analysis across the four developmental time points revealed 2809 differentially expressed proteins, predominantly enriched in starch synthesis and flavonoid biosynthesis pathways. Through multiomics analysis, key candidate enzymes involved in starch synthesis—ADP-glucose pyrophosphorylase small subunit (APS), ADP-glucose pyrophosphorylase large subunit (APL), starch-branching enzyme (SBE), soluble starch synthase (SSS), and granule-bound starch synthase (GBSS)—were identified, and their expression dynamics across seed development were characterized. Simultaneously, key enzymes in the flavonoid biosynthesis pathway—phenylalanine ammonia lyase (PAL), chalcone synthase (CHS), flavanone 3-hydroxylase (F3H), and anthocyanidin synthase (ANS)—were also identified, establishing a proteomic framework linking *E. ferox* seed development to bioactive flavonoid accumulation (Wu et al., 2022). It has been found that more proteins are involved in the later stages of development than in the early stages.

Abscisic acid (ABA) is an important plant hormone that is essential for plant growth and development. This proteomic study also characterized the ABA signal transduction pathway in developing makhana seeds, identifying Pyrabactin Resistance (PYR)/PYR1-Like proteins (PYL)/Regulatory Component of ABA Receptor (RCAR) proteins, type 2C protein phosphatase (PP2C), and SNF1-related kinase 2 (SnRK2)—components of the core ABA signaling module that regulate seed maturation, dormancy, and stress responses (Wu et al., 2022). In this study, changes in ABA content directly correlate with changes in starch content. The identification of these signaling proteins provides an understanding of the developmental programs that govern bioactive compound accumulation in makhana seeds.

The comprehensive metabolomic characterization of *E. ferox* seeds using liquid chromatography–high resolution mass spectrometry (LC-HRMS) combined with Global Natural Products Social Molecular Networking (GNPS) has identified a total of 438 compounds, of which 283 were identified through searching databases and 155 through spectral network propagation (Xie et al., 2025). These compounds span multiple chemical classes, including glycerophospholipids, carbohydrates and their conjugates, fatty acids and their derivatives, fatty amines, glycerolipids, cinnamic acid derivatives, sphingolipids, lignans, terpenoids and their derivatives, and flavonoids and their derivatives (Xie et al., 2025).

Statistical analysis of raw versus processed (dry-fried and bran-fried) seeds revealed 32 and 38 significantly different compounds, respectively, with most showing increased levels after thermal processing, a finding that has important implications for understanding how traditional processing methods modulate the bioactive composition of makhana (Xie et al., 2025).

Polyphenolic compounds represent the most extensively studied class of bioactive metabolites in *E. ferox* and are predominantly concentrated in the seed coat. For example, 2β-hydroxybetulinic acid 3β-oleiate (HBAO), a pentacyclic triterpene derived from *E. ferox* seeds, may help prevent the important complications in diabetic rats, and it might be a potential therapeutic candidate to combat diabetes (Ahmed et al., 2019).

Gas chromatography–mass spectrometry (GC-MS) has complemented LC-based methods by characterizing the volatile and semi-volatile metabolome, capturing fatty acids, organic acids, and terpenoids that may be missed by LC methods (Sinha and Jha, 2025). Through GC-MS, applied to map the volatile and semivolatile metabolites of both raw and popped *E. ferox* seeds, 36 metabolites were detected, including fatty acids, esters, and phytosterols. Major metabolites identified include glycerol, decanoic acid, butanoic acid, gallic acid, myristic acid, palmitic acid, linoleic acid, tricosanoic acid, myoinositol, and farnesol. The metabolomics profiling identified a mixture of primary metabolites (fatty acids, sugar alcohols) and secondary metabolites used to treat problems related to spermatogenesis, rheumatism, polyurea, parturition, and biliary dysfunction. It has been found that the essential fatty acids present in makhana seeds are important for preventing cancer, cardiovascular protection, promoting healthy brain and vision development, preventing diabetes mellitus, reducing hypertension, reducing arthritis, and preventing neurological and neuropsychiatric diseases (Sinha and Jha, 2025).

Additional techniques that have contributed to makhana bioactive research include HPLC-UV for routine quantification of known phenolics and isoflavones; Fourier-transform infrared spectroscopy (FTIR) for functional group characterization; nuclear magnetic resonance (NMR) spectroscopy for definitive structural elucidation of isolated compounds; and scanning electron microscopy (SEM) for visualization of structural changes during processing (Sinha and Jha, 2025; Xie et al., 2025).

13.3.1.3 Impact of processing on the makhana bioactive proteome and metabolome

Traditional makhana processing involves a multistep sequence of initial roasting (270–280°C), tempering (approximately 48 h), followed by final roasting (260–270°C) to produce the porous kernels consumed as popped makhana (Nehal et al., 2015). This thermal processing fundamentally transforms the seed matrix from a loose crystalline structure to a highly porous configuration, with dramatic increases in water absorption capacity (376%) and oil absorption capacity (317%). The popping process significantly increases total sugar content (by 154%), degree of gelatinization (by over 1100%), available carbohydrates (by 27%), and starch digestibility, while reducing phytic acid content from 281 to 168 mg/100 g and eliminating detectable tannins (Nehal et al., 2015).

Recent data showed that roasting significantly improved mineral content (from 0.58% to 0.66%), protein content (from 11.40% to 14.57%), total phenolic content (from 346.02 to 470.62 mg GAE/100 g), flavonoid content (from 4.15 to 4.43 mg CE/g), 2,2-diphenyl-1-picrylhydrazyl (DPPH) radical scavenging activity (from 48.54% to 79.13% inhibition), and ferric reducing antioxidant power (from 662.46 to 957.14 μmol $FeSO_4$/g) (Liaquat et al., 2022). Roasted fox nuts were found to contain higher levels of phenolics and flavonoids, along with enhanced antioxidant activity, compared with unroasted nuts. Roasting also increased protein and mineral contents, thereby improving their overall nutrient density (Liaquat et al., 2022). In addition, antioxidant activity significantly increased, antinutritional substances, such as phytates and tannins, decreased, and mineral bioavailability increased as a result of germination (Islam et al., 2025).

Despite significant advances, the proteomic and bioactive characterization of makhana is still at an early stage compared with that of major crops. Several critical limitations constrain the field. First, the makhana seed proteome has been characterized using iTRAQ technology (Fig. 13.5), and independent validation using orthogonal proteomic approaches is required. Second, the functional annotation of many identified proteins through gene knockout, overexpression, functional characterization, or enzyme activity assays in *E. ferox* remains largely unexplored. Third, most identified proteins are abundant and soluble, while membrane-associated, low-abundance, and post-translationally modified proteins have not yet been investigated. Future studies should focus on the following directions to advance the translation of makhana's bioactive and medicinal components:

1. Several promising research directions can be identified for advancing makhana bioactive science. The application of deep shotgun proteomics using data-independent acquisition (DIA) methods, combined with storage protein depletion strategies analogous to those developed for soybean, could substantially expand the characterized makhana seed proteome (Krishnan and Natarajan, 2009).
2. Phosphoproteomics and glycoproteomics studies can provide insights into the posttranslational regulation of key biosynthetic enzymes and the glycosylation patterns of seed proteins.

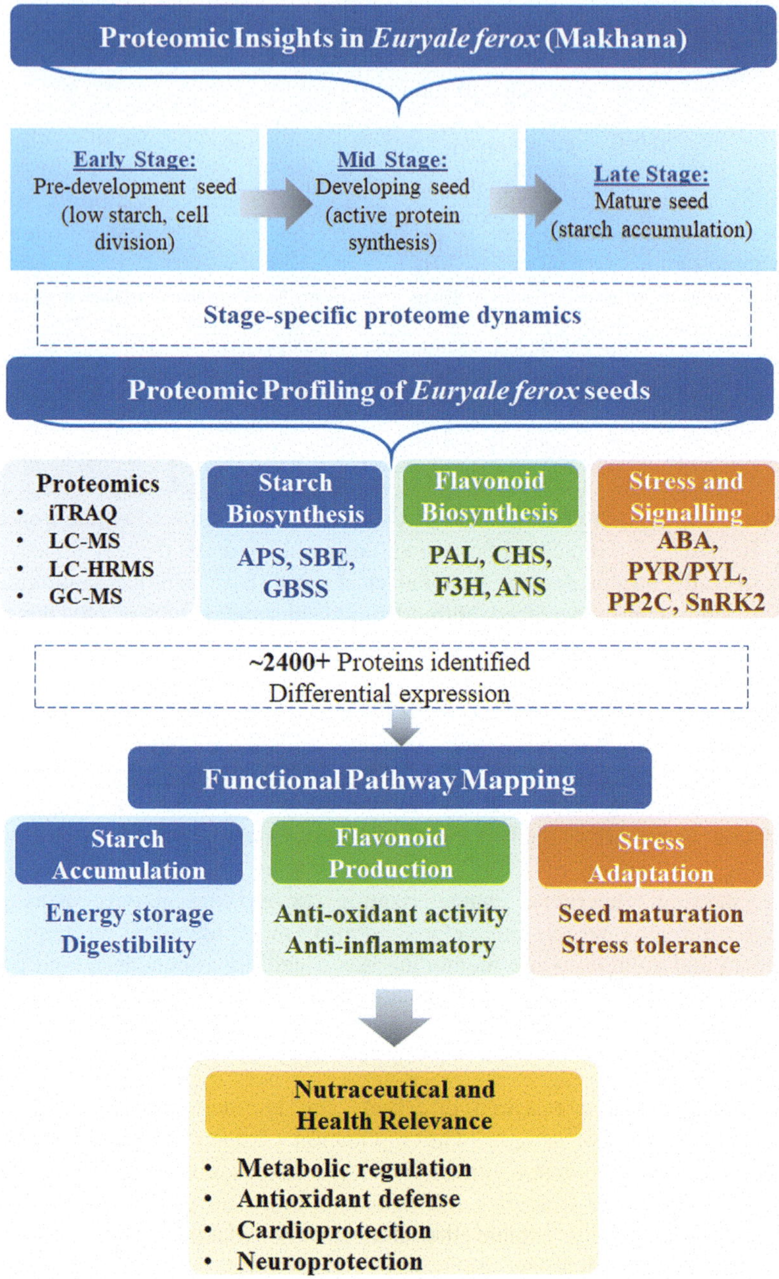

FIGURE 13.5 Stage-specific proteome dynamics in *E. ferox* seeds illustrating key pathways-starch biosynthesis, flavonoid metabolism, and ABA-mediated stress signaling, identified by iTRAQ and LC–MS/HRMS. Differential protein expression across developmental stages drives bioactive synthesis and underpins the functional and nutraceutical relevance of makhana.

3. Development of standardized extracts from the currently discarded polyphenol-rich seed coat bran could yield functional food ingredients and nutraceutical products with applications in antioxidant supplementation, glycemic control, and antiaging cosmeceuticals.
4. The mechanisms and pathways involved in the treatment of diabetes, gastrointestinal disorders, and cancer remain unclear. Randomized controlled trials examining the effects of standardized makhana consumption on glycemic control, lipid profiles, antioxidant status, and antiaging processes could strengthen its scientific credibility and facilitate its acceptance by global consumers.

13.4 Physiological relevance of the makhana bioactive molecule

Makhana exerts a wide range of physiological effects on the human body due to its bioactive compounds. Makhana seeds are pharmacologically important due to their diverse phytoconstituents, including polyphenols, flavonoids, tocopherols, triterpenoids, polysaccharides, and cyclic dipeptides (Jiang et al., 2023). Phenolic compounds and flavonoids are well known for their strong antioxidant activity in makhana seeds. Raw makhana is a high-carbohydrate, low-fat seed with moderate protein content and useful amounts of minerals and essential amino acids. It contains approximately 10.8%–11.8% moisture, 9.15% protein, only 0.46% fat, 2.55% dietary fiber, and nearly 87.5% carbohydrates, providing around 362 kcal per 100 g. Its mineral profile indicates appreciable levels of P/PO_4^{3-} (≈132.5 mg/100 g), K^+ (≈50.3 mg/100 g), and Mg^{2+} (≈13.8 mg/100 g), along with smaller but relevant amounts of Ca^{2+}, Fe^{2+}/Fe^{3+}, manganese (Mn^{2+}), Zn^{2+}, Cu^{2+}, and trace molybdenum ($Mo^{4+/6+}$), supporting its role as a micronutrient source (Jha et al., 2018) as shown in Table 13.3. The essential amino acid analysis shows the presence of

TABLE 13.3 Properties of makhana (Jha et al., 2018; Mahesh Kumar et al., 2023).

Nutrient	Concentration	Physiological benefit
Protein	10%–15%	Muscle recovery, cell metabolism
Mg^{2+}	11.2–20 mg/100 g	Antiobesity, glucose regulation
Ca^{2+}	149 mg/100 g	Bone health, metabolic control
Fiber	7.6%–14.5%	Digestive health, satiety, and low GI
Essential AA	High Arginine, Methionine	Antiaging, immune function

lysine, histidine, isoleucine, leucine, methionine, phenylalanine, threonine, and valine in measurable quantities, suggesting that makhana protein is qualitatively better than most cereal proteins, which are often lysine-deficient. This is reflected in the comparative Essential Amino Acid Index (EAAI)/ Biological Value (BV) data (Devi, 2022). This is particularly rare in plant-based foods, making it a critical resource for vegetarians and vegans. Human health is greatly influenced by the EAAI and BV, which are essential for assessing the quality of protein in food. While BV indicates how well the protein is used for tissue maintenance, EAAI refers to the balance of essential amino acids, which is critical for the development, repair, and general health of the body (Mittal et al., 2020). This protein-rich seed, with a high EAAI and a favorable lysine-to-proline and arginine-to-proline ratio, is superior to other cereals (Jha et al., 2018). Overall, these findings support the notion that makhana is an energy-dense food with low fat and good-quality protein, enriched with key minerals, which justifies its traditional use as a strengthening and health-promoting ingredient and its potential as a superfood (Kapoor et al., 2022).

Makhana seed extracts exhibit free radical–scavenging potential and enhance the production of antioxidant enzymes, including superoxide dismutase, catalase, and glutathione peroxidase, thereby protecting the body (Lee et al., 2002).

Additionally, makhana-derived polysaccharides and proteins modulate the body's immune responses and improve metabolic regulation (Jha et al., 2018). High Mg^{2+} levels are particularly notable for their effects on metabolic health. Mg^{2+} acts as a cofactor for enzymes, such as thiamine pyrophosphokinase, which is essential for energy metabolism. Deficiency in serum Mg^{2+} is closely linked to general obesity, making makhana an effective dietary intervention for weight management (Rosique-Esteban et al., 2018). Furthermore, the combination of lysine and methionine in makhana may support the formation of carnitine in cells, which facilitates the burning of fat for energy (Rebouche, 2004). These combined effects support the potential use of makhana as a dietary supplement to promote the overall health and prevent chronic diseases (Jha et al., 2018).

The nutraceutical value of makhana can be further improved through processing methods, such as roasting, which enhances the levels of phenolics, flavonoids, and antioxidant capacity. Owing to its rich phytochemical and nutritional profile, makhana exhibits multiple medicinal properties, including antioxidant, antiinflammatory, antidiabetic, cardioprotective, hepatoprotective, antimicrobial, and potential anticancer activities, largely mediated by the modulation of oxidative stress and key cellular signaling pathways (Shaw et al., 2025). Collectively, these attributes establish makhana as a potent functional food with significant applications in health promotion, disease prevention, and therapeutic research.

13.5 Report on the medicinal potential of *E. ferox*

E. ferox is increasingly recognized for its medicinal potential owing to a diverse array of bioactive compounds, including flavonoids, phenolic acids, polysaccharides, and triterpenoids. These constituents exhibit strong antioxidant, antiinflammatory, antidiabetic, and anticancer activities by modulating key molecular pathways, including oxidative stress regulation, inflammation signaling, and cellular metabolism (Table 13.4). Consequently, makhana-derived bioactive molecules play a crucial role in the prevention and management of chronic diseases, providing a promising foundation for further exploration of their therapeutic properties.

13.5.1 Antioxidant and antiinflammatory properties

Makhana exhibits significant antioxidant and antiinflammatory properties owing to its rich reservoir of bioactive phytochemicals, particularly polyphenols, flavonoids, phenolic acids, tannins, and other secondary metabolites (Jiang et al., 2023). Metabolomic studies have identified a remarkably diverse phytochemical profile in *E. ferox*, with as many as 589 secondary metabolites reported across different tissues, among which polyphenols constitute nearly 87%, highlighting its strong antioxidant potential (Xie et al., 2025). Major antioxidant constituents reported in makhana include gallic acid, chlorogenic acid, catechin, epicatechin, rutin, pyrogallol, ferulic acid, and corilagin, many of which are known to scavenge free radicals and reduce oxidative damage effectively. Makhana seeds are a potent source of natural antioxidants, thanks to their high phenolic content, which neutralizes free radicals and reduces inflammation, thereby protecting tissues from oxidative damage (Jiang et al., 2023). Several studies on makhana seeds and their seed coats have shown that these phenolic extracts neutralize reactive oxygen species (ROS). In lung fibroblast cells (V79-4), makhana extracts improved cell viability under oxidative stress by inducing superoxide dismutase (SOD), catalase (CAT), and glutathione peroxidase (GPx), with GPx showing the most pronounced increase in activity (Lee et al., 2002; Zhang et al., 2023).

Makhana shell extracts contain corilagin, a monomer shown to inhibit excessive macrophage activation, reflecting makhana's antiinflammatory effects. It functions by downregulating protein phosphorylation in the nuclear factor kappa B (NF-κB) and toll-like receptor 4 (TLR4) signaling pathways, leading to a reduction in the secretion of proinflammatory cytokines, such as interleukin 1 Beta (IL-1β), tumor necrosis factor (TNF-α), and interleukin 6 (IL-6) (Zhao et al., 2008). This presents a promising strategy for therapeutics of chronic inflammatory diseases, such as arthritis.

TABLE 13.4 Therapeutic relevance of bioactive compounds in makhana across various diseases (Idris and Jana, 2018; Jha et al., 2018; Jiang et al., 2023; Mahesh Kumar et al., 2023; Qin et al., 2025; Zhang et al., 2023).

Disease/condition	Pathophysiology	Bioactive components	Therapeutic significance
Cardiovascular disease	Oxidative stress, endothelial dysfunction, lipid peroxidation, vascular inflammation, and hypertension	Polyphenols, flavonoids, tocopherols, K^+, Mg^{2+}, and arginine	May reduce ROS, improve vascular homeostasis, suppress inflammation, and support blood pressure and endothelial regulation
Diabetes mellitus	Hyperglycemia, insulin resistance, oxidative stress, β-cell dysfunction, and inflammation	Polyphenols, flavonoids, polysaccharides, dietary fiber, and Mg^{2+}	May improve glucose metabolism, enhance insulin sensitivity, protect pancreatic cells, and reduce diabetes-associated oxidative stress
Metabolic syndrome	Dyslipidemia, insulin resistance, obesity, and chronic low-grade inflammation	Polyphenols, flavonoids, minerals, and low-fat nutritional profile	May assist in metabolic regulation, reduce inflammatory burden, and support overall metabolic balance
Obesity	Adipose tissue expansion, chronic inflammation, altered lipid metabolism, and oxidative stress	Low-fat nutrient profile, fiber, and antioxidant phytochemicals	May support weight management, improve metabolic efficiency, and reduce obesity-associated oxidative and inflammatory stress
Neurodegenerative disorders	Neuroinflammation, oxidative neuronal injury, mitochondrial dysfunction, and neuronal degeneration	Polyphenols, flavonoids, amino acids, and antioxidant compounds	May exert neuroprotective effects by reducing oxidative damage, suppressing neuroinflammation, and supporting neuronal survival
Neuropathic pain/ neuroinflammation	Inflammatory signaling, oxidative stress, neuronal sensitization, and altered neural signaling	Polyphenols, flavonoids, and antiinflammatory phytochemicals	May attenuate neuroinflammatory responses, reduce oxidative neuronal stress, and support neuronal homeostasis

Cancer	Uncontrolled proliferation, oxidative stress, inflammation, evasion of apoptosis, and altered signaling	Polyphenols, flavonoids, tannins, alkaloids, and saponins	May exhibit antiproliferative, proapoptotic, antiinflammatory, and redox-modulatory properties
Chronic inflammation	Persistent cytokine activation, inflammatory mediator release, and oxidative tissue damage	Flavonoids, phenolic acids, tannins, and corilagin-like phytochemicals	May inhibit proinflammatory mediators, suppress NF-κB-associated signaling, and restore inflammatory balance
Oxidative stress–related disorders	Excess free-radical generation, cellular redox imbalance, and biomolecular damage	Gallic acid, chlorogenic acid, catechin, epicatechin, rutin, and flavonoids	May enhance antioxidant defense systems and protect cells from oxidative injury
Liver dysfunction/ hepatotoxicity	Toxin-induced oxidative stress, inflammatory injury, and hepatocellular damage	Polyphenols, flavonoids, and antioxidant molecules	May support hepatocellular protection and improve oxidative-inflammatory defense mechanisms
Kidney dysfunction/ nephrotoxicity	Oxidative injury, inflammatory damage, and renal cellular stress	Polyphenols, flavonoids, minerals, and antioxidant compounds	May reduce renal oxidative burden and help preserve kidney tissue integrity
Aging and age-related decline	Cumulative oxidative stress, chronic inflammation, reduced repair capacity, and mitochondrial dysfunction	Antioxidants, flavonoids, amino acids, and minerals	May support healthy aging, reduce cellular stress, and promote physiological resilience
Immune dysfunction	Dysregulated immune signaling, inflammatory imbalance, and poor cellular defense	Polyphenols, flavonoids, and micronutrients	May contribute to immune modulation and help maintain immune-inflammatory homeostasis
Gastrointestinal disorders	Gut oxidative stress, inflammation, mucosal injury, and impaired digestion	Fiber, polyphenols, flavonoids, and micronutrients	May support digestive health, reduce gut inflammation, and contribute to mucosal protection

13.5.2 Cardiovascular disease

Makhana seeds contribute to cardiovascular health through their rich content of polyphenols, flavonoids, tocopherols, minerals, and essential amino acids, which collectively help regulate major cardiovascular risk factors (Jiang et al., 2023). Its antioxidant bioactives reduce oxidative stress and lipid peroxidation, thereby protecting vascular endothelium and cardiac tissues from damage, while its antiinflammatory compounds help suppress chronic low-grade inflammation associated with atherosclerosis, hypertension, and endothelial dysfunction (Jiang et al., 2023; Liu et al., 2023). In addition, makhana is a low-fat, low-sodium, and low-glycemic index food, making it beneficial for controlling obesity, insulin resistance, and metabolic syndrome, all of which are strongly linked to cardiovascular diseases (Jha et al., 2018). The presence of minerals, such as Mg^{2+} and K^{+}, along with amino acids like arginine, may further support vascular tone, nitric oxide production, electrolyte balance, and blood pressure regulation (Zhang et al., 2023).

Additionally, makhana seeds contain compounds, for example, flavonoids, phenolic compounds, and phytosterols that improve blood circulation, reduce cholesterol levels, and prevent oxidative stress to blood vessels. Furthermore, in vitro studies confirmed that makhana extracts possess potent ROS-scavenging activity, mitigating the free radical surge during reperfusion (Zhang et al., 2023). This reflects that makhana consumption can be associated with improved cardiovascular health. Thus, makhana shows promise as a cardioprotective functional food that may aid in the prevention and dietary management of cardiovascular disease.

13.5.3 Neuroprotective potential

The neuroprotective potential of makhana is largely contributed to by its effects on oxidative stress and autophagy in the brain (Jiang et al., 2023). Neurodegenerative diseases, such as Alzheimer's (AD) and Parkinson's (PD), are associated with cellular energy imbalance and the buildup of damaged proteins inside the brain. Makhana extracts contain the petroleum ether fraction (ES-PE), which upregulates the adenosine monophosphate–activated protein kinase (AMPK) pathway, which, in turn, activates the unc-51-like autophagy activating kinase 1 (ULK1), leading to mTORC1 inhibition and the induction of autophagosome formation to restore ATP and recycle nutrients (Kim et al., 2011). The activation of the AMPK–ULK1–mTOR axis promotes autophagosome formation, facilitating the clearance of toxic protein aggregates and restoration of cellular ATP levels. These mechanisms are highly relevant to AD and PD pathology, where defective autophagy contributes to neuronal degeneration; thus, AMPK-mediated autophagy induction (Mary et al., 2025; Park and Kim, 2024) by makhana bioactives may offer a potential neuroprotective strategy.

In addition, its antiinflammatory phytochemicals may suppress inflammatory mediators and signaling pathways involved in microglial activation and neuroinflammatory responses, which are key contributors to neuronal dysfunction and cell death. The presence of amino acids and micronutrients in makhana may further support neuronal metabolism, synaptic function, and cellular repair mechanisms (Xie et al., 2025). Vitamin E acetate in the extracts provides antioxidant protection for hippocampal neurons. Collectively, makhana shows potential as a dietary neuroprotective agent and may be considered a valuable nutraceutical for supporting brain health and preventing neurodegenerative and neuroinflammatory conditions.

13.5.4 Antidiabetic and metabolic regulation

Makhana extracts are recognized both traditionally and scientifically for their role in treating type 2 diabetes. The extracts contain flavonoids and polysaccharides, which improve glycemic control and reflect on diabetic management in patients (Idris and Jana, 2018). Its bioactive compounds help improve insulin sensitivity and glucose uptake while reducing oxidative stress, which plays a critical role in the pathogenesis of diabetes and its complications (Idris and Jana, 2018; Zhang et al., 2023). The antioxidant constituents of makhana help protect pancreatic β-cells from oxidative damage, thereby supporting insulin secretion and maintaining glycemic control (Zhang et al., 2023). Additionally, makhana has a low-glycemic index, which helps slow glucose release and improve postprandial blood sugar regulation (Mahesh Kumar et al., 2023). Its antiinflammatory properties further help in mitigating chronic low-grade inflammation associated with insulin resistance and metabolic syndrome. Moreover, the presence of dietary fiber and Mg^{2+} supports lipid metabolism, weight management, and overall metabolic homeostasis, reducing the risk of obesity and associated disorders (Jha et al., 2018; Jiang et al., 2023). Collectively, these properties highlight makhana as a promising functional food for the prevention and management of diabetes and metabolic disorders.

E. ferox seed extracts are highly effective in managing metabolic disorders associated with blood glucose regulation. Ethanol extracts of these extracts promote insulin secretion and protect pancreatic β-cells from oxidative damage (Ahmed et al., 2015). In insulin-resistant cells, the polysaccharide fraction EFSP-1 enhances glucose uptake by upregulating Glucose Transporter Type 4 (GLUT-4) expression via the Phosphoinositide 3-kinase (PI3K)/Protein Kinase B (AKT) signaling pathway (Wang et al., 2023). Additionally, seed shell extracts inhibit key carbohydrate-digesting enzymes, including α-amylase and α-glucosidase, which play crucial roles in controlling postprandial blood glucose levels. Furthermore, specific triterpenoids, such as 2β-hydroxybetulinic acid caprylate (HBAC) and HBAO, have demonstrated the ability to normalize plasma glucose and glycosylated hemoglobin (HbA1c) levels while restoring renal and hepatic histology in diabetic animal models (Ahmed et al., 2019).

13.5.5 Other major diseases

In recent times, makhana extracts have shown significant antitumor activity. A study on lung carcinoma (A549 cells) showed that makhana extracts induced apoptosis by inhibiting AKT signaling and activating tumor protein 53 (TP53), leading to ROS generation and disruption of mitochondrial membrane potential in cancer cells (Zhang et al., 2023). Human gastric cancer (SGC7901) and hepatoma (HepG2) cell lines have shown that seed shell extracts can arrest the cell cycle and inhibit growth (Nam et al., 2019). Moreover, extracts have shown dose-dependent inhibition of *E. coli* and other pathogens, suggesting potential for treating infectious diseases (Devi et al., 2022; Kumar et al., 2024). Additionally, several preclinical studies on the pharmacological effects of *E. ferox* underscore its importance for the body's well-being (see Table 13.5).

13.6 Nutraceutical and functional food relevance

Makhana is increasingly recognized as a functional food due to its nutritional richness and health-promoting properties. Makhana seeds are low in fat, rich in essential nutrients, and gluten-free, which makes them ideal for health-conscious people (Jiang et al., 2023). Nowadays, makhana is added to a variety of food products, including snacks, breakfast cereals, and nutraceutical formulations, to promote health and wellness. This transition is primarily influenced by the unique physical and nutritional transformations that occur during processing. It is generally consumed in popped form; the popping process involves roasting the seeds at 335°C for approximately 6 minutes, followed by the mechanical impact, which causes the starchy perisperm to expand (Vikash et al., 2013; Khadatkar et al., 2020). This process enhances the digestibility by breaking down complex protein molecules and converting starch into more easily digestible forms. Popping also increases the concentrations of minerals (Na^{+}, Ca^{2+}, K^{+}, Mg^{2+}, Mn^{2+}, and Cu^{2+}) and dietary fiber (Shukla et al., 2024). The popping process alters the nutritional composition of makhana. Studies show that the total mineral content and calorific value generally decrease during the popping process due to thermal processing. An increase in protein content is analyzed, and it is described as denaturation, with complex protein structures reduced to simpler, more digestible forms (Farooqi et al., 2025). Roasting at high temperatures (270–280°C) significantly increases total phenolic and total flavonoid content, thereby enhancing bioavailability (Mahesh Kumar et al., 2023). This change in the composition is primarily due to structural modification in the kernel that increases mineral availability. However, the phytochemical content, including phenols,

TABLE 13.5 Representative preclinical studies on the pharmacological effects of *E. ferox* (Jiang et al., 2023; Lee et al., 2002).

Activity	Animal/Cell line	Doses	Key effects
Antioxidation and antiinflammation	V79-4 and RAW 264.7 cell lines	0.8–400 μg/mL	DPPH radical scavenging and inhibition of iNOS, Cox-2, and NO production
	D-galactose-induced aging in Kunming mice	100, 200, and 400 mg/kg (oral)	Increased SOD, CAT, and GSH-Px activities; decreased MDA content in liver and kidneys
Antidiabetic and hypoglycemic	Streptozotocin-induced diabetic Wistar rats	100–400 mg/kg (oral)	Decreased blood glucose; increased insulin levels; and restored hepatic enzymes
	HepG2 and 3T3-L1 cells	25, 100, and 400 μg/mL	Increased glucose consumption by upregulating GLUT-4 via the PI3K/Akt pathway
Hepatoprotective and cardioprotective	High-fat diet-induced ICR mice	15 and 30 mg/kg (oral)	Reduced body weight and liver lipid deposition; increased SOD activity
	Ischemia/reperfusion injury model (Rats)	50 and 500 mg/kg (oral)	Reduced myocardial infarct size and improved ventricular function via ROS scavenging
Anticancer	A549 Lung Carcinoma cells and Balb/c nu/nu mice	50–150 μg/mL (in vitro); 100 mg/kg (in vivo)	Induced p53-dependent apoptosis; suppressed tumor growth in mice
	Gastric Cancer SGC7901 and Hepatoma HepG2 cells	50–800 μg/mL	Strong inhibitory effect on proliferation (up to 92.63% in SGC7901 cells)
Cytotoxicity	Brine shrimp lethality bioassay	62.5–1000 μg/mL	Showed marginal toxicity with LC50 values of 0.17 and 0.20 mM
	Hybridoma N18RE-105 cells	10 μg/mL	Dose-dependent protection against glutamate-induced neuronal cell death
Antifatigue	Kunming mice (Swimming test)	100, 200, and 400 mg/kg (oral)	Significantly prolonged exhaustive swimming time
Antidepressant	CUMS mouse model	0.1, 0.15 g/kg (oral)	Attenuated depressive behavior via AMPK-ULK1 pathway-mediated autophagy

flavonoids, and antioxidants, decreases significantly because these bioactive compounds are mainly located in the bran, which is removed during popping. The popping process enhances digestibility and certain nutritional content while decreasing some bioactive constituents (Zhao et al., 1989).

Makhana, being gluten-free, is a safe and nutritious option for individuals with celiac disease or gluten sensitivity, aiding better overall digestion. Furthermore, its low-glycemic index (37%) makes it an ideal evening snack for health-conscious people and those with metabolic disorders, as it provides a slow-release energy profile and prevents sharp spikes in blood glucose.

13.7 Storage and handling

In harvesting *E. ferox*, the seeds are sieved, cleaned, and mechanically dried to reduce moisture content to a threshold below 8%–10%, which is important for inhibiting bacterial growth. Less drying is the major cause of fungal contamination and damaging of kernel quality. Studies show that both popped and raw seeds absorb moisture, which can lead to microbial growth (Gaurav et al., 2022). A low relative humidity environment (below 50%–60%) is maintained during plant storage to prevent microbial growth. Modern storage techniques include packaging systems, such as polyethylene terephthalate materials and multilayer laminated pouches, which reduce oxidative reactions and moisture uptake, thereby improving shelf life (Sharma et al., 2024). Temperature also plays a vital role in preserving the quality of makhana. Studies show that storage at cool temperatures (4–10°C) reduces physicochemical changes, including starch discoloration and lipid rancidity (Sharma et al., 2024). Temperature fluctuations can degrade bioactive compounds, including flavonoids and phenolics, and reduce the product's antioxidant properties (Sharma et al., 2024). Wang et al. studied the effect of γ-aminobutyric acid (GABA) treatment on the quality of fresh-cut makhana stems during cold storage. The findings revealed that 5 mM GABA treatment significantly delayed spoilage by maintaining stem firmness and color, improving nutrient levels, reducing oxidative damage and lignin deposition, and enhancing protective enzyme activities (SOD, peroxidase (POD), and ascorbate peroxidase (APX)) (Wang et al., 2025).

Handling techniques during the postharvest process are crucial. The popping method involves tempering and rapid heating, and recent research shows the need for controlled heating reactions to remove nutrient loss (Sharma et al., 2024). Hygiene control is another important factor, as *E. ferox* preparation is labor-intensive and prone to contamination if sanitation is not maintained effectively. Current good manufacturing practices and enhanced mechanization have been recommended to reduce microbial contamination and enable crop safety. Mechanical damage is minimized during transportation and handling of the crop, which is essential, as damaged kernels are more sensitive to moisture absorption and quality degradation (Kumar and Kumar, 2015).

13.8 Future perspectives

Future studies should focus on the molecular mechanisms underlying the bioactivity of makhana compounds, on large-scale clinical studies to validate health benefits, and on the development of innovative functional foods and nutraceutical products. Advances in omics technologies, such as proteomics and metabolomics, will further enhance our understanding of the therapeutic potential of *E. ferox*. Future research should prioritize:

- Clinical standardization: Transitioning from rodent models to human clinical trials to establish precise dosages for the management of diabetes, depression, and cardiovascular disease.
- Genomic optimization: Leveraging the sequenced genome to develop varieties that are more resilient to climate change and have even higher concentrations of thioredoxins and essential minerals.
- Waste valorization: Developing industrial-scale extraction methods for corilagin and other polyphenols from the seed shell, which is currently a neglected by-product.
- Nanoencapsulation: Using advanced biotechnological approaches to encapsulate makhana bioactives to improve their oral bioavailability and stability in functional food formulations.

13.9 Conclusion

Makhana (*E. ferox* Salisbury) is a nutritionally rich aquatic crop with significant potential as a functional food and nutraceutical. Traditionally valued in Ayurveda and TCM, it is now supported by modern scientific evidence highlighting its diverse bioactive composition, including polyphenols, flavonoids, polysaccharides, and essential nutrients. Advances in proteomic and metabolomic studies have improved understanding of its molecular composition and bioactive potential, linking makhana to important physiological effects, including antioxidant activity, metabolic regulation, cardiovascular support, and neuroprotection. These properties underscore its relevance in the prevention and management of chronic lifestyle-associated disorders. In addition to its health benefits, makhana is important for sustainable agriculture and the development of functional foods, with growing applications in value-added dietary products. Essentially, further research focusing on mechanistic validation, clinical studies, and bioactive standardization is essential to fully establish its therapeutic and commercial potential. Overall, *E. ferox* serves as a promising food–medicine interface, bridging traditional knowledge with modern nutraceutical innovation.

Declaration

Artificial Intelligence (AI) tools were used only for English language editing in the preparation of this manuscript. No AI tools were used to generate any kind of content, including text, tables, or figures.

Authorship contribution statement

SDS: Conceptualization, Investigation, Writing Original Draft, Validation, Methodology, Data curation, and Revision. SSS: Conceptualization, Investigation, Writing Original draft, Validation, Methodology, Data curation, and Revision. ID: Writing Original Draft and Revision. P, ST, and CD: Assisted with Writing the Original Draft and Revision. MC: Writing Original Draft, Revision, Conceptualization. GP: Conceptualization, Investigation, Writing Original Draft, Validation, Methodology, Data curation, Supervision, and Revision. SP: Conceptualization, Writing Original Draft, Investigation, Supervision, Validation, Project administration, Methodology, Resources, Data curation, and Revision.

Acknowledgments

SP is supported by the Science and Engineering Research Board (SERB) Start-up Research Grant (SRG) (SRG/2022/000067) and the Indian Institute of Science Education and Research Kolkata (IISER Kolkata). GP is supported by the Seed Money Scheme (DRIE/SMS/DRIE-10918/2024–2025/5545) at the Birla Institute of Technology, Mesra, Jharkhand. SDS and SSS are supported by the IIT Guwahati institutional fellowship. ID is supported by the Institutional Research Fellowship by Birla Institute of Technology, Mesra, Jharkhand.

References

Ahmad, A., 2020. A study on constraints of Makhana cultivation and suitable measures for its better development. J. Pharmacogn. Phytochem. 9 (1), 984–987.

Ahmed, D., Sharma, M., Kumar, V., Bajaj, H.K., Verma, A., 2015. 2β-hydroxybetulinic acid 3β-caprylate: an active principle from *Euryale Ferox* Salisb. seeds with antidiabetic, antioxidant, pancreas & hepatoprotective potential in streptozotocin induced diabetic rats. J. Food Sci. Technol. 52 (9), 5427–5441. https://doi.org/10.1007/s13197-014-1676-0.

Ahmed, D., Khan, M.I., Sharma, M., Khan, M.F., 2019. Novel pentacyclic triterpene isolated from seeds of *Euryale Ferox* Salisb. ameliorates diabetes in streptozotocin induced diabetic rats. Interdiscip. Toxicol. 11 (4), 275–288. https://doi.org/10.2478/intox-2018-0027.

Aman Nisar, Md, Sharma, N.K., Panda, C.K., Sinha, S.P., Behera, S.K., Kumari, M., 2024. Socio-economic and socio-psychological traits of Makhana growers of Kosi region. J Community Mobiliz Sustain. Dev. 19 (2), 496–501. https://doi.org/10.5958/2231-6736.2024.00133.6.

Bhatt, B.P., Gupta, V.K., Kumar, L., Singh, I.S., Sarkar, B., 2017. *Euryale ferox* (Salisb): promising aquatic food crop of eastern Indo gangetic plains. Int. J. Curr. Microbiol. Appl. Sci. 6 (6), 1914–1921. https://doi.org/10.20546/ijcmas.2017.606.223.

BioRender. (2026). Created in BioRender. DashSharma, S. https://BioRender.com/3w2yql4.

Biswas, P., Das, M., Boral, S., Mukherjee, G., Chaudhury, K., Banerjee, R., 2020. Enzyme mediated resistant starch production from Indian Fox Nut (*Euryale ferox*) and studies on digestibility and functional properties. Carbohydr. Polym. 237, 116158. https://doi.org/10.1016/j.carbpol.2020.116158.

Das, J.K., Saikia, K., Handique, G.K., Handique, A.K., 2021. Nutritive values and dietary anti-oxidant of some under-utilized seeds and nuts from ethnic sources of North East India. Crop Res. 56 (1–2), 60–66. https://doi.org/10.31830/2454-1761.2021.011.

Devi, K., 2022. A comprehensive review on nutraceutical properties of *Euryale ferox* Salisb. J. Plant Sci. Res. 9 (2), 222.

Devi, M., Sharma, K., Narayan Jha, S., Arora, S., Patel, S., Kumar, Y., Kumar Vishwakarma, R., 2020. Effect of popping on physicochemical, technological, antioxidant, and microstructural properties of makhana seed. J. Food Process. Preserv. 44 (10). https://doi.org/10.1111/jfpp.14787.

Devi, N., Rani, K., Kharb, P., Kaushik, P., 2022. Bio-fabrication of euryale ferox (makhana) leaf silver nanoparticles and their antibacterial, antioxidant and cytotoxic potential. Plants 11 (20), 2766. https://doi.org/10.3390/plants11202766.

Devi, O.R., Moirangthem, L., Keithellakpam, O.S., Sharma, N., Singh, K.B., 2024. Phytochemical screening and evaluation of antioxidant potential in *Euryale ferox* Salisb. and eupatorium birmanicum DC. of Manipur, India. Pharmacogn. J. 16 (6), 1231–1237. https://doi.org/10.5530/pj.2024.16.201.

Ekal, V., Kumbamoorthy, S., 2024. Formulation and proximate analysis of makhana-enriched laddoos. Int. J. Nutr. Pharmacol. Neurol. Dis. 14 (2), 205–212. https://doi.org/10.4103/ijnpnd.ijnpnd_92_23.

Farooqi, R., Maryam, H., Jaffar, H.M., Malik, T., 2025. Therapeutic potential of fox nut (Euryale ferox) in osteoarthritis management: a nutraceutical approach. Food Rev. Int. 1–27. https://www.tandfonline.com/doi/full/10.1080/87559129.87552025.82541850.

Fern, K., 2026. Useful tropical plants. Tropical plants database. http://tropical.theferns.info/viewtropical.php.

Gaurav, A., Nath, D., Barbhuiya, R.I., Pradhan, R.C., 2022. Physical characterization and mass modeling of traditionally popped makhana (*Euryale ferox* Salisbury) variety: sabour-1 makhana. J. Food Process Eng. 45 (6). https://doi.org/10.1111/jfpe.13927.

Guo, Z., Min, J., 2020. Characterization the complete chloroplast genome of *Euryale ferox* (Nymphaeaceae), an medicinal plant species in China. Mitochondrial DNA Part B 5 (3), 2389–2390. https://doi.org/10.1080/23802359.2020.1775516.

Gupta, V.K., Kumar, L., Jha, B., Jha, I., Bhatt, B., Singh, A., 2022. Status of Makhana Cultivation in India.

Idris, Md., Jana, B.R., 2018. Anti-aging amino acids in *Euryale ferox* (Salisb.): a review. Adv. Plants Agric. Res. 8, 1 %U. https://medcraveonline.com/APAR/anti-aging-amino-acids-in-euryale-ferox-salisb-a-review.html.

Islam, Z., Bashir, K., Singh, S., Siddiqui, S., 2025. Effect of germination on physicochemical, functional and antinutritional characteristics of makhana seed (*Euryale ferox*) flour. Food Human. 5, 100717. https://doi.org/10.1016/j.foohum.2025.100717.

Jain, A., Birkumar Singh, H., Kanjilal, P.B., 2010. Economics of Foxnut (*Euryale ferox* Salisb.) cultivation: a case study from Manipur in North eastern India. Indian J.Nat. Prod. Res 1 (1), 63–67.

Jana, B.R., Srivastava, A., Idris, M., 2019. New makhana (*Euryale ferox* Salisb.) processed products for health benefit. J. Food Process. Technol. 10 (4).

Jana, B.R., Kumar, M., Raut, S.M., 2024. Antidiabetic biomolecules and nutrient elements in Makhana (*Euryale ferox* Salisb.). Ann Phytomed. 13 (2). https://doi.org/10.54085/ap.2024.13.2.52.

Jha, V., Shalini, R., Kumari, A., Jha, P., Sah, N.K., 2018. Aquacultural, nutritional and therapeutic biology of delicious seeds of *Euryale ferox* salisb.: a minireview. Curr. Pharm. Biotechnol. 19 (7), 545–555. https://doi.org/10.2174/1389201019666180808160058.

Jiang, J., Ou, H., Chen, R., Lu, H., Zhou, L., Yang, Z., Ethnopharmacological, T., 2023. Phytochemical, and pharmacological review of *Euryale ferox* Salisb.: a Chinese medicine food homology. Molecules 28 (11), 4399. https://doi.org/10.3390/molecules28114399.

Joseph, A., Ramesh, G., Analysis, N., 2023. Phytochemical and antioxidant activity of a food product formulated with fox nuts (*Euryale ferox*). Asian J. Biol. Life Sci. 12 (2), 279–285. https://doi.org/10.5530/ajbls.2023.12.38.

Kapoor, S., Kaur, A., Kaur, R., Kumar, V., Choudhary, M., 2022. *Euryale ferox* , a prominent superfood: nutritional, pharmaceutical, and its economical importance. J. Food Biochem. 46 (12). https://doi.org/10.1111/jfbc.14435.

Khadatkar, A., Mehta, C.R., Gite, L.P., 2020. Makhana (*Euryle ferox* Salisb.): a high-valued aquatic food crop with emphasis on its agronomic management—A review. Sci. Hortic. 261, 108995. https://doi.org/10.1016/j.scienta.2019.108995.

Kim, J., Kundu, M., Viollet, B., Guan, K.L., 2011. AMPK and mTOR regulate autophagy through direct phosphorylation of Ulk1. Nat. Cell Biol. 13 (2), 132–141. https://doi.org/10.1038/ncb2152.

Krishnan, H.B., Natarajan, S.S., 2009. A rapid method for depletion of Rubisco from soybean (Glycine max) leaf for proteomic analysis of lower abundance proteins. Phytochemistry 70 (17–18), 1958–1964. https://doi.org/10.1016/j.phytochem.2009.08.020.

Kumar, A., 2025. Present status, prospects and future scope of makhana in Bihar. Int. J. Innov. Hortic. 14 (2), 187–191. https://doi.org/10.5958/2582-2527.2025.00020.8.

Kumar, S., Kumar Jha, K., 2026. Unveiling regional disparities in makhana farming: a multidimensional analysis from Bihar. Indian J. Ext. Ed. 62 (1), 15–20. https://doi.org/10.48165/ijee.2026.62103.

Kumar, S., Prakash Kumar, S., 2015. Effect of different grades on physical and mechanical properties of popped makhana. Int. J. Process. Post Harvest Technol. 6 (1), 101–108. https://doi.org/10.15740/HAS/IJPPHT/6.1/101-108.

Kumar, R., Sanduja, D.M., painuly, N., 2024. A research on phytochemical and antimicrobial investigation of *Euryale ferox* seeds. Educ. Admin. Theor. Pract. https://doi.org/10.53555/kuey.v30i6.5884.

Kumari, A., Jha, V., 2017. Ethnic uses of Makhana (*Euryale ferox* Salisb.) in Mithila (north Bihar) and other parts of India. J. Tradit. Folk Practices 5 (1). https://doi.org/10.25173/jtfp.83.

Lee, S.E., Ju, E.M., Kim, J.H., 2002. Antioxidant activity of extracts from *Euryale ferox* seed. Exp. Mol. Med. 34 (2), 100–106. https://doi.org/10.1038/emm.2002.15.

Lee, M.R., Kim, J.H., Son, E.S., Park, H.R., 2009. Protective effect of extracts from *Euryale ferox* against glutamate-induced cytotoxicity in neuronal cells. Nat. Prod. Sci. 15 (3), 162–166.

Li, X., Guo, H., Liu, S., Luan, X., Yang, Y., Zhang, X., 2026. Structural characterization of starch-like α-glucan from the seeds of *Euryale ferox* Salisb and evaluation on the anti-inflammatory activity after sulfated modification. Carbohydr. Polym. 373. https://doi.org/10.1016/j.carbpol.2025.124659.

Liaquat, M., Pasha, I., Ahsin, M., Salik, A., fox nuts, R., Euryale, F.L., 2022. Contain higher concentration of phenolics, flavonoids, minerals and antioxidants, and exhibit lower Glycemic Index (GI) in human subjects. Food Prod. Process. Nutr. 4 (1). https://doi.org/10.1186/s43014-021-00081-x.

Liu, A.L., Wang, Y.H., Wang, T.Y., Zhu, Y., Wu, P., Li, L.J., 2023. Comparative metabolomic profiling of secondary metabolites in different tissues of *Euryale ferox* and functional characterization of phenylalanine ammonia-lyase. Ind. Crops Prod. 195. https://doi.org/10.1016/j.indcrop.2023.116450.

Mahesh Kumar, M., Goudar, G., Sharma, P., Vishwakarma, R., Gogoi, P., Mahajan, A., Chandragiri, A., Kalpuri, S., Babu Geddam, J.J., Suresh, C., Chary, M., Venketrajireddy, G., Radhika, M.S., Sreenu, P., Tulja, B., 2023. Differentiating the nutrient composition, in-vitro starch digestibility, individual polyphenols and anti-oxidant properties of raw and popped makhana (*Euryale ferox*). J. Food Meas. Char. 17 (6), 5828–5844. https://doi.org/10.1007/s11694-023-02074-5.

Mary, A., Barale, S., Eysert, F., Valverde, A., Lacas-Gervais, S., Bauer, C., Eddarkaoui, S., Buée, L., Buée-Scherrer, V., Checler, F., Chami, M., 2025. Hampered AMPK-ULK1 cascade in Alzheimer's disease (AD) instigates mitochondria dysfunctions and AD-related alterations which are alleviated by metformin. Alzheimers Res. Ther. (1), 17. https://doi.org/10.1186/s13195-025-01772-0.

Mittal, R., Sharma, S., Mittal, A., 2020. A critical review on ethnobotanical and pharmacological aspects of *Euryale ferox* Salisb. Pharmacogn. J. 12 (6), 1444–1454. https://doi.org/10.5530/pj.2020.12.199.

Mondal, T., Sarkar, T., Sengupta, S., Kundu, S., 2024. Nutritional and pharmacological aspects of Trapa natans: an underutilized boon crop of West Bengal. Int. J. Res. Agron. 7 (4), 69–72. https://doi.org/10.33545/2618060x.2024.v7.i4b.522.

Nam, G.-H., Jo, K.-J., Park, Y.-S., Kawk, H.W., Kim, S.-Y., Kim, Y.-M., 2019. In vitro and in vivo induction of p53-dependent apoptosis by extract of euryale ferox Salisb in A549 human caucasian lung carcinoma cancer cells is mediated through Akt signaling pathway. Front. Oncol. 9. https://doi.org/10.3389/fonc.2019.00406.

Nehal, N., Mann, S., Gupta, R.K., 2015. Two promising under-utilized grains: a review. Indian J. Tradit. Knowl. 14 (3), 416–422.

Park, J.M., Kim, D.H., 2024. A paradigm shift: AMPK negatively regulates ULK1 activity. Autophagy 20 (4), 960–962. https://doi.org/10.1080/15548627.2023.2223465.

Qian, L.-H., Wu, J.-Y., Wang, Y., Zou, X., Zhou, G.-C., Sun, X.-Q., 2022. Genome-wide analysis of NBS-LRR genes from an early-diverging angiosperm *Euryale ferox*. Front. Genet. 13, 880071. https://www.frontiersin.org/articles/880010.883389/fgene.882022.880071/full.

Qin, J., Zheng, X., Wang, D., Song, C., Shi, W., Lu, Y., 2025. Degradation polysaccharide of *Euryale ferox* Salisb. Seeds significantly enhanced the hypoglycemic efficiency of type 2 diabetes mellitus. Int. J. Biol. Macromol. 318 (Pt 3), 145238. https://doi.org/10.1016/j.ijbiomac.2025.145238.

Qu, C., Yu, D., Jing, Z., Gu, S., Wang, Y., Xie, W., Wu, Q., 2025. Comparative study on structural characterization, physicochemical properties, and in vitro probiotic activities of resistant starch from different varieties of Euryale ferox. Food Chem. X 28, 102572. https://doi.org/10.1016/j.fochx.2025.102572.

Ramesh, K., Singh, I.S., Jha, V., 2026. Bioactive compounds in *Euryale ferox* Salisb. and their nutritional and therapeutic actions-a review. Vegetos 39 (2), 480–488. https://doi.org/10.1007/s42535-025-01209-x.

Rebouche, C.J., 2004. Kinetics, pharmacokinetics, and regulation of L-Carnitine and acetyl-L-carnitine metabolism. Ann. N Y Acad. Sci. 30–41. https://doi.org/10.1196/annals.1320.003. http://onlinelibrary.wiley.com/journal/10.1111/(ISSN)1749-6632 1033.

Rehman, A.U., Khan, A.U., Sohaib, M., Rehman, H., 2024. Comparative analysis of nutritional properties, phytochemical profile, and antioxidant activities between red and green water chestnut (Trapa natans) fruits. Foods 13 (12). https://doi.org/10.3390/foods13121883.

Rosique-Esteban, N., Guasch-Ferré, M., Hernández-Alonso, P., Salas-Salvadó, J., 2018. Dietary magnesium and cardiovascular disease: a review with emphasis in epidemiological studies. Nutrients 10 (2), 168. https://doi.org/10.3390/nu10020168.

Semwal, D.P., Pandey, A., Gore, P.G., Ahlawat, S.P., Yadav, S.K., Kumar, A., 2021. Habitat prediction mapping using BioClim model for prioritizing germplasm collection and conservation of an aquatic cash crop "makhana" (*Euryale ferox* Salisb.) in India. Genet. Resour. Crop Evol. 68 (8), 3445–3456. https://doi.org/10.1007/s10722-021-01265-7.

Sharma, K., Patel, S., Jha, S.N., Devi, M., Nandan, S., Vishwakarma, R.K., 2024. Change in popping, structural, functional, and crystalline properties of roasted makhana seeds in prolong storage: an FTIR, GC-MS, SEM, XRD study. Res. Square. https://doi.org/10.21203/rs.3.rs-3910663/v1.

Sharma, K., Patel, S., Devi, M., Vishwakarma, R.K., 2024. Effect of packaging material and storage period on popping behaviour of roasted makhana seeds. J. Agric. Eng. (4), 61. https://doi.org/10.52151/jae2024614.1870.

Shaw, R., Das, S., Chaube, R., 2025. Reproductive health benefits of *Euryale ferox* Salisb.: a short comprehensive review. J. Endocrinol. Reprod. 183–190. https://doi.org/10.18311/jer/2024/44756.

Shukla, N., Sharma, J., Sharma, P.K., 2024. A review on: *Euryale ferox* (Fox nut). J. Pharmacogn. Phytochem. 13 (5), 50–52. https://doi.org/10.22271/phyto.2024.v13.i5a.15057.

Singh, D.K., Singh, I.S., Kumar, U., Kumar, A., Bhatt, B.P., 2018. Traditional wisdom of Mallah community regarding makhana production and processing in North Bihar. Indian J. Ext. Educ. 54 (2), 76–82. https://epubs.icar.org.in/index.php/IJEE/article/view/143741.

Singh, D.K., Kumar, A., Singh, I.S., Kumar, U., Chandra, N., Bhatt, B.P., 2020. Value Chain Analysis of Makhana. Technical Bulletin No. R-67/Patna-38.

Sinha, C.B.K., Jha, V., 2025. Identification of nutraceutically important metabolites in raw and popped makhana (*Euryale ferox* Salisb.) seeds using GCMS technique. Int. J. Plant Environ. 11 (03), 514–524.

Song, C.W., Wang, S.M., Zhou, L.L., Hou, F.F., Wang, K.J., Han, Q.B., Li, N., Cheng, Y.X., 2011. Isolation and identification of compounds responsible for antioxidant capacity of *Euryale ferox* seeds. J. Agric. Food Chem. 59 (4), 1199–1204. https://doi.org/10.1021/jf1041933.

Sonu, K., Jha, K.K., 2026. Unveiling regional disparities in makhana farming: a multidimensional analysis from Bihar. Ind. J. Exten. Educ. 62 (1), 15–20. https://journals.acspublisher.com/index.php/ijee/article/view/23263.

Suraiya, S., Ria, S.J., Tanzim Riya, M.U., Ritu, F.Y., Sumona, A.A., Rodela, A.B., Akter, L., Uddin, M.S., Hasan, M.N., 2024. Nutritional and biofunctional characterizations of four novel edible aquatic plants of Bangladesh. Heliyon 10 (15). https://doi.org/10.1016/j.heliyon.2024.e35538.

Vikash, K., Raushan, K., Dinesh, R., Vishal, K., Anjani, K., 2013. Influence of moisture content and grade on engineering properties of makhana. Adv. Appl. Sci. Res. 4 (6), 29–35.

Wang, Y., Yuan, H., Wang, Y., 2023. Treatment of diabetes nephropathy in mice by germinating seeds of *Euryale ferox* through improving oxidative stress. Foods 12 (4), 767. https://doi.org/10.3390/foods12040767.

Wang, Y., Wei, H., Wang, S., Yu, F., Zhai, L., Yang, J., Wei, Z., 2025. γ-Aminobutyric acid treatment enhances quality and improves antioxidant activities of fresh-cut *Euryale ferox* stems during postharvest storage. Front. Nutrit. 12. https://doi.org/10.3389/fnut.2025.1527555.

Wardhan, H., Suntwal, T., Gulati, A., 2025. Strategies to Boost India's Makhana Exports.

Wen, X., Liang, Y., Shan, H., Chang, X., Song, X., Shen, S., Fu, Y., Chen, D., Chen, F., Li, Y., Guan, Q., Gao, Q., Qi, W., Li, Y., Wang, Z., Kong, H., Liu, H., Gao, X., Zhou, X.,

Thorogood, C., Zhang, L., 2025. The genome of giant waterlily provides insights into the origin of angiosperms, leaf gigantism, and stamen function innovation. Plant Commun. 6 (6), 101342. https://doi.org/10.1016/j.xplc.2025.101342.

Wu, C.Y., Chen, R., Wang, X., Shen, B., Yue, W., Wu, Q., 2013. Antioxidant and anti-fatigue activities of phenolic extract from the seed coat of *Euryale ferox* Salisb. and identification of three phenolic compounds by LC-ESI-MS/MS. Molecules 18 (9), 11003–11021. https://doi.org/10.3390/molecules180911003.

Wu, P., Liu, A., Li, L.J., 2021. Metabolomics and transcriptome analysis of the biosynthesis mechanism of flavonoids in the seeds of *Euryale ferox* Salisb at different developmental stages. Mol. Genet. Genom. 296 (4), 953–970. https://doi.org/10.1007/s00438-021-01790-1.

Wu, P., Liu, A.L., Zhu, Y., Xiang, L., Wang, Y.H., Li, L.J., 2022. Proteomic analysis of *Euryale ferox* Salisb seeds at different developmental stages. Gene 834, 146645. https://doi.org/10.1016/j.gene.2022.146645.

Xie, X., Zeng, C., Zhang, R., Zhu, W., Li, H., Huang, Z., 2025. Profiling and discrimination of *Euryale ferox* seeds from different processing methods using liquid chromatography high-resolution mass spectrometry combined with molecular networking and statistical analysis. Metabolites 15 (4), 225. https://doi.org/10.3390/metabo15040225.

Yang, Y., Sun, P., Lv, L., Wang, D., Ru, D., Li, Y., Ma, T., Zhang, L., Shen, X., Meng, F., Jiao, B., Shan, L., Liu, M., Wang, Q., Qin, Z., Xi, Z., Wang, X., Davis, C.C., Liu, J., 2020. Prickly waterlily and rigid hornwort genomes shed light on early angiosperm evolution. Nat. Plants 6 (3), 215–222. https://doi.org/10.1038/s41477-020-0594-6.

Zhang, L., Zeng, J., Yuan, E., Chen, J., Zhang, Q., Wang, Z., Yin, Z., 2023. Extraction, identification, and starch-digestion inhibition of phenolics from *Euryale ferox* seed coat. J. Sci. Food Agric. 103 (7), 3437–3446. https://doi.org/10.1002/jsfa.12460.

Zhang, Y., Huang, J., Qin, Y., Chen, X., Xu, F., Zhu, K., Zhang, X., Wu, Z., Zhang, J., Zhang, Y., 2025. Type 3 resistant starch prepared from *Euryale ferox* alleviated the obesity symptoms via gut microbiota regulation. Int. J. Biol. Macromol. 311 (Pt 3), 143906. https://doi.org/10.1016/j.ijbiomac.2025.143906.

Zhao, H., Zhao, S., Sun, C., Guillaume, D., 1989. Glucosterols in extracts of *Euryale ferox* identified by high resolution NMR and mass spectrometry. JLR (J. Lipid Res.) 30 (10), 1633–1637.

Zhao, L., Zhang, S.L., Tao, J.Y., Pang, R., Jin, F., Guo, Y.J., Dong, J.H., Ye, P., Zhao, H.Y., Zheng, G.H., 2008. Preliminary exploration on anti-inflammatory mechanism of Corilagin (beta-1-O-galloyl-3,6-(R)-hexahydroxydiphenoyl-d-glucose) in vitro. Int. Immunopharmacol. 8 (7), 1059–1064. https://doi.org/10.1016/j.intimp.2008.03.003.

Chapter 14

Value added product development using makhana (snacks, flour, bars, etc.)

Ritu Kukde[1], Vaishali Thakre[2], Pakkiranna Sivamma[3], Thongam Sunita[1], Priyabrat Kapri[1], Pratiksha Puranik[4], Vikram Kad[4] and Ganesh Shelke[4]
[1]*ICAR-Central Institute of Post Harvest Engineering and Technology, Ludhiana, Punjab, India;* [2]*ICAR-DoCA Pulse Storage Project, Dr. PDKV, Akola, Maharashtra, India;* [3]*ICAR-National Institute of Seed Science & Technology, Mau, Uttar Pradesh, India;* [4]*Dr. A.S.C.A.E & T, MPKV, Rahuri, Maharashtra, India*

14.1 Introduction

Makhana (*Euryale ferox* Salisb.), commonly known as foxnut or gorgon nut, is an aquatic food crop belonging to the family Nymphaeaceae. It is typically cultivated in ponds, lakes, swamps, ditches, and other stagnant water bodies. The crop is native to and widely cultivated in the tropical and subtropical regions of Southeast and East Asia, including India, China, Japan, Korea, Bangladesh, and neighboring countries.

In India, makhana is commercially cultivated mainly in the states of Bihar, West Bengal, Assam, Tripura, and Manipur, with Bihar being the largest producer. The crop occupies an estimated area of over 27,000 acres, contributing nearly 90% of the global production. Makhana is obtained by popping the dried seeds of *E. ferox* at high temperatures, which converts the hard seed into a light, edible, and expanded product.

Four high-quality varieties of pop are identified in traditional makhana markets: lava, murha, turi, and mix. The magnitude of the pop is almost entirely responsible for the variations in quality.

Gunny bags are used for makhana transactions in these ancient marketplaces. The weight of these gunny sacks, which have a uniform size, is a good indicator of the makhana's quality.

A bag with a low weight is therefore a sign of high quality because well-produced makhana pops are bigger and lighter. Generally speaking, a bag weighing 8 kilograms indicates high-quality lava makhana, while a bag

Makhana (Euryale ferox) as a Functional Food. https://doi.org/10.1016/B978-0-443-44421-0.00018-4

weighing more than 10 kilograms typically contains lower-quality makhana (Minten et al., 2010).

Popped makhana is highly valued for its nutritional attributes. It is a rich source of easily digestible protein and essential minerals such as potassium, phosphorus, and magnesium, while being low in fat and cholesterol (Kumar et al., 2023). Due to these characteristics, makhana is increasingly being recognized as a functional and health-promoting food. It is also nutritionally characterized by low fat and cholesterol content, moderate protein levels, and a favorable mineral profile, particularly magnesium, potassium, phosphorus, zinc, and copper. The popping process significantly influences the biochemical composition of makhana seeds. High-temperature, short-time popping modifies starch and protein structures, thereby enhancing their digestibility and improving overall nutrient availability. This thermal process not only improves sensory attributes such as texture and flavor but also reduces antinutritional factors, leading to enhanced bioavailability of nutrients. Additionally, popping helps retain thermolabile nutrients, resulting in a ready-to-eat product with minimal nutrient loss.

Makhana is naturally gluten-free, making it suitable for individuals with gluten intolerance or celiac disease. Several studies have reported its health-promoting properties, including antioxidant, hepatoprotective, cardioprotective, antifatigue, antiaging, and antimelanogenic effects. The low sodium and high potassium content of makhana may also contribute to blood pressure regulation and cardiovascular health.

Although numerous studies have investigated the nutritive value and antioxidant activity of makhana, limited research has focused on the impact of the popping process on its nutrients and phytochemicals, highlighting the need for further systematic studies (Table 14.1).

Commercial cultivation of makhana in India is largely confined to selected districts of North Bihar, West Bengal, Assam, and Manipur. The produce reaches domestic markets through a well-established supply chain involving producers, processors, local traders, wholesalers, commission agents, retailers, and consumers.

Despite being the world's leading producer, India currently exports only about 1% of its total makhana production, amounting to nearly 100 t of popped makhana annually. Major export destinations include the United States, the United Kingdom, Australia, Bangladesh, Pakistan, and Gulf countries such as the UAE, Qatar, Saudi Arabia, Kuwait, and Oman (Singh et al., 2020).

In recent years, global demand for makhana has increased significantly, particularly in countries such as the Philippines, China, Thailand, Malaysia, Bangladesh, and Nepal, owing to its nutritional value, therapeutic benefits, and growing popularity as a healthy snack. Companies such as Shakti Sudha Industries, Patna, have played a key role in promoting makhana exports through organized processing and online marketing platforms.

TABLE 14.1 Comparative nutritional composition of makhana and selected snack foods.

Component	Unit	Cashew	Almond	Popcorn	Makhana (raw)	Makhana (popped)
Proximate composition						
Moisture	g	4.4	4.37	3.32	**11.3**	**12**
Ash	g	2.25	2.62	1.42	**0.45**	**0.38**
Protein	g	18.78	18.41	12.9	**9.15**	**11.03**
Fiber	g	3.86	13.06	14.5	**2.56**	**3.26**
Fats	g	45.2	58.49	4.54	**0.46**	**0.33**
Carbohydrates (g)	g	25.46	3.04	77.8	**87.52**	**84.87**
Minerals						
Calcium	mg	34.0	228.0	7.0	**14.50**	**20.94**
Sodium	mg	9.0	1.5	8.0	**2.92**	**4.06**
Potassium	mg	635.0	699.0	329.0	**50.25**	**48.39**
Magnesium	mg	307.0	318.0	144.0	**13.83**	**12.71**
Phosphorus (mg)	mg	500.0	446.0	358.0	**132.53**	**124.01**
Iron (mg)	mg	5.95	4.59	3.19	**5.73**	**2.67**
Manganese (mg)	mg	1.78	2.54	1.11	**1.30**	**1.24**
Zinc (mg)	mg	5.34	3.5	3.08	**1.50**	**1.04**
Copper (mg)	mg	2.23	1.08	0.26	**0.71**	**0.76**

Values are expressed per 100 g edible portion. Compiled and adapted from multiple published sources.

14.2 Need for value addition

Makhana (*E. ferox* Salisb.), commonly known as foxnuts or gorgon nuts, represents one of India's most significant aquatic cash crops with immense potential for value addition and commercial development. Makhana has transitioned from a traditional food ingredient to an emerging superfood in both domestic and international markets (Kapoor et al., 2022; Kumar et al., 2016). The makhana industry continues to operate primarily in raw and minimally processed forms despite its rich nutritional profile and rising market demand, leaving significant opportunities for value-added product development unexplored. The makhana trade currently has a highly biased economic structure that favors middlemen. Farmers collectively make less than ₹250 crore, while traders make about ₹550 crore, despite the industry's thousands of crores.

Mithila makhana, a highly nutritious superfood, serves as a vital engine for economic growth in Bihar. According to nutritional analyses, makhana has a better nutrient profile than many traditional snack options. According to Kumar et al. (2016) and Rathod et al. (2023), the seed endosperm is rich in essential minerals such as phosphorus (79.1 mg/100 g), potassium (56.0 mg/100 g), magnesium (20.00 mg/100 g), and zinc (5.6 mg/100 g). It also contains about 9.7% protein, 76.9% carbs, and 0.1% fat. Understanding how processing processes affect nutritional and functional qualities is essential to turning raw makhana into products with additional value. Recent studies have provided understanding of the important changes that take place during makhana processing, offering vital information for strategies related to product development. The popping process causes significant changes in the nutritional composition and is still the predominant processing procedure. The nutritional value and functional qualities of makhana can be improved using advanced processing procedures, especially germination, which have emerged as potential bioprocessing technologies. In addition to improving water absorption capacity from 2.48 to 5.65 g/g and oil absorption capacity from 1.33 to 2.63 g/g, Islam et al. (2025) showed that 72-hour germination dramatically improves moisture, protein, and ash content. Crucially, germination significantly lowers antinutritional factors, reducing tannins from 3.69 to 2.65 mg/100 g and phytic acid from 168.64 to 82.94 mg/100 g. This improves protein digestibility and mineral bioavailability. Makhana flour development is a particularly attractive option for value addition, since it may be used in a variety of product categories such as bakery goods, infant meals, and gluten-free formulations. As a gluten-free alternative, makhana flour addresses the rising market segment of consumers with celiac disease and gluten intolerance, which is anticipated to affect millions worldwide.

Current market trends indicate rapid growth in the derivatives segment, which encompasses makhana flour, powder, and energy bars. These processed forms find applications in baby food formulations, diabetic-friendly diets, and protein-rich snack bars, catering to the expanding functional and health-oriented foods market. The development of export-oriented value-added products, including flavored snacks, makhana flour, and ready-to-eat bars tailored to international taste preferences, can significantly enhance export revenues and market penetration in high-value markets, including the United States, the United Kingdom, Canada, Australia, and Middle Eastern countries. The absence of state-of-the-art processing units and cold storage facilities leads to significant postharvest losses and quality deterioration. Value addition through "time utility" (storage) and "form utility" (postharvest processing) is essential to maintain quality standards and reduce wastage, making Bihar's makhana more competitive in national and international markets.

As the world's population becomes more health-conscious and people's socioeconomic status improves due to their increasingly busy lifestyles, there is an increasing need for quick and nutrient-dense snacks. In this regard, makhana, or foxnut, shows promise as a component that fits the changing dietary choices of people who are health-conscious (Kumar and Kumar, 2025).

Foxnuts, also known as makhana or *E. ferox*, have gained popularity as a superfood due to their high fiber, low calories, and plenty of antioxidant qualities. Because it is rich in macro and micronutrients, it is referred to as a functional food. Carbohydrates (80% starch), high-quality protein, vitamins (C and E), and minerals (magnesium, potassium, phosphorus, iron, and zinc) are the key nutritional components found in *E. ferox* seed (Kumar et al., 2016a, 2016b).

About 61% of the raw *E. ferox* is made up of carbs, 12.1% moisture, 15.6% protein, 1.35% fat, 7.6% fiber, 1.8% minerals, and 362 kcal/100 g. With an amino acid index of roughly 89%–93%, it is extremely nutritious. The amounts of glutamic acid, leucine, and isoleucine are approximately 16.64%, 8.34%, and 4.18%, respectively, while the ratio of arginine + lysine/proline is 4.74–7.6 (Jana and Idris, 2018). Cu (0.3–0.71 mg/kg), Zn (0.9–1.50 mg/kg), Fe (0.8–5.73 mg/kg), and Mn (11.3–13.83 mg/kg) were among the trace metals found in *E. ferox,* according to several studies (Devi and Deka, 2021; Kumar et al., 2016a, 2016b). It has been noted that this plant is a good source of minerals, which are necessary to maintain healthy bones, muscles, the heart, and the brain. In addition to these, *E. ferox* is rich in carotenoids, flavonoids, and vitamins (Kumar et al., 2020).

About 0.6 lakh MT of makhana are produced in India, of which 40% (0.23–0.25 lakh MT) are exported, and the remaining amount is used domestically. Since 2021–22 and 2024–25, the market has expanded by 17%–18% a year due to premiumization, growing branded players, and increased health consciousness. Due to its manufacturing leadership and the rising demand for plant-based, clean-label superfoods around the world. India is in a good position to grow makhana into a high-end, international snack. In the United States, EU, Middle East, and Southeast Asia, where health-conscious snacking is quickly gaining popularity, value-added formats such as roasted, flavored, organic, and functional varieties provide greater margins and excellent export potential (APEDA, 2025). Foxnuts expand a lot in volume when processed, making them costly to transport despite low weight. This high transportation expense persists, and there is an eager want to create value-added products from them.

14.3 Types of value-added products from makhana

Popped makhana is the most commonly used and most highly demanded processed food from the Gorgon nut. It is used mostly in the auspicious rituals of North Bihar and other eastern states of India. This popped makhana is also consumed by roasting in ghee and sprinkling rock salt to add to the savory taste. Moreover, these are also used as an ingredient in dal makhni, kofta, dum aloo, palak makhana, makhana curry, makhana pulao, matar makhana, etc. As an additive, popped makhana flour improves the baking properties of cookies and buns. It is also mixed with water chestnut (*Trapa natans*) flour to prepare a traditional dish called *Pua*. Newly developed companies are developing flavored makhana pops using popular and widely accepted flavors like onion, tomato, pudina, chilli, butter, etc. A company named Shakti Sudha Industries in Patna is preparing a blend of wheat and makhana flour, naming it "makhana atta," along with many more products like makhana cookies, makhana flakes, makhana bhujia, etc. Various ICAR institutes like ICAR—CIPHET, ICAR—NRC for makhana, ICAR—RCER, Patna, have attempted research activities for technology development and preparing value-added products from makhana, namely makhana kheer mix, makhana powder, makhana barfi, makhana kalakand, makhana curry, makhana chapati, etc. (Singh et al., 2020). Some of the value-added products are given in Fig. 14.1. The use of foxnuts in cooking has advanced beyond conventional methods in recent years. Foxnuts are being roasted, flavored, and incorporated into energy bars, morning cereals, and even desserts by food entrepreneurs and health brands (Kumar and Kumar, 2025). In this chapter, various value-added makhana products are given, such as roasted foxnuts, makhana snacks, makhana flour, makhana bread/bun, makhana cookies/biscuits, instant makhana kheer mix, makhana chapatti, makhana pakora, makhana kalakand, makhana barfi, and baby food.

14.3.1 Ready-to-eat snacks

The most common attributes associated with consumer reluctance to consume this nut as a snack are its soggy texture and bland flavor; therefore, foxnuts require roasting as a pretreatment prior to consumption to improve their textural and sensory acceptability.

FIGURE 14.1 Value-added products from makhana: (a) popped makhana, (b) makhana *kheer*, (c) makhana powder, (d) roasted makhana snacks, (e) makhana *chapati*, (f) makhana *barfi*, (g) makhana *halwa*, (h) makhana pickle, (i) makhana weaning mix, (j) makhana pasta, (k) ready-to-eat makhana snacks, (l) fat-free flavored makhana.

14.3.1.1 Roasted foxnuts

The popped foxnut preparation method has been mentioned. The seeds are dried to remove the seed kernel from the seed coat. The seeds are passed through the sieves of different sizes to differentiate them into different grades. The graded seeds are heated in a pan with continued stirring at a temperature of about 230–335°C for 6 min. The seeds are stored in the open baskets for about 40–50 h. This helps loosen the seed coat from the kernel. These seeds

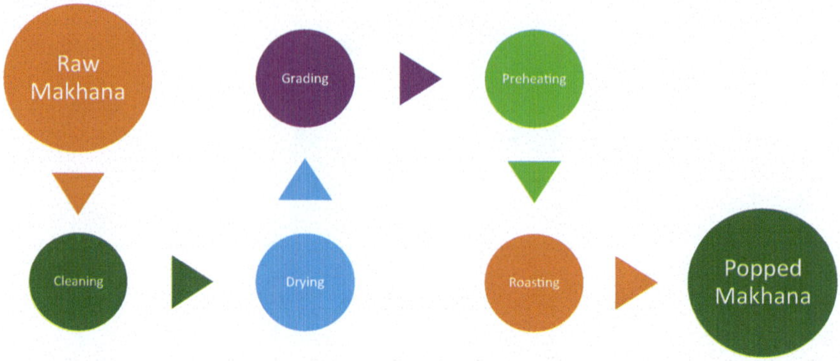

FIGURE 14.2 Flowchart for popping process of foxnut

are then roasted at approximately 230–335°C. When the seeds start to crack, 6–7 seeds are kept on some hard surface and are hit with a hammer. In this process, the seed coat breaks and the seed kernel comes out in an expanded form. The edible part is then polished by rubbing it against a bamboo basket immediately to avoid moisture absorption. The final grading of the seeds is done on the basis of the size and the color of the seed. The final product is packed and sent to the markets (Ahmad et al., 2020; Kumar et al., 2020) (Fig. 14.2).

14.3.1.2 Makhana snacks

Popped makhana are first cleaned, weighed, and then roasted, followed by slurry flavoring in flavor (masala, tomato, chilli, chocolate, onion, and others) drum, depending upon the market requirements, and then packed into different pack sizes (PMFME-makhana, foxnut).

14.3.1.3 Salted makhana

Salted makhana is prepared by dry roasting raw makhana in a nonstick pan until crisp, then sprinkling with salt according to taste and tossing to coat evenly before cooling and storing in an airtight container.

14.3.1.4 Makhana ready-to-eat breakfast cereal

Breakfast cereals that are ready to eat fall into the group of foods that are generally stable, have a long shelf life, and are quite popular right now. Almonds, cashew nuts, tutti frutti, and other dry fruits are combined with the makhana after it has been flaked or reduced in size. It could be basic or have flavors like honey, malt, chocolate, or fruit. It could be a product with an additional component, coated, or uncoated. Depending on the needs of the final consumer, it can be served with milk, water, sugar, or salt (PMFME-makhana, foxnut).

Makhana has a lot of protein, which is very good for the body. Magnesium, a strong channel blocker, is plentiful in lotus seeds and greatly enhances blood and oxygen levels. It is extensively consumed on many occasions in India and offers numerous other nutritional benefits. With a compound annual growth rate of over 13%, the market for ready-to-eat snacks and namkeens is projected to be worth over Rs. 50,000 crore. Changes in lifestyle, increasing urbanization, the rise in nuclear families, and rising disposable income are the segment's main development factors.

Other factors that have contributed to its growth include product innovations, strong marketing initiatives, and low-cost pricing strategies, making products more affordable for consumers. The sale of the snack food category (ready-to-eat packed foods like chips, extracted food, etc.) alone has grown more than sixfold. The global demand for nutritious and organically produced foods is growing rapidly in developed countries like Europe, the United States, Japan, and Australia. Therefore, there is a huge potential for marketing makhana as a ready-to-eat snack in the domestic as well as international markets. The product positioning should be done carefully to target the defined customer segment and promote it as a high-value item.

14.3.2 Makhana flour products

14.3.2.1 Makhana flour

The made makhana flour can be used in various product-making. The procedure for making makhana flour is given below (Jana et al., 2019) (Fig. 14.3).

14.3.2.2 Makhana chapatti

The recipe for the makhana chapatti is made with raw makhana powder (250 g), bread wheat (atta: 250 g), and pure ghee or oil (10 mL). The dough is made with the help of water. Chapatti is then fried in a nonstick pan at about 150°C (for roasting) (Fig. 14.4).

14.3.2.3 Makhana bread/bun

It is an innovative product made from a blend of wheat and makhana flour or powder, suitable for both gluten-rich and gluten-free variants (Fig. 14.5).

14.3.2.4 Makhana pakora

Makhana pakora is made from makhana raw powder (250 g), pure besan (150 g), and soybean oil (250 g). Pakora is fried and scanned on a spoon at about 150°C (Fig. 14.6).

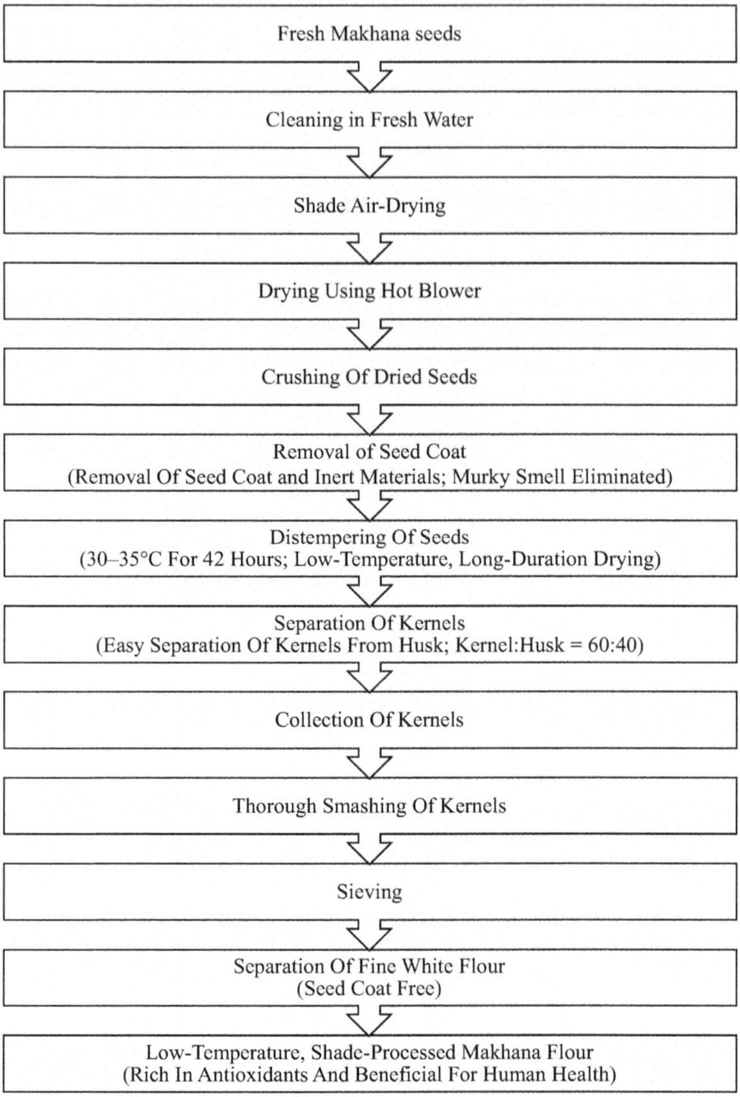

FIGURE 14.3 Flowchart for preparation of makhana flour.

14.3.3 Gluten-free bakery applications

14.3.3.1 Makhana bread/bun

The makhana flour can be used to make various bakery products. The ingredients required for the preparation of the sponge and dough formula, along with their respective quantities and percentages, are given as follows. The sponge portion consists of flour (70 kg, 35%), compressed yeast (3 kg, 1.50%),

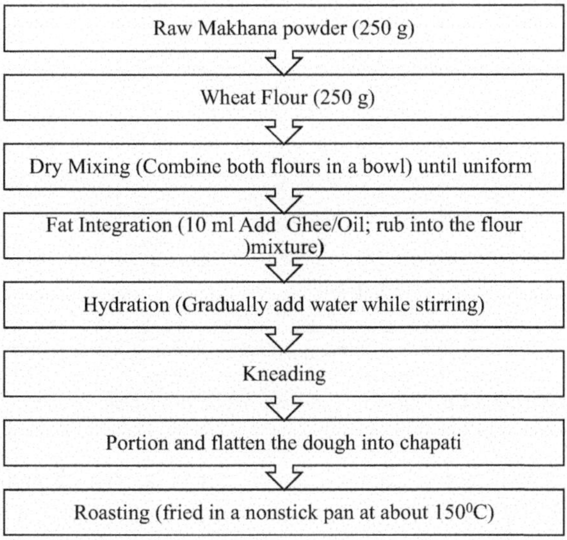

FIGURE 14.4 Flowchart for preparation of makhana chapati.

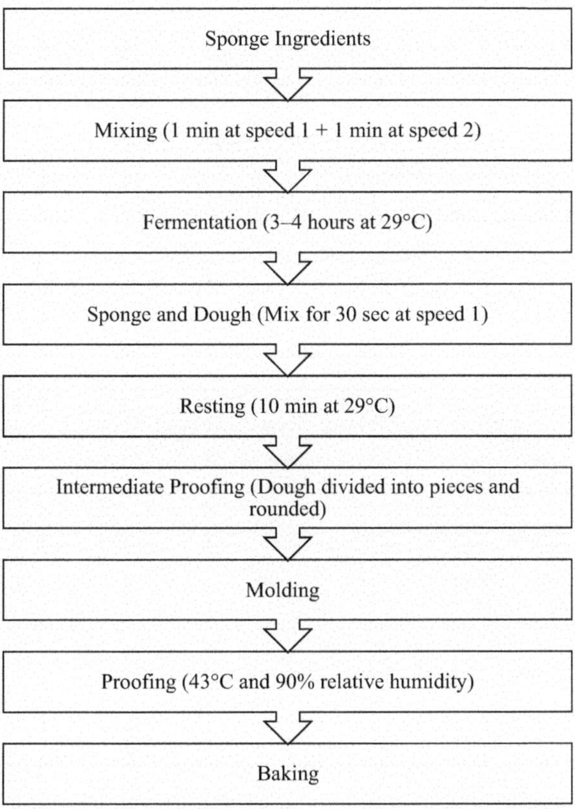

FIGURE 14.5 Flowchart for preparation of makhana bread/bun.

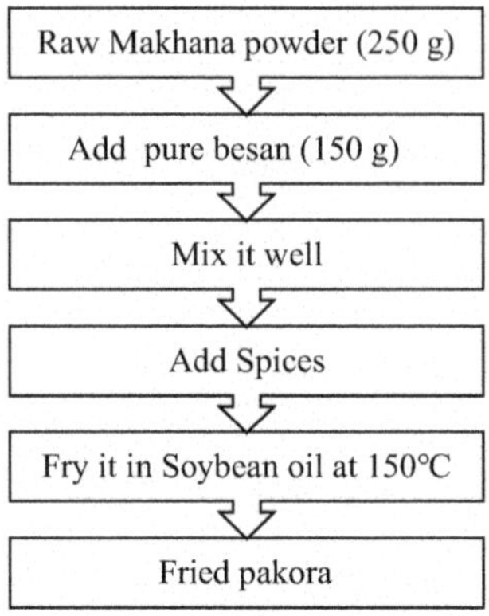

FIGURE 14.6 Flowchart for preparation of makhana pakora.

water (46 kg, 23%), and yeast food (0.3 kg, 0.15%). The dough portion includes flour (30 kg, 15%), high fructose corn syrup (42%) at 18 kg (9%), shortening (6 kg, 3%), salt (2 kg, 1%), ascorbic acid at 100–200 ppm (0.05%), water in variable quantity (12.50%), and calcium propionate (0.12 kg, 0.06%) (Fig. 14.7).

FIGURE 14.7 Flowchart for preparation of makhana bread or bun.

14.3.3.2 Makhana cookies/biscuits

Cookies can be made from makhana (PMFME-makhana, foxnut). The ingredients required for making makhana cookies are makhana flour (100 kg), sucrose (42 kg), all-purpose shortening (40 kg), high-fructose corn syrup (1.5 kg), nonfat dry milk (1 kg), sodium chloride (1.25 kg), sodium bicarbonate (1 kg), ammonium bicarbonate (0.50 kg), and deionized water (variable) (20.75 kg). The flowchart for the method of preparation of makhana cookies/biscuits is given as follows (Fig. 14.8).

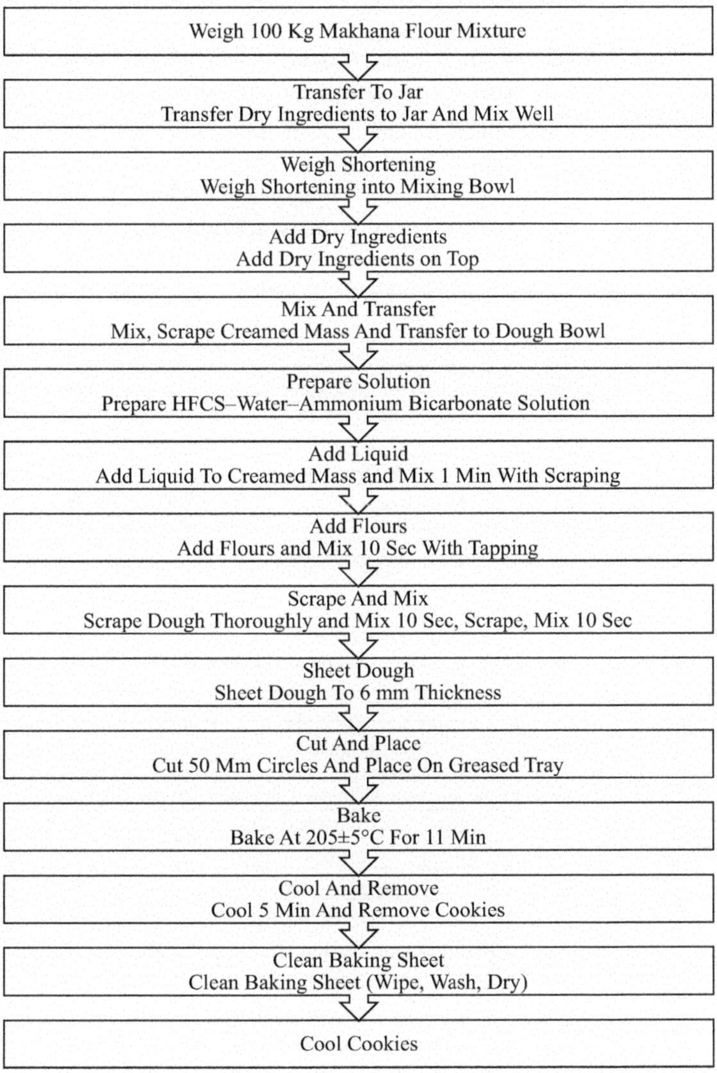

FIGURE 14.8 Flowchart for preparation of makhana cookies.

14.3.3.3 Makhana-based vegetable cookies

Madhubani Makhana Pvt. Ltd. in Bihar, India, was the source of spent makhana flour. Additional raw materials, including refined wheat flour, salted butter, sugar, baking powder, *Chenopodium album*, and *Spinacia oleracea*, also purchased from the market. *C. album* and *S. oleracea* were properly cleansed with tap water to get rid of dust, unwanted objects, etc. They were blanched for 2 to 3 minutes at 95 ± 5 °C in hot water, quickly chilled in cold water, and then pureed. Anand et al. (2020) made dough balls weighing 15 g were prepared for each cookie formulation, flattened to 5 mm thickness, and baked in a single-deck electric oven at 180 °C for 15–20 min (Fig. 14.9).

*14.3.3.4 Nutri cookies (makhana (*Euryale ferox*) and ragi (*Eleusine coracana*) supplemented and incorporated with dates)*

Raw materials such as popped makhana, ragi flour, wheat flour, sugar, dates powder, milk, butter, baking powder, baking soda, and pistachios were procured from the local market, while analytical-grade chemicals were available in the industry's analytical laboratory. Processing and analytical equipment, including a hot air oven, Kjeldahl apparatus, muffle furnace, and moisture meter, were used for analysis. Makhana powder was prepared by roasting

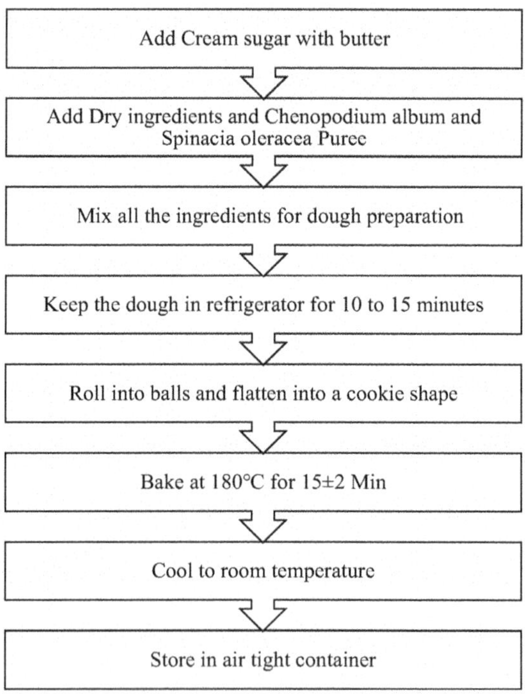

FIGURE 14.9 Flowchart for preparation of makhana-based vegetable cookies.

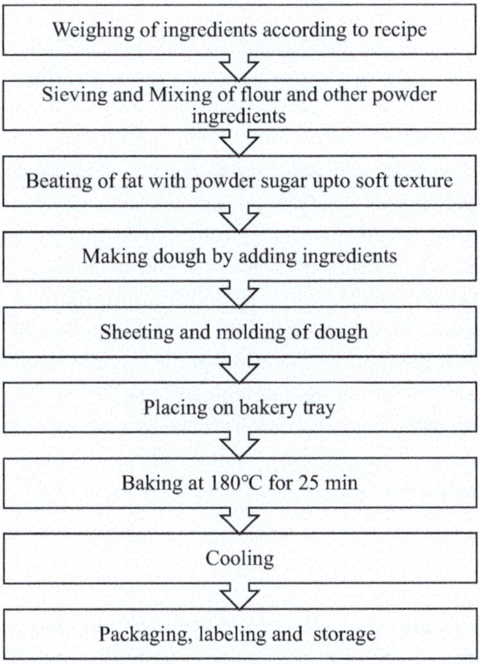

FIGURE 14.10 Flowchart for the preparation of makhana-based nutri cookies.

popped makhana, followed by grinding and sieving. For cookie preparation, butter was creamed with powdered sugar, after which makhana powder, ragi flour, wheat flour, dates powder, baking powder, and baking soda were mixed in, and milk was added gradually to obtain a uniform dough. The dough was sheeted, molded, placed on baking trays, and baked at 180°C for 25 min. After baking, the cookies were cooled at room temperature and packed for further analysis (Fig. 14.10).

14.3.4 Health and nutrition bars

Numerous varieties of bars exist in the market, classified into specific categories according to their principal nutritional function, ingredients, and intended consumers. This category encompasses energy bars, delivering rapid carbohydrate-derived energy; granola bars, abundant in fiber and whole grains for prolonged energy; low-sugar or sugar-free bars, designed for glycemic control and weight management; and functional or fortified bars, which provide supplementary health advantages via vitamins, minerals, or bioactive compounds.

14.3.4.1 Energy bars

Adequate nutritional consumption is essential for enhancing physical activity, optimizing athletic performance, and facilitating postexercise recovery. Professional athletes must carefully meet their nutritional requirements by ingesting appropriate volumes and varieties of food. Athletes require snacks between main meals for a healthy lifestyle and continuous training sessions. Energy bars are a rapid, handy meal that is ready for use and possesses an extended shelf life without refrigeration requirements. Energy bars supply the body with carbohydrates to sustain energy during exercise or training sessions. Energy bars are offered in many flavors and textures, typically comprising a combination of energy-dense nutrients, vitamins, minerals, and a significant amount of sugar. Energy bars have developed to accommodate a wider consumer demographic with diverse dietary habits and nutritional requirements. Despite their convenience and popularity, energy bars have been questioned for their nutritional quality, since certain brands may include elevated quantities of added sugars, artificial additives, and preservatives (Kanya et al., 2025).

14.3.4.2 Granola bars

A granola bar typically consists of rolled oats, nuts, seeds, dried fruits, and a binding sweetness like honey or syrup. These ingredients are generally baked or compressed into a bar form, facilitating portability and convenience for consumption. Granola bars serve as a rapid energy source, supplying carbohydrates, fiber, and occasionally protein, depending upon the ingredients utilized. They are often ingested as a snack, a light breakfast, or as fuel before and after workouts. Variations encompass low-sugar bars, protein-fortified bars, and bars containing supplementary functional components such as vitamins, minerals, or superfoods, addressing diverse dietary requirements and preferences.

14.3.4.3 Low-sugar and functional food bars

Low-sugar and functional food bars are a developing category of convenience meals aimed at offering essential nutrients while also providing specific health advantages beyond just energy enhancement. In contrast to traditional energy or granola bars designed for quick caloric intake, low-sugar functional bars prioritize alternative sweeteners (such as honey or natural syrups), minimized added sugars, and bioactive components to enhance particular physiological functions. These bars are typically composed of functional elements such as dietary fiber, plant-based proteins, minerals, antioxidants, and conventional medicinal constituents to meet population-specific requirements, including maternal nutrition, gastrointestinal health, glycemic regulation, and metabolic wellness.

14.3.4.4 Makhana powder and dates bar

Makhana powder and dates offer the potential to develop a specialized product designed for athletes' nutritional needs. Makhana, a protein-dense food, functions as a wholesome snack or component in diverse goods, providing low sugar and elevated protein levels, making it suitable for a balanced athletic diet. The raw components for the energy bar comprise makhana powder (16.66 g), rice crisps (8.33 g), watermelon seeds (4.16 g) and dried fruits such as dates (21.6 g), figs (12.5 g) and walnuts (7.5 g), amalgamated with jaggery (20.82 g), and honey (8.33 g).

The process begins by heating jaggery and honey in a pan until a homogeneous syrup is achieved, which acts as the binding agent. Makhana powder is gently roasted to improve its flavor and nutritional stability. Chopped dates, figs, walnuts, and watermelon seeds are sequentially incorporated into the syrup, while the mixture is continuously stirred in order to ensure uniform coating and adequate binding. Rice chips are added last to preserve their crunchy texture. The amalgamated mixture is immediately placed into a greased mold, and consistent pressure is exerted to form the mixture into a dense slab. The slab is permitted to cool at ambient temperature until it solidifies. Subsequent to setting, the slab is segmented into uniform portions to create individual energy bars. Fig. 14.1 illustrates the comprehensive flow chart for the preparation of makhana-dates bar, as shown by Kanya et al. (2025) (Fig. 14.11).

The makhana energy bar contains 383.12 kcal per 100 g, with carbohydrates at 75.5% (25.16% dry value [DV]), protein at 4.78% (6.37% DV), fat at 6.89% (10.28% DV), moisture at 11.48%, dietary fiber at 10.41% (26.02% DV), and ash at 1.35 g/100 g, illustrating its nutrient-rich profile appropriate for active lifestyles. The bar contains 383.12 kcal per 100 g, providing approximately 20% of the necessary daily calorie intake and serving as an instant and sustainable energy source. The elevated glucose content facilitates glycogen restoration, improving endurance and postponing exhaustion during physical activity (Jeukendrup, 2014). Dietary fiber enhances digestive health, controls blood glucose levels, and aids in achieving satiety (Anderson et al., 2009). Adequate protein levels facilitate muscle repair and regeneration (Pasiakos et al., 2015), while beneficial unsaturated fats enhance cardiovascular health and regulate hunger (Ros, 2010). The ash content signifies vital minerals that promote bone health, muscle function, and cellular activities. The sensory evaluation employing a nine-point Hedonic scale reveals bar scores for taste (7.7 0.7), odor (7.2 1.2), color (7.0 1.5), texture (7.5 1.0), and overall acceptability (7.6 0.9), signifying positive sensory characteristics and attractive physical features.

The overall cost for manufacturing the bar amounts to Rs. 77.94 per 120 g. A comparison with a commercial energy bar indicates that the makhana energy bar contains elevated levels of carbohydrates and dietary fiber, has

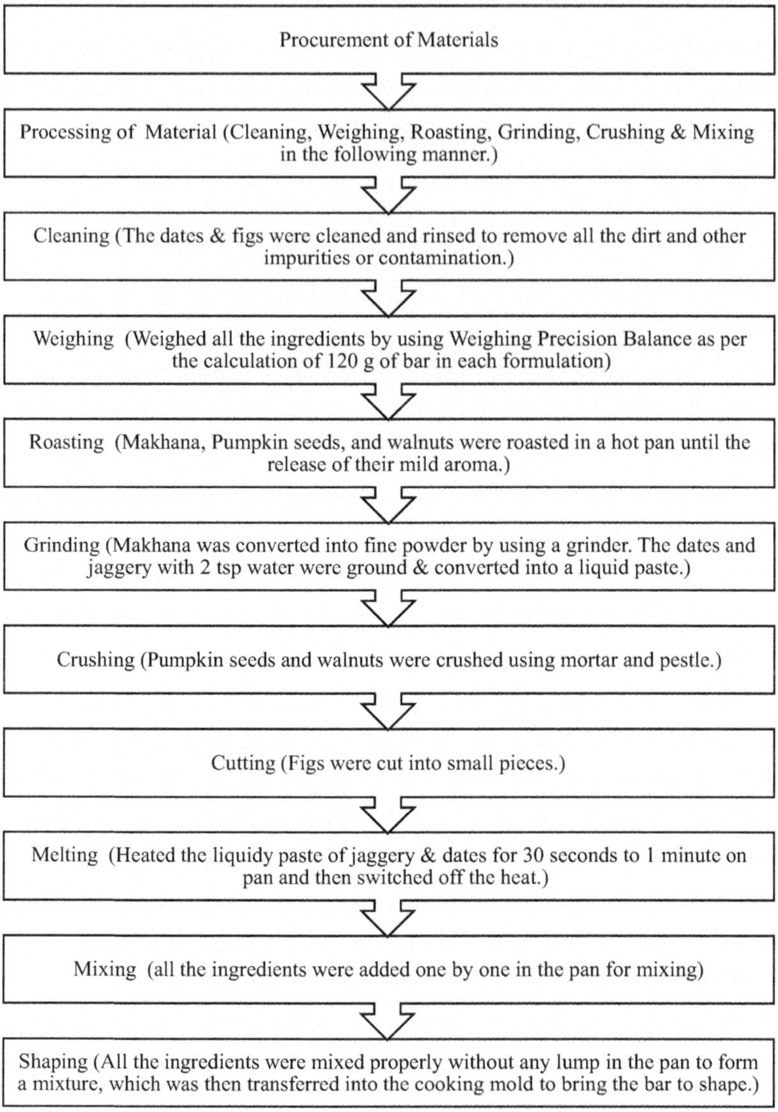

FIGURE 14.11 Preparation of makhana-dates bar.

reduced fat content, and is devoid of significant allergens, hence enhancing digestibility and rendering it an acceptable allergen-free alternative. This energy bar offers a well-rounded combination of macronutrients and vital minerals, facilitating immediate energy requirements, enhancing athletic performance, aiding recovery, and promoting long-term health advantages.

14.3.4.5 Fero nutri bar

The ingredients utilized in the formulation of the fero nutri bar consist of popped makhana (150 g), mixed nuts (25 g), pitted dates (50 g), honey (35 g), jaggery (50 g), and butter (50 g). These ingredients offer a suitable equilibrium of carbohydrates, fats, moderate protein, and vital micronutrients, while also providing favorable sensory characteristics.

To prepare nutri bars, melt the butter in a pan, add makhana, and roast over moderate heat until a light brown color and crunchy texture are attained. This process improved both the flavor and structural integrity of the bars. Chopped almonds and dried fruits were then incorporated into the roasted mixture, followed by the incorporation of honey and jaggery, which served as natural sweeteners and binding agents. The mixture was agitated continuously to guarantee uniform distribution of components and to achieve a cohesive, granular mass. Subsequently, pour the produced hot mixture into a greased tray and spread it evenly to achieve a uniform thickness. Refrigerate the mixture in the tray for approximately 1 hour to ensure proper setting. After setting, detach the nutri bar slab and slice it into homogeneous rectangular bars (Sahithi et al., 2022).

The fero nutri bar demonstrates a balanced and nutritionally optimized formulation. The nutri bar has a moisture content of 6.21%, signifying excellent product stability and an appealing crisp texture. Carbohydrates are the primary component at 70.87 g/100 g, providing sufficient energy supply. The protein level is 6.42 g per 100 g, indicating a moderate contribution from whole food sources like makhana, almonds, and dried fruits. While this level does not classify the product as a high-protein bar, it significantly contributes to daily protein consumption. The fat level is 18.36 g per 100 g, predominantly derived from butter and nuts, which improves flavor and energy density. The ash concentration is 2.14%, indicating an enhanced mineral profile attributed to the naturally elevated calcium and iron levels in makhana. The dietary fiber content of 3.18 g/100 g increases the functional value of the bar by aiding digestive health and enhancing satiety. The nutritional profile of the fero nutri bar indicates that makhana serves as an effective cereal substitute in granola-style nutri bars, with balanced macronutrients and enhanced mineral density. The fero nutri bar is deemed satisfactory by consumers, achieving scores of 8.06 for color, 7.93 for texture, 7.66 for aroma, and 8.13 for taste, culminating in an overall acceptability rating of 8.2.

14.3.4.6 Makhana nutri bar for lactating women

The raw materials utilized in the formulation of the nutri bar comprised semolina (suji) (92.5 g), makhana (7.5 g), honey (42 g), dried ginger (0.1 g), gum acacia (5 g), almonds (5 g), cashew nuts (5 g), coconut (50 g), fenugreek (0.1 g), ghee (50 g), and chocolate flavor. These nutrients possess nutritional value and functional attributes, especially for maternal health and lactogenic potential.

To prepare the nutri bar, heat ghee in a flat-bottomed pan to 80°C, then progressively add chopped almonds, cashew nuts, and coconut, roasting until a consistent brown color is achieved. Subsequently, sauté dried ginger and fenugreek individually, ensuring to eliminate excess oil to prevent bitterness. Fry the semolina-makhana mixture in the remaining ghee until a pleasing aroma and color are attained, then add gum acacia. Continue stirring until the crystals expand and the sputtering ceases. Thereafter, incorporate honey into the roasted mixture and stir frequently to achieve uniform integration. Subsequently, integrate the roasted dry fruit combination, chocolate flavor, and 0.1% preservative, then spread the homogeneous mixture onto trays, mold into bars of specified dimensions, and allow to cool (Jan et al., 2012).

Nutri bar using 7.5% makhana fortification and honey as a natural sweetener demonstrates a balanced and nutritionally enriched formulation. The nutri bar comprises 83.53% carbohydrates, providing sufficient energy, while substituting refined sugar with honey improves digestion and functional value. The bar contains approximately 18.29% protein, indicating a moderate yet nutritionally substantial level appropriate for lactating women, aiding in tissue repair and metabolic activities. The fat content was comparatively low at 5.45%. A dietary fiber level increase of 2.80% may enhance digestion and promote gastrointestinal health. The 2.49% ash percentage in the bar signifies an enhanced mineral profile due to the inclusion of makhana and other nutrient-rich components. The sensory profile indicates elevated scores for color (8.10), texture (7.80), aroma (8.10), taste (8.10), and overall acceptability (8.0), illustrating that the incorporation of 7.5% makhana preserves favorable sensory attributes while augmenting nutritional value, thus rendering it a highly acceptable functional nutribar.

14.3.4.7 Functional nutri bar (foxnut–barnyard millet-based)

The nutri bar, composed of foxnut and barnyard millet, is formulated with broken makhana flour and broken barnyard millet powder as primary ingredients, supplemented by coconut milk, sucrose, palm sugar, and liquid glucose for binding and sweetness.

To prepare a 100 g bar, heat 30 g of coconut milk (w/v) in a pan over medium heat for 3 min, then incorporate sucrose and palm sugar, stirring continuously for approximately 15 min to achieve a homogeneous syrup. 15 g of roasted makhana flour and 15 g of broken barnyard millet powder are subsequently added to the mixture and mixed thoroughly until a brown color is achieved. Liquid glucose is included at the final stage to improve adhesion, and the heated mixture is subsequently poured into molds. The molded bars are cooled to ambient temperature. The overall sweetener concentration is restricted at 35% to ensure satisfactory taste and texture (Kumar et al., 2024).

The foxnut-barnyard millet nutri bar comprises many bioactive metabolites that enhance its functional attributes. Principal constituents comprise palmitoyl serinol, palmitoleic acid, oleamide, docosahexaenoic acid, ferulic acid, glycitein, apigenin, oleanolic acid, kammogenin, bufogargarizin, and carvone, among others. These metabolites encompass classes including lipids, fatty acids, phenolics, flavonoids, sterols, saponins, and alkaloids and are linked to antioxidant, antiinflammatory, cardioprotective, antidiabetic, antimicrobial, and anticancer properties, underscoring the bar's potential as a functional, bioactive-rich snack. Texturally, the foxnut-barnyard millet-based nutri bar demonstrates a hardness of **42,723.39 ± 268.49 g s** and a fracturability of **19.05 ± 0.31 s**, indicating good structural integrity with acceptable bite and chewiness. Sensory evaluation with a high overall acceptability score of **8.1 ± 1.21** on a nine-point hedonic scale, confirming that the balanced ingredient ratio and controlled sweetener level (≤35%) result in a nutri bar with favorable nutritional quality, pleasant texture, and consumer acceptability.

14.3.4.8 Allergen-free nutri bar

The allergen-free nutri bar comprises 40 g of roasted foxnut (makhana) flour, 50 g of coconut milk, 30 g of palm sugar, 15 g of sucrose, 10 g of liquid glucose, and 2 g of amaranth seeds. Fresh makhana seeds undergo cleaning, air-drying, and subsequent drying at 30–35°C for 42 h prior to the removal of seed coatings and grinding of the kernels into fine flour. Fresh coconut is grated and immersed in hot water for 5–7 min, thereafter pureed and filtered to get coconut milk, which can be refrigerated for 2–3 days or frozen for future use.

To prepare the bars, coconut milk is heated in a pan over medium heat, to which sucrose and palm sugar are added and stirred for 15 min until fully dissolved. Roasted foxnut flour is blended until the mixture attains a brown color, followed by the addition of liquid glucose to bind the components. The resulting mixture is thereafter poured into molds, permitted to cool to ambient temperature, and optionally coated with amaranth seeds to augment texture. Ultimately, the bars are compacted and preserved, resulting in a nutrient-rich, high-fiber snack that is devoid of prevalent allergens and appropriate for sensitive individuals (Kumar, 2022) (Fig. 14.12).

The nutrition bar comprises 10.9 g of protein, 8.25 g of fat, and 34.65 g of carbohydrates per 100 g. The moisture content is 37.29%, while the total ash content is 2.65%. The mineral composition of the bar includes calcium at 90 mg, magnesium at 2.85 mg, phosphorus at 5.65 mg, potassium at 218 mg, sodium at 48 mg, copper at 0.12 mg, iron at 3.29 mg, and zinc at 1.45 mg. The nutrition bar has an antioxidant activity of 46.8% and a total phenolic content of 7.79 μg/mL. This foxnut-based nutrition bar is allergen-free and has a high caloric content of around 261 Kcal per 100 g.

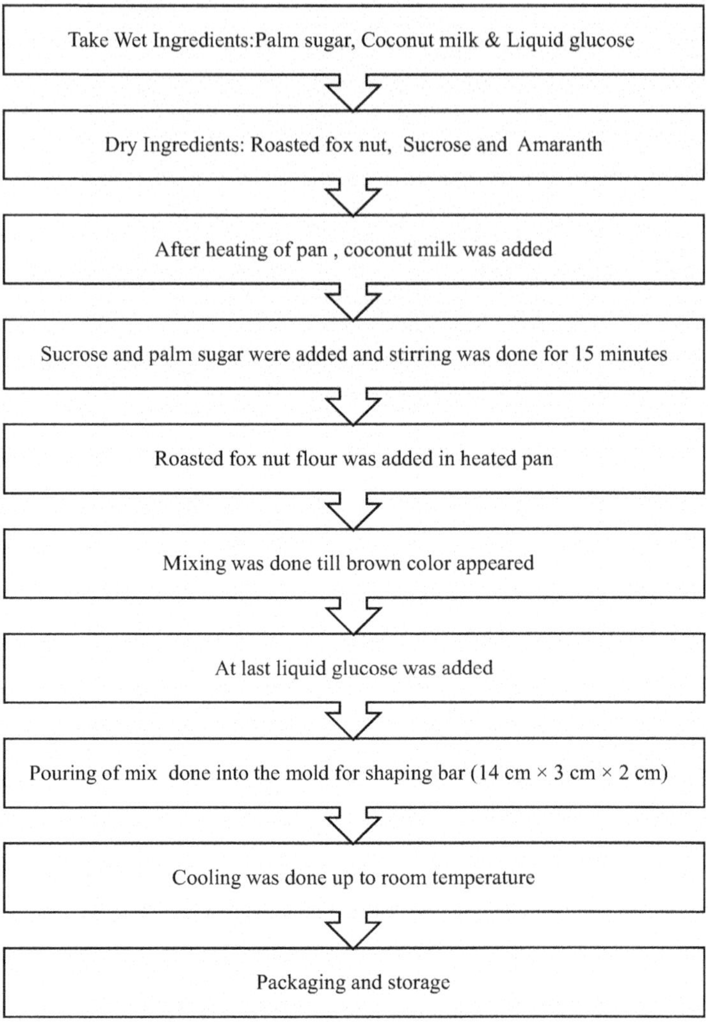

FIGURE 14.12 Preparation of allergen-free nutri bar.

14.3.4.9 Gorgon nut and millet snack bar

The nutri bar is formulated with 5% gorgon nut (makhana) and a 10% composite blend of pearl millet, finger millet, and amaranth. The additional ingredients consist of oats (10%), skim milk powder (3%), cocoa powder (3%), watermelon seeds (3%), honey (35%), butter (5%), jaggery (5%), and liquid glucose (1%). The process commences with the roasting of premium gorgon nuts at 100°C for 2–3 min to eliminate moisture, thereafter cooling them to room temperature and grinding them into a fine powder. Butter and jaggery are melted in a pan at 100°C, followed by the addition of liquid glucose and honey. All dry components, comprising gorgon nut powder, millets, oats, cocoa powder, skim milk powder, and watermelon seeds, are combined and blended manually to create a homogeneous combination. The amalgamation is subsequently shaped into greased molds and permitted to cool, resulting in bars, each weighing roughly 50 g (Adhav et al., 2023). The snack bar has 8.31% protein, 1.11% fiber, 5.27% fat, 76.77% carbs, and 387.57 kcal per 100 g. It exhibits significant antioxidant activity (59.47% 2,2-diphenyl-1-picrylhydrazyl [DPPH] inhibition) and is abundant in minerals: iron 3.75 mg, calcium 76.25 mg, phosphorus 167.97 mg, magnesium 69.82 mg, and potassium 234.87 mg per 100 g. This renders it a nutrient-rich, functional energy bar appropriate for all age categories.

14.3.5 Other value-added products

14.3.5.1 Makhana vermicelli

Makhana vermicelli is prepared by roasting makhana until crisp, grinding into a coarse powder, mixing with semolina, water, and seasoning, forming into thin noodles, and cooking until tender.

14.3.5.2 Makhana halva

To make makhana halwa, roast makhana until crisp and then crush into pieces. Cook in milk until thickened, sweeten with sugar and cardamom, and garnish with nuts and saffron before serving hot or at room temperature.

14.3.5.3 Makhana pickle

To prepare makhana pickle, roast makhana until crisp. Then, mix with pickling spices, vinegar, and salt. Let it marinate for a few hours before serving. Adjust seasoning according to taste preferences.

14.3.5.4 Makhana baby foods or porridge

Popped makhana is first cleaned, the quality is checked, and then the makhana is roasted on medium heat until golden. Once golden, check if ready by crushing one makhana piece between your fingers. If it cracks into pieces and, when further crushed, turns into powder. Then cool the makhana for 20 min before grinding. Grind the makhana into a fine powder. Add the powder to a colander or strainer and sieve the flour and packed into airtight container. It may be mixed with bengal gram flour (sattu), banana powder to prepare different types of baby foods (PMFME-foxnut, makhana).

14.3.5.5 Makhana kalakand

Makhana kalakand can be prepared from raw makhana powder (300 g), sugar (100 g), channa (600 g), milk (1.0 L), pure ghee (5 g), and elaichi (cardamom) powder (2 gm). The boiling temperature was 130–140°C for 30 min to a thick consistency and then cut into small pieces after cooling of the resultant mass (PMFME-foxnut, makhana) (Fig. 14.13).

14.3.5.6 Makhana barfi

Makhana barfi is made using raw makhana powder (300 g), sugar (200 g), milk (1.5 L), pure ghee (5 g), elaichi (cardamom) powder (2 g), and arrowroot powder (2 g). These ingredients are then thoroughly combined in preparation for boiling. The boiling temperature is kept at 130–140°C for 30 min to achieve a thick consistency, then cut into small pieces after cooling (PMFME-foxnut, makhana) (Fig. 14.14).

14.3.5.7 Instant makhana kheer mix

It is rich and creamy sweet with delicious flavor. Given below is the procedure for making makhana kheer mix (PMFME-makhana, foxnut) (Fig. 14.15).

14.3.5.8 Makhana churna

Dried makhana (foxnut/*E. ferox*) seeds are the primary component in makhana churn.

Makhana seeds are cleaned, dried, and then gently roasted till crisp in order to prepare churna. After cooling, they are ground into a fine powder in a grinder and sieved to create a smooth churna. For use, the manufactured powder is kept in an airtight container. With Madhura Rasa, Sheeta Virya, and Snigdha Guna, makhana churna is a nourishing Ayurvedic formulation that is

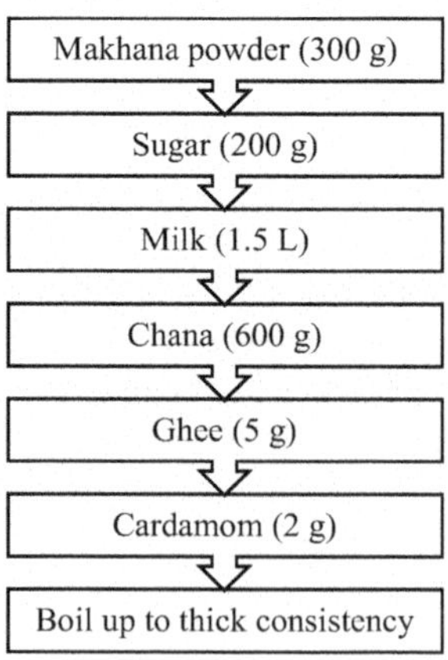

FIGURE 14.13 Flowchart for preparation of makhana kalakand.

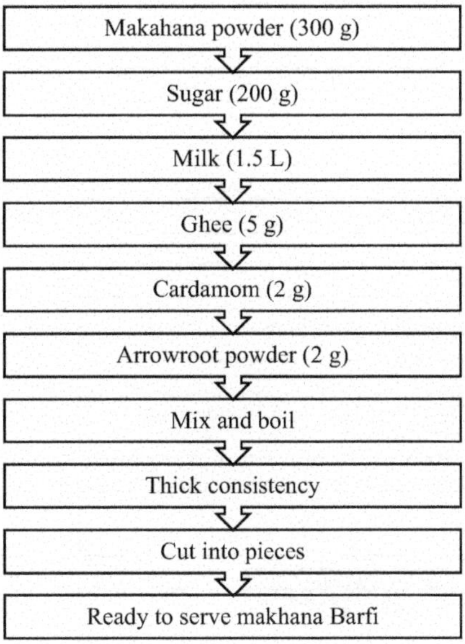

FIGURE 14.14 Flowchart for preparation of makhana barfi.

particularly helpful for Vata and Pitta balance. As a mild Rasayana, it promotes uterine health, strength, and energy, especially during pregnancy and the postpartum phase. It has high levels of magnesium, potassium, antioxidants, and dietary fiber, all of which benefit maternal health by enhancing digestion, lowering stress and fatigue, controlling blood pressure, and more. Additionally, makhana churna enhances Ojas, encourages fetal feeding (Garbha Pushti), and facilitates postpartum recuperation. Vigor, which is crucial for the healing process after giving birth (Rao et al., 2025) (Table 14.2).

14.4 Waste valorization for gorgon nut

Agro-industrial waste poses a significant challenge globally, spurring researchers to find cost-effective ways to extract bioactive compounds such as polyphenols. These compounds exhibit various pharmacological activities like antioxidant, antibacterial, and antiinflammatory properties, offering potential applications in food additives, nutraceuticals, and cosmetics.

Apart from the seeds, which are roasted and popped to get the edible endosperm, called "makhana *lawa*," other parts of the plants also have high nutritional composition. Dutta et al. (1986) have conducted studies to assess the plant contents and nutritional quality of various plant parts of makhana. It was observed that the potassium (K) content showed a decreasing trend toward upper plant parts. Rhizome gave the maximum K content of 2120 mg/kg, followed by leaves (1940 mg/kg) and fruit sheaths (1180.9 mg/kg). However,

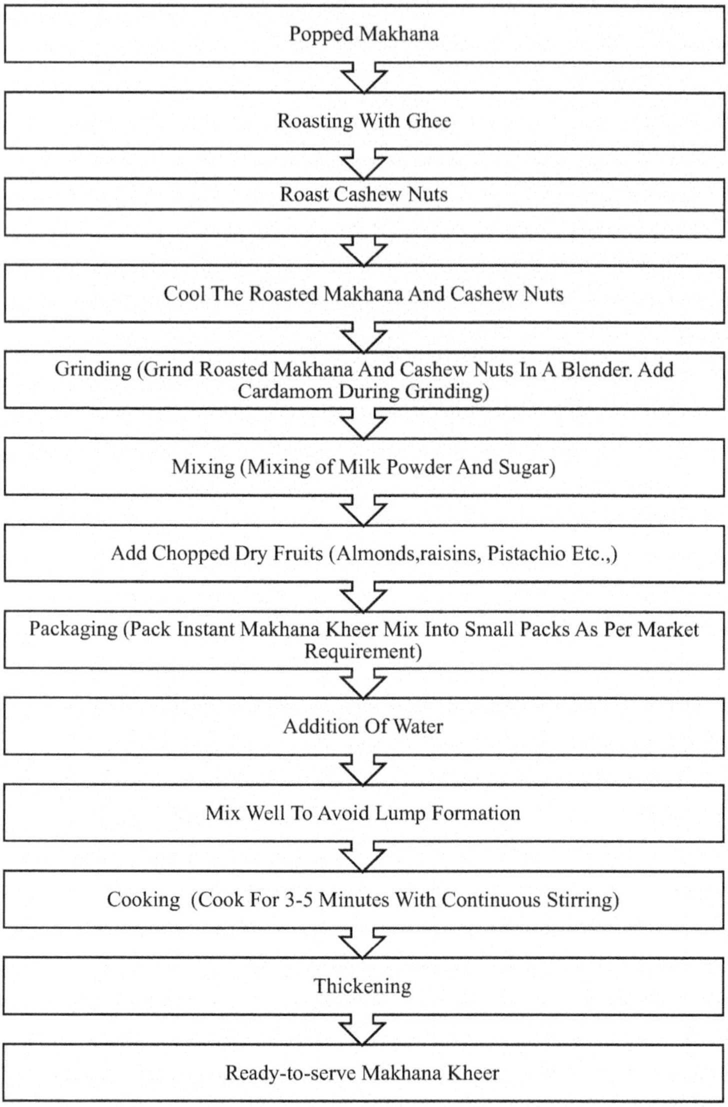

FIGURE 14.15 Flowchart for preparation of instant makhana kheer mix.

sodium (Na) distribution showed a reverse trend. Makhana leaves gave the highest Na content of 1480 mg/kg, followed by rhizome (1440 mg/kg) and fruit sheath (1320 mg/kg). The seeds also showed high Na accumulation, yielding in the range of 180–240 mg/kg. This explains the reason why makhana plants grow well in low Na soils. Phosphorus (P) content was more concentrated in the fruit portions. Inflorescence buds contained the maximum

TABLE 14.2 Product categories, processing techniques, and shelf life of makhana-based products.

Product category	Product type	Processing techniques	Major ingredients	Key quality attributes	Shelf life (approx.)
Ready-to-eat snacks	Roasted makhana	Cleaning, grading, roasting, seasoning	Whole makhana, edible oil, spices	Crisp texture, uniform roasting, low oil absorption	3–6 months
	Flavored makhana	Roasting, flavor coating, cooling, packaging	Makhana, oil, spice blends, herbs	Enhanced sensory appeal, flavor retention	3–5 months
	Salted makhana	Dry roasting raw makhana, Flavoring	Makhana oil, salt	Enhance appearance and taste	3–5 Months
Flour-based products	Makhana flour	Dry roasting, Distempering, grinding, sieving	Makhana seeds	Fine particle size, high digestibility, gluten-free	6–9 months
Bakery products	Makhana cookies/ crackers	Mixing, dough forming, baking	Makhana flour, wheat flour, fat, sugar	Crispness, light color, nutritional enrichment	2–4 months
Convenience foods	Makhana porridge mix	Roasting, grinding, blending	Makhana flour, milk solids, sweeteners	Easy reconstitution, smooth mouthfeel	6–8 months
Nutrition bars	Makhana-based bars	Roasting, binding, molding, setting	Makhana, nuts, honey/ jaggery, seeds	Compact structure, chewiness, nutrient density	4–6 months
Dairy sweets	Makhana kalakand	Mixing, boiling, thickening, setting, cooling, and cutting	Makhana powder (300 g)	Gluten-free, dense, chewy, slightly grainy texture	5–6 days or up to a week
			Sugar (200 g)		
			Milk (1.5 L)		
			Chana (600 g)		
			Ghee (5 g)		
			Cardamom (2 g)		

Continued

TABLE 14.2 Product categories, processing techniques, and shelf life of makhana-based products.—cont'd

Product category	Product type	Processing techniques	Major ingredients	Key quality attributes	Shelf life (approx.)
	Makhana barfi	Roasting, grinding, cooking, and setting: Sweetening, cooking, setting, cooling, and cutting	Makahana powder,	Dense, melt-in-mouth flavor, creaminess	30–45 days
			Sugar,		
			Milk,		
			Ghee,		
			Cardamom, Arrowroot powder		
	Instant makhana kheer mix	Popping, roasting, cooling,	Popped makhana, ghee, cashew nuts, cardamom, milk powder, dry fruits	Good flowability, solubility, creamy, smooth, etc.	4–12 months
		Grinding,			
		Mixing			
		Cooking, thickening			

P content of 5715 mg/kg, followed by fruit sheaths (5318 mg/kg), while the same in vegetative parts ranged between 1863 and 2715 mg/kg. The P content of common fruits like litchi, guava, and mango varies in the range of 300–800 mg/kg. Therefore, makhana is a richer source of P than common fruits. Makhana plant is a good source of micronutrients as well. Fruits contain a high zinc (Zn) content of 42.9–66 mg/kg, which is much higher than common fruits like cucumber (23.3 mg/kg), mango (7.3 mg/kg), banana (5.3 mg/kg), and colocasia (15.5 mg/kg). Moreover, the critical limit for Zn in food crops ranges between 11 and 35 mg/kg. This shows that Zn is not a limiting factor for makhana. It is also a rich source of iron (Fe), varying in the range of 1994–2236 mg/kg against critical limits of 35–69 mg/kg for common crops. It also has higher manganese (Mn) content (25–35 mg/kg) than common fruits (7.3–29.6 mg/kg). The crude protein in makhana is about 10 times higher than that of common fruits. Therefore, the plant parts of makhana show a promising contribution in minerals and micronutrients, making it a superior crop over other staple crops, fruits, and vegetables.

Makhana processing by-product is bran outer coat. Makhana bran may be used as poultry feed ingredients. Makhana outer coat may be used for fuel purposes. Recent findings suggest its untapped potential as a source of bioactive compounds with antioxidant and antibacterial properties. This agro-industrial waste holds promise for further exploration, potentially leading to the development of commercially viable products.

14.5 Challenges and future prospects

14.5.1 Challenges

Despite increased interest in makhana-based value-added goods, a number of linked hurdles impede their development and commercialization. Raw material variability due to different cultivation practices and processing conditions affects product consistency, while makhana's naturally bland taste and unique soggy texture necessitate careful formulation with sweeteners, binders, and flavoring agents to achieve consumer acceptability while maintaining nutritional benefits. Shelf-life stability is a key problem since products are subject to lipid oxidation, moisture absorption, and textural deterioration, and maintaining microbiological safety without artificial preservatives is especially difficult. Furthermore, processing processes such as popping, roasting, and grinding raise production costs, while volume expansion after processing increases transportation costs, restricting mass-market affordability and scalability.

Infrastructure and systemic challenges impede commercialization attempts. The lack of sophisticated processing units and cold storage facilities causes severe postharvest losses and quality deterioration, hurting competitiveness in both national and international markets. Farmers earn less (around ₹250 crore) than middlemen (about ₹550 crore), despite the industry's significant worth. Limited systematic study on how processing affects nutrients and phytochemicals, along with rigorous regulatory compliance requirements for labeling, health claim verification, and international food safety standards, creates further impediments to large-scale manufacturing and export promotion. Addressing these multiple difficulties necessitates collaborative efforts in technological development, infrastructural investment, equitable value chain restructuring, and extensive scientific validation.

14.5.2 Future prospects

The future of makhana-based value-added goods seems very promising, thanks to the convergence of technological innovation, industry expansion, and increasing consumer preferences toward functional and plant-based foods. Processing technology advancements, such as extrusion, microencapsulation, and enhanced germination procedures, improve texture, shelf life, and nutrient bioavailability while decreasing antinutritional elements. The potential for broad product diversification is enormous, encompassing ready-to-eat flavored snacks, gluten-free bakery items, traditional sweets, nutrition bars, infant foods, and makhana flour for specialty uses. These products can be strategically formulated for specific populations, such as children, athletes, diabetic patients, and celiac disease patients, with functional enhancements such as fortification, probiotics, and bioactive compounds that have been shown to have antioxidant, cardioprotective, and antiaging properties.

The commercial landscape presents extraordinary growth opportunities, particularly in export markets where India currently ships only 1% of its production despite rising global demand across the United States, EU, Southeast Asia, and Gulf countries. Value-added formats—roasted, flavored, organic, and functional varieties—command premium margins in health-conscious international markets. Apart from commercial benefits, the makhana value chain promotes climate-resilient agriculture, boosts regional economies in Bihar, West Bengal, Assam, and Manipur, and provides avenues for fair economic empowerment of farmers who now receive limited industry earnings. With continued scientific validation of health claims, infrastructure investment in processing facilities and quality control systems, supportive government policies such as PMFME, and integration with clean-label and nutraceutical trends, makhana has the potential to transform from a traditional ingredient to a globally recognized superfood, driving sustainable agricultural development while meeting contemporary nutritional needs.

14.6 Conclusion

Makhana-based value-added products represent a game-changing prospect in the functional food industry, combining great nutritional benefits with high economic viability. These goods are rich in protein, dietary fiber, critical minerals, and bioactive components, as well as naturally gluten-free and low in fat, making them ideal for today's health-conscious consumers. Makhana's adaptability allows for the development of a wide range of products, including ready-to-eat snacks, bakery goods, nutrition bars, and infant foods, with processing techniques such as popping and germination improving nutritional content and digestibility. The market's excellent 17%–18% annual growth rate, driven by premiumization and health awareness, displays healthy demand, while India's dominant position as the world's largest producer (contributing 90% of worldwide output) gives a strong basis for expansion.

Despite obstacles in raw material consistency, sensory optimization, and infrastructural development, the sector has vast untapped potential, especially given its current 1% export rate. Strategic value addition not only increases shelf stability and lowers transportation costs, but it also generates considerable job possibilities and economic empowerment at the local level. As consumers worldwide prefer plant-based, allergen-free, and nutrient-dense meals, makhana-based products are perfectly positioned to grab both local and international markets, particularly in health-conscious developed countries and Southeast Asia. With continued research, technological advancement, and infrastructure investment, makhana has the potential to transform from a traditional ingredient into a premium international superfood that supports sustainable agriculture, preventive healthcare, and rural economic development while addressing current nutritional challenges.

References

Adhav, G.A., Agrawal, R.S., Ghodke, S.V., Shere, P.D., 2023. Quality evaluation of gorgon nut and millet snack bar: a health-conscious approach. Int. J. Nephrol. Renovascular Dis. 8 (10), 237–2342.

Ahmad, A., Azad, C.S., Kumar, B., 2020. A study on constraints of makhana cultivation and suitable measures for its better development. J. Pharmacogn. Phytochem. 9 (1), 984–987. https://doi.org/10.22271/phyto.2020.v9.i1q.10580.

APEDA, 2025. Market Intelligence Report: Makhana. https://apeda.gov.in/sites/default/files/202512/MIC_Monthly_dashboard_Makhana_15122025.pdf.

Anderson, J.W., Baird, P., Davis Jr, R.H., Ferreri, S., Knudtson, M., Koraym, A., Williams, C.L., 2009. Health benefits of dietary fiber. Nutr. Rev. 67 (4), 188–205.

Devi, M.B., Deka, S.C., 2021. Physicochemical properties and structure of starches of foxnut (*Euryale ferox* Salisb.) from India and its application. J. Food Process. Preserv. 46 (2). https://doi.org/10.1111/jfpp.16262. Article e16262.

Islam, Z., Bashir, K., Singh, S., Siddiqui, S., 2025. Effect of germination on physicochemical, functional and antinutritional characteristics of makhana seed (*Euryale ferox*) flour. Food Human. 100717.

Jan, K., Jairajpuri, D.S., Jan, S., 2012. Preparation of nutri bar for lactating women. IOSR J. Environ. Sci. Toxicol. Food Technol. 1 (5), 10–14.

Jeukendrup, A.E., 2014. Carbohydrate intake during exercise and performance. Nutrition 30 (7–8), 769–775.

Jana, B.R., Idris, M.D., 2018. Anti-aging amino acids in *Euryale ferox* (Salisb.): a review. Adv Plants Agric Res 8 (1), 39–43. https://doi.org/10.15406/apar.2018.08.00289.

Jana, B.R., Srivastava, A., Idris, M., 2019. New makhana (*Euryale ferox* Salisb.) processed products for health benefit. J. Pharmacogn. Phytochem. 8 (2), 1662–1666.

Kapoor, S., Kaur, A., Kaur, R., Kumar, V., Choudhary, M., 2022. *Euryale ferox*, a prominent superfood: nutritional, pharmaceutical, and its economical importance. J. Food Biochem. 46 (12), e14435.

Kumar, A., Yadav, P.K., Paras, N., Nath, S., Kumari, U., 2016. *Euryale ferox* Salisb. Is an immense potential aquatic cash fruit crop: a review. Int. J. Adv. Agric. Sci. Technol. 3 (3), 35–49.

Kumar, L., Singh, A.K., Bhatt, B.P., 2016. Nutritional status of recently developed makhana (gorgon nut) variety-Swarna Vaidehi. J. AgriSearch 3 (4), 199–205.

Kumar, M., Shekhar, D., Kumari, A., Prasad, R.P., Bhatt, B.P., 2020. Makhana farming for maximizing farm income: a success story. Biotica Res. Today 2 (9), 962–965.

Kumar, A.V., 2022. A Novel Development of Allergen-free Nutrition Bar. Doctoral Diss., Department of Dairy Science and Food Technology, Institute of Agricultural Sciences. Banaras Hindu University.

Kumar, M., Goudar, G., Sharma, P., Vishwakarma, R., Gogoi, P., Mahajan, A., Tulja, B., 2023. Differentiating the nutrient composition, in-vitro starch digestibility, individual polyphenols and anti-oxidant properties of raw and popped makhana (*Euryale ferox*). J. Food Meas. Char. 17 (6), 5828–5844.

Kumar, A., Kumar, R., Chanchal, 2025. A study on culinary innovation with fox nut: a nutritional approach to healthy snacks. TMIMT Int. J. 1–15 (ISSN: 2348-988X).

Kumar, V., Kumar, A., Singh, M.K., Dhyani, P., Mishra, H., Rai, D.C., 2024. Bioactive metabolites identification of the foxnut and broken millet-based nutritional bar using HR-MS. Food Chem. Mol. Sci. 9, 100214.

Kanya, M., Pavani, P., Rakshit, M., Taliwal, C., Mekam, M., 2025. Development and evaluation of energy bar using makhana powder and dates. In: Revolutionizing Agri-Food Systems: Sustainability, Health, and Innovation. Springer Nature Switzerland, Cham, pp. 167–184.

Minten, B., Singh, K.M., Sutradhar, R.K., 2010. The Makhana Value Chain and the Fast Emergence of Branding in Food Retail: Evidence from Bihar (India). *Available at SSRN 2406566*.

Pasiakos, S.M., McLellan, T.M., Lieberman, H.R., 2015. The effects of protein supplements on muscle mass, strength, and aerobic and anaerobic power in healthy adults: a systematic review. Sports Med. 45 (1), 111–131.

Rao, S., Patil, A., Jadav, M., 2025. Therapeutic potential of makhana Churna and Sharkarayukta Dugdha: an ayurvedic perspective. J. Ayurveda Integr. Med. Sci. 10 (5), 198–202.

Rathod, R.V., Neve, G., Jain, A., Giri, P., 2023. A comprehensive review on health benefits and nutritional aspects of foxnut (Makhana). Pharm. Innov. 12 (6), 4432–4438.

Ros, E., 2010. Health benefits of nut consumption. Nutrients 2 (7), 652–682.

Sahithi, G., Anees Fathima Thabassum, Z., Madhavi, R., Shiva Kumara, C.S., Satish, A., 2022. Estimation of nutritional composition and sensory evaluation of fero nutri bars developed using *Euryale ferox* (makhana seeds). Int. J. Food Nutr. Sci. 11 (7), 774–785.

Singh, D.K., Kumar, A., Kumar, U., Singh, I.S., Chandra, N., Bhatt, B.P., 2020. Value chain analysis of makhana. Tech. Bull. No. 3.

Chapter 15

Packaging technologies and smart packaging for makhana-based products

Aastha Bhardwaj[1], Bisma Jan[2], Shazan Khan[1], Nitya Sharma[3], Praveen Kumar Dubey[4] and Tanweer Alam[1]

[1]*Indian Institute of Packaging (IIP), New Delhi, India;* [2]*Department of Biotechnology, IILM University, Greater Noida, Uttar Pradesh, India;* [3]*Food, Land and Water Program, WRI India, New Delhi, India;* [4]*Indian Institute of Packaging (IIP), Lucknow, Uttar Pradesh, India*

15.1 Introduction

Makhana a non-cereal food crop, also known as Fox Nut or *Euryale ferox*, is an important crop of the Nymphaeaceae family often consumed in popped and/or roasted form. Worldwide, it is known by various names such as "Gorgon nut" and "Prickly water lily", however, in the Indian subcontinent, it is commonly called "Makhana," and its seeds are typically consumed in popped or roasted form (Liaquat et al., 2022). Some researchers describe makhana seeds as "Black Diamonds" or "Black Gems of Wetlands" due to the distinctive black color of their outer seed coat (Liaquat et al., 2022). It is consumed either as a stand-alone snack product, or added to other food products as a nutritional enrichment ingredient. It is cultivated in the flood-prone regions of Bihar, Assam, and West Bengal in India and is known for its high nutritional value. India is the world's largest producer of Makhana, with Bihar accounting for about 80%–85% of global production. Improved farming techniques, conventional pond-based and field-based culture, and integrated Makhana-fish systems drove a 40%–50% increase in cultivated area (25,000–27,000 to 35,000–40,000 ha) and an almost twofold increase in production (32,000–35,000 to 60,000–63,000 MT) between 2020 and 2025 (MIC, 2025). The crop also gained the prestigious geographical indication (GI) tag (Mithila Makhana: 696) in year 2022 and is gaining export demand in markets like the US, Middle East, Japan, Canada, and Australia. Although India dominates makhana production, its contribution to global exports remains comparatively low with only about 1%–2% of the total production being

Makhana (Euryale ferox) as a Functional Food. https://doi.org/10.1016/B978-0-443-44421-0.00009-3

exported. With the increasing global demand for plant-based, gluten-free, and functional foods, India has a substantial opportunity to expand its presence and strengthen its position in the international makhana market. Consumers prefer products that are uniformly sized, properly processed, and attractively packaged, along with a rising preference for flavored and value-added varieties.

Makhana is considered a superfood due to its rich nutritional composition and also serves as an excellent source of protein for vegetarian diets (Liaquat et al., 2022). Nutritional composition of makhana seeds and popped makhana is given in Table 15.1 Fox nuts contain around 9.7% easily digestible protein, comprising all essential amino acids, with essential amino acid index of 89%–93%, and optimum arginine + lysine/proline ratio (4.74–7.6) (Singh et al., 2021). This makes them an excellent source of essential amino acids that the human body cannot synthesize and therefore must obtain through the diet. Makhana is considered a natural detoxifying agent and is traditionally used in managing chronic conditions such as kidney disorders, diarrhea, excessive

TABLE 15.1 Nutritional content of makhana seed and popped makhana.

Parameters	Makhana seed	Popped makhana
Carbohydrate (% by wt.)	57.0	79.8
Protein (% by wt.)	7.2	8.7
Fat (% by wt.)	0.3	0.5
Moisture (% by wt.)	34.7	10.4
Total ash (% by wt.)	0.3	0.4
Crude fiber (% by wt.)	0.5	0.2
Amylose (%)	19.0	18.2
Calorific value (K·cal/100g)	259	358
Phosphorus (mg/100g)	66.1	53.2
Pottasium (mg/100g)	35.6	42.0
Iron (mg/100g)	0.8	1.4
Calcium (mg/100g)	9.5	18.5
Magnesium (mg/100g)	11.3	13.9
Sodium (mg/100g	48.2	71.0
Copper (mg/100g)	0.3	0.5
Manganese (mg/100g)	0.9	1.3
Zinc (mg/100g)	0.9	1.1

leucorrhea, and spleen dysfunction (Vikram and Mishra, 2021). In addition to their nutritional value, makhana possess several medicinal properties that have been documented in the ancient literature of Chinese and Indian medicine. Beyond their macro and micronutrient composition, the presence of phenolic compounds and strong antioxidant activity contributes significantly to their therapeutic potential. Furthermore, these nuts are undoubtedly convenient and palatable snacks, and regular consumption of one serving per day has been associated with a reduced risk of developing chronic diseases such as cardiovascular disorders, type II diabetes, cancer, hypertension, and neurodegenerative diseases (Islam et al., 2023; Chang et al., 2016).

Popped makhana is highly hygroscopic in nature that is, it readily absorbs moisture from the environment. This often leads to loss of crispiness, soggy texture, and increased susceptibility to microbial growth (Islam et al., 2025). Therefore, optimum packaging is required that can guard against moisture, oxygen, and environmental contaminants in order to preserve the sensorial attributes of makhana and makhana-based products (Sharma et al., 2024). Additional challenges faced by the small-scale exporters concern meeting the sanitary and phytosanitary (SPS) requirements with importing nations. As a result, the product is often rejected in international markets due to inconsistencies in product quality and poor packaging. Through this chapter, authors have made an effort to explore the conventional packaging aspects of makhana and role of advanced packaging technologies in improving its safety, quality and shelf life until it reaches the end consumer, along with challenges and future trends in this direction.

15.2 Quality loss mechanisms in makhana

Makhana is a delicate food product whose inherent composition and outside environmental conditions of storage have a significant impact on its sensory quality. Its porous structure is ascribed to roasting and popping treatments, making it highly susceptible to humid conditions. Makhana tends to take up water vapor from atmosphere when exposed to high relative humidity, thereby losing its crispiness and becoming softer in texture. The imbibed moisture acts as a plasticizer that makes the starch-protein matrix lose its rigidity and stiffness. Another theory suggests that the makhana loses its residual moisture under extreme dry conditions, making it brittle and vulnerable to breakage or powdering. Moisture sorption isotherms play an important role in understanding the hygroscopic behavior of makhana and its susceptibility to quality degradation. Since makhana partly resembles the category of puffed cereal-based snacks, popcorn and extruded products, it may exhibit a Type II sigmoidal sorption isotherm that describes the relationship between its equilibrium moisture content (EMC) and surrounding relative humidity at constant temperature. The sigmoidal curve can be divided into three prominent regions; the first region is the low water activity region, where water is strongly bound to specific sites within the food matrix. The second region is the intermediate water activity region where water molecules accumulate in layers leading to an increase in moisture content and softening of its texture. The third region is high water activity region, characterized by sharp increase in moisture uptake which is attributed to capillary condensation within the porous structure. This stage leads to loss of texture and enhanced risk of microbial growth in makhana. The critical water activity for maintaining optimum crispiness generally lies between 0.4 and 0.5; once exceeded weakens the starch-protein matrix causing characteristic loss of its crunch. Mathematical models for example, Guggenheim-Anderson-de Boer (GAB) which has been widely used for cereal-based snack products, may also be applied to makhana and makhana-based products to estimate critical moisture content and understand moisture uptake behavior. Additionally, empirical validation must also be done through storage observation studies that is, by monitoring texture properties at different RH levels. Thus, from a packaging perspective, sorption isotherms are essential for designing packaging systems that cater to effective moisture barrier.

15.3 Packaging requirements for makhana and makhana-based products

Packaging is an amalgamation of art, science, and technology intended to ensure product performance, preserve purity, and reliability of the supply chain. It involves preparing products for transit, distribution, storage, marketing, and consumption. Appropriate packaging also helps minimize waste while preserving the product's intended quality (Dani et al., 2024). As evident from recent studies, packaging trends have been progressively evolving to match consumer preferences. Resealable closure systems, recyclable materials, and sustainable and biodegradable materials are being encouraged. Consumers are seeking snack items with prolonged freshness and shelf life while also being environmentally sustainable. Also, there is a growing demand for portion controlled formats such as single serve pouches or single serve mini cups, common among students, working professionals and fitness enthusiasts. For makhana and its products, packaging materials must cater to specific functional and regulatory compliances to ensure its safety, stability, and shelf life. The material must be affordable, safe, non-toxic, and provide barrier protection against all environmental influences. Additionally, packaging must be durable and strong enough to withstand wear and tear during transit, handling and distribution.

It should also be compatible with packaging machinery for efficient processing and sealing. Oxygen in the presence of moisture can cause oxidative deterioration and quality loss especially in roasted and flavored makhana hence, low oxygen levels in packaged products can be achieved through several methods, including vacuum packaging, flushing the package with inert gases such as nitrogen (N_2) or carbon dioxide (CO_2), or incorporating oxygen absorbers along with packaging materials that have a low oxygen transmission rate (OTR) (Cummins and Lyng, 2016). Similarly, water vapor barrier properties are crucial in the packaging makhana and makhana-based products because these products are generally dry and their quality largely depends on maintaining crispness. Therefore, packaging materials with high resistance to water vapor transmission are required to prevent moisture absorption. In some cases, moisture-absorbing sachets containing desiccants are included in packages to remove excess humidity.

Packaging of popped makhana is generally carried out in bulk sacks of approximately 8–10 kg, which are mainly used for wholesale distribution. For direct consumer sales, the product is also packed in smaller retail packages, typically around 250 g, which are marketed through local markets as well as modern retail outlets. According to reports from Agricultural and Processed Food Products Export Development Authority (APEDA), a farmer's share in the consumer rupee for makhana is estimated to be approximately 55%, suggesting that a considerable portion of the final retail price is absorbed by intermediaries, processors, and retailers along the value chain. However, this share may vary depending on the structure and efficiency of the supply chain (Min et al., 2010).

15.4 Packaging materials used for makhana products

Packaging of makhana and its products is primarily carried out to protect the product from external environmental factors, particularly air and moisture. Proper packaging aids in maintaining aroma and freshness of makhana for comparatively longer duration. Storage life of makhana and makhana-based products mainly depends on the type and quality of packaging material used (Kashyap and Shukla, 2025). Additionally, by employing eco-friendly procedures that reduce the consumption of packaging films and use sustainable packaging materials, snack processors are investigating innovative ways to minimize carbon impact. Snacks like makhana must be protected from physical damage, textural alterations, and flavor degradation by using effective packaging materials. Selection of packaging depends on the quality attributes of snack and its characteristics, like water activity, moisture, and fat content (Serna-Saldivar, 2022). Packaging materials for makhana are as below (Fig. 15.1).

15.4.1 Polyethylene (PE)

Polyethylene is considered as most commonly packaging material for dry food products because it has good moisture barrier properties, durability, and flexibility. Makhana is highly hygroscopic and can easily absorb moisture from surroundings. Since PE packaging offers moisture barrier properties it therefore maintains crispness, texture, and shelf life of makhana. Among all PE materials, low-density polyethylene (LDPE) is commonly used for packaging makhana due to its good moisture barrier properties, flexibility, sealing properties which help to maintain the crispness and other sensory quality attributes of the product. Another commonly used packaging material is linear low-density polyethylene (LLDPE) particularly in laminated packaging structures such as PET/LLDPE or BOPP/PE. This kind of packaging material provides good puncture resistance and stronger sealing characteristics. Additionally, high-density polyethylene (HDPE) is mainly used for bulk storage bags or liners rather than small retail packs. Overall, LDPE and LLDPE are the most preferred PE materials for makhana packaging because they protect the product from moisture absorption and help extend its shelf life. Recently, effect of storage period and packaging materials on the popping behavior of roasted makhana seeds was evaluated by storing roasted seeds in different materials such as LDPE, polypropylene (PP), biaxially oriented polypropylene (BOPP), and silver laminate for 6 months. The results showed that thicker LDPE (80 μm), BOPP, PP, and silver laminate packaging effectively maintained popping efficiency, expansion ratio, and recovery of popped makhana for up to about 135 days, whereas 15 μm LDPE maintained quality only up to 60 days. Overall, roasted makhana seeds stored at about 15.5% moisture content in airtight packaging could be popped effectively with acceptable color and quality for approximately **120 days** (Sharma et al., 2024).

15.4.2 Polyethylene terephtalate

Polyethylene terephthalate (PET) is a type of polyester characterized by high tensile strength, which is achieved after orientation. It exhibits good chemical resistance, is lightweight, flexible, and maintains stability over a wide range of temperatures. In case of makhana PET also provides effective barrier properties against gases, odors, and lipids, although its permeability to water vapor is relatively higher compared to some other packaging materials (Awulachew, 2022). This type of packaging material is used popped makhana and its products packaging because of excellent odor free and gas barrier properties. Blow molded bottles made up of PET or PVC is used.

15.4.3 Tin cans and composite cans

Tin cans are widely used for packaging of makhana in customized sizes and capacities depending on market requirements. These cans are lined with a protective coating that avoids direct contact with the metal, therefore minimizing the risk of corrosion and undesirable chemical reactions that could compromise product quality. Lacquering also helps to maintain aroma, and flavor of the product during storage. Tin cans also offer advantages like excellent mechanical strength, durability during transportation, and good barrier properties against moisture, air, and light. Additionally, tin cans are resistant to physical damage, making them ideal for long-term storage. Tin cans are recyclable and can be reused after processing, thereby promoting environmental sustainability (Deshwal and Panjagari, 2020). Apart from providing excellent barrier protection, **tinplate** also provide good thermal stability, which makes it withstand high temperature hermetic sealing. Lastly, tinplate offer high ductility and formability, allowing it to be molded into containers of various sizes and designs for different packaging applications (Sarkar and Aparna, 2020).

Another form of rigid packaging for makhana are the composite cans or containers that are comprised of paperboard, aluminum foil, and plastic. These layers impart durability, strength as well as barrier to the product stored inside—all of which have a significant impact on the shelf life of popped and roasted makhana.

15.4.4 Glass bottles

Glass bottles and jars are commonly used for packaging makhana in retail quantities, ranging from 50 to 250 g. They offer excellent protection against moisture, gases, and contaminants, thereby, maintaining freshness of the product. Glass is non-reactive and chemically inert which ensures that the taste, aroma, and nutritional value of makhana remain unchanged. The transparency of glass containers allows easy visualization of the product. However, glass based materials are not widely used for makhana because they are prone to breakage during transportation and handling.

15.4.5 Lined cartons

Another packaging material used for makhana and makhana-based products is lined cartons or stand-alone paper. These packs are present in different sizes and shapes making them ideal for different retail quantities and marketing requirements. Flexible pouches made from laminated materials or aluminum foil is kept inside the carton to give an extra resistance against light, moisture, and oxygen hence, aiding in retaining the freshness and other quality attributes of the product. Inner flexible packaging and outer carton provides excellent protection, easy handling, and enhanced shelf life. Lastly, lined cartons are considered appealing to consumers and comparatively economical packaging options, making them popular for commercial distribution and retail marketing.

15.4.6 Flexible films/pouches/laminates

Flexible films, laminates and pouches are widely used packaging materials made from laminated or multilayer films. Lamination is the process of bonding two or more layers of films together to increase barrier properties, appearance, and tensile strength of the packaging. These packages are present in varied forms, like pillow pouches or self-standing pouches, which offer convenience and good visual appeal. Moreover, flexible packaging systems are considered economical when compared to other types of packaging materials. Selection of a suitable laminate largely depends on factors such as the compatibility of the food contact layer, heat-sealing ability, heat-sealing strength, and the shelf life requirements of the packaged product (Wagne, 2016).

15.4.7 Aluminum

Aluminum foil is most versatile kind of material used in laminated packaging and has broad applications in the packaging of food products, including makhana. This kind of packaging material offers excellent barrier properties against oxygen, other gases, volatile aromas, moisture, and light, making it better than other plastic laminate materials. These exceptionally strong barrier characteristics makes aluminum foil ideal for products like makhana, which significantly requires protection from moisture and oxidation to preserve its crispy texture and other quality parameters during storage. So, aluminum foil is usually incorporated into laminates when relatively longer shelf-life stability is required. Even though aluminum-coated plastic laminates can also be used as an alternative, their barrier properties are generally lower than that of aluminum foil laminates. In addition, aluminum foil is available in rigid, semi-rigid, and flexible packaging forms that makes them withstand thermal processing with minimal heat damage to the product. For such applications, seal strength, along with good mechanical stability, is very important. Stability of aluminum foil when in direct contact with food largely depends on the chemical composition of the product. Aluminum used in food packaging is usually considered safe, although in few cases an inner protective coating on the foil is advised (Kerry, 2012). The relationship between key physico-chemical properties of makhana, packaging requirements, and suitable packaging materials used to meet these requirements are summarized in Table 15.2, Fig. 15.1.

15.5 Modified atmospheric packaging and active packaging for makhana and makhana-based products

Modified atmospheric packaging (MAP) has evolved as crucial technology in food preservation and processing and plays an important role in minimizing food waste, which accounts for nearly 30%–40% of total global food production. In MAP, composition of gases such as nitrogen, carbon dioxide, and oxygen are changed within the package, thereby retarding microbial growth, oxidative reactions, and sensory deterioration in food products. As a result, the storage life of commodities can be significantly extended. As per many

shelf life but also for enhancing sustainability and reducing losses across a wide range of food systems. Aseptic packaging can also be used to take advantage of high-temperature processing, extend the shelf life of food products at normal storage temperatures, and ensure effective sterilization within the package. However, no literature study exists related to aseptic packaging of makhana as a whole. Infact, a majority of research focus on moisture barrier packaging for makhana-based instant mixes, snacks, bakery products, beverages, and ready-to-eat formulations (Kapoor et al., 2024; Saxena et al., 2022; Mridula et al., 2024).

Active Packaging (AP), that has been widely used for dry fruits, may be effectively correlated and adapted for makhana due to similarities in their water activity, sensitivity to RH, and oxidative deterioration, especially in flavored variants such as roasted, buttered, or spiced makhana. Though they structurally differ, the mechanisms of quality loss in makhana and dry fruits are comparable. Oxygen is one of the critical parameters; oxygen scavengers are used in dry fruits like almonds and cashews to prevent lipid oxidation. Thus, shelf life of makhana and its flavored variants may be enhanced by utilizing nitrogen flushing and adding oxygen scavengers to maintain its sensory properties. Antioxidant-based active packaging may also be used in this direction to delay rancidification reactions in makhana variants. Antioxidants added in the form of sachet formats or directly to packaging materials can help reduce lipid oxidation and stabilize the shelf life, particularly for export grade makhana. Likewise, moisture is another crucial variable that influences the textural properties of makhana just like the dry fruits. As previously mentioned, Makhana is highly hygroscopic, thus loses its crispiness even with a small amount of moisture absorption. Therefore, use of moisture absorbers (such as silica gel sachets) in bulk or multi-serve makhana packages is utmost essential.

Hence, from the standpoint of packaging design, an integrated active packaging systems combining dessicants, high moisture barrier materials (e.g., metalized laminates), and oxygen scavengers for flavored or oil coated makhana shall be appropriate. This shall tackle the very hygroscopic character of makhana while addressing the quality deterioration mechanisms effectively.

15.6 Smart Packaging for makhana

Although direct studies related to smart and intelligent packaging of makhana is very limited, the concept can be applied to products such as makhana because it is a quality sensitive snack product. Makhana shares many traits with other dry snack foods like potato chips, popcorn, and roasted nuts that have been extensively researched in the context of intelligent and active packaging systems. Therefore, smart packaging technologies developed for snack foods may also be applicable to makhana packaging. In order to monitor or enhance food quality during storage and distribution, smart packaging technologies integrate intelligent and active systems. Sensors, indicators, RFID tags, and oxygen scavengers are examples of smart packaging technologies that can give real-time information regarding the safety and freshness of products by accurately monitoring changes occurring within the packaged food product.

Smart packaging systems are broadly classified into two main categories. **Intelligent packaging** combines conventional packaging materials with advanced electronic sensing technologies, such as sensors and indicators, to monitor changes in food quality and safety during storage and distribution. These systems enable detection of variations in temperature, gas composition, or microbial activity, thereby ensuring safety and acceptability of the food until it reaches the end consumer. In contrast, **active packaging** involves the incorporation of functional components, such as antioxidants, antimicrobial agents, or oxygen scavengers, into conventional packaging materials. These active components interact with the packaged food or its surrounding environment (headspace) to enhance product stability, extend shelf life, and maintain overall quality during storage (Thirupathi Vasuki et al., 2023).

Numerous intelligent packaging methods have been presented in snack packaging research to track food quality and storage conditions. In nitrogen-flushed snack packaging, for instance, oxygen indicators and gas sensors are utilized to identify oxygen entrance or package leakage, preserving product freshness and preventing oxidative deterioration. Another crucial smart packaging element is humidity indicators, which are especially useful for dry snacks since they can identify increase in relative humidity within the container that could cause microbial deterioration and crispness loss. Integrating humidity sensors or moisture indicators into makhana packaging could offer real-time information on product stability because moisture absorption is the main factor causing quality deterioration in makhana. Additionally, snack packaging has employed smart labels using radio-frequency identification (RFID), near-field communication (NFC), and QR codes to improve consumer engagement, supply chain monitoring, and product traceability (Biji et al., 2015). These technologies could be used to support traceability in makhana packaging, particularly in export markets where quality control and transparency are becoming more and more crucial.

In view of the recent advancements in snack packaging research, these options may also be applicable to makhana and related products. Although not validated for makhana, their successful application in snack packaging hints at strong potential to enhance shelf life and facilitate real time quality monitoring of makhana. Given the hygroscopic nature of makhana and its susceptibility to moist conditions, the integration of humidity sensors, oxygen sensors, freshness indicators and IoT enabled systems could significantly improve its shelf life.

15.7 Labeling aspects of makhana packaging

Labeling is a critical component of food packaging that asserts safety and quality of the food and ensured its compliance as per regulations. In India, Food Safety and Standards Authority of India (FSSAI) oversees food labeling regulations under the Food Safety and Standards (Labeling and Display) Regulations, 2020. For packaged makhana, packaging is utmost essential to communicate nutritional information, product storage conditions, product safety aspects, in addition to looking aesthetically pleasing. As per FSSAI, all pre-packaged foods must carry labels with these mandatory details, typically in English or Hindi or other languages. Labels must be clear, prominent, and legible. In case of exports to other countries, compliance requirements extend to APEDA registration, SPS certification, HACCP or ISO 22000, etc. Current industry practices rely on moisture-proof laminated pouches, vacuum-sealed barrier films, and corrugated shipping containers to preserve crispness and mitigate contamination risk during the supply chain (Kapoor et al., 2022; Sharma et al., 2024; Kumar et al., 2024).

One of the most important requirements is the correct name of the food that must also indicate the nature of the product such as "Roasted Makhana" or "Makhana Snack (Peri-Peri/Masala etc.). For plain Makhana, this may be simple but for all flavored variants, all added components namely edible oil, salt, spices, falvaoring agents and/or preservatives, if any, must be clearly listed. The nutritional labeling, net quantity declaration, batch number/lot number/code number, date of manufacture/packing, name and complete address of the manufacturer/packer/importer, FSSAI license number, storage instructions, allergen declaration (in case of flavored makhana), vegetarian logo (green dot), barcode/QR code (for traceability), and certifications (e.g., organic, if applicable) must be mentioned on the packaging to fulfill regulatory requirements while also addressing product specific needs. It may carry nutritional or health claims such as "Gluten free" or "Low fat", however, such claims must be scientifically substantiated to avoid misleading the consumer.

In order to enable standardized packaging for export quality makhana, APEDA and Indian Institute of Packaging (IIP) developed a packaging prototype design for "Mithila Makhana," with Geographical Indication (GI) tag (696). The primary packaging includes a polymeric monomaterial liner with an outer corrugated fiberboard box (CFB). The label design integrates product branding, consumer communication regulatory information and aesthetics. Fig. 15.2 shows the design label for developed makhana packaging box. The front panel clearly shows:

- Product name: Mithila Makhana - Fox Nut
- Product visuals: Makhana bowl and a backdrop of an aquatic setting
- Color combination: Yellow and earthy tones
- Net amount: 200 g

FIGURE 15.2 Prototype label design for Mithila makhana packaging developed by IIP for APEDA.

- Tagline: Healthy-Crispy-Tasty
- Logo: APEDA
- QR code: To scan and get more details

The back panel comprises of a table with nutrition facts that are mandatory to declare under FSSAI regulations and help consumers know about the nutritional profile of the makhana. FSSAI logo along with FSSAI license number denotes regulatory conformity with Indian food safety regulations. The back panel also includes the GI tag no. 696, batch number, date of packaging, maximum retail price (MRP) and best before date to ensure transparency about product's shelf life. The label also mentions storage conditions—Keep in cool, dry & hygienic place. This is important as makhana is a highly hygroscopic material and is susceptible to loss of crispiness and freshness when exposed to moist conditions. The label also includes manufacturer/packer details, address information, customer care contact details and email contact for consumer queries. 1-D barcodes are commonly used for retail and logistics purposes. Thus, the packaging label promotes Mithila Makhana as a desirable snack with eye-catching aesthetics, in addition to providing information about the product and conforming to regulations (FSSAI, 2018) ensuring product safety.

15.8 Research gaps

Besides technological advancements in food packaging techniques, a number of research gaps pertaining to safety of packaging, sustainability, and functionality of packaging for makhana exists. One major issue is the migration/leaching of undesirable packaging components such as additives, chemicals etc. into food. Very few reliable techniques for measuring these migrations are available, particularly in the case of complex food matrices. Thus, there is an urgent need to concentrate on development of affordable and scalable technologies along with bringing concurrence between intelligent packaging technologies and regulatory frameworks in makhana packaging sector (Gupta et al., 2024; Wang et al., 2025). Some of the research gaps are mentioned as follows:

15.8.1 Limited research available on makhana-specific packaging systems

Limited research studies are available that have evaluated the performance of packaging materials during makhana storage. For instance, (Sharma et al., 2024), demonstrated the effect of various packaging materials with optimum barrier properties for maintaining makhana quality. However, it focused mainly on makhana storage behavior rather than on the optimization of packaging technology used. Being porous in nature, makhana may interact with the packaging material, particularly polymer-based films and multi-layer laminates, leading to migration of specific packaging components which presents another significant research gap. Further to this, insufficient systematic studies examining migration aspects for makhana packaging is available. In India, flexible pouches, owing to their good moisture barrier and convenient handling properties, currently dominate the makhana packaging market. Nevertheless, these pouches are adapted for general snack packaging rather than specifically for makhana. Thus, future research efforts should focus on the development and design of packaging systems customized to physicochemical and hygroscopic properties of makhana, thereby ensuring product quality and safety. Sustainable materials, biodegradable films, and modified barrier materials may aid in improving the quality and shelf stability of makhana and related products, while lowering overall environmental impacts.

15.8.2 High cost of biodegradable and sustainable materials

Conventional petroleum-based plastics have undoubtedly long dominated food packaging industry due to their versatility, low cost and excellent mechanical and barrier properties. Owing to growing concerns over plastic waste and long term ecological damage, food packaging sector is gradually transitioning to biodegradable and sustainable packaging materials (Marsh and Bugusu, 2007; Siracusa et al., 2008). In comparison to conventional plastics, biopolymers including polylactic acid (PLA), polyhydroxyalkanoates (PHA), starch-based polymers, and cellulose derivatives frequently require more expensive raw materials, intricate processing techniques, and smaller manufacturing volumes. Together, these elements raise the total cost of production, which discourages food producers and processors from using them. For instance, because PLA and PHA materials require fermentation and purification during manufacturing; their production costs can be many times greater than those of polyethylene or polypropylene. Small and medium-sized businesses (MSMEs), which control the majority of the makhana processing and packaging industry in India, are most affected by this cost barrier. To stay competitive in both domestic and international markets, many MSMEs rely on inexpensive packaging materials and operate with little capital expenditure. Adoption of biodegradable packaging frequently necessitates further expenditures in new equipment, bio-based material supply networks, and adherence to changing sustainability regulations. As a result, despite their negative effects on the environment, many processors still rely on low-cost multilayer plastic laminates.

15.8.3 Lack of LCA studies

The scarcity of a comprehensive life cycle assessment (LCA) assessing environmental impacts of various packaging materials (including harvest of raw material, production, transportation, consumption until disposal/recycling) used for makhana is another research gap in makhana packaging industry. Current literature elaborates on the functional performance of packaging materials, such as moisture barrier qualities, shelf-life extension, and storage stability. While these factors certainly are important for preserving food quality, they seldom explain how employing packaging technologies impact the environment. For example, conventional multilayer plastic laminates that are commonly used for snacks packaging have good barrier attributes; difficulty arises during their collection, sorting and recycling (Siracusa et al., 2014). In absence of thorough environmental assessment, researchers find it challenging to conduct reliable LCA and address specific impacts.

LCA has been extensively used to evaluate food packaging solutions including dairy, meat and beverages. The studies demonstrate well how much energy and resources are consumed throughout the supply chain from extraction to end-of-life disposal. For example, flexible materials consume lesser material and energy during production in comparison to rigid materials; however, owing to its low recyclability rates, it presents more waste management problems (Williams et al., 2018). Likewise, the overall sustainability of biodegradable materials is dependent on sourcing of raw materials, availability of infrastructure for composting or degradation, amount of energy consumed etc. Currently no research demonstrates the LCA of makhana packaging. Thus, policymakers and industry stakeholders are unable to make well-informed judgments about sustainable packaging solutions. Replacing biodegradable materials with traditional plastics may lessen plastic waste burden, but their overall environmental value may be diminished if these substitutes require greater energy inputs during production. In order to assess trade-offs between material performance, cost, and environmental impact, LCA's are crucial. LCA findings may also facilitate development and laying down evidence-based regulations promoting eco-friendly packaging solutions in the makhana sector.

15.8.4 Lack of technical know-how & limited research & development (R & D)

The quantification of non-intentionally added stimulants (NIAS) is one of the most difficult hurdles to deal with in the modern food packaging industry. NIAS may arise in packaging from recycled plastics and adhesives, impurities in raw materials, polymer degradation or interaction between food and material. Advanced analytical techniques such as gas chromatography—mass spectrometry (GC-MS) and liquid chromatography—mass spectrometry (LC-MS) can detect and quantify NIAS even in trace amounts. However, there is still uncertainty in safety assessments because globally optimized methods for quantifying NIAS are still evolving. Regulatory compliance brings another challenge. In India, Food Safety and Standards Authority of India (FSSAI) oversees food packaging safety under the Food Safety and Standards (Packaging) Regulations, 2018. These regulations specify approved list of packaging materials, specific as well as overall migration limits as well as testing procedures clearly. Migration testing often uses food stimulants to mimic properties of different kinds of foods. While these are suitable for standardized testing, they may fail to represent the behavior of complex food matrices like that of makhana, resulting in an inaccurate detection result.

This raises serious concerns about repurcussions of product interaction with the packaging materials during prolonged storage and transit. Further to this, MSMEs and small businesses lack the technical know-how needed to comprehend and implement the regulatory requirements. Interdisciplinary expertise in food science, material science and engineering, and polymer chemistry, and knowledge of regulatory frameworks is necessary. In the absence of sufficient capacity building programs, industry-academia partnership, and domestic R & D activities, it would be difficult to cater to the gaps. In this direction, technology adoption must be greatly increased by creating specialized packaging research laboratories, encouraging industry-academia cooperation, hand-holding training programs, creating pilot-scale packaging research facilities, and obtaining funding through government initiatives. Future developments are increasingly focused on AI enabled intelligent packaging systems and sustainability. Emerging trends highlight the use of machine learning algorithms integrated with sensors to enable real time detection of contamination and spoilage as well as shelf-life predictions with over 90% accuracy (Alrashdi and Taloba, 2026). Also, there is a growing shift toward indigenous biodegradable polymers namely shellac and algae-based films for food products (Sagar and Rani, 2026). At the same time, the makhana packaging sector is adopting digital traceability systems including QR codes and block chain technologies that may further ensure transparency and facilitate meeting international export standards. Together, these technological innovations and advancements are revolutionizing packaging from passive to an active aspect of digital food ecosystem.

15.9 Roadmap for advancing makhana packaging

To close existing gaps in scientific research related to packaging of makhana and makhana-based products, a well-organized roadmap is the need of the hour. This roadmap must include short and long term priorities, catering to makhana's extreme sensitivity to moisture and its widespread commercialization. In the short term, the focus must be on methodical packaging material identification and evaluation; in absence of scientific data on moisture sorption isotherms and critical water activity, results on standardized packaging design for makhana is still lacking. Some studies revealed that high barrier materials like BOPP, PP, and laminated structures can extend the shelf life and stability of makhana up to 120–135 days, however LDPE can only preserve product quality till 60 days. Further research is required to establish critical moisture limits specific to makhana and controlled shelf life studies along with WVTR based material comparisons must also be included in the short term study goal. Furthermore, the focus must also be on integration of smart and active packaging technology with optimized use of desiccants and oxygen scavengers in makhana packaging. Simultaneously, adherence to food safety regulations needs to be strengthened via consistent labeling, traceability and quality control.

Long term priorities must focus on eco-friendly sustainable packaging solutions for makhana packaging. The development of biodegradable or recyclable high barrier materials appropriate for low moisture snacks like makhana is necessary in light of growing environmental concerns. Standardized packaging procedures, such as recommended materials, storage conditions, and shelf-life benchmarks, will also be essential for guaranteeing uniformity, particularly for export markets. Institutional support from R & D organizations can facilitate this transition through technology development and industry requirements. To sum up, developing makhana packaging necessitates a sequential strategy that incorporates scientific findings into a coherent framework for high barrier packaging.

15.10 Conclusion

The packaging of makhana and makhana-based products has gradually transitioned from conventional wrappers to more sophisticated and advanced systems. Owing to its highly hygroscopic and porous nature, selection of optimized packaging material is much critical to preserve its nutritional and sensorial properties including texture, taste and smell as well as maintaining its shelf life. Traditional polymeric materials such as PE, PP, PET and multilayered laminates exhibit excellent barrier properties however; growing concerns over plastic wastes undermine their role. The chapter highlights the significance of chemical migration in foods through food packaging materials, and the need for advanced analytical technologies for their accurate detection and quantification. Looking at the emerging trends in packaging, future advancements are expected to involve incorporation of intelligent packaging and AI driven predictive systems to ensure product quality throughout the supply chain. Development of sustainable, biodegradable and eco-friendly packaging materials align well with the circular economy goals as well as modern consumer demands. To strengthen the export competitiveness of makhana and its products, QR codes and blockchain technology may be utilized like other food products. In a nutshell, combining interdisciplinary collaboration strategy, innovation in materials, and upgradation of regulatory frameworks are essential to drive and shape the future of makhana packaging.

References

Alrashdi, I., Taloba, A.I., 2026. AI-driven smart packaging: enhancing sustainability and cybersecurity in food supply chains. Front. Sustain. Food Syst. 9. https://doi.org/10.3389/fsufs.2025.1712080, 2571–581X.

Awulachew, M.T., 2022. A review of food packaging materials and active packaging system. Int. J. Health Policy Plan 1 (1).

Biji, K.B., Ravishankar, C.N., Mohan, C.O., Srinivasa Gopal, T.K., 2015. Smart packaging systems for food applications: a review. J. Food Sci. Technol. 52 (10), 6125–6135. https://doi.org/10.1007/s13197-015-1766-7, 0022–1155.

Chang, S.K., Alasalvar, C., Bolling, B.W., Shahidi, F., 2016. Nuts and their co-products: the impact of processing (roasting) on phenolics, bioavailability, and health benefits - a comprehensive review. J. Funct.Foods 26, 88–122. https://doi.org/10.1016/j.jff.2016.06.029.

Cummins, E.J., Lyng, J.G., 2016. Packaging systems and materials used for meat products with particular emphasis on the use of oxygen scavenging systems. In: Emerging Technologies in Meat Processing: Production, Processing and Technology. Wiley, pp. 231–263. https://doi.org/10.1002/9781118350676.ch9.

Dani, R., Tiwari, K., Prabhakar, A.P.B., 2024. A review of food packaging materials and its impact on environment. AIP Conf. Proc. 2978 (1). https://doi.org/10.1063/5.0182889.

Deshwal, G.K., Panjagari, N.R., 2020. Review on metal packaging: materials, forms, food applications, safety and recyclability. J. Food Sci. Technol. 57 (7), 2377–2392. https://doi.org/10.1007/s13197-019-04172-z.

FSSAI, 2018. Food Safety and Standards (Packaging) Regulations. Food Safety and Standards Authority of India.

Gupta, R.K., Pipliya, S., Karunanithi, S., Eswaran U, G.M., Kumar, S., Mandliya, S., Srivastav, P. P., Suthar, T., Shaikh, A.M., Harsányi, E., Kovács, B., 2024. Migration of chemical compounds from packaging materials into packaged foods: interaction, mechanism, assessment, and regulations. Foods 13 (19). https://doi.org/10.3390/foods13193125.

Habiba, U., Bajpai, A., Shafi, Z., Pandey, V.K., Singh, R., 2025. Advancing sustainability through modified atmosphere packaging (MAP) for fresh food preservation: a critical review. J. Stored Prod. Res. 112. https://doi.org/10.1016/j.jspr.2025.102657.

Islam, Z., Bashir, K., Shaida, B., Siddiqui, S., 2023. A comprehensive examination of the nutritional value and health advantages of makhana (*Euryale ferox*). J Sci Tech Res 13 (2).

Islam, Z., Bashir, K., Singh, S., Rastogi, M., Siddiqui, S., Punia Bangar, S., 2025. Structural and functional properties of starch from popped seeds of makhana (*Euryale ferox*). Starch - Stärke 77 (7). https://doi.org/10.1002/star.70030, 0038–9056.

Kapoor, S., Kaur, A., Kaur, R., Kumar, V., Choudhary, M., 2022. Euryale ferox, a prominent superfood: nutritional, pharmaceutical, and its economical importance. J. Food Biochem. 46 (12). https://doi.org/10.1111/jfbc.14435.

Kapoor, S., Kaur, A., Kumar, V., Choudhary, M., 2024. Chemical, textural, antioxidant potential and sensory characterization of Foxnut powder (*Euryale ferox*) added bakery products. Nutr. Food Sci. 54 (3), 547–561. https://doi.org/10.1108/nfs-03-2023-0047, 0034–6659.

Kashyap, A., Ji Shukla, O., 2025. Sustainable food supply chain: exploration, identification, and analysis of the critical drivers for the foxnut (Makhana) industry. J. Glob. Oper. Strateg. Sourc. 18 (3), 559–582. https://doi.org/10.1108/jgoss-05-2023-0042, 2398–5364.

Kerry, J., 2012. Aluminium Foil Packaging. Elsevier BV, pp. 163–177. https://doi.org/10.1533/9780857095701.2.163.

Kumar, V., Kumar, A., Kumar Singh, M., Dhyani, P., Mishra, H., Chandra Rai, D., 2024. Bioactive metabolites identification of the foxnut and broken millet-based nutritional bar using HR-MS. Food Chem. Mole Sci 9. https://doi.org/10.1016/j.fochms.2024.100214.

Liaquat, M., Pasha, I., Ahsin, M., Salik, A., fox nuts, R., Euryale, F.L., 2022. Contain higher concentration of phenolics, flavonoids, minerals and antioxidants, and exhibit lower Glycemic Index (GI) in human subjects. Food Prod Process Nutr 4 (1). https://doi.org/10.1186/s43014-021-00081-x, 2661–8974.

Market Intelligence Cell (MIC) for APEDA, Monthly dashboard –Makhana (Fox-Nut) Sep-2025, https://apeda.gov.in/sites/default/files/2025-10/MIC_Monthly_dashboard_Makhana_30102025.pdf.

Marsh, K., Bugusu, B., 2007. Food packaging - roles, materials, and environmental issues. J. Food Sci. 72 (3), R39–R55.

Min, S.C., Kim, Y.T., Han, J.H., 2010. 18 Packaging and the shelf life of cereals and snack foods. Food Packag. Shelf Life 339.

Mridula, D., Vishwakarma, R.K., Arora, S., Bala, M., 2024. Formulation and characterization of novel cereal gluten-free pasta from semi-popped makhana, water chestnut, and potato. Food Bioprocess Technol. 17 (11), 4262–4277. https://doi.org/10.1007/s11947-024-03349-5, 1935–5130.

Sagar, N.A., Rani, N., 2026. Recent trends and innovations in smart and AI-based food packaging: a review. Front. Food Sci. Technol. 5. https://doi.org/10.3389/frfst.2025.1665055.

Sarkar, S., Aparna, K., 2020. Food packaging and storage. Res Trends Home Sci Ext 3, 27–51.

Saxena, D., Hussain, I., Singh, S., Kumar, S., Garg, N., 2022. Optimization and storage study of rabri enriched with date syrup and makhana. Trends Sci. 19 (9), 3669. https://doi.org/10.48048/tis.2022.3669, 2774–0226.

Serna-Saldivar, S.O., 2022. Packaging of snack foods. In: Snack Foods. CRC Press, pp. 83–112.

Sharma, K., Patel, S., Devi, M., Vishwakarma, R.K., 2024. Effect of packaging material and storage period on popping behaviour of roasted makhana seeds. J. Agric. Eng. 61 (4). https://doi.org/10.52151/jae2024614.1870, 0976–2418.

Singh, D.K., Kumar, A., Kumar, U., Singh, I.S., 2021. Production and profitability of makhana (*Euryale ferox* salisb.) in aquatic ecosystem of North Bihar, India. Indian J Ext Educ 57 (2), 62–66. https://doi.org/10.5958/2454-552x.2021.00056.6, 0537–1996.

Siracusa, V., Rocculi, P., Romani, S., Rosa, M.D., 2008. Biodegradable polymers for food packaging: a review. Trends Food Sci. Technol. 19 (12), 634–643. https://doi.org/10.1016/j.tifs.2008.07.003, 09242244.

Siracusa, V., Ingrao, C., Lo Giudice, A., Mbohwa, C., Dalla Rosa, M., 2014. Environmental assessment of a multilayer polymer bag for food packaging and preservation: an LCA approach. Food Res. Int. 62, 151–161. https://doi.org/10.1016/j.foodres.2014.02.010.

Thirupathi Vasuki, M., Kadirvel, V., Pejavara Narayana, G., 2023. Smart packaging—an overview of concepts and applications in various food industries. Food Bioeng 2 (1), 25–41. https://doi.org/10.1002/fbe2.12038, 2770–2081.

Vikram, D.V., Mishra, D.S., 2021. A review on Euryale Ferox: daily dietary regime for Covid -19 patients. Int J Appl Res 7 (6S), 29–34. https://doi.org/10.22271/allresearch.2021.v7.i6sa.8607, 23947500.

Wagner, J.,J.R. (Ed.), 2016. Multilayer Flexible Packaging. William Andrew.

Wang, J., Liang, X., Hou, H., Deng, J., Lin, Q., Wen, L., Bai, J., 2025. A review on migration behavior and safety regulation of inorganic nanoparticles in food packaging: multi-scale mechanisms, risk assessment, and innovative strategies. Food Control 178, 111490. https://doi.org/10.1016/j.foodcont.2025.111490, 09567135.

Williams, H., Wikström, F., Wetter-Edman, K., Kristensson, P., 2018. Decisions on recycling or waste: how packaging functions affect the fate of used packaging in selected Swedish households. Sustainability 10 (12), 4794. https://doi.org/10.3390/su10124794, 2071–1050.

Chapter 16

Safety, quality standards, regulatory aspects, and value chain opportunities in makhana processing and export

Nizwa Itrat[1], Javaria Shabbir[2], Beenish Israr[1], Abdul Momin Rizwan Ahmad[3], Muhammad Rizwan Majeed[4] and Walid Elfalleh[5]

[1]*Institute of Home Sciences, Faculty of Food Nutrition and Home Sciences, University of Agriculture, Faisalabad, Pakistan;* [2]*Department of Nutrition and Dietetics, The University of Faisalabad, Faisalabad, Pakistan;* [3]*Department of Health Sciences, University of York, York, United Kingdom;* [4]*Department of Nutrition, University of Tennessee, Knoxville, TN, United States;* [5]*Department of Biology, College of Science, Imam Mohammad Ibn Saud Islamic University (IMSIU), Riyadh, Saudi Arabia*

16.1 Introduction

Fox nut or gorgon nut (Makhana) is a unique water crop that is made of the seeds of the plant of the genus *Euryale ferox*. It is long held to possess nutritional values and curative qualities and has long been cultivated in the select areas of India, Nepal, and China (Atal et al., 2019). Over the past few decades, makhana has been researched in other countries as a superfood due to its excellent protein value, low-fat makeup, antioxidant properties, and the fact that it is gluten-free (Atal et al., 2019). On one hand, the newfound awareness has established new prospects of processing and exporting, but on the other hand, it has reiterated the need to observe safety, high-quality standards and the regulations that are emerging (Garg et al., 2021).

Food commodities to be exported to the international markets must meet high standards, thus intended to protect the health of the consumer and build confidence. In the case of makhana, this requires careful control of all the stages of the value chain, including the process of cultivation and harvesting, as well as processing, packaging, and distribution. The quality of the product

Makhana (Euryale ferox) as a Functional Food. https://doi.org/10.1016/B978-0-443-44421-0.00017-2

and export opportunities can be impaired by potential safety hazards, such as microbial contamination, pesticides, and ineffective storage (Jha, 1999). Regulatory factors are also critical. The domestic authorities, for example, the Food Safety and Standards Authority of India (FSSAI) or other authorities in other manufacturing countries, have procedures that regulate the processing, labeling, and marketing of makhana (Pahwa et al., 2025). Another requirement is that exporters should comply with international regulations that often contain highly specified requirements on labeling, disclosure of allergens, organic certification, and country-of-origin records (Kumar and Jha 2024; Swetha and Agalya 2024).

The export environment is also influenced by trade policies, tariffs, and nontariff barriers. As an example, the sanitary and phytosanitary regulations that are part of the World Trade Organization (WTO) framework affect makhana, which is accepted in the foreign market (Pahwa et al., 2025). The producers who have the insight to understand and adjust to such regulatory architectures are in a better position to expand their coverage and prevent the high-priced rejections during entry points (Khadatkar et al., 2015). In addition to safety and regulatory consent, makhana presents great opportunities in the value chain (Khadatkar et al., 2015). On the production level, farmers can gain training on improved harvesting methods, mechanization, and sustainable agronomic methods (Singh et al., 2020a,b).

The monotypic genus of water lilies is a miracle food, snack, or even a therapeutic meal option for breakfast in cases of chronic diseases (Dér, 2008). The low sodium and high potassium make it a perfect combination for patients with cardiovascular diseases and hypertension (Singh, n.d.). Majorly, India produces 90% of the global production; therefore, major safety issues can significantly impact the export of this aquatic crop (Singh, n.d.). In Ayurvedic history, makhana was called a divine food and an important part of Traditional Chinese Medicine due to its nephroprotective properties. Traditionally, it was only consumed as roasted makhana, but now a lot of preparations of makhana have been developed, including premixes for breakfast, flour for bread making, flavored makhana, and energy bars (Dega and Barbhai, 2023). Another therapeutic application is due to its being a very good source of plant-based protein (Kumar et al., 2016a,b). The glycemic index is only 22, due to the resistant starch present in great amounts, which makes it a perfect option for diabetics. The kaempferol lowers the inflammation in the body and reduces the risk of complications of these diseases (Vallepu et al., 2023). This production can be increased by adopting modern cultivation methods in freshwater ponds and wetlands. The temperature for its production is 20–35° in neutral pH water (Devi et al., 2022).

Supply chain improvements include cold storage, logistics optimization, and digital traceability systems, which enhance operational efficiency and reduce postharvest losses (Kashyap and Shukla, 2023). Diversification on the market side to snack foods, functional foods, and health supplements presents new growth opportunities, especially in areas where consumers are becoming more interested in plant-based, gluten-free, and high-protein foods (Kashyap and Shukla, 2025). Entrepreneurship and rural development are also prospects of the makhana industry. Small and medium enterprises (SMEs), female-owned businesses, and cooperative models can be crucial in strengthening the value chain (Singh et al. 2020a,b).

Makhana can be involved in employment, the generation of income, and the empowerment of rural people by linking them to the global supply networks. Meanwhile, by positioning makhana as a culturally iconic superfood, the popularity of the product can be increased in other countries, which will create a niche (Singh et al., 2020a,b). However, there are still difficulties. The chain of supply is often disjointed, infrastructure is limited, and awareness of the producers of the international standards is still in its early phases (Minten et al., 2010a,b). To solve these problems, concerted efforts involving research, innovation, and policy assistance are needed. Subsidies, training, and facilitation of exports by governments and industry associations can be used to reinforce the sector (Minten et al., 2010a,b). The researchers can also make contributions by designing new technologies for processing, increasing shelf life, and designing environmentally friendly packaging solutions (Kumari and Kumar, n.d.; Jana et al., 2019).

Consequently, safety and quality assurance play a role as a technical requirement and as a strategic instrument for enhancing competitiveness in global trade. Establishing a robot quality standard is vital for positioning (Pahwa et al., 2025). Makhana, as a reliable and premium product, in the international standards of market consumers in regions, such as North America and Europe, expects imported food to comply with globally recognized benchmarks, including cortex element areas and ISO Certified National regulations enforced by authorities, such as the United States Food and Drug Administration and the food safety authority (Patil et al., 2022; Sonu and Jha, 2025). With the combination of these factors, the makhana industry can become a reliable provider of quality, safe, and innovative products that will establish its future in the local and global markets.

16.2 Food safety in makhana processing

16.2.1 Safety risks in makhana

Makhana, like many other food products, undergoes a range of risks during its production to consumption path (Table 16.1). Microbial contamination is common, notably when the seeds have not been dried properly or when they are stored under moist conditions (Ahmad et al., 2020). The growth of fungi, such as the *Aspergillus* species, may lead to a decrease in both quality and safety. Chemical risks come not only from crops in the form of fertilizers or pesticides residues, but also from the cultivation as residue of these chemicals

TABLE 16.1 Hazards and control measures in makhana processing (Kumar, 2018; Nag and Gite, 2020; Saha, 2020).

Stage	Potential hazards	Control measures	Examples/Details	References
Cultivation	Chemical residues	Adoption of good agricultural practices (GAP), use of biofertilizers, and monitoring water quality	Excessive use of pesticides and fertilizers	Sinha and Jha (2025)
	Heavy metals	Regular testing of water sources and shifting cultivation away from polluted areas.	Contaminated ponds or irrigation water	Rai et al. (2002)
	Biological hazards	Periodic cleaning of ponds, controlled water management	Growth of algae or pathogens in stagnant water	Devi et al. (2022)
Harvesting	Physical contamination	Manual sorting, mechanical cleaning, use of sieves	Stones, husk fragments, mud	Gaurav et al. (2022)
	Microbial contamination	Training workers in hygiene, sanitizing tools	Handling unclean tools or hands	
Drying	Microbial growth	Controlled drying to $<12\%$ moisture, use of mechanical dryers	High moisture content leading to mold	Gehija et al. (2019)
	Physical hazards	Covered drying platforms, insect-proof nets	Dust, insects, during open sun drying	
Roasting	Incomplete roasting	Standardized roasting temperatures and times, calibrated equipment	Survival of microbes due to low temperature	Kumar et al. (2014)
	Chemical hazards	Monitoring roasting conditions, avoiding excessive heat	Formation of acrylamide at very high temperatures	

Packaging	Moisture and oxygen exposure	Use of vacuum-sealed or nitrogen-flushed packs, food-grade packaging	Poor sealing, permeable materials	Kashyap and Shukla (2023)
	Pest infestation	Pest-proof packaging, regular inspection	Entry of insects during storage	
	Labeling errors	Consent with labeling regulations, double-checking documentation	Missing allergen or origin information	
Storage	Fungal growth	Temperature and humidity control, use of dehumidifiers	High-humidity environments	Saxena et al. (2022)
	Insect infestation	Regular fumigation, integrated pest management	Beetles and moths in storage facilities	
	Cross-contamination	Segregated storage, proper handling protocols	Mixing with other products	
Distribution/ Export	Physical damage	Use of sturdy cartons, cushioning materials	Breakage during transport	Gaurav et al. (2022)
	Temperature fluctuations	Cold chain logistics, insulated containers	Spoilage during long transit	Gaurav et al. (2022)
	Regulatory nonconsent	Strict adherence to export regulations and traceability systems	Missing export certificates, improper documentation	Pahwa et al. (2025)

(Priyadarshi and Ahmad, 2025). The polluted water contributes to heavy metals such as arsenic in the makhana seeds. Furthermore, physical hazards also become a threat to the makhana final product, such as stones, dust, foreign materials, and packaging materials (Nag and Gite, 2020; Pahwa et al., 2025).

16.2.2 Processing environment

Food safety is the most important factor in gaining the trust of people and the hazard free environment (Fig. 16.1). The safety begins with the hygiene of workers working in the processing unit under strict regulations and safety rules. The sanitization of all equipment used for cleaning, roasting, sieving, and packing is compulsory on an interval basis for the safety of makhana for consumer consumption. The storage spaces must also be monitored for hazard analysis. These storehouses must be a dry, ventilated, and microbial contamination-free environment for consumer safety. Physical hazards also contained a pest control system for ensuring the prevention of the vector borne diseases (Sharma et al., 2023; Sinha and Srinivasan, 2023).

16.2.3 Risk management systems

A well-structured food safety system can ensure hazard-free makhana for consumer health. Food safety systems regulate these rules in processing

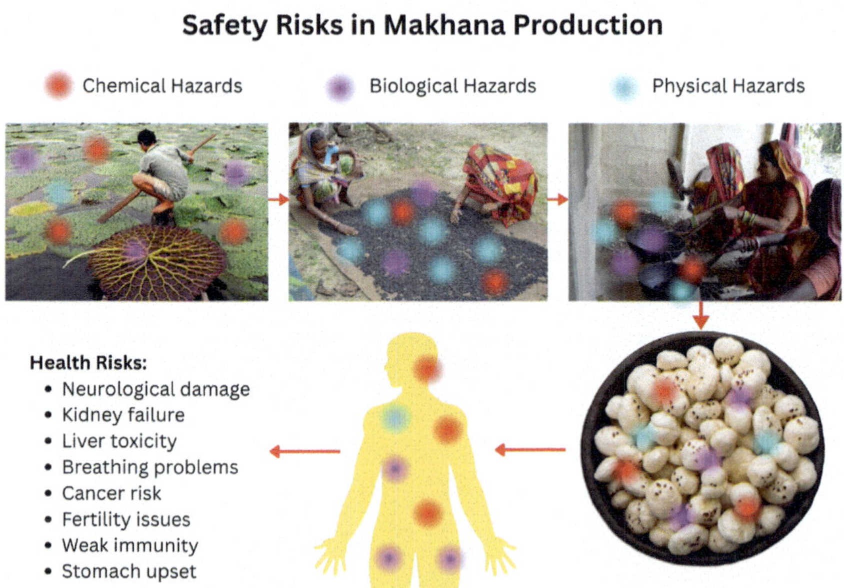

FIGURE 16.1 Safety risks in makhana cultivation, harvesting, and production.

units for the prevention of contamination. The most important and famous system among all food safety systems is Hazard Analysis and Critical Control Points (HACCP) used for microbial control (Khadatkar et al., 2020). Good manufacturing practices (GMPs) also control safe handling and hygiene of workers at the workspace. All these systems ensure the trading of makhana for international exports (Khadatkar et al., 2015; Nag and Gite, 2020).

16.2.4 Postharvest handling

To prevent the growth of microbes, seeds should be adequately dry to decrease the moisture content to harmless levels, usually, to less than 12%. Although the traditional sun drying is still common, mechanical dryers have more stable performance and are not reliant upon weather phenomena. The sensitive nature of makhana makes it prone to postharvest nutritional losses. But with proper roasting, the makhana not only becomes safer but also more delicious for consumption. Another safety assurance can be obtained by the selection of safe and good-quality packaging material. Good storage facilities are helpful for temperature and humidity to avoid spoilage (Thakur et al., n.d.; Sharma et al., 2023).

16.2.5 Training and awareness

Farmers, processors, and workers should be trained in how to understand the value of hygiene and safe handling and meet the standards. Both awareness campaigns and industry associations can be used to help small producers embrace good practices, and they can provide guidelines and advice. The culture of safety can be cultivated so that all the stakeholders, including the farmer up to the exporter, can play a role in keeping the products intact (Khadatkar et al., 2020). This shared responsibility cannot be done without maintaining the reputation of makhana in domestic and international markets through standards and quality assurance. Food safety depends on both knowledge and technology systems (Nag and Gite, 2020; Sonu and Jha, 2025).

16.2.6 Transparency of the food chain

The advanced food system ensures the in-depth management of food safety aspects (Fig. 16.2). The use of all the advanced digital tools, such as the QR code barcode and blocks in the platform, also ensures the end strength and the transparency of food safety. The transparent food system, must verify, the regulations from the source to the final product of Makhana, as this is important for consumer trust and buyers in international trade (Sharma et al., n.d.; Minten et al., 2010a,b).

FIGURE 16.2 Supply chain transparency for food safety.

16.3 Quality standards and assurance

16.3.1 Physical quality parameters

Among the most important quality indicators, the physical properties of makhana should be noted (Table 16.2). Consumers and buyers often judge makhana by the size of the kernel (Fig. 16.3), expansion ratio, color, and texture. Bigger kernels with equal growth are usually favored both in the domestic and foreign markets (Khadatkar et al., 2020). Another critical aspect is the whiteness of the popped seeds because the consumer might lose interest due to discoloration (Fig. 16.4). The texture is also a factor; crispness is a characteristic that is always desirable in the consumption of snacks. Achieving consistency of these parameters requires close attention to these parameters through the processing process, particularly at the roasting phase, in which temperature and time have a direct effect on expansion and appearance (Gaurav et al., 2022; Priyadarshi and Ahmad, 2025).

16.3.2 Nutritional attributes

The Makhana is in nutrient rich snack that is packed with fiber, protein, and essential minerals, such as calcium, potassium, and magnesium. As it is low fat and low in calories with a low glycemic index, Makhana is a very good option for gluten-sensitive people. Being low fat and low calorie makes it a perfect snack for diabetic people. The powerful antioxidant potential of Makhana makes it a super food that helps in reducing inflammation and protecting the digestive system and heart health. The calcium in Makhana makes it a perfect

TABLE 16.2 Quality standards and assurance in makhana (Gaurav et al., 2022; Islam et al., 2023; Priyadarshi and Ahmad, 2025).

Dimension Gaurav et al. (2022)	Quality parameters Devi et al. (2020)	Assurance measures Sonu and Jha (2026)	Examples/ Indicators Islam et al. (2025)
Physical quality	Kernel size, expansion ratio, color, texture	Standardized roasting conditions, mechanical grading, and visual inspection	Large, uniformly expanded kernels; bright white color; crisp texture
Nutritional quality	Protein, amino acids, antioxidants, low-fat	Laboratory nutrient profiling, in compliance with codex and ISO standards	Protein ~10–12%; lysine and methionine content; flavonoids and polyphenols
Sensory attributes	Taste, aroma, mouthfeel	Sensory evaluation panels, consumer acceptance testing	Mild nutty flavor, crunchy texture, fresh aroma
Safety parameters	Microbial load, pesticide residues, heavy metals	HACCP, accredited lab testing, certification audits	Absence of mold, bacteria, or contaminants
International benchmarks	Codex alimentarius, ISO 22000, FDA, EFSA	Certification audits, documentation, and traceability systems	Consent with global food safety and labeling requirements
Packaging and shelf life	Moisture control, pest resistance, and labeling accuracy	Food-grade packaging, shelf-life testing, labeling consent	Vacuum-sealed packs, nitrogen-flushed packaging, and clear labeling
Processing consistency	Uniform roasting, minimal breakage, reduced defects	Standard operating procedures, calibrated roasting equipment	Even expansion, a low percentage of broken kernels
Export documentation	Certificates of origin, organic certification, allergen disclosure	Verified documentation, third-party audits, digital traceability	Properly documented shipments, in line with WTO and importing country rules

Continued

TABLE 16.2 Quality standards and assurance in makhana (Gaurav et al., 2022; Islam et al., 2023; Priyadarshi and Ahmad, 2025).—cont'd

Dimension Gaurav et al. (2022)	Quality parameters Devi et al. (2020)	Assurance measures Sonu and Jha (2026)	Examples/ Indicators Islam et al. (2025)
Sustainability and ethics	Eco-friendly packaging, fair trade practices	Environmental certifications, fair trade labels, and CSR initiatives	Biodegradable packs, reduced carbon footprint, ethical sourcing
Consumer expectations	Convenience, innovation, branding	Product diversification, branding strategies, consent with marketing claims	Ready-to-eat packs, flavored makhana, gluten-free claims

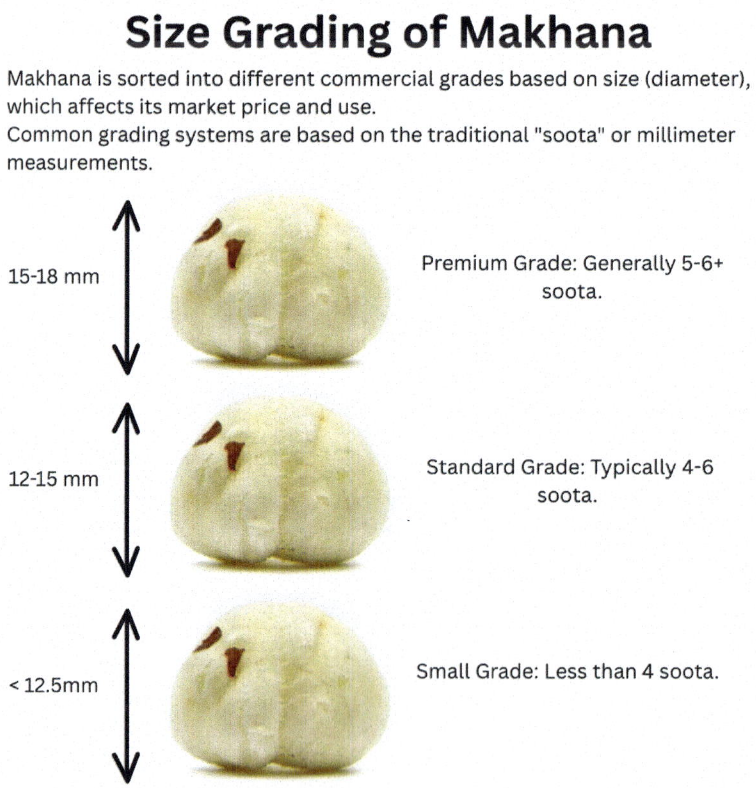

FIGURE 16.3 Size grading of makhana.

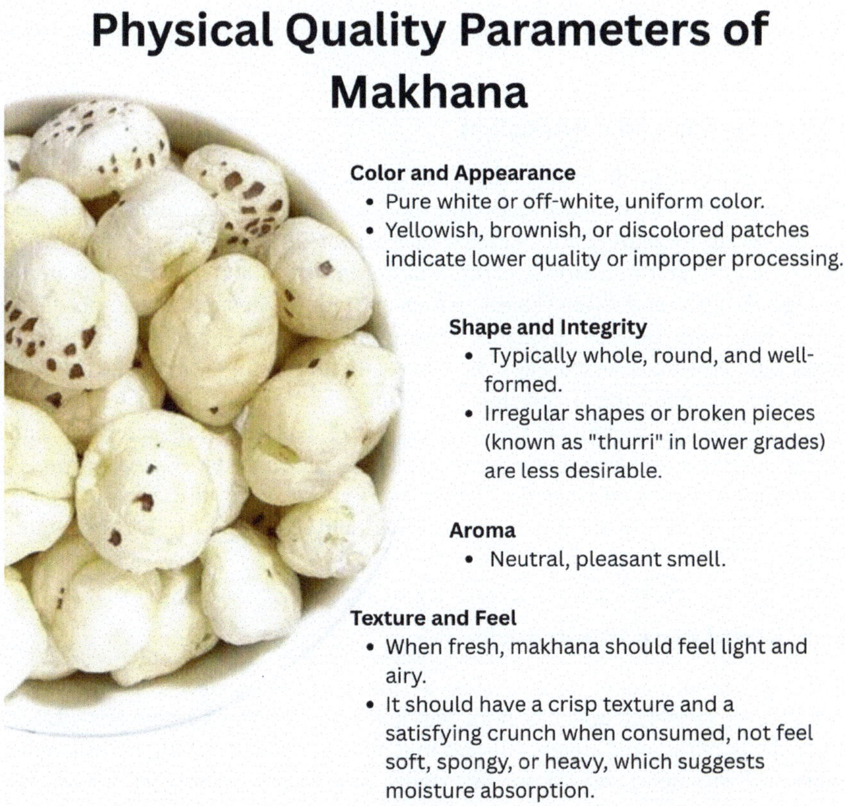

FIGURE 16.4 Physical and sensory quality parameters of makhana.

option for good and healthy bones. Traditionally, it helps calm the mind and improves sleep quality. This is a good option for weight management in weight-conscious population (Singh, n.d.; Mahesh Kumar et al., 2023; Jana et al., 2025a,b).

16.3.3 International quality benchmarks

There are other requirements enforced by national laws; the U.S. Food and Drug Administration (FDA) focuses on the accuracy of labeling and disclosure of allergens, whereas the European Food Safety Authority (EFSA) enforces a strict limit on the content of contaminants, such as pesticide residues and heavy metals. Procurement of such benchmarks will require investments in facility analysis, certification process, and record-keeping. Export companies that follow these standards earn confidence and achieve an entry into the high-end market segments. ISO certification for international quality management is important for ensuring the quality of fox nuts. This is a good snack option for weight management in weight-conscious population (Nag and Gite, 2020; Sharma, 2022).

16.3.4 Testing and certification

Testing and certification are also some of the basic components of quality assurance in the makhana processing industry. Certified testing laboratories can be set up to analyze the samples of microbial safety, chemical residues, heavy metals, and nutritional value. Purchasers have a certification system, such as HACCP and ISO systems, which provide evidence that the products are in line with the set standards. Traceability mechanisms are additional items that ensure transparency because suppliers can be tracked by the purchaser to their origin through barcoding or blockchain-enabled layers. All of this helps to reduce the chances of rejection of the economy and create a sense of consumer confidence in the market. Certification is becoming a competitive edge, allowing exporters to differentiate their products and to obtain high prices (Nag and Gite, 2020; Kashyap and Shukla, 2024).

16.4 Regulatory frameworks

The legal environment that relates to the processing and exportation of makhana is complex and has both national and global requirements (Table 16.3). These structures are established to defend consumer health, maintain the quality of products, and secure fair trade. Regulatory consent is not only a legal limitation to producers and exporters of goods and services but also a strategic necessity to enter the global market. Regulatory provisions lie across various fields, such as food safety, labeling, documentation, trade policy, and sustainability. Understanding and consent to these structures will make the makhana industry competent enough to establish trustworthiness, reduce risks of rejection, and strengthen its global trade stance (Nag and Gite, 2020; Authority, 2022).

The blockchain system, bar coding, and other traceability mechanisms are very important to enhance transparency by enabling the purchasing and tracking of the suppliers along with their origins. These mechanisms help consumers trust that they will have safe food products without any hazards or risks to their health. The scientific certificate differentiators help the consumer to get healthy, nutritious food with premium pricing and availability throughout the year. International regulatory bodies, which are the main force for the regulation of processing and exportation rules and regulations, both for National and International standards, are important to safeguard consumer health. Global market rules with the provision of food safety, food labeling documentation, trading policy, and sustainable systems (Kumar et al., 2011; Kumari et al., 2024).

TABLE 16.3 Regulatory frameworks for makhana processing and export (Minten et al., 2011; Nag and Gite, 2020; Authority, 2022).

Regulatory area	Domestic requirements	Key assurance measures	International requirements
Food safety	FSSAI (India), PSQCA (Pakistan) standards for hygiene, contaminants, and permissible pesticide levels	HACCP, ISO 22000 certification, accredited laboratory testing, consent audits	FDA (US), EFSA (EU), ASEAN limits on microbial load, pesticide residues, and heavy metals
Labeling	Mandatory nutritional labeling, allergen disclosure, and country of origin	Verified documentation, consent audits, and third-party certification	FDA nutrition facts, EFSA allergen rules, and organic labeling standards
Documentation	Certificates of analysis, domestic quality assurance	Third-party audits, digital traceability, and blockchain-enabled documentation	Certificates of origin, phytosanitary certificates, organic certification, and WTO-compliant trade documents
Trade policies	National export promotion schemes, tariff structures	Monitoring tariff schedules, adapting to nontariff barriers, and trade facilitation programs	WTO sanitary and phytosanitary measures, free trade agreements, and import restrictions
Sustainability and ethics	Encouragement of eco-friendly packaging, fair wages, and community development	Environmental audits, CSR initiatives, ethical sourcing, and sustainability certifications	Fair trade, organic, eco-label certifications, and carbon footprint reporting
Consumer protection laws	Domestic consumer rights laws ensure safe and truthful products	Consent with consumer rights legislation, transparent labeling, and grievance redressal mechanisms	International consumer protection frameworks require transparency and safety
Certification schemes	Local certifications for food safety and quality	Regular certification renewals, independent audits, and alignment with global benchmarks	ISO, HACCP, GlobalG.A.P., fair trade, and organic certifications
Import restrictions	Country-specific bans or limits on contaminants	Preexport testing, consent with importing country standards, and proactive risk management	EU maximum residue levels, US FDA import alerts

16.4.1 Domestic regulations

The safety and quality assurance of makhana is based on the underlying structure of domestic food regulatory bodies. In India, the limits of pesticide residues, microbial safety parameters, and labeling requirements are prescribed by the Food Safety and Standards Authority of India (FSSAI). These laws ensure that makhana that is sold in the country is of safe quality and meets minimum quality criteria. Similar organizations in other producing countries, such as the Pakistan Standards and Quality Control Authority (PSQCA), issue orders related to hygiene standards, packaging, and marketing. Local laws are usually the first step to export preparation, and local manufacturing is standardized to conform to the global standards (Kumar et al., 2011; Kumar and Jha, 2024).

16.4.2 International consent

Exporting of makhana requires adherence to the international regulations to adhere to, which is usually stricter than the domestic ones. The United States Food and Drug Administration (FDA) has very strict standards that include the accuracy of labels, disclosure of allergens, and the levels. Maximum residue content of pesticides and heavy metals is set by the European Food Safety Authority (EFSA), and comprehensive provisions on food safety management. The standards that are practiced by the member states of ASEAN are also unique, and the exporters are required to meet these standards to access markets in the region. These international frameworks usually require certification according to an internationally accepted framework, for example, ISO 22,000 or HACCP, and thus demonstrate that food safety management has been followed (Nag and Gite, 2020; Saha, 2020).

16.4.3 Labeling and documentation

Labeling is a very important aspect of regulatory consent, as it forms the channel through which the imperative information is passed to the consumers and regulatory agencies. Makhana exported must have nutritional information, warnings about allergies, an indication of the country of origin, and other relevant certification logos. Organic claims require to be proven by accredited certifying bodies, where outbound shipments still require ancillary documents, such as certificates of analysis, phytosanitary certificates, and certificates of origin. Mislabeling or incomplete documentation may trigger shipment delays, fines, or even outright rejection at points of entry (Saha, 2020; Gaurav et al., 2022).

16.4.4 Trade policies and tariffs

The tariff regimes and nontariff barriers shape international trade in makhana. In the WTO, sanitary and phytosanitary provisions have been established to protect food safety in international trade. On one hand, the free trade agreements can reduce the tariff burden and open new markets, but on the other hand, complex certification requirements tend to be nontariff barriers. To maximize the opportunities and mitigate the risks, exporters must ensure that they have situational awareness of the changing trade policies and tariff schedules, and preferential trade accords (Singh et al., 2020a,b; Swetha and Agalya, 2024).

16.4.5 Responsibility and ethical Conformance

In addition to the traditional regulatory requirements, sustainability and ethical levels are becoming more prominent in international trade. Stakeholders, the consumers, and regulatory bodies are paying closer attention to eco-friendly packs, consent with fair trade standards, and carbon footprint declarations. Incorporating sustainable practices could enable exporters to enter niche markets and develop better brand images. Certifications Fair Trade, Organic, and eco-labels are further guarantees to consumers that the production of makhana is responsible. Ethical conformity also includes fair remuneration, occupational health and safety, and community improvement programs, which further increase the value of the social aspect of the makhana industry (Saha 2020; Kashyap and Shukla, 2023).

16.5 Value chain opportunities

The makhana industry has significant potential in all its value chain, including cultivation and primary production, and extends to processing, marketing, and exportation (Table 16.4). Considering the growing worldwide need for plant-based, gluten-free, and nutritionally nutritious foods, makhana is gradually being recognized as a superfood. The strengthening of the value chain also increases profitability and the survival of rural livelihoods, entrepreneurship, and sustainable development. Along the chain, there are prospects at every level, and they include farmer training, technological innovation, supply chain administration, market diversification, and branding (Minten et al. 2010a,b; Singh et al. 2020a,b).

16.5.1 Primary production

At the production level, there are some areas of improvement that can be identified, and they include optimization of cultivation practices and improvement of training programs for farmers. Traditional ways of harvesting the makhana are labor-intensive and in many cases inefficient. Manual work and increasing productivity can be relieved by the introduction of mechanization and the implementation of innovative harvesting tools (Kashyap and Shukla, 2025).

TABLE 16.4 Value chain opportunities in makhana (Minten et al., 2010a,b; Singh et al., 2020a,b; Kashyap and Shukla, 2023).

Stage	Examples/ Details	Strategic measures	Opportunities
Primary production	GAP adoption, water management, and pest control	Farmer cooperatives, training programs, and mechanized harvesting tools	Improved cultivation, farmer training, and mechanization
Processing innovations	Flavored makhana, makhana flour, and nutraceuticals	Energy-efficient roasting, automated grading, and R&D investment	Value addition and product diversification
Supply chain management	Cold storage, humidity-controlled warehouses, and digital tracking	Logistics modernization, blockchain traceability, and inventory management systems	Efficiency, transparency, and reduced losses
Market diversification	Snacks, functional foods, and gluten-free alternatives	Branding strategies, targeted marketing, and consumer awareness campaigns	Expansion into new categories and regions
Export potential	North America, Europe, and the Middle East	Certification (organic, fair trade), e-commerce platforms, and international retail partnerships	Access to premium global markets
Entrepreneurship and SMEs	Women-led enterprises, rural cooperatives, and startups	Microfinance support, incubation programs, and niche product development	Local innovation and employment generation
Sustainability integration	Biodegradable packaging and waste valorization	Environmental certifications, CSR initiatives, and green technologies	Eco-friendly practices and circular economy
Policy support	Export facilitation and rural development schemes	Public–private partnerships, policy advocacy, and capacity building	Government incentives, subsidies, and training

16.5.2 Processing innovations

Processing is a vital step for value-added products. The innovative semi-automatic popping machines make the roasting process quick and safe. The machines reduce the processing time from 3 days to some hours with quite accuracy. With the help of these processing units, flavored makhana is also made with perfection in flavor and taste quality. The protein-based makhana powders, energy bars, and flavored makhana snacks of export quality become possible with help of these new technologies (Gem, n.d.; Kashyap and Shukla, 2025).

16.5.3 Supply chain management

Advanced digital systems enhance transparency and buyer confidence, helping exporters to reduce the waste, boost efficiency, and improve competitiveness. The cold chain and controlled storage maintain appropriate temperatures and humidity from harvest through transit to preserve quality and shelf-life. Efficient routing, timely pickups, and minimized dwell times further reduce delays. The temperature, humidity, location, and tracking enable protective interventions and improve the trust of buyers. Quality Assurance with the help of clear quality parameter at procurement minimizes rejections and downstream losses (Gem, n.d.; Jana, 2017).

16.5.4 Market diversification

International retail chains and e-commerce websites offer new avenues of reaching consumers across the world. The market of makhana aligns with the superfood trend in the international market, where consumers seek low-calorie options that are rich in protein and antioxidants. In America and Europe, plant-based nutritionally rich food supplement and snacks are in demand for health-conscious buyers. Promoting makhana as a premium superfood by a targeted branding and highlighting the Heritage and health benefits, such as heart support and weight management, is a very good strategy for enabling food system. Diversification reduces on a single market by balancing local and export volumes against fluctuations. Good branding focuses on authenticity, while supply chain improvements from prior discussions minimize losses and sustain quality (Kashyap and Shukla, 2023; Kumar et al., 2025).

16.5.5 Entrepreneurship and SMEs

SMEs are very important in strengthening the makhana value chain. Innovation can be triggered through women-owned enterprises, rural cooperatives, and startups, which will create job opportunities. Entrepreneurship provides a new generation for the development of the local economies of makhana. With the focus on niche markets, for example, flavored makhana or eco-friendly packaging, the SMEs can acquire a unique position in the sphere (Singh et al., 2021; Kashyap and Shukla, 2023).

16.6 Challenges and strategic interventions

Makhana industry is growing day-by-day, but still faces multiple challenges that decrease its potential for development and competitiveness (Table 16.5).

TABLE 16.5 Challenges and strategic interventions in makhana (Kumar et al., 2016a,b; Ahmad et al., 2020; Khadatkar et al., 2020).

Challenge	Details/Examples	Strategic interventions
Fragmented supply chain	Small-scale farmers, lack of coordination, and inconsistent quality	Formation of cooperatives, digital platforms for coordination, and farmer-producer organizations
Limited infrastructure	Inefficient drying/roasting, inadequate storage, and poor logistics	Investment in mechanical dryers, energy-efficient roasting units, cold storage facilities, and public–private partnerships
Lack of awareness of standards	Nonconsent with ISO, HACCP, and codex, leading to export rejections	Training programs, workshops, certification support, and industry association guidance
Research and innovation needs	Limited product diversification, short shelf life, and traditional packaging	R&D in nutraceuticals, eco-friendly packaging, new processing technologies, and collaboration with universities
Policy support	Limited subsidies, complex export documentation, and tariff barriers	Targeted subsidies, simplified documentation, export facilitation, and rural entrepreneurship programs
Global positioning strategies	Low awareness in international markets, weak branding	Branding as a superfood, cultural storytelling, e-commerce expansion, and digital marketing campaigns
Financial constraints	Limited access to credit for small farmers and SMEs	Microfinance schemes, low-interest loans, and incubation programs for startups
Sustainability challenges	Environmental concerns and waste management issues	Adoption of eco-friendly packaging, waste valorization, and carbon footprint reduction initiatives

16.6.1 Fragmented supply chain

There are a lot of difficulties in the makhana food chain. Currently, a fragmented makhana chain is a big challenge. This is initiated by small-scale farmers, who lack the new technologies for harvesting and processing. Digital technologies can serve as a solution to these kinds of challenges (Kumar et al., 2022).

16.6.2 Poor infrastructure

Conventional drying and roasting methods are often ineffective, and poor storage systems promote postharvest wastage. There is a need to invest in modern infrastructure, such as mechanical dryers, energy-efficient roasting units, cold storage, and humidity-controlled warehouses. The solution to these gaps can be found in the partnerships between the public and the private in such a way that infrastructure would be accessible to the small-scale producers (Khadatkar et al., 2020; Kashyap and Shukla, 2023).

16.6.3 Deficiency of standards Consciousness

Lots of manufacturers and processors are characterized by low levels of awareness about international quality and safety requirements. This knowledge gap results in nonconsent, and this may culminate in the rejection of shipments in the export markets. These stakeholders must be educated on the world standards, ISO, HACCP, and the Codex Alimentarius through training programs, workshops, and awareness. Certification can be conducted through industry associations and government agencies that can offer technical support in the process of consent.

16.6.4 Research and innovation requirement

To remain competitive in the makhana industry, there is a need to continually research and innovate. The existing threats are poor product diversification, limited shelf life, and the use of conventional packaging techniques (Sinha and Srinivasan, 2023; Jana et al., 2024, 2025a,b).

16.6.5 Policy support

Government policies can positively influence shaping the Makhana industry. The subsidies and export promotion program can be helpful for the new farmers to grow and build the Makhana industry using innovative technologies. Policies can also encourage mechanisms that offer financial incentives for certification-free or less costly innovative technology supply for new farmers to enhance the Makhana industry, additionally simplifying the export process. Users can also help farmers to access the Global market of Makhana. Although makhana has several nutritional benefits, this product is rather a novelty in many foreign markets. The use of e-business networks, foreign innovation chains, and online marketing efforts will increase its exposure (Kumar et al., 2011; Saha, 2020).

16.7 Possible side effects of overconsumption

16.7.1 Bad digestion

The lotus seeds can create digestive issues when consumed in large quantities. They might be hard to digest because of their thick texture and high content of starch, particularly for people who have poor digestive enzymes (Mahesh Kumar et al., 2023). The lotus seeds have relatively poor digestibility, which may cause the following symptoms: heaviness, stomach pain, nausea, or indigestion (Islam et al., 2023). Lotus seeds should also not be eaten too late in the night or without sufficient chewing, as this can also slow down the digestion process. The lotus seeds are regarded as heavy and cooling within traditional medicine, and when taken in excess, may disrupt the digestive fire. Poor digestion in the long run may lead to a decrease in nutrient content and result in overall discomfort (Islam et al., 2023). Lotus seeds are better consumed in moderate amounts and in the daytime to prevent digestive complications. They can be roasted lightly with little oil to make it easier to digest. Therefore, it is essential to have balance and mindful eating to avoid problems with digestion (Kumar et al., 2025).

16.7.2 Allergic reactions

Lotus seeds may also result in an allergic reaction in a few people, even though this is very rare. The symptoms can be itching, skin rashes, lips or face swelling, nausea, or difficulty breathing in severe cases. These responses are a result of an allergy to some proteins in lotus seeds. Individuals who have a history of being allergic to food ought to be very careful when taking lotus seeds for the first. The allergic reactions can occur within a few hours or immediately after consumption. In extreme situations, anaphylaxis can ensue, and this needs emergency treatment. Children and those who have weak immunity might be susceptible. In case of any strange symptoms following the consumption of lotus seeds, the intake of the same should be discontinued at once and medical counsel obtained. Thus, lotus seeds are mostly safe, but allergic reactions should also be noted.

16.7.3 Bloating and flatulence

Excessive consumption of lotus seeds may result in gaseousness and bloating in others. This is because lotus seeds have complex carbohydrates that might not be easily digested by all. When these carbohydrates are undigested and get to the large intestine, they are fermented by bacteria in the gut, resulting in the production of gas (Ansari et al., 2026). The result of this gas build-up leads to bloating, stomach aches, and flatulence. Individuals who have weak digestion or other diseases, such as irritable bowel syndrome (IBS), are more prone to these effects. Baked or spiced lotus seeds can also be an added irritant to the digestive system because of additional spices or oil (Rao et al., 2025). A lotus seed eaten in large quantities or consumed excessively may increase gastrointestinal upsets (Islam et al., 2025). Lotus seeds are recommended in small amounts to be chewed to minimize bloating. It is also possible to reduce the amount of gas formed by avoiding heavy seasoning and combining it with light foods (Süfer, 2025).

16.7.4 Blood sugar spikes

Lotus seeds are rich in carbohydrates and this can have an impact on blood glucose (Jana et al. 2025a,b). When taken in large amounts with sweetened syrups or sugar-containing bars, they can lead to rapid blood sugar spikes, particularly in individuals who have diabetes or resistance to insulin (Paduva et al., 2025). During the digestion of lotus seeds, some of their carbohydrates are oxidized to glucose, hence contributing to high levels of blood sugar (Kumar, 2025). Even though lotus seeds contain a moderate GI, excessive intake of this food may lead to inadequate GI. This effect may be increased by sweetened batches of lotus seed or roasted lotus seed cooked in oil. Regular fluctuations in the blood sugar level can cause the development of long-term risks of insulin resistance, fatigue, and hunger (Jana et al., 2024). The lotus seeds are supposed to be eaten with care and supervision by diabetic people in terms of their portion. They could be combined with fiber-rich foods or protein to slow down glucose absorption (Jana et al., 2024). Thus, the consumption of lotus seed-based bars is to be regulated, especially among people who control their sugar levels (Joshi and Nande, 2024).

16.7.5 Weight gain

Lotus seeds (makhana) are considered a healthy snack, yet consuming too many may lead to weight gain. Lotus seeds are fat-free but contain a lot of carbohydrates and calories. They may add to the overall daily caloric intake when consumed in large amounts, particularly in roasted or sweetened forms. If these additional calories are not used during physical activity, they are stored as fat, and the body gradually gains weight (Wakodikar and Bodhankar, n.d.). Individuals who eat lotus seeds as snacks and fail to control their portions may end up consuming more energy than needed (Shukla et al., 2024). Moreover, lotus seeds contain a lot of glycemic load when used in large quantities, which may lead to high insulin secretion and accumulation of fats (Sinha and Jha, 2025). There is a greater risk of weight gain when the lotus seeds are cooked in oil or butter, sugar, or salt because they add a further level of calories. Thus, lotus seeds may be a part of a balanced diet, but it is necessary to consume them in moderation (Gaurav et al., 2022). People who attempt to lose weight or maintain a healthy body weight need to be conscious of the size of the portions and how they are prepared to prevent the accumulation of unwanted weight (Gaurav et al., 2022).

16.8 Constipation

Lotus seeds are regarded as helpful in moderate quantities in terms of digestion; overeating might result in constipation. Lotus seeds are dry in nature, and they have starch, which slows bowel movements, particularly when ingested in large quantities (Dega and Barbhai, 2023). When an individual fails to consume adequate water and lotus seeds, the likelihood of constipating feces and constipation is high (Rao et al., 2025). The issue is more prevalent in patients with sensitive intestinal systems or those who constipate in the first place (Sinha and Jha, 2025). Lotus seeds have a low moisture value, which can absorb water in the intestines, thereby causing the stool to become dry and hard (Gavhane et al., n.d.). Lotus seeds need to be taken in moderation with a lot of fluid intake to avoid constipation. Normal bowel movements can also be maintained by ensuring that the diet is in the form of fruits, vegetables, and other foodstuffs that contain a lot of fiber. Thus, lotus seeds can be healthy, but in excess, when they are not properly hydrated, they can have adverse impacts on bowel health (Singh et al., 2023).

16.9 Conclusion

The makhana industry is evolving very rapidly, influenced by global demand for gluten-free and nutrient-dense food products. The safety is nonnegotiable, and strict control of chemical, microbiological, and physical hazards is needed to be checked by HACCP and Codex Alimentarius, FDA, and EFSA standards. Global regulations also cover the certifications, labeling, and documentation. The value chain is the backbone of the makhana industry for better supplying the global demands. Makhana, being a superfood, is a great opportunity for entrepreneur, new farers, and sustainable food systems.

Self declaration

Artificial Intelligence (AI) tools were used only for English language editing in the preparation of this manuscript. No AI tools were used to generate any kind of content, including text, tables, or figures.

References

Ahmad, A., Azad, C., Kumar, B., 2020. A study on constraints of Makhana cultivation and suitable measures for its better development. J. Pharmacogn. Phytochem. 9 (1), 984–987.

Ansari, A.P., Anwar, N., Ahmed, N.Z., Makbul, S.A.A., 2026. Phytochemistry and Pharmacology of Aquatic Plants Used in Unani System of Medicine. Crude Drugs of Unani Medicine. Apple Academic Press, pp. 231–279.

Atal, R., Patil, S., Sohane, R., Datt, R., Roy Choudhary, S., Singh, S., 2019. Economics of makhana production and socio-economic profile of makhana growers in Bihar. Curr. J. Appl. Sci. Technol. 38 (6), 1–5.

Authority, E.F.S., 2022. Technical Report on the notification of roasted and popped kernels from the seeds of *Euryale ferox* Salisb. In: As a Traditional Food from a Third Country Pursuant to Article 14 of Regulation (EU) 2015/2283. Wiley Online Library.

Dega, V., Barbhai, M.D., 2023. Exploring the underutilized novel foods and starches for formulation of low glycemic therapeutic foods: a review. Front. Nutr. 10, 1162462.

Dér, P., 2008. Alternatives for the Treatment of Ischemic Heart. Univ. Debrecen. Doctoral dissertation.

Devi, M., Sharma, K., Narayan Jha, S., Arora, S., Patel, S., Kumar, Y., Kumar Vishwakarma, R., 2020. Effect of popping on physicochemical, technological, antioxidant, and microstructural properties of makhana seed. J. Food Process. Preserv. 44 (10), e14787.

Devi, N., Rani, K., Kharb, P., Kaushik, P., 2022. Bio-fabrication of *Euryale ferox* (Makhana) leaf silver nanoparticles and their antibacterial, antioxidant and cytotoxic potential. Plants 11 (20), 2766.

Garg, N., Chauhan, K., Baria, K., Mehan, M.B., 2021. Foxnut (Makhana): The Powerpack Seed. Everyman's Science, pp. 165–168.

Gaurav, A., Nath, D., Barbhuiya, R.I., Pradhan, R.C., 2022. Physical characterization and mass modeling of traditionally popped makhana (*Euryale ferox* salisbury) variety: Sabour-1 makhana. J. Food Process. Eng. 45 (6), e13927.

Gavhane, M. N. P., S. Nagarale, V. Babar, V. D. Tambe, M. Nisha and S. Mhaske Review on: Prickly Water Lily.

Gehija, J., Cholera, M., Patel, K., Kori, V., Bhaskar, C.L., 2019. Stability Study of an Ayurvedic Formulation Paushtika Biscuit for Undernutrition in Pediatric Age Group With Respect to Baseline Microbial Profile.

Gem, A.C. The Makhana Codex: The Ultimate Interactive Guide to its Nutrition, Benefits & Science.

Islam, Z., Bashir, K., Shaida, B., Siddiqui, S., 2023. A comprehensive examination of the nutritional value and health advantages of makhana (*Euryale ferox*). J. Sci. Techn. Res. 13 (2).

Islam, Z., Bashir, K., Singh, S., Rastogi, M., Siddiqui, S., Bangar, S.P., 2025. Structural and functional properties of starch from popped seeds of makhana (*Euryale ferox*). Starch-Stärke, e70030.

Jana, B., 2017. Organic cultivation of makhana in low land during rainy season for enhancing farm profitability. Int. J. Ecosyst. Ecol. Sci.-IJEES 7 (3), 419–424.

Jana, B., Bhatt, B., Srivastava, A., 2025a. Biochemical components of makhana (*Euryale ferox* salisb) and its relevance to anti-obesity effect in human. Asian J. Dairy Food Res. 44 (4), 614–619.

Jana, B., Bhatt, B., Srivastava, A., 2025b. Biochemical components of makhana (*Euryale ferox* salisb) and its relevance to anti-obesity effect in human. Asian J. Dairy Food Res. Int. 44 (4).

Jana, B., Kumar, M., Raut, S., 2024. Antidiabetic biomolecules and nutrient elements in makhana (Euryale ferox Salisb.). Ann. Phytomed. 13 (2), 521–528.

Jana, B., Srivastava, A., Idris, M., 2019. New makhana (*Euryale ferox* Salisb.) processed products for health benefit. J. Food Process. Technol. 10 (4), 789.

Jha, S., 1999. Physical and hygroscopic properties of makhana. J. Agric. Eng. Res. 72 (2), 145–150.

Joshi, P.N., Nande, P.J., 2024. Diabetes Management: A Journey to Wellness through Nutrition and Lifestyle Choices. Notion Press.

Kashyap, A., Shukla, O.J., 2023. Analysis of critical barriers in the sustainable supply chain of MSMEs: a case of Makhana (Foxnut) industry. Benchmark Int. J. 30 (6), 2040–2061.

Kashyap, A., Shukla, O.J., 2024. A SD approach for optimizing food supply chain cost: a case of foxnut (makhana) industry. In: Responsible Production and Consumption. CRC Press, pp. 77–82.

Kashyap, A., Shukla, O.J., 2025. Sustainable food supply chain: exploration, identification, and analysis of the critical drivers for the foxnut (Makhana) industry. J. Glob. Oper. Strateg. Sourc. 18 (3), 559–582.

Khadatkar, A., Gite, L., Gupta, V., 2015. Interventions to reduce drudgery of workers in the traditional method of harvesting Makhana (*Euryale ferox* salisb.) seeds from ponds. Curr. Sci. 1332–1337.

Khadatkar, A., Mehta, C., Gite, L., 2020. Makhana (*Euryale ferox* Salisb.): a high-valued aquatic food crop with emphasis on its agronomic management–A review. Sci. Hortic. 261, 108995.

Kumar, A., 2018. A diagnostic study of makhana production technology among the growers of Madhubani district (M.Sc. (Ag.) Thesis/Doctoral dissertation, depending on the degree awarded). Dr. Rajendra Prasad Central Agricultural University, Pusa, Samastipur, Bihar, India.

Kumar, A. (2025). Goodbye PCOS, hello life: the complete guide to understanding PCOS & a 90-day natural blueprint to balance hormones, boost fertility & regain control without Pills or pain, Dr. Ankit Kumar Publishing House.

Kumar, A., Jha, A.K., 2024. Makhana: a valuable aquatic food crop. Magnesium 11 (13), 19.

Kumar, A., Yadav, P., Nat, P., Nath, S., Kumari, U., 2016a. *Euryale ferox* Salisb. Is an immense potential aquatic cash fruit crop: a review. Int. J. Adv. Agricult. Sci. Technol. 3 (5), 35–49.

Kumar, L., Singh, A.K., Bhatt, B., 2016b. Nutritional status of recently developed makhana (gorgon nut) variety-Swarna Vaidehi. J. AgriSearch 3 (4), 199–205.

Kumar, P., Kumari, R., Singh, A.K., 2025. Makhana (*Euryale ferox*): the super snack from water. In: Fasting Superfoods: Cultivation, Nutrition & Market Potential. Springer, pp. 115–124.

Kumar, S., Akhtar, M.N., Patel, D.K., Saha, T., Sah, S., 2022. Status of alternaria leaf blight and leaf spot diseases of makhana in North Bihar and their management strategies under integrated approach. Indian Phytopathol. 75 (3), 781–787.

Kumar, U., Kumar, A., Singh, K.M., 2011. Constraints and drudgery in makhana cultivation. Int. J. Extens. Educ. 7, 47–51.

Kumar, V., Rajak, D., Kumar, R., Kumar, V., Devi Sharma, P., 2014. Design and development of low-cost makhana grading and roasting machine. Int. J. Food Eng. 10 (3), 357–366.

Kumari, N., Dana, S., Mondal, A., Sau, S., Sarkar, M.R., Hazmt, N., 2024. Constraints encountered by Makhana-cum-fish culturist in adoption of improved practices. Indian J. Anim. Health 63 (2), 334–340.

Kumari, P. and G. Kumar. Boosting Bihar's Economy Through Mithila Makhana: Opportunities and Challenges.

Mahesh Kumar, M., Goudar, G., Sharma, P., Vishwakarma, R., Gogoi, P., Mahajan, A., Chandragiri, A., Kalpuri, S., Babu Geddam, J., Suresh, C., 2023. Differentiating the nutrient composition, in-vitro starch digestibility, individual polyphenols and anti-oxidant properties of raw and popped makhana (*Euryale ferox*). J. Food Meas. Char. 17 (6), 5828–5844.

Minten, B., Singh, K., Sutradhar, R., 2011. Branding in Food Retail of High Value Crops in Asia: Case of Makhana from Bihar (India).

Minten, B., Singh, K.M., Sutradhar, R.K., 2010a. The Makhana Value Chain and the Fast Emergence of Branding in Food Retail: Evidence from Bihar (India). Available at SSRN 2406566.

Minten, B., Singh, K.M., Sutradhar, R.K., 2010b. The Makhana Value Chain in Bihar: An Assessment and Policy Implications. Available at SSRN 2406538.

Nag, P., Gite, L., 2020. Health hazards in farming. In: Human-Centered Agriculture: Ergonomics and Human Factors Applied. Springer, pp. 205–237.

Paduva, A., Baby, N.J., and Niba, P. F. (2025). Nutrient packed digestive crackers: HARNESSING THE POWER OF Cannabis sativa L. AND Euryale ferox, St. Teresa's College (autonomous) Ernakulam.

Pahwa, V., Sirsikar, I., Singh, R., Shukla, N.P., Verma, K.K., Sachdev, A., 2025. Proceedings from the national symposium on Makhana: unlocking India's aquatic superfood for rural prosperity. Intl. Food Policy Res. Inst.

Patil, S., Patel, V., Sohane, R., Rehaman, S.M., 2022. Challenges in In-situ processing of makhana and integration of farmers with makhana value chain. In: International Conference on Agriculture and Applied Sciences for Food Security (AAPS-2017).

Priyadarshi, A., Ahmad, H., 2025. Makhana and the sustainable development goals: a pathway to nutrition, livelihoods, and environmental sustainability. Int. J. Adv. Multidiscip. Res. Stud 5, 1481–1483.

Rai, U., Tripathi, R., Vajpayee, P., Jha, V., Ali, M., 2002. Bioaccumulation of toxic metals (Cr, Cd, Pb and Cu) by seeds of *Euryale ferox* Salisb.(Makhana). Chemosphere 46 (2), 267–272.

Rao, S., Patil, A., Jadav, M., 2025. Therapeutic potential of makhana Churna and Sharkarayukta Dugdha: an Ayurvedic perspective. J. Ayurv. Integr. Med. Sci. 10 (5), 198–202.

Saha, D., 2020. Food Processing: Understanding Common Threads. Economics of the Food Processing Industry: Lessons from Bihar. Springer, India, pp. 13–52.

Saxena, D., Hussain, I., Singh, S., Kumar, S., Garg, N., 2022. Optimization and storage study of rabri enriched with date syrup and makhana. Trends Sci. 19 (9), 3669.

Sharma, K., 2022. Studies on Process Mechanization and Storage of Gorgan Nut (*Euryale ferox*) for Enhanced Recovery of Popped Makhana.

Sharma, K., Vishwakarma, R.K., Patel, S., Jha, S.N., Devi, M., 2023. Organic production, post-harvest processing, and value-added intervention of fox nuts (*Euryale ferox*) for export. In: Transforming Organic Agri-Produce into Processed Food Products. Apple Academic Press, pp. 363–393.

Sharma, S. K., M. Kumari, Bairwa, S. L., Rahman, M., Panda, C.K., Homa, F., and Kushwaha, C. Economic analysis of makhana cultivation in darbhanga district of Bihar, India. Seed. 3688: 5.85.

Shukla, N., Sharma, J., Sharma, P., 2024. A review on: *Euryale ferox* (fox nut). J. Pharmacogn. Phytochem. 13 (5), 50–52.

Singh, D., Kumar, A., Kumar, U., Singh, I., Chandra, N., Bhatt, B., 2020a. Value chain analysis of makhana. Techn. Bull. 3.

Singh, D.K., Kumar, A., Kumar, U., Chandra, N., Singh, I., 2021. Value addition in makhana through post-harvest operations. Food Sci. Rep. 2, 29–32.

Singh, D.K., Kumar, U., Kumar, A., Singh, I., 2020b. Strategies for export promotion of popped makhana in India. Food Sci. Rep. 1, 30–33.

Singh, I. S. Dynamics of Nutrients of Aquatic Crop (Gorgon Nut or Makhana) under Different Systems of Cultivation-A.

Singh, J., Chauhan, H., Kumari, V., Singh, R., 2023. Aquatic vegetables for nutrition and entrepreneurship. In: Vegetables for Nutrition and Entrepreneurship. Springer, pp. 389–408.

Sinha, C.B.K., Jha, V., 2025. Identification of nutraceutically important metabolites in raw and popped makhana (*Euryale ferox* salisb.) seeds using GCMS technique. Int. J. Plant Environ. 11 (03), 514–524.

Sinha, S., Srinivasan, G., 2023. Redesigning tools for fox nut (makhana) farming and processing: an ergonomic study. In: International Conference on Humanizing Work and Work Environment. Springer.

Sonu, K., Jha, K., 2025. Assessing Regional Disparities in Makhana Farming: A Study from Bihar.

Sonu, K., Jha, K.K., 2026. Unveiling regional disparities in makhana farming: a multidimensional analysis from Bihar. Indian J. Ext. Educ. 62 (1), 15–20.

Süfer, Ö., 2025. Roasting Technology for Production of Ready-To-Eat Snacks. Ready-To-Eat Snacks. Apple Academic Press, pp. 35–70.

Swetha, A., Agalya, E., 2024. Makhana: as a modern day superfood. Today's Multidiscip. Res. Persp. Volume-1 (9).

Thakur, A. K., I. S. Singh and R. Sharma Harvesting and Processing of Makhana (Euryale ferox salisb.)–Research Aspect and Prospect.

Vallepu, S., Shaik, K.B., Shaik, N., Vadde, R., Ragimekula, N., Patan, S.S.V.K., 2023. Influences of household processing methods on nutritional composition, anti-nutritional factors and antioxidant activities of foxtail millet. Recent Progr. Nutrit. 3 (4), 1–27.

Wakodikar, R. D. and H. Bodhankar. Formulation of Makhana (Euryale ferox) and Ragi (Eleusine Coracana) Supplemented Nutri Cookies Incorporated with Dates Powder.

Chapter 17

Makhana industry landscape: Market trends, sustainability, innovation, and future prospects

Isabella de Andrade Rezende[1], Flaviana Coelho Pacheco[1, 2], Wilson de Almeida Orlando Júnior[1], Luiza Oliveira Mello[1], Ana Flávia Coelho Pacheco[1] and Paulo Henrique Costa Paiva[1]
[1]*Cândido Tostes Dairy Institute (EPAMIG), Lieutenant Luis de Freitas, Juiz de Fora, Brazil;*
[2]*Federal University of Viçosa, Viçosa, Brazil*

17.1 Introduction

Euryale ferox Salisb., widely known as fox nut, is a monotypic genus of the Nymphaeaceae family. Fox nuts are found mainly in wetlands, such as lakes, ponds, and puddles to shallow bodies of water. In addition to fox nut, it is known globally by several other names, such as gorgon nut and lotus seeds, although the true lotus seed comes from another aquatic species, *Nelumbo nucifera* Gaertn) (Kumar et al., 2011).

It is a well-known aquatic crop for its edible seeds with high economic value and potential throughout the country, since it has a high nutritional value (Mandal et al., 2010). Makhana contains approximately 9.7% highly digestible protein, about 76% carbohydrates, 12.8% moisture, 0.1% fat, and 0.5% total minerals (Arjun et al., 2016).

Beyond its nutritional relevance, makhana has gained prominence in the global agri-food landscape due to shifting consumer preferences toward plant-based, minimally processed, and functional foods (Sinha and Jha, 2025). This trend has stimulated growing international demand, driving diversification of products derived from *E. ferox*, such as roasted snacks, flours, protein-enriched formulations, and nutraceutical ingredients. Market analyses indicate rapid expansion of the makhana sector, although estimates of market size and growth rates vary considerably across sources, underscoring the need for

Makhana (Euryale ferox) as a Functional Food. https://doi.org/10.1016/B978-0-443-44421-0.01017-9

methodological transparency and standardized data collection within this emerging industry (Jiang et al., 2023; Sinha and Jha, 2025).

From a food science and postharvest technology perspective, the value of makhana is strongly influenced by processing conditions, particularly drying, roasting, and popping operations, which affect starch structure, protein functionality, bioactive compound retention, and shelf stability (Mandal et al., 2010; Kumar et al., 2016). These processing stages are critical for determining product quality, safety, and consumer acceptance, as well as for enabling industrial scalability and export potential (Kumar et al., 2016).

Sustainability considerations further reinforce the relevance of makhana as a strategic crop within the bioeconomy (Garrett et al., 2025). Cultivation systems based on shallow wetlands and integrated aquaculture models have been described as low-input and biodiversity-friendly, contributing to ecosystem services such as water regulation, carbon sequestration, and livelihood resilience for smallholder farmers, particularly in Bihar, India, which accounts for the majority of global production (Jiang et al., 2023). Nevertheless, the sector faces significant challenges related to climate change, wetland degradation, labor-intensive harvesting practices, and limited mechanization, which may constrain future growth if not addressed through coordinated innovation and policy support.

In this context, the present chapter provides a comprehensive overview of the makhana (*Euryale ferox*) value chain, integrating perspectives from food science, postharvest technology, agricultural sustainability, and bioeconomy. By synthesizing current knowledge on market dynamics, nutritional and functional properties, processing technologies, sustainability frameworks, and innovation pathways, this chapter aims to position makhana as an emerging, multifunctional agri-food platform with substantial potential for scientific advancement, industrial development, and socioeconomic impact.

17.2 Makhana (*Euryale ferox*)

17.2.1 Presentation of makhana (*Euryale ferox*)

Makhana is scientifically known as *Euryale ferox* Salisb (EFS) (Fig. 17.1) and belongs to the Nymphaeaceae family. It is distinguished as a species of the genus Euryale, widely cultivated in Asia, especially in countries such as India, China, Bangladesh, Japan, and parts of Southeast Asia (Kapoor et al., 2022). Classified as a hydrophyte, it is predominantly distributed in warm and humid environments. It preferentially develops in lakes, ponds, or shallow wetland areas (depths around 0.2–2 m). It exhibits excellent tolerance to temperature variations and requires soils rich in organic matter (Jiang et al., 2023).

Euryale ferox represents a fundamental constituent of traditional food and medicinal practices used in East Asia for over 2000 years. In Traditional Chinese Medicine, its seeds serve as a functional food and are prescribed for

FIGURE 17.1 The flower (A), fruit (B, C, D) shell (E) and seed kernel (F) of *E. ferox* Salisb. *Modified from Zhang, L., Chen, Y., Zeng, J., Zang, J., Liang, Q., Tang, D., Wang, Z., Yin, Z., 2022. Digestive and physicochemical properties of small granular starch from Euryale ferox seeds growing in Yugan of China. Food Biophys. 17 (1), 126–135. https://doi.org/10.1007/s11483-021-09706-7.*

kidney and intestinal dysfunctions. They were first mentioned in classic texts, such as Shen Nong's book on materia medica, while in Indian Ayurvedic medicine, EFS are prescribed for various biliary, rheumatic, and urinary dysfunctions. EFS is an unfinished multifunctional plant; Japanese and Korean ethnopharmacology reveal its unified popular use to increase vitality and longevity and support digestive health (Jiang et al., 2023).

Therefore, the recent expansion of Makhana into global markets is a result of the growing recognition of its nutritional and functional value, and versatility in industrial applications. These qualities have encouraged the use of Makhana in the formulation of healthy snack lines, baking ingredients, nutraceutical formulations, and plant-based diet products in Europe, North America, and Oceania (Jiang et al., 2023).

17.2.2 Socioeconomic and agri-food importance

EFS is an important source of income and livelihood in producing communities in the state of Bihar, India, which holds approximately 90% of the world's production and is mostly characterized by family farms in the regions of Madhubani, Darbhanga, Purnia, and Katihar. The socioeconomic profile of these families is mostly made up of small producers who own up to 2 hectares of cultivated land and employ, in most cases, family labor, with a strong interconnection of their production stages and the need to sell the harvest to cover the expenses of the vicious cycle that tends to last as long as they are producers. In addition, there are several structural difficulties in accessing technological assistance, financing, and variations in market prices can significantly influence the decision to continue production (Garrett et al., 2025).

Traditional Makhana cultivation systems in shallow wetlands contribute significantly to the resilience of local food systems by diversifying agricultural practices with lower demand for chemical inputs and reduced environmental impact (Jiang et al., 2023; Garrett et al., 2025). In addition to being a staple food, Makhana can strengthen production chains, making them more resilient to climate change. This happens through sustainable management practices and combined cultivation with fish, as has already been done in Bihar, where producers cultivate the plant along with fish farming in semiintensive systems. This combination not only increases the income of producing families but also helps to reduce the social and environmental risks of the region (Garrett et al., 2025).

Thanks to its nutritional profile, with low caloric density and high content of minerals and bioactive compounds, Makhana has gained prominence, boosting its value as a functional ingredient in lines of healthy snacks, flours, and nutraceutical supplements. This recognition not only differentiates the product in the market, increasing its commercial competitiveness, but also promotes the social and economic inclusion of the communities involved in its production chain (Jiang et al., 2023).

17.2.3 Goals

This chapter aims to demonstrate the current landscape of the Makhana (*Euryale ferox*) industry, showcasing the socioeconomic, production, technological, and scientific aspects that underpin the emergence of this food as a key ingredient in global food systems. It seeks to provide a detailed context of cultivation practices, industrial processing methods, sustainability and innovation topics, and to systematize gaps and opportunities for scientific development related to the use of Makhana in different production chains and segments of the bioeconomy.

In addition to providing an updated overview of the Makhana sector, this chapter seeks to critically examine how agronomic, technological, and governance advances effectively support the consolidation of *E. ferox* as a strategic crop in the global bioeconomy. It specifically aims to contextualize the current market expansion of Makhana in comparison with other unconventional food plants and plant-based functional foods, analyze how production systems, postharvest technologies, and industrial processing models contribute to or limit scale-up and value addition, discuss the robustness and gaps of available evidence on nutritional, functional, and clinical benefits and identify major research gaps and innovation priorities that can guide future work in breeding, processing engineering, digital traceability, and sustainability assessments.

17.3 Global market overview

17.3.1 Current economic size of the industry

The global Makhana (*Euryale ferox*) market reached an estimated value of US $44.4 million in 2023 and could reach over US$90 million by 2033. This is due to the growing demand for functional foods in the Asian, European, and

North American markets (Sinha and Jha, 2025). Production is heavily concentrated in India, mainly in the state of Bihar, which accounts for approximately 80% of the world's supply, followed by China, Nepal, and, to a lesser extent, some Southeast Asian countries (Jiang et al., 2023). Consumption has also shown significant growth in the United States, Canada, the United Kingdom, and Australia, as Makhana is incorporated as an ingredient in various types of snacks, cereal mixes, baked goods, and nutraceuticals, among other applications.

Nevertheless, these figures stem mainly from private market intelligence reports and online dashboards, for which the underlying sampling procedures and coverage are not always transparent; therefore, they should be interpreted as indicative orders of magnitude rather than precise, audited values.

This market segmentation can be divided into four major categories: ready-to- eat snacks, ingredients for the food industry (such as flours and other extracts), industrialized/processed foods, and nutraceutical products. The snack segment is currently the most representative, driven by consumption trends linked to convenience, healthiness, and innovation in textures (Sinha and Jha, 2025).

17.3.2 Growth dynamics and outlook for the coming years

There are records of quite encouraging projections for the coming years, reaching estimates of compound annual growth between 8 and 11% until 2030 (Jiang et al., 2023; Sinha and Jha, 2025). The expansion of Makhana follows trends that are also observed in other Unconventional Food Plants (UNFPPs) and plant nootropics, such as microgreens, adaptogenic mushrooms, and functional seeds (Jiang et al., 2023; Singh et al., 2019).

The main reasons for this growth include greater public awareness of health, the pursuit of healthy eating habits, the increase in the number of vegetarians and vegans, innovation in plant-based products, and the expansion of digital distribution channels. The spread of gourmet lines, protein blends, functional foods, and dietary supplements also boosts demand prospects at the international level.

The global Makhana industry presents a competitive configuration dominated mainly by large Indian companies, which concentrate on the distribution and processing of seeds on a commercial scale and expand finished product lines to international markets. While India leads the chain in the supply of raw materials, China stands out for its high investment in intensive processing technologies and industrial diversification focused on its medicinal use. Nepal, still in its infancy, has highlighted local initiatives aimed at exporting and developing producer cooperatives (Jiang et al., 2023; Garrett et al., 2025).

In recent years, there has been an emergence of startups and companies specializing in premium and functional products, such as gourmet snacks, enriched flours, and baking ingredients, which explore differentiation through quality, nutrition, and territorial appeal. Territorial branding strategies, such as the designation of origin "Bihar Makhana," have been employed to add commercial value and improve the perception of authenticity in very demanding markets, in addition to favoring initiatives for organic certification and traceability, ensuring greater transparency. Regarding barriers to entry into new international markets, we have strict sanitary regulations, appropriate postharvest practices, food safety certifications, and high logistics costs, especially in cold chains and exports to continents such as Europe and North America. Aspects such as exchange rate fluctuations, tariff differences, and the need for industrial standardization also represent significant challenges for new entrants.

17.3.3 Considerations on market methodologies

Market analysis for agri-food products such as Makhana demands methodological approaches that reconcile statistical rigor, geographic representativeness, and the possibility of data traceability. Internationally, it is common to use private sources (specialized consultants), institutional databases (governmental and multilateral organizations), as well as academic databases that compile information on production, consumption, and export/import dynamics (Atal et al., 2020). The use of long historical series, structured questionnaires, random sampling, direct interviews, and sensory analyses of processed products are recurring practices to guide value estimates and market segmentation.

It should be noted that there are statistical limitations that impact the accuracy of projections, mainly due to the scarcity of standardization of data and information collection methods, divergences between criteria adopted by each source, and a lack of traceability mechanisms in fragmented production chains (Jiang et al., 2023). Discrepancies between private, governmental, and academic studies can be observed, especially in the volumes of national production and exports, pointing to the need for methodological harmonization and transparency in sampling processes, as already discussed in other sectors of the agri-food market (Benzing et al., 2021).

In the case of Makhana, such limitations are even more pronounced because official agricultural statistics rarely distinguish *E. ferox* from other aquatic crops, while private consultancy reports often rely on short time series and proprietary extrapolation models. As a result, discrepancies between estimates for production,

export volumes, and market size are common, and cross-verification among independent datasets is seldom possible. Rather than attempting to reconcile all numbers into a single "true" value, this chapter adopts a conservative stance that highlights ranges and trends, explicitly acknowledging that robust, harmonized, and traceable market statistics for Makhana are still lacking.

17.4 Botanical, agronomic, and ecological aspects

17.4.1 Morphological and ecophysiological aspects

Euryale ferox is a perennial aquatic plant belonging to the Nymphaeaceae family. It exhibits striking morphological characteristics, including floating, spiny leaves, purplish-blue flowers, and rounded seeds encased in a resistant seed coat. Its reproductive cycle is annual, beginning with seed germination in shallow water, followed by leaf development, epigynous flowers, and subsequent fruiting, culminating in seed dispersal between April and October in temperate environments. Seed morphogenesis is complex, involving maturation processes under the influence of plant hormones and environmental variations. Physiologically, the plant requires temperatures between 22°C and 30°C, high nutrient availability, and a neutral to slightly acidic pH for optimal growth and productivity. Fibrous root architecture favors efficient fixation in waterlogged soils, while ecophysiological adaptations confer tolerance to water variations and promote robustness in wetland areas (Jiang et al., 2023).

17.4.2 Cultivation systems

Makhana cultivation in Bihar, India, is traditionally carried out in shallow wetlands and freshwater ponds, where small producers manage the crop extensively and manually, using practices empirically passed down through generations (Garrett et al., 2025). The production cycle begins with the preparation of the ponds between October and November, followed by the planting of the seeds between December and January. During growth, cleaning management, control of competing weeds, and water monitoring are carried out to ensure successful germination and plant development.

With the growth of global demand and the need to increase productivity, emerging intensive systems are being implemented, such as excavated ponds, semiintensive nurseries, and hybrid agroecosystems, in which rotation techniques, co-cultivation of fish, and nutrient optimization are integrated into management. In more technologically advanced areas, controlled irrigation, specific fertilization, and automated monitoring of environmental parameters are used (Annual Report 2024–25). These systems can improve seed quality and increase germination rates, providing greater uniformity and reducing postharvest losses. The final quality of the seeds is closely related to the agricultural practices used, being influenced at this stage by factors such as proper tank cleaning, disease control, water management, and harvest time (Garrett et al., 2025).

17.4.3 Agronomic challenges

Its cultivation faces a series of agronomic challenges that directly impact its productivity and final product quality. Among the main obstacles are phytopathological diseases caused by fungi and bacteria that affect aquatic plants, as well as the occurrence of specific pests, such as aquatic insects and competing organisms that can reduce seed germination (Garrett et al., 2025). Water variations, associated with the fluctuation of water levels in tanks and wetlands, also represent considerable risks, especially in regions susceptible to prolonged droughts or flash floods. These variabilities can result in failures in root fixation and affect the reproductive cycle of *Euryale ferox*.

Another aggravating factor refers to the costs of implementing and managing production systems, which are still high due to the predominance of manual management, the demand for constant cleaning of tanks, and the need for continuous monitoring of environmental parameters. Intensive systems require additional investments in infrastructure and technology, which limits adoption by small municipal producers (Garrett et al., 2025). Therefore, the impact of climate change becomes increasingly relevant in the context of Makhana farming, threatening the stability of the production cycle through changes in average temperature, increased frequency of extreme events, and reduced natural water reserves (Jana et al., 2018). These trends necessitate the adoption of adaptation strategies, such as integrated pest management, diversification of water resource use, and research on genotypes more resilient to environmental variations.

17.5 Chemical composition, nutritional quality, and Functional Properties

17.5.1 Detailed nutritional composition

Makhana seeds are primarily characterized by their high carbohydrate content, which accounts for approximately 76.9% of their total composition. Starch represents the predominant fraction, exceeding 70% of the seed's dry weight and serving as its primary nutritional component. Native starch (ES) extracted from *E. ferox* kernel starch (EFKS) typically exhibits an amylose content of approximately 20.82%, although values can reach 27.13% depending on the specific cultivar.

The proximate composition values compiled in Table 17.1 synthesize findings from diverse regions of India and China, reflecting variations in cultivars, analytical protocols, and moisture bases. Consequently, these reported ranges should be interpreted as biologically plausible intervals rather than exact universal constants; much of this variability likely stems from methodological discrepancies rather than inherent nutritional contrasts among the samples.

A critical functional attribute of Makhana starch is its significant content of resistant starch (RS). Native starch extracted from *E. ferox* typically contains substantial amounts.

TABLE 17.1 Proximate composition and energy value.

Component	Quantity (%) or value	Scientific notes	Source
Carbohydrates	75.4%–76.9%	Primary macronutrient and most abundant component.	(Mittal et al., 2020; Shukla et al., 2023; Kumar et al., 2016)
Carbohydrates (starch- based)	78%	Proportion of carbohydrates present in starch form within the seeds.	(Sinha and Jha, 2025)
Proteins	9.7%–11.72%	Protein content typically ranges between 10% and 12%.	(Shukla et al., 2023; Mittal et al., 2020; Kumar et al., 2016)
Fat (lipids)	0.1%–0.51%	Classified as nearly fat-free.	(Shukla et al., 2023; Mittal et al., 2020; Kumar et al., 2016)
Moisture	11.82%–12.8%	Moisture content measured in seeds or grains.	(Sinha and Jha, 2025; Mittal et al., 2020; Shukla et al., 2023; Kumar et al., 2016)
Minerals (ash)	0.5%–0.62%	Total ash content represents mineral composition.	(Shukla et al., 2023; Mittal et al., 2020; Kumar et al., 2016)
Starch (kernel/ Isolated starch)	68%–71%	Starch concentration in kernel of cultivars from Assam and Manipur.	(Devi and Deka, 2022)
Amylose (EFKS starch)	25.3%–27.13%	Maximum amylose content comparable to rice starch.	(Devi and Deka, 2022; Kumar et al., 2016)
Resistant starch (RS) (native starch)	59.58%–68.76%	Significantly higher than most cereal starches.	(Devi and Deka, 2022)
Caloric value (kcal/100 g)	Raw: 362; Popped: 328	Popped products exhibit slightly reduced caloric value.	(Mittal et al., 2020)

Source: From authors.

RS levels, with specific cultivars reaching between 59.58% and 68.76% in their native state. RS functions similarly to dietary fiber; it evades enzymatic digestion in the small intestine and undergoes fermentation by colonic microbiota, such as Bifidobacterium.

This fermentation process yields short-chain fatty acids (SCFAs), which optimize the intestinal environment and have been linked to improved postprandial glycemic responses and a reduced risk of insulin resistance.

However, most available data on the glycemic index (GI) of Makhana are derived from in vitro models or purified starch fractions. While these standardized assays provide valuable comparative insights, they may not fully capture the complexity of human digestion or the influence of the food matrix in commercial products. Consequently, while current evidence suggests a moderately low glycemic impact, particularly for minimally processed kernels, rigorous human clinical trials remain essential to substantiate specific dietary and therapeutic recommendations.

Regarding its macronutrient profile, Makhana is notable for its negligible lipid content (0.1%–0.51%), making it an ideal component for calorie-restricted diets. The protein content is moderate (9.7%–12.0%), and the energy density of raw seeds is approximately 362 kcal/100 g. Furthermore, Makhana serves as a relevant source of essential minerals, including phosphorus (0.9%), calcium (0.02%), and iron (~1.4 mg/100 g).

As detailed in Table 17.2, these micronutrient concentrations are subject to environmental and technical variables, such as soil fertility, water quality, and analytical methodology. Therefore, these values should be interpreted as indicative of the species' nutritional potential rather than fixed regulatory constants.

TABLE 17.2 Vitamins and minerals.

Micronutrient	Quantity (%) or IU/g	Scientific notes	Source
Phosphorus (P)	0.028%–0.9%	Phosphorus content in market samples typically ranges from 0.0769% to 0.0899%.	(Sinha and Jha, 2025; Mittal et al., 2020; Shukla et al., 2023; Kumar et al., 2016)
Calcium (Ca)	0.02%–0.0392%	Calcium content is measured in seeds.	(Sinha and Jha, 2025; Shukla et al., 2023; Kumar et al., 2016)
Iron (Fe)	0.0014%–0.0043%	Equivalent to approximately 1.4 mg per 100 g.	(Sinha and Jha, 2025; Shukla et al., 2023; Kumar et al., 2016)

TABLE 17.2 Vitamins and minerals.—cont'd

Micronutrient	Quantity (%) or IU/g	Scientific notes	Source
Sodium (Na)	Detected (e.g., Sodium (Na) 800–1000 ppm in unrefined starch)	The plant can absorb sodium even from low-sodium soils.	(Sinha and Jha, 2025; Kumar et al., 2016)
Magnesium (Mg)	Detected (e.g., >2000 ppm in bran)	Essential minerals are involved in blood pressure regulation.	(Shukla et al., 2023; Kumar et al., 2016)
Potassium (K)	Seeds: 159–240 mg/kg; Rhizome: 2170 mg/kg	High potassium content makes makhana suitable for individuals with hypertension.	(Vikram and Mishra, 2021)
Vitamin A	62.23–63.84 IU/g	Quantified in market samples.	(Kumar et al., 2016)
Vitamin C (ascorbic acid)	0.183%–0.203%	Present in relatively low concentrations.	(Sinha and Jha, 2025b; Kumar et al., 2016)

From authors.

From a nutraceutical point of view, it is notable for being rich in Magnesium and Potassium and having a low Sodium content, which makes it an ideal food for patients with hypertension. However, the plant can absorb sufficient Sodium (Na) even in low-content soils, with a content ranging from 180 to 200 mg/kg in the seeds. Regarding vitamins and other compounds, the seed contains B complex, Vitamin C (ascorbic acid), as well as phenol and sugar.

The protein quality of *E. ferox* stands out as one of its most distinctive nutritional attributes. Its Essential Amino Acid Index (EAAI) is remarkably high, ranging from 89% to 93%. In its raw state, the EAAI reaches 93.63%, surpassing major staple crops such as soybean (85.59%), rice (82.88%), and wheat (65.18%).

The amino acid profile is comprehensive (Table 17.3), with glutamic acid and arginine being the most prevalent, reaching levels of 16.64 g/16 g N and 15.19 g/16 g N, respectively. The high concentration of these specific amino acids, alongside lysine (3.79 g/16 g N), methionine (3.06 g/16 g N), and leucine (8.34 g/16 g N), provides a biochemical rationale for Makhana's traditional use in promoting reproductive health and maintaining tissue integrity, particularly regarding skin and hair.

TABLE 17.3 Aminoácidos e Qualidade da Proteína.

Property/Amino acid	Value (g/16 g N)	Scientific significance and comparison source	(References)
Essential amino acid index (EAAI)	93.63% (Raw); 89.97% (Fried)	Higher than staple foods such as rice (82.88%), wheat (65.18%), and cow's milk (88.8%).	(Sinha and Jha, 2025; Kumar et al., 2016; Mittal et al., 2020)
Arginine	15.19 (Raw); 16.07 (Fried)	Highly abundant amino acid and a biochemical precursor of creatine.	(Shaw et al., 2025; Kumar et al., 2016; Mittal et al., 2020)
Glutamic acid	16.64 (Raw); 17.06 (Fried)	Most abundant amino acid in the protein profile.	(Kumar et al., 2016)
Lysine	3.79 (Raw); 4.69 (Fried)	Essential amino acid with significant nutritional importance.	(Kumar et al., 2016)
Methionine	3.06 (Raw); 2.95 (Fried)	Essential sulfur-containing amino acid and precursor of creatine, important for skin and hair metabolism.	(Shaw et al., 2025; Kumar et al., 2016; Mittal et al., 2020)
Leucine/ Isoleucine ratio	1.7 (Raw); 1.66 (Rice/W heat)	A ratio of approximately 1.7 is considered nutritionally favorable.	(Mittal et al., 2020; Kumar et al., 2016)
Arginine + lysine/ Proline	6.3%–9.6%	High proportion indicating good overall protein quality.	(Mittal et al., 2020; Kumar et al., 2016)
Biological value (BV)	55 (Fried); 84.5 (Cow's milk)	Comparable to the biological value reported for most cereals.	(Kumar et al., 2016)
Estimated glycemic index of purified resistant starch (RS3)	0.5491	Represents the estimated glycemic index (eGI) of purified resistant starch (RS3) from *Euryale ferox*, calculated using the hydrolysis index ($eGI = 39.71 + 0.5491 \times HI$).	(Yang et al., 2025)

From authors.

Despite this robust amino acid profile, the Biological Value (BV) of popped Makhana seeds is comparatively lower (~55). This discrepancy is primarily attributed to an amino acid imbalance, specifically a high leucine-to-isoleucine ratio (1.84–1.99), which exceeds the FAO/WHO recommended standard of 1.75. This imbalance can interfere with the absorption and metabolic utilization of other branched-chain amino acids (BCAAs).

Table 17.3 synthesizes these protein quality indices based on current literature, which focuses primarily on Indian cultivars. However, given the limited number of datasets and the variability across processing methods (raw vs. fried vs. popped), further systematic studies are required to fully map the protein quality diversity across different genotypes and industrial segments.

Makhana has a moderately low glycemic index (GI 50–55), allowing for a mild postprandial effect on blood glucose, a characteristic valued in glycemic control diets and in the prevention of metabolic syndromes (Liaquat et al., 2022). Comparatively, Makhana has a similar energy value to traditional popcorn, but is superior in protein, dietary fiber, and essential minerals. Its competitive composition positions the food as a healthy alternative to ultra-processed snacks and common plant-based foods.

17.5.2 Variables that influence quality

The quality of Makhana seeds is directly influenced by aspects related to the plant's genotype, geographic origin, environmental conditions during cultivation, and harvest time (Jiang et al., 2023; Saha et al., 2022). Genetic varieties may exhibit varying resistance to diseases, changes in nutritional composition, and concentrations of bioactive compounds. The location of cultivation, especially in areas like Bihar, affects the physical parameters and consistency of the seeds due to climatic conditions and local management (Garrett et al., 2025).

Environmental factors, such as water availability, average temperature, and soil fertility, directly influence the physiological development of seeds and their germination capacity. The ideal harvest time is crucial: seeds harvested early tend to have lower vigor and alterations in nutritional composition, while late harvests can increase the risks of deterioration and pathogen infections.

Postharvest processing encompasses steps such as roasting, drying, and maturation, which are essential for preserving the functional, sensory, and physiological quality of seeds. Inadequate drying and storage methods can lead to the loss of bioactive compounds, germination potential, and susceptibility to pest attacks (Garrett et al., 2025). The adoption of modern techniques, such as controlled drying, proper maturation, and storage in climate-controlled environments, contributes to extending shelf life and maintaining physiological qualities.

17.6 Postharvest, technology, and industrial processing

The processing of Makhana seeds involves a series of fundamental operations to guarantee the quality, food safety, and industrial standardization of the final product. Initially, harvesting can be done manually, especially in traditional Bihar systems, or mechanically in larger industrial units, which allows for speed and reduced losses due to deterioration (Garrett et al., 2025). Fig. 17.2 illustrates the harvesting process:

The processing of *Euryale ferox* begins with the manual harvesting of mature and dehiscent seeds, which takes place between August and September (Mandal et al., 2010). Next, the seeds undergo rigorous washing to remove mucilage and debris. The raw material, with an initial moisture content of 37% (wet basis) (Kumar et al., 2016), goes through controlled drying processes, either solar or mechanical, until it reaches a safe level of 25% (wet basis) (Devi and Deka, 2022). Then, they are subjected to granulometric classification (into 5 to 7 classes or up to 15–16°), an essential step to ensure uniformity in heat transfer in subsequent thermal processes (Kumar et al., 2016).

The crucial step in the transformation involves preheating the seeds to a temperature between 280°C and 335°C, followed by a quenching period of 40–50 h under ambient conditions. This process favors the relaxation of internal tensions and the detachment of the pericarp, resulting in an increase in popping yield. Popping is achieved by roasting small batches at high temperatures (230°C–335°C). Under these conditions, the internal vapor pressure, stabilized at about 11% moisture (wet basis), results in the explosive rupture of the shell when exposed to an immediate mechanical impact (striking on a hard surface with a wooden hammer) (Kumar et al., 2016; Mandal et al., 2010).

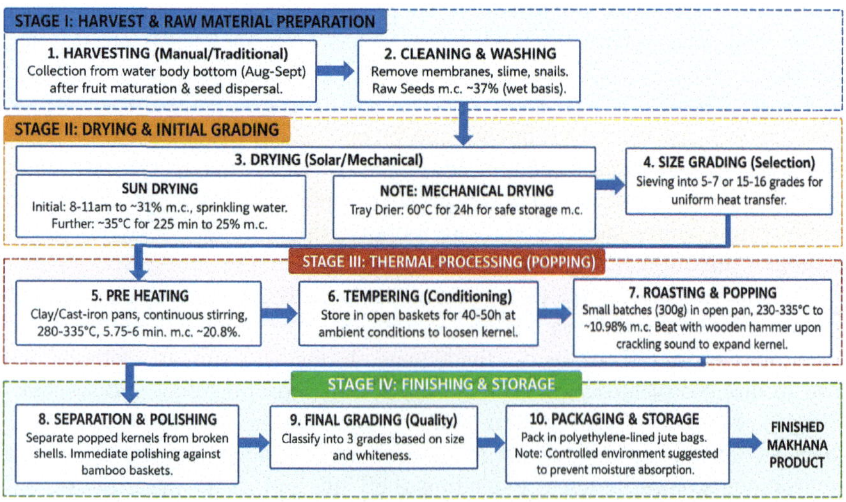

FIGURE 17.2 Technical flowchart: Makhana processing method. *From authors.*

In the final finishing step, the expanded kernels are separated and polished by friction (in bamboo baskets) to reduce moisture absorption (Mandal et al., 2010). The product is then classified according to criteria of whiteness and physical integrity (in 3 grades) before being packaged in burlap sacks lined with polyethylene (Kumar et al., 2016). It is recommended that storage be carried out in hygrometrically controlled environments to preserve stability, since the dried seeds are stored for safe preservation (Devi and Deka, 2022).

17.6.1 Technological innovations

In recent years, there have been significant advances in the industrial processing of Makhana, especially regarding alternative methods that seek to maintain its bioactive compounds and improve the functional properties of the seeds. The application of infrared and microwave technology, such as industrial fluidized bed assisted heating, demonstrates a significant reduction in density, grain weight, and hardness, in addition to promoting an increase in fissures and porous structures. These factors facilitate the expansion and release of previously retained nutrients, expanding the possibilities for using the product in industrial processes (Wang et al., 2022; PMC9407493).

Dry-frying and bran-frying processes show differences in retention and bioactive compounds, such as polysaccharides, polyphenols, and glycolipids. Research in metabolomics indicates that, depending on the method used, there is a considerable change in the lipid profile and in the presence of differential compounds (32–38 markers between techniques), which can be strategically used for the development of functional products, according to the growing market demand (Yang et al., 2025; MDPI, 2025; PMC12028916).

In addition, investment in minimal processing, with modified atmosphere (MAP) and the use of active packaging that incorporates natural antioxidants, seeks to preserve thermolabile compounds and prolong product stability, serving gourmet and nutraceutical segments with high added value.

17.6.2 Food safety

Food safety in Makhana processing requires attention to several hazards, especially microbiological ones, such as contamination by bacteria, fungi, and deterioration by environmental agents during postharvest and storage (Saha et al., 2022; Garrett et al., 2025). The use of controlled drying, moisture monitoring, temperature control, and the adoption of good manufacturing practices are essential to reduce the risk of proliferation of Salmonella spp., Bacillus cereus, and coliforms, which can affect seed lots in inadequately managed environments (Alonso et al., 2021).

Another challenge relates to lipid oxidation, which causes sensory changes and loss of bioactive compounds. Prevention strategies include the use of active packaging with oxygen barriers, modified atmosphere packaging (MAP), and the use of natural antioxidants. These measures extend shelf life, maintaining the sensory and functional quality of the processed seeds (Wang et al., 2022). The industrial segment of Makhana must comply with international food safety regulations, such as the recommendations of the Food Safety and Standards Authority of India (FSSAI), the US FDA, and European Union standards, which require traceability, sanitary records, maximum contaminant limits, and the adoption of Hazard Analysis and Critical Control Points (HACCP). Compliance with these requirements is essential for export and global marketing, as well as adding value to the supply chain and ensuring consumer confidence (Garrett et al., 2025).

17.7 Sustainability and bioeconomy

17.7.1 Contributions of makhana to the sustainability of aquatic environments

Makhana farming stands out for its strategic role in aquatic ecosystems, especially in wetland regions of India. These areas offer ideal conditions for the development of *Euryale ferox*, promoting ecosystem services such as maintaining aquatic biodiversity, regulating water resources, and protecting the soil. The presence of Makhana contributes to the ecological structure of wetlands, serving as food and shelter for different aquatic species and birds, and favoring nutrient cycles in natural reservoirs (Raut et al., 2025).

The sustainable cultivation of Makhana, especially when associated with integrated aquaculture techniques such as semiintensive fish co-culture systems, enables the rational use of water resources, reduces the use of industrial inputs, and enhances the conservation of water quality (Raut et al., 2025). Plants help keep the water clean, acting as natural filters against the accumulation of debris, and capture carbon in wetlands, which reduces greenhouse gas emissions (Garrett et al., 2025).

In addition, Makhana can be part of local and regional bioeconomy strategies, generating sustainable income and contributing to environmental preservation. Carbon capture, habitat conservation, and increased biodiversity in cultivated areas make Makhana an essential element in new proposals for sustainable aquatic production.

17.7.2 Relationships between community, environment, and makhana cultivation

Growing *Euryale ferox* in shallow wetlands and integrated aquaculture systems constitutes an efficient strategy to mitigate dependence on exogenous inputs, notably synthetic fertilizers and agrochemicals. By capitalizing on sedimentary fertility and the trophic balance inherent in these ecosystems

(Raut et al., 2025), the practice favors nutrient cycling and reduces the load of water pollutants.

From a socioeconomic point of view, Makhana production continues to be a force that promotes economic stability for smallholder farmers, mainly in the Bihar region and Nepal. Evidence shows that the incorporation of rural development and extension policies directly improves average income and regional food security (Garrett et al., 2025). Regarding their adaptability to environmental conditions, these systems demonstrate greater buffering capacity than conventional terrestrial crops, except in cases of extreme hydrological events (Jiang et al., 2023). Thus, there is certainly still room for improvement in this value chain, which represents an appropriate corrective measure against intensive production, thereby reinforcing agroecological models that prioritize environmental resilience along with social integration.

17.7.3 Life cycle assessment (LCA)

The Life Cycle Assessment (LCA) of Makhana considers direct and indirect emissions throughout the production stages, from cultivation in wetlands and industrial processing to distribution and marketing. Studies indicate that Makhana cultivation has significantly lower emissions compared with high-intensity terrestrial crops, such as rice, corn, and beans, due to the lower use of synthetic inputs, lower energy demand in semiintensive systems, and the ability of wetland systems to sequester carbon, mitigating the climate impact (Hoekstra, 2016).

In addition to low greenhouse gas emissions, there is a reduction in post-harvest losses and less solid waste generation in the processing stages, reflecting the full use of seeds and the adoption of circular economy practices in some production chains. Mitigation strategies, such as the use of clean technologies like solar drying and biodegradable packaging, as well as the adoption of renewable energies in industrial stages, are being incorporated to enhance the environmental sustainability of the sector (Jalaj et al., 2019).

At the time of writing this study, there is still a need for standardization of environmental indicators and the establishment of specific LCA protocols for aquatic products, including factors for carbon sequestration, water balance, and impact on biodiversity, to increase the robustness of assessments and facilitate international comparisons.

To date, no comprehensive life cycle inventory has been developed explicitly for Makhana systems, and most environmental assessments extrapolate emission factors and water use coefficients from better-studied crops such as rice or generic wetland agriculture. While these analogies offer a first approximation, they may overlook unique features of *E. ferox* cultivation, including its dual role in providing food and ecosystem services in shallow wetlands. Developing dedicated LCA protocols that integrate carbon sequestration dynamics, biodiversity indicators, and water-quality regulation is, therefore, a priority for accurately benchmarking Makhana against other agri-food commodities.

17.8 Innovation and technological trends

17.8.1 New product development

Euryale ferox is recognized as a prominent superfood, abundant in macronutrients, micronutrients, and bioactive compounds, giving it vast potential for both the food and nonfood industries. The preferred form of consumption of this food is the popped seed, which is the crucial base product driving its growing industrial use (Wang et al., 2022).

The process of transforming raw foxnuts into Popped Foxnut is a procedure traditionally considered skillful and time-consuming. This processing cycle, schematically illustrated in Fig. 17.3, encompasses several phases, such as drying, grading, preheating, roasting, and cleaning. Roasting is carried out at high temperatures, ranging approximately between 230 and 335°C (Kapoor et al., 2022).

The heat treatment performed during popping is crucial for the physicochemical characteristics of the final product. As shown in Fig. 17.1, this process results in key nutritional changes: there is an increase (↑) in protein content and soluble dietary fiber. The increase in protein occurs through the conversion of complex protein molecules into simpler structures. Simultaneously, there is a reduction (↓) in fat content, mainly due to the removal of the bran portion during popping, and in the concentration of carbohydrates (starch), which degrade into smaller molecules, increasing the total sugar

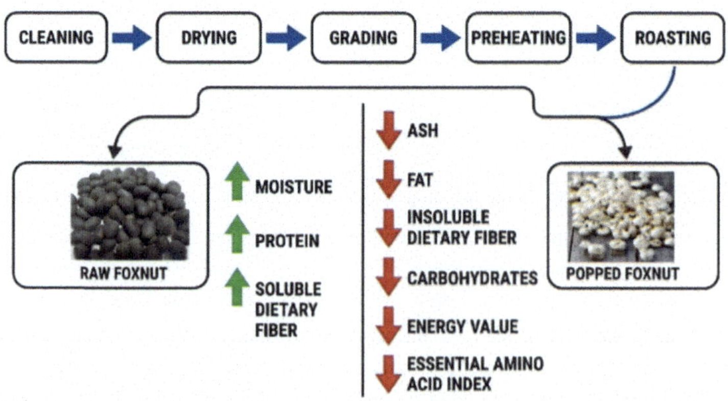

FIGURE 17.3 Schematic illustration of the popping process of foxnut (*Euryale ferox*). *From authors.*

content. The popped product is easily digestible and considered healthy for human consumption.

This enhanced nutritional profile, combined with its functional food nature, is the basis for the growing diversification of the Makhana chain. There is an investment in technological expansion and texturization processes aimed at developing premium products with high nutritional appeal. Puffed seeds and their derivatives are widely used as gourmet and functional snacks, in the baking industry for formulating biscuits, breads, gluten-free pasta, and protein mixes, as well as in the production of powdered extracts and functional blends intended for the plant-based beverage and food supplement industry (Wang et al., 2022).

17.8.2 Digitization of the supply chain

The modernization of Makhana's production is gaining momentum with the innovations of Industry 4.0, which bring more advanced resources for traceability control and intelligent production management. The incorporation of IoT (Internet of Things) sensors allows for real-time monitoring of critical environmental parameters, such as humidity and temperature, during cultivation and postharvest, generating data that supports quick and accurate decisions to Shahab et al. (2025).

The digital traceability of the chain can be enhanced by blockchain systems Fig. 17.4, ensuring transparency, authenticity of origin, and reliability in the data shared between producers, distributors, and consumers worldwide. Cloud computing enables the integrated management of production stages, where it is possible to connect the different links in the chain and facilitate remote access to information on batch, quality, sustainability, and certifications, strengthening governance practices and adding commercial value to the product (Queiroz et al., 2019).

The advancement of digitalization in agribusiness provides greater food security, helps respond to environmental variables, and fosters sustainable practices, adapting the Makhana supply chain to overcome challenges of scale, market, and regulatory compliance (Gurtu and Johny, 2019).

17.8.3 Genetic and genomic innovation in EFS: Molecular foundations and breeding perspectives

As already mentioned, despite its potential as a functional food, *E. ferox* cultivation is often neglected and has received little attention in terms of crop optimization and management. However, the growing demand for functional foods and the use of Makhana in the development of new products require an effort to disseminate traditional knowledge and modernize cultivation and processing. Thus, genetic and genomic innovation are crucial areas for improving production and product quality (Kapoor et al., 2022).

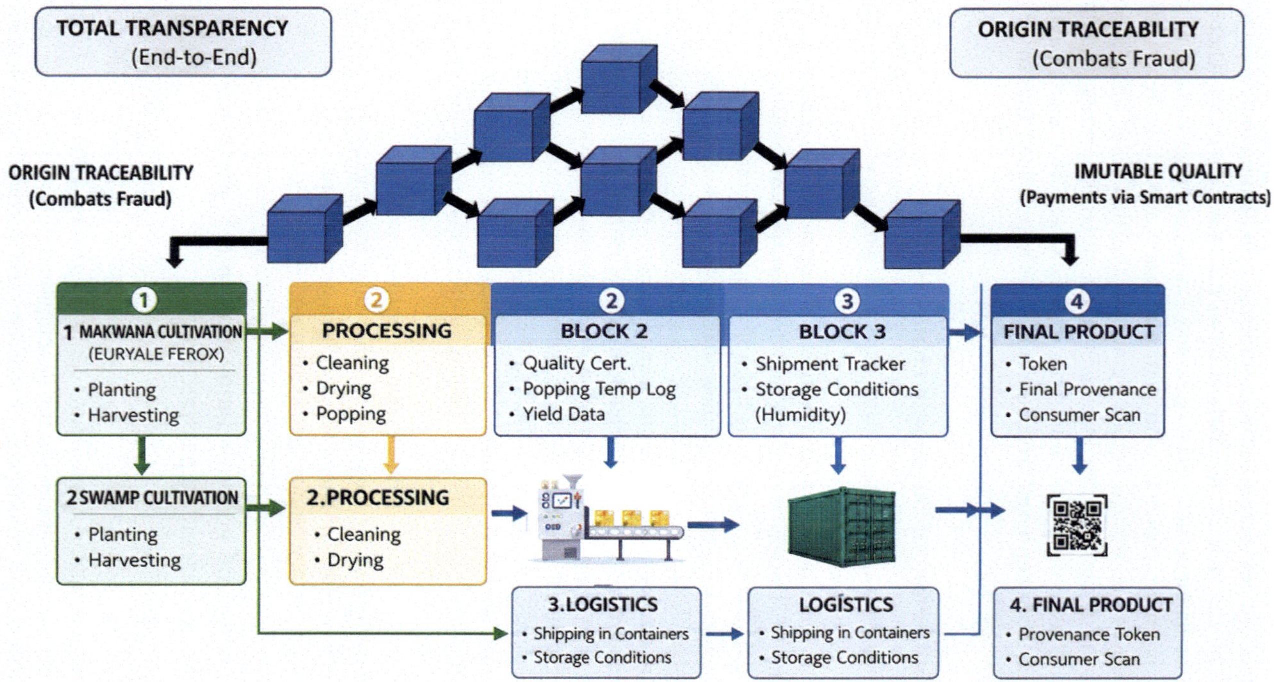

FIGURE 17.4 Schematic representation of the integration of blockchain technology into the *Euryale ferox* (Makhana) value chain. The diagram demonstrates the data flow from cultivation in the swamps to the end consumer, highlighting the correlation between the physical processing steps and the generation of immutable digital records to guarantee traceability and quality. *From authors.*

Recent molecular studies have focused on understanding how the seed develops and its phytochemical composition, explaining important biosynthesis pathways. Detailed research used iTRAQ (Isobaric Tags for Relative and Absolute Quantitation) technology to analyze the proteins of *E. ferox* seeds at different growth stages (10, 20, 30, and 40 days). This analysis found more than 2800 different proteins (Wu et al., 2023). The study of differential protein enrichment demonstrated that the starch biosynthesis pathways and the flavonoid biosynthesis pathways were highly enriched (Kumar et al., 2016). This information is crucial, since the *E. ferox* seed has a high content of polyhedral starch, and flavonoids are responsible for a large part of its antioxidant properties.

In addition, transcriptome analysis and metabolomics were applied to investigate the biosynthesis mechanism of flavonoids in seeds at different stages of development. The presence of 129 types of flavonoid metabolites in *E. ferox* was reported, including proanthocyanidins, flavonols, flavones, chalcones, isoflavones, flavanones, flavonoid carbonates, dihydroflavanols, flavanols, and anthocyanins. The detailed understanding of these molecular and metabolic pathways allows future genetic improvement efforts to focus on gene manipulation to increase the concentration of specific bioactive compounds, such as flavonoids and triterpenes (hypoglycemic agents) (Wu et al., 2023).

Innovation in *E. ferox* has been manifesting itself, in an incipient way, in the development of new varieties. A recently developed Makhana variety was named Swarna Vaidehi. In addition, the market recognizes four types of Makhana pop based on size: lava, murha, turi, and mix (Semwal et al., 2021).

However, innovation must go beyond the development of limited varieties. Given the difficulty and labor-intensive nature of traditional harvesting, there is an urgent need for technological development in production, including improved harvesting practices (Khadatkar et al., 2015). Understanding engineering properties, such as fracture properties, is desirable for optimizing processing and adding value to *E. ferox* products.

In conclusion, although Makhana is still mostly cultivated and processed using traditional methods, the evidence of its biochemical complexity, revealed by proteomic and metabolomic studies, opens a vast field for biotechnology and genetic improvement. The National Research Center for Makhana, Darbhanga, is already active in this area and has the potential to lead the increase in production and the development of new technologies that benefit the production chain (Kapoor et al., 2022).

17.8.4 Farmers, economic potential, and new business flows

The resilience of *Euryale ferox* cultivation is rapidly changing due to the improvement of the global socioeconomic position and the consequent increase in the search for healthier foods (Kapoor et al., 2022). This market expansion gives the plant great potential to create new food products, opening new possibilities for the sector (Kumar et al., 2016). According to the National Makhana Research Center, this production extends over 15,000 ha in India, generating 120,000 tons of seeds, which are transformed into 40,000 tons of processed Makhana. There is a noticeable increase in economic value, going from 250 crore rupees for the farmer to a revenue of 550 crore rupees for the trader (Apeda, 2020), where proper processing and value aggregation are fundamental to increasing producers' income, resulting in distinct market classifications such as lava, murha, turi, and mix (Kumar et al., 2020). On the contrary, commercial expansion faces challenges, since the high volume of foxnut expansion considerably increases transportation costs, requiring a deeper study of the engineering and fracture properties of the material to optimize logistics (Kumar et al., 2016). It is also possible to observe that the sector has come into the focus of attention of the Indian government and private companies (Kapoor et al., 2022), an institutional intervention necessary especially to reshape harvesting practices, which remain dangerous and labor-intensive, causing health problems for workers, such as skin rashes and thorn injuries (Khadatkar et al., 2015; Khadatkar et al., 2020). To meet future demand, efforts are needed to disseminate production technologies that can increase yields in the field (Kapoor et al., 2022).

17.9 Technoeconomic analyses and industrial feasibility

17.9.1 Cost structure and margins

The cost structure of the *Euryale ferox* (Makhana) value chain is significantly affected by low production efficiency and a strong need for manual labor. Overcoming these problems is essential to take advantage of the enormous market potential of this superfood.

The cost of production (cultivation) depends heavily on the yield per hectare. Kapoor et al. (2022) show an important difference in efficiency: the traditional method of cultivation in still waters (tanks, lakes, ponds) offers a reasonable yield of raw seed. On the contrary, the field system, which represents progress in crop management, manages to increase the yield to a range of 2.6–3.0 tons per hectare. Although the exact values per hectare are not available, this yield difference shows that the field system is more efficient in converting cultivated areas into raw material. This is a crucial advantage to meet the growing demand.

The harvesting of *E. ferox* is the main factor in operational costs and a point of great social risk. It is a risky and labor-intensive task, as shown by Kapoor et al. (2022) and Khadatkar et al. (2015). Workers are constantly subject to skin problems, dirt from the growing environment, and the danger of plant thorns. This difficulty requires urgent action from governments and public policy makers, aimed at improving harvesting practices and developing production technology that reduces this burden.

Transforming Makhana into pop (final product) is a very artisanal and time-consuming process, still very much tied to old practices. Seed preparation, considered a crucial step, takes a long time (about 40–50 h to facilitate the loosening of the grain shell), which impacts a high labor cost, hindering the increase in large-scale production (Kapoor et al., 2022; Ahmad and Kumar, 2020).

In transportation, the cost of logistics represents a major challenge, reducing the viability of industrializing foxnut pop. Due to its large volume, the final product makes transportation expensive (Kapoor et al., 2022; Kumar et al., 2016). To improve logistics and processing, it is essential to thoroughly study the characteristics of the seed and how it can be transported (Kapoor et al., 2022; Kumar et al., 2016).

When analyzing the gains in Makhana production, we realize that most of the money comes in the final stage, when the raw seed becomes the pop product (Kapoor et al., 2022). The field cultivation system results in a higher yield of raw seeds compared with the traditional system, which uses permanent bodies of water (Kapoor et al., 2022; Khadatkar et al., 2020).

17.9.2 Industrial scale models

The ability to increase Makhana's industrial production depends on how different planting and preparation methods mix and evolve, directly affecting the amount of raw material available and the practicality of transportation (Kapoor et al., 2022).

The structure of the production chain is defined by two forms of cultivation that determine how efficient agricultural production is:

- Traditional Planting in Water: Being the most common throughout history, this type of planting uses areas of water that do not dry up, such as lakes and ponds. Even though it is common practice, space is not well utilized, generating a raw seed production of only 1.8–2.0 tons per hectare. Harvesting in this system is dangerous and requires more labor. Workers face social and health risks due to contact with water and the plant's thorns (Khadatkar et al., 2015). It is evident that this method limits production growth.
- Intensive Field Planting: This method is the most suitable for industrialization and increased production. There is agreement among experts that the planting area of *E. ferox* needs to be increased and, above all, intensified, especially in the field system, where water use is better utilized. The field system is much more efficient, increasing production by between 2.6 and 3.0 tons per hectare (Kapoor et al., 2022).

The processing stage is where the added value is most tangible, and the distinction between scales becomes more acute:

- Traditional Processing Model (Decentralized and Manual): The transformation of raw seed (black diamond) into pop (white diamond) is a laborious process that requires high artisanal skill. It involves a series of manual phases: drying, sorting, preheating, and, critically, tempering, where the seeds are stored for long hours so that the seed coat loosens. The final popping step is performed by striking a few seeds (6–7) with a hammer on a hard surface. Due to its dependence on individual skill and the long processing cycle, this model is inherently small-scale and decentralized (Kapoor et al., 2022; Ahmad and Kumar, 2020).
- Industrial Processing Model (Mechanized and Centralized): The transition to industrial scale requires the complete mechanization of the process. The high consumption of *E. ferox* in popped form requires the mechanical transformation of the roasted seed into a lightweight, high-volume product. The high volumetric expansion of foxnut popping becomes a challenging logistical factor, increasing transportation costs (Kumar et al., 2016). To optimize industrial efficiency, a deeper understanding of the seed's engineering properties, such as its fracture characteristics, is fundamental and crucial for the development of improved processing and logistics equipment (Kumar et al., 2016). Therefore, mechanization of popping is the main driver for maximizing production and efficiency (Kapoor et al., 2022).

The versatility of large-scale production is evident in the numerous applications of *E. ferox* across multiple sectors, indicating high value-added diversification:

- Functional Food Sector: Recognized for its beneficial qualities, Makhana stands out as a superfood, opening up opportunities in the food industry. It is a key component in the creation of functional foods and nutraceutical products. *E. ferox* starch, with its high capacity to withstand paste temperature, is particularly promising in both food and nonfood industries.
- Pharmaceutical and Nutraceutical Sector: The application of Makhana on an industrial scale extends to the health sector. It is a traditional component in Ayurvedic formulations and, more recently, its bioactive compounds (glycosyl sterols, polyphenols, and triterpenes) demonstrate antidiabetic, cardioprotective, and antioxidant activities, validating its inclusion in high-value pharmaceutical and nutraceutical products.
- Non-Food Sector (Waste Valorization): Industrial production demands the use of by-products. The seed husk, for example, can be transformed into activated carbon with a large surface area, used to remove pollutants from industrial waste, completing the cycle of sustainability and valorization (Kumar et al., 2016).

Thus, the Makhana value chain is experiencing a crucial moment, migrating from limited aquatic production to intensified cultivation, coupled with the indispensable mechanization of processing and broad diversification into high-value sectors such as nutraceuticals and functional foods (Kapoor et al., 2022).

17.9.3 Barriers and opportunities for global expansion

Makhana's journey to conquer the global market is not without obstacles, typical of agri-food products. It is crucial to meet the demanding import standards of large markets such as the European Union, the USA, and Japan, which range from food safety to proof of sustainable practices (Garrett et al., 2025). Compliance with regulations such as HACCP, Fairtrade, and organic labels implies additional costs, requiring investment in new technologies, bureaucracy, and third-party inspections.

Fluctuations in exchange rates and political turbulence in exporting nations can destabilize prices in the international market, compromising profit margins and making it challenging to establish long-term agreements. To compete with other healthy options such as soy, quinoa, and Brazil nuts, Makhana needs to stand out for its nutritional value, innovations, and by focusing on niche consumers who value products with environmental labels and proven origin (Jiang et al., 2023).

However, the scenario is promising, driven by the growing global interest in "functional foods" and the demand for products that benefit public health. Policies to encourage exports, online sales platforms, and the strengthening of production chains linked to origin labels and digital management systems facilitate the entry of small and medium-sized enterprises (Garrett et al., 2025).

17.10 Research gaps and scientific agenda for the future

17.10.1 Metabolomics and advanced characterization

Research advances in Makhana have provided innovative approaches in metabolomics and advanced characterization of the phytochemical components of seeds. Using techniques such as high-resolution mass spectrometry, liquid chromatography coupled to spectrometry (LC-HRMS), and multivariate statistical analysis, between 32 and 38 differential compounds were identified in dry frying and bran-frying techniques of the seeds, including polysaccharides, polyphenols, and glycolipids that exhibit distinct lipid profiles and metabolic markers, according to the industrial processing applied (Yang et al., 2025).

Recent studies in Darbhanga, Bihar, investigated the carbohydrate and protein profile and the impact of environmental factors on the concentrations of these nutrients, promoting comparisons between Makhana and other functional foods such as ramdana (*Schoenoplectiella articulata*) and highlighting the importance of characterization to enhance the use of these native systems and diversify the local bioeconomy (Raut et al., 2025).

Identifying and analyzing these new phytochemicals from seeds is essential to understand their bioactivity, molecular interaction mechanism, and potential use as nutraceutical ingredients. Deepening these techniques reinforces the need to expand specialized databases and integrate modern methods into the innovation process in the sector.

17.10.2 Clinical and functional studies

The research has primarily focused on its metabolic effects in humans, especially regarding glycemic control, reduction in oxidative stress, and cardiovascular benefits. Controlled studies show improvements in metabolic markers and antioxidant capacity in groups that regularly consume Makhana preparations, highlighting an impact on the lipid profile, postprandial glycemia control, and mitigation of inflammatory biomarkers (Jiang et al., 2023; Saha et al., 2022). It is crucial to standardize how we measure the bioavailability of active compounds, as the results depend on how the product is processed and the individual characteristics of each person.

In animal models of induced diabetes, Makhana extracts demonstrated a significant reduction in fasting blood glucose, improved glucose tolerance, and increased plasma insulin concentrations, along with positive changes in liver enzymes involved in gluconeogenesis. These effects were accompanied by an improvement in the lipid profile, reducing triglyceride and total cholesterol levels. Histopathological signs of tissue protection were also observed in the pancreas, liver, and kidney, suggesting an adjuvant role in the prevention of chronic complications associated with diabetes (Kumar et al., 2022).

The literature points to the need for larger-scale clinical trials and in-depth study of the mechanisms of action of the phytochemicals present in the seeds, especially in the context of preventive nutrition and functional supplementation, in addition to reinforcing the opportunities for the application of Makhana in lines of nutraceutical supplements and functional foods intended for target audiences, such as diabetics, the elderly, and athletes (Saha et al., 2022).

Despite this, the clinical evidence base remains relatively narrow: most human studies involve small cohorts, short intervention periods, and heterogeneous preparations ranging from crude seed flour to concentrated extracts. Control groups, randomization procedures, and blinding are not always clearly reported, and dose– response relationships are seldom explored systematically. These constraints limit the generalizability of current findings and underline the need for larger, rigorously designed trials before firmly attributing disease-modifying effects to routine Makhana consumption.

The integration of food chemistry studies, bioavailability assays, and well-designed clinical trials will allow for the definition of dose profiles, presentation forms, and more precise indications for the use of Makhana as a functional ingredient in products intended for vulnerable populations or those with high physiological performance (Ahmed et al., 2015).

17.10.3 Postharvest innovations

To reduce waste and maintain the nutritional and functional properties of Makhana seeds, there has been progress in new postharvest techniques. It is of utmost importance to have combined methods that bring together proper handling, new storage technologies, and sustainable actions to mitigate losses in quality and quantity during preprocessing, which in Brazil can reach more than 50% of losses, and which are also important for Makhana in new productions (Observatório Latinoamericano, 2024).

Among the techniques used, noteworthy are the methods for verifying moisture and temperature, the use of biodegradable and special packaging (MAP), and also advances in mechanical and light drying, which attempt to protect the heat-sensitive compounds that exist in the seeds. The application of these techniques, often necessary to obtain organic certification and sell in more demanding markets, is directly related to guaranteeing the safety and quality of the final product (Centro-Oeste Brasileiro, 2010).

In the area of simpler drying and processing, there are initiatives to improve the combinations of actions such as predrying, tempering, roasting, and “popping,” using lower temperatures and shorter times, to increase the production of good-quality beans while reducing darkening, loss of heat-sensitive nutrients, and the emergence of bad flavors. The way the processes are done with Makhana shows that the correct technological sequences can significantly increase the harvest of first-rate beans and decrease the total processing time compared with older methods (Tracedata Research, 2025).

Together, the modernization of postharvest operations and the use of clean and sustainable technologies strengthen productive sustainability, reduce waste throughout the chain, and increase the competitiveness of Makhana in the market (Sinha et al., 2025).

17.10.4 Socioeconomic and governance studies

The analysis shows how Makhana is important for providing job opportunities and increasing income in rural areas, especially in Bihar, where most producers are small farmers working with their families. This production helps strengthen the local economy, create jobs, maintain ancient customs, encouraging constant progress and crop diversity (Ahmad and Kumar, 2020). However, there are social problems such as financial fragility, difficulty in obtaining technologies, and the dangers caused by climate change, which require government actions that encourage everyone's participation, learning, and honesty in management.

The way the sector is managed has improved with the introduction of certifications, government projects, and ways to monitor the product, which ensures that cultivation follows international food safety and environmental care rules. Agreements between institutions and the use of digital tools increase clarity and social commitment, which are fundamental to adding value to production and preventing it from becoming just another commodity, which would diminish its value (Jiang et al., 2023; Garrett et al., 2025).

This integrated approach aims to foster local economies, preserve traditional knowledge, and reduce social vulnerabilities associated with unequal access to technologies and climate change. Strengthening the production chain with a focus on added value, food security, and sustainability has the potential to increase income generation, improve the living conditions of farmers, and give Makhana strategic relevance in the national and international scenario (Rai et al., 2024).

17.11 Conclusions

This chapter delineates *Euryale ferox* (Makhana) as an emerging aquatic crop of growing relevance within global food systems, positioned at the intersection of agronomic tradition, functional nutritional value, and socio-ecological sustainability. The analysis demonstrates that its value chain, while still largely artisanal and regionally concentrated, holds significant potential to contribute to food security, bioeconomic development, and wetland conservation. However, realizing this potential requires overcoming structural constraints and building a more robust, critical scientific foundation.

From an economic perspective, market projections suggest an expansion trajectory driven by demand for functional and plant-based foods. Nevertheless, the scarcity of harmonized primary data, the lack of transparent statistical methodologies, and the high production concentration in India introduce volatility and limit growth predictability. The commercial valorization of Makhana, therefore, depends not only on consumption trends but also on the institutionalization of reliable information systems and the geographical diversification of production.

Agronomically and technologically, persistent operational bottlenecks compromise efficiency and scalability. Cultivation remains vulnerable to hydrological variability, pest and disease incidence, and labor-intensive practices, particularly during harvesting. Industrial processing, in turn, still relies on semiartisanal methods, with limited automation and standardized quality controls. The transition to intensive, mechanized systems represents a necessary condition for enhancing productivity and reducing costs, yet it demands investment in applied research and technology transfer adapted to local conditions.

In the realm of nutrition and health, Makhana's chemical profile, rich in resistant starch, high-quality proteins, and bioactive compounds, justifies its recognition as a functional food. However, the available evidence derives largely from in vitro studies, animal models, or preliminary clinical trials with small sample sizes and heterogeneous methodological designs. Claims regarding glycemic impact, metabolic modulation, and phytochemical bioavailability lack robust human substantiation, limiting their applicability in dietary recommendations and regulatory approvals.

The sustainability of the Makhana chain, often highlighted as an inherent virtue of wetland cultivation, has yet to be systematically quantified. Life-cycle assessments tailored to aquatic cropping systems, accounting for ecosystem services, water balance, carbon sequestration, and impacts on aquatic biodiversity, are notably absent. Similarly, discussions on digital traceability and genomic innovations remain largely conceptual, with limited translation into operational tools or improved varieties accessible to producers.

Considering this landscape, clear priorities emerge for a dedicated scientific and innovation agenda centered on *Euryale ferox*. There is an urgent need for randomized controlled clinical trials that validate health benefits in human populations, alongside the development of standardized analytical protocols for characterizing their bioactive compounds. In agronomy, breeding programs should focus on developing resilient, high-yielding cultivars adapted to diverse cropping systems, while process engineering must optimize postharvest and popping techniques that preserve functional attributes and enable industrial scale-up.

Concurrently, chain governance requires stronger articulation among public, private, and academic actors, with emphasis on clear regulatory frameworks, accessible certification schemes, and mechanisms for equitable value distribution. Digitalization, if properly contextualized, can enhance transparency and efficiency, but its implementation must consider the reality of smallholders and local infrastructure.

Declaration of generative AI and AI-assisted technologies

Artificial intelligence (AI) tools were used only for reviewing the English text in the preparation of this manuscript. No AI tools were used to generate any type of content, including text, tables, or figures.

Funding

The authors would like to thank the Minas Gerais Research Foundation (FAPEMIG) for the financial support provided under project APQ-01782-23, PPE-00090-23/PPE-00079-24/PPE-00045-24 and the Minas Gerais Agricultural Research Corporation (EPAMIG).

References

Ahmad, A., Kumar, B., 2020. A study on socio-economic condition of farmers cultivating makhana in Darbhanga district of Bihar, India. Int. J. Curr. Microbiol. Appl. Sci. 9 (2), 1318–1325. https://doi.org/10.20546/ijcmas.2020.902.155.

Ahmed, D., Sharma, M., Kumar, V., Bajaj, H.K., Verma, A., 2015. 2β-hydroxybetulinic acid 3β-caprylate: an active principle from *Euryale Ferox* Salisb. seeds with antidiabetic, antioxidant, pancreas & hepatoprotective potential in streptozotocin induced diabetic rats. J. Food Sci. Technol. 52 (9), 5427–5441. https://doi.org/10.1007/s13197-014-1676-0.

Annual Report 2024-25. ICAR-Regional Center for Eastern Region, Patna (Bihar). Available at https://icarrcer.in/wp-content/uploads/Annual-Report-2024-25.pdf.

Alonso, V.P.P., de Oliveira Morais, J., Kabuki, D.Y., 2021. Incidence of Bacillus cereus, Bacillus sporothermodurans and Geobacillus stearothermophilus in ultra-high temperature milk and biofilm formation capacity of isolates. Int. J. Food Microbiol. 354, 109318.

Apeda, 2020. Market Intelligence Report: Makhana. Disponível em: https://apeda.gov.in/sites/default/files/2025- 10/MIC_Monthly_dashboard_Makhana_30102025.pdf https://apeda.gov.in/sites/default/files/2025- 10/MIC_Monthly_dashboard_Makhana_30102025.pdf. Acesso em: 11 out. 2025.

Atal, R., Patil, S., Kumar, S., Lata, S., Roy, S., 2020. Processing of Makhana and value chain financing in Bihar. Int. J. Curr. Microbiol. Appl. Sci. 9 (5), 1322–1326.

Benzing, V., Siegwart, V., Spitzhüttl, J., Schmid, J., Grotzer, M., Roebers, C.M., Schmidt, M., 2021. Motor ability, physical self-concept and health-related quality of life in pediatric cancer survivors. Cancer Med. 10 (5), 1860–1871.

Centro-Oeste Brasileiro, 2010. Technological Innovation in Organic Agriculture: A Case Study of Post-harvest Processing Certification.

Devi, M.B., Deka, S.C., 2022. Physicochemical properties and structure of starches of foxnut (*Euryale ferox* Salisb.) from India and its application. J. Food Process. Preserv. 46 (2). https://doi.org/10.1111/jfpp.16262.

Garrett, M., et al., 2025. Garrett ranking of constraints and satisfaction of makhana (*Euryale ferox*) growers in Bihar. Int. J. Ext. Ed. 17 (2), 136–143.

Gurtu, A., Johny, J., 2019. Potential of blockchain technology in supply chain management: a literature review. Int. J. Phys. Distrib. Logist. Manag. 49 (9), 881–900. https://doi.org/10.1108/ijpdlm-11-2018-0371.

Hoekstra, A.Y., 2016. A critique on the water-scarcity weighted water footprint in LCA. Ecol. Indic. 66, 564–573. https://doi.org/10.1016/j.ecolind.2016.02.026.

Jalaj, R.K., Jalaj, V.K., Kumar, P., 2019. Traditional Methods of Harvesting and Processing of Makhana. *Euryale ferox* Salisb, pp. 321–326.

Jana, B.R., Srivastava, A., Idris, M., 2018. Climate change effects and quality makhana (*Euryale ferox* Salisb.) production under wetland ecosystem of North Bihar. Int. J. Curr. Microbiol. Appl. Sci. 7 (12), 578–587. https://doi.org/10.20546/ijcmas.2018.712.072.

Jiang, J., Ou, H., Chen, R., Lu, H., Zhou, L., Yang, Z., 2023. The ethnopharmacological, phytochemical, and pharmacological review of *Euryale ferox* Salisb.: a Chinese medicine food homology. Molecules 28 (11), 4399. https://doi.org/10.3390/molecules28114399, 1420–3049.

Kapoor, S., Kaur, A., Kaur, R., Kumar, V., Choudhary, M., 2022. *Euryale ferox*, a prominent superfood: nutritional, pharmaceutical, and its economical importance. J. Food Biochem. 46 (12). https://doi.org/10.1111/jfbc.14435.

Khadatkar, A., Gite, L.P., Gupta, V.K., 2015. Interventions to reduce drudgery of workers in the traditional method of harvesting Makhana (*Euryale ferox* salisb.) seeds from ponds. Curr. Sci. 109 (7), 1332–1337. http://www.currentscience.ac.in/Volumes/109/07/1332.pdf.

Khadatkar, A., Mehta, C.R., Gite, L.P., 2020. Makhana (*Euryle ferox* Salisb.): a high-valued aquatic food crop with emphasis on its agronomic management – a review. Sci. Hortic. 261, 108995. https://doi.org/10.1016/j.scienta.2019.108995.

Kumar, L., Gupta, V.K., Jha, B.K., Singh, I.S., Bhatt, B.P., Singh, A.K., 2011. Status of Makhana (Euryale ferox Salisb.) Cultivation in India. Technical Bulletin No. R-32/PAT-21. ICAR Research Complex for Eastern Region, Patna, India.

Kumar, A., et al., 2016. Nutritional and medicinal value of fox nut (*Euryale ferox* Salisb.) and its chemical composition. Int. J. Adv. Agricul. Sci. Technol. 3 (5), 35–49.

Kumar, R., et al., 2022. Antidiabetic biomolecules and nutrient elements in Makhana (*Euryale ferox*): a nutraceutical perspective. Ann. Phytomed. 11 (2), 45–54.

Kumar, M., Shekhar, D., Kumari, A., Prasad, R.P., Bhatt, B.P., 2020. Makhana farming for maximizing farm income: a success story. Biot. Res. Today 2 (9), 962–965.

Latinoamericano, O., 2024. Perdas nas etapas pós-colheita do pré- processamento: um panorama atual da logística brasileira.

Liaquat, M., Pasha, I., Ahsin, M., Salik, A., 2022. Roasted fox nuts (Euryale Ferox L.) contain higher concentration of phenolics, flavonoids, minerals and antioxidants, and exhibit lower Glycemic Index (GI) in human subjects. Food Prod. Process. Nutr. 4 (1), 1.

Mandal, S.M., Chakraborty, D., Dey, S., 2010. Phenolic acids act as signaling molecules in plant-microbe symbioses. Plant Signal. Behav. 5 (4), 359–368.

Mittal, R., Sharma, S., Mittal, A., 2020. A critical review on ethnobotanical and pharmacological aspects of *Euryale ferox* Salisb. Pharmacogn. J. 12 (6), 1444–1454. https://doi.org/10.5530/pj.2020.12.199.

Queiroz, M.M., Telles, R., Bonilla, S.H., 2019. Blockchain and supply chain management integration: a systematic review of the literature. Supply Chain Manag.: Int. J. 25 (2), 241–254. https://doi.org/10.1108/scm-03-2018-0143.

Rai, R.S., et al., 2024. Assessing Regional Disparities in Makhana Farming: A Study from Bihar, India. Research Square, [S. l.].

Raut, S., Sarma, K., Singh, I.S., Singh, J., Kumar Padala, V., Kumar Ahirwal, S., Kumar, T., Kumar, R., Jana, B.R., Prasad Bhatt, B., 2025. Performance and sustainability of the makhana (*Euryale ferox* Salisbury) cum fish culture system in India. Indian J. Fish. 72 (2). https://doi.org/10.21077/ijf.2025.72.2.147354-10, 0970–6011.

Saha, S., Sharma, A., Sarkar, T., 2022. Nutritional and nutraceutical potentials of Makhana (*Euryale ferox* Salisb.): a review. Int. J. Food Prop. 25 (1), 1142–1157.

Semwal, D.P., Pandey, A., Gore, P.G., Ahlawat, S.P., Yadav, S.K., Kumar, A., 2021. Habitat prediction mapping using BioClim model for prioritizing germplasm collection and conservation of an aquatic cash crop 'makhana' (*Euryale ferox* Salisb.) in India. Genet. Resour. Crop Evol. 68 (8), 3445–3456. https://doi.org/10.1007/s10722-021-01265-7, 0925–9864.

Shahab, H., Naeem, M., Iqbal, M., Aqeel, M., Ullah, S.S., 2025. IoT-driven smart agricultural technology for real-time soil and crop optimization. Smart Agric. Technol. 10, 100847.

Shaw, R., Das, S., Chaube, R., 2025. Reproductive health benefits of *Euryale ferox* Salisb.: a short comprehensive review. J. Endocrinol. Reprod. 183–190. https://doi.org/10.18311/jer/2024/44756, 2455–1244.

Shukla, N., et al., 2023. The importance of *Euryale ferox* in different contexts. Asian J. Innov. Res. 2 (4), 1–3.

Sinha, C.B.K., Jha, V., Sinha, C.B.K., 2025. JHA, V. GC-MS-based metabolomic profiling of makhana (*Euryale ferox* Salisb.). Int. J. Plant Environ. 11 (3), 514–524. https://doi.org/10.18811/ijpen.v11i03.09.

Sinha, C.B.K., Jha, V., 2025. Identification of nutraceutically important metabolites in raw and popped makhana (*Euryale ferox* Salisb.) seeds using GCMS technique. Int. J. Plant Environ. 11 (03), 514–524. https://doi.org/10.18811/ijpen.v11i03.09, 2455–202X.

Singh, S., et al., 2019. New Makhana (*Euryale ferox* Salisb.) processed products for health and nutrition. J. Pharmacogn. Phytochem. 8 (2), 190–196.

Tracedata Research, 2025. India Makhana (Fox Nut) Market Outlook to 2030 – by Market Structure, by Product Type, by Size/grade, by Distribution Channel, by Application, and by Region. [S.l.]: Tracedata Research, 2025.

Vikram, D.V., Mishra, D.S., 2021. A review on *Euryale Ferox*: daily dietary regime for Covid -19 patients. Int. J. Appl. Res. 7 (6S), 29–34. https://doi.org/10.22271/allresearch.2021.v7.i6sa.8607.

Yang, X., et al., 2025. Profiling and discrimination of *Euryale ferox* seeds from different processing methods using liquid chromatography high-resolution mass spectrometry combined with molecular networking and statistical analysis. Metabolites 15 (4), 225.

Wang, S., et al., 2022. Effect of thermal treatment on the physicochemical, ultrastructural, and antioxidant characteristics of *Euryale ferox* seeds and flour. Front. Nutr. 9, 1007493.

Wu, M., et al., 2023. The effect and mechanism of corilagin from *Euryale ferox* Salisb shell on LPS-induced inflammation in Raw264.7 cells. Foods 12 (5), 979.

Index

Note: Page numbers followed by "f" indicate figures, "t" indicate tables.

D

E

F

N

T

U

V

W

Printed and bound by CPI Group (UK) Ltd, Croydon, CR0 4YY
17/09/2026
02218533-0006